T0182153

UNITEXT for Physics

UNITEXT for Physics series, formerly UNITEXT Collana di Fisica e Astronomia, publishes textbooks and monographs in Physics and Astronomy, mainly in English language, characterized of a didactic style and comprehensiveness. The books published in UNITEXT for Physics series are addressed to graduate and advanced graduate students, but also to scientists and researchers as important resources for their education, knowledge and teaching.

More information about this series at http://www.springer.com/series/13351

Oliver Piattella

Lecture Notes in Cosmology

 Springer

Oliver Piattella
Núcleo Cosmo-UFES and Department
   of Physics
Federal University of Espírito Santo
Vitória, Espírito Santo
Brazil

ISSN 2198-7882   ISSN 2198-7890   (electronic)
UNITEXT for Physics
ISBN 978-3-030-07060-1   ISBN 978-3-319-95570-4   (eBook)
https://doi.org/10.1007/978-3-319-95570-4

This Springer imprint is published by the registered company Springer Nature Switzerland AG
The registered company address is: Gewerbestrasse 11, 6330 Cham, Switzerland

To Giuseppina, Lilli e Dorotea
Together we are gold

Preface

Considerate la vostra semenza: fatti non foste a viver come bruti ma per seguir virtute e canoscenza

(Consider well the seed that gave you birth: you were not made to live as brutes, but to follow virtue and knowledge)

Dante Alighieri, Divina Commedia, Canto XXVI

These lecture notes are based on the handwritten notes which I prepared for the cosmology course taught to graduate students of PPGFis and PPGCosmo at the Federal University of Esprito Santo (UFES), starting from 2014.

This course covers topics ranging from the evidence of the expanding universe to Cosmic Microwave Background anisotropies. In particular, Chap. 1 commences with a bird's-eye view of cosmology, showing its tremendous evolution during the last fifty years and the open problems on which cosmologists are working today. From Chap. 2 on starts the conventional content, hopefully exposed in a not too conventional manner. The main topics are as follows: the expansion of the universe, relativistic cosmology and Friedmann equations (Chap. 2); thermal history, Big Bang Nucleosynthesis, recombination and cold relic abundance (Chap. 3); cosmological perturbation theory and perturbed Einstein equations (Chap. 4); perturbed Boltzmann equations (Chap. 5); primordial modes of perturbations (Chap. 6); random variables and stochastic character of cosmological perturbations (Chap. 7); inflationary paradigm (Chap. 8); evolution of perturbations (Chap. 9); anisotropies in the Cosmic Microwave Background sky (Chap. 10). A selection of extra topics is offered in Chap. 11, whereas Chap. 12 collects some important and well-known physical and mathematical results.

Scattered throughout the text are many exercises. I chose not to put them at the end of a chapter or at the end of the book because I want the reader to stop and do some work in the moment in which this is needed. Most of the exercises are not extensions of the material covered, but consist in developing calculations necessary to the topics addressed. The idea here is, since these are lecture notes, not only to provide a book for consultation but to put the reader-student to work.

When I prepared these lecture notes, I heavily relied on the following three famous textbooks in cosmology:

- S. Dodelson, Modern Cosmology, (Dodelson 2003),
- V. Mukhanov, Principles of Physical Cosmology, (Mukhanov 2005),
- S. Weinberg, Cosmology, (Weinberg 2008),

but also on many other books and papers, which are cited throughout the text. I tried not to simply develop the calculations there contained, but to provide an original presentation. I hope to have succeeded, but of course the final word is up to the reader.

I want now to make three recommendations to the reader-student.

First, read the original papers in order to make contact with the original ideas in their primeval forms and to appreciate the geniuses of their authors.

Second, the technological advance has provided us with sophisticated instruments such as Inspire, http://inspirehep.net/, a powerful tool for researching papers. One of the features I like most about it is the possibility to check which papers have cited the one in which we are interested, thus allowing us to rapidly get up to date on a given topic. I sometimes joke about this by rephrasing a sentence which we all have many times read in papers and books when referring to a certain work: *see ... and references therein*. Thanks to the above-mentioned citation tool we can now add also the sentence *see ... and references thereout*. These notes will be up to date when published, but cosmology is a very vivacious research field so they will probably become outdated in a few years. My recommendation is thus to take advantage of the *references thereout* tools and keep yourself always updated.

Third and final one, in preparing these notes I have intensively employed the CLASS code. This is very user-friendly, so it was not as hard as it may sound. My point in doing so is that analytic calculations are very stimulating and very useful in order to understand the physics ruling the cosmological phenomena, so they have an enormous didactical value. On the other hand, observation needs precise calculations, and these can be done only numerically. So, my recommendation here is the following: do not be afraid of numerical codes and learn how to use them proficiently.

I thank all my students and colleagues who have helped me, through questions and suggestions, in the challenging but rewarding task of writing these notes. Extra thanks to Rodrigo Von Marttens, whose help with CLASS has been crucial in order for me to rapidly grasp how to run the code. Special thanks are due to my editor Aldo Rampioni and his assistant Kirsten Kley-Theunissen for their kind help and encouragement throughout the realisation of the project.

Vitória, Brazil   Oliver Piattella
February 2018

References

Dodelson, S.: Modern cosmology. Academic Press, Amsterdam, Netherlands (2003)
Mukhanov, V.: Physical foundations of cosmology. Cambridge University Press, Cambridge, UK (2005)
Weinberg, S.: Cosmology. Oxford University Press, Oxford, UK (2008)

Contents

1 Cosmology 1
  1.1   The Expanding Universe and its Content 1
      1.1.1   Olbers's Paradox 3
      1.1.2   The Accelerated Expansion of the Universe
            and Dark Energy 4
      1.1.3   Dark Matter 4
  1.2   Cosmological Observations 7
      1.2.1   The Cosmic Microwave Background 7
      1.2.2   Redshift Surveys 8
      1.2.3   Gravitational Waves Observatories 9
      1.2.4   Neutrino Observation 9
      1.2.5   Dark Matter Searches 10
  1.3   Redshift 10
  1.4   Open Problems in Cosmology 10
      1.4.1   Cosmological Constant and Dark Energy 11
      1.4.2   Dark Matter and Small-Scale Anomalies 12
      1.4.3   Other Problems 13
  References 13

2 The Universe in Expansion 17
  2.1   Newtonian Cosmology 17
  2.2   Relativistic Cosmology 18
      2.2.1   Friedmann–Lemaître–Robertson–Walker Metric 19
      2.2.2   The Conformal Time 22
      2.2.3   FLRW Metric Written with Proper Radius 22
      2.2.4   Light-Cone Structure of the FLRW Space 23
      2.2.5   Christoffel Symbols and Geodesics 25
  2.3   Friedmann Equations 28
      2.3.1   The Hubble Constant and the Deceleration
            Parameter 30

	2.3.2	Critical Density and Density Parameters	32
	2.3.3	The Energy Conservation Equation	33
	2.3.4	The ΛCDM Model	35
2.4		Solutions of the Friedmann Equations	37
	2.4.1	The Einstein Static Universe	37
	2.4.2	The de Sitter Universe	38
	2.4.3	Radiation-Dominated Universe	40
	2.4.4	Cold Matter-Dominated Universe	41
	2.4.5	Radiation Plus Dust Universe	43
2.5		Distances in Cosmology	44
	2.5.1	Comoving Distance and Proper Distance	45
	2.5.2	The Lookback Time	45
	2.5.3	Distances and Horizons	47
	2.5.4	The Luminosity Distance	48
	2.5.5	Angular Diameter Distance	50
References			52

3 Thermal History   55
3.1		Thermal Equilibrium and Boltzmann Equation	55
3.2		Short Summary of Thermal History	57
3.3		The Distribution Function	60
	3.3.1	Volume of the Fundamental Cell	61
	3.3.2	Integrals of the Distribution Function	62
3.4		The Entropy Density	64
3.5		Photons	67
3.6		Neutrinos	70
	3.6.1	Temperature of the Massless Neutrino Thermal Bath	71
	3.6.2	Massive Neutrinos	74
	3.6.3	Matter-Radiation Equality	76
3.7		Boltzmann Equation	78
	3.7.1	Proof of Liouville Theorem	79
	3.7.2	Example: The One-Dimensional Harmonic Oscillator	79
	3.7.3	Boltzmann Equation in General Relativity and Cosmology	81
3.8		Boltzmann Equation with a Collisional Term	84
	3.8.1	Detailed Calculation of the Equilibrium Number Density	87
	3.8.2	Saha Equation	89
3.9		Big-Bang Nucleosynthesis	90
	3.9.1	The Baryon-to-Photon Ratio	91
	3.9.2	The Deuterium Bottleneck	92

3.9.3   Neutron Abundance 94
3.10 Recombination and Decoupling 100
3.10.1 Decoupling 104
3.11 Thermal Relics 106
3.11.1 The Effective Numbers of Relativistic Degrees
of Freedom 110
3.11.2 Relic Abundance of DM and the WIMP Miracle 113
References 114

4 **Cosmological Perturbations** 117
4.1 From the Perturbations of the FLRW Metric to the Linearised
Einstein Tensor 117
4.1.1 The Perturbed Christoffel Symbols 120
4.1.2 The Perturbed Ricci Tensor and Einstein Tensor 123
4.2 Perturbation of the Energy-Momentum Tensor 125
4.3 The Problem of the Gauge and Gauge Transformations 131
4.3.1 Coordinates and Gauge Transformations 132
4.3.2 The Scalar-Vector-Tensor Decomposition 134
4.3.3 Gauges 140
4.4 Normal Mode Decomposition 142
4.5 Einstein Equations for Scalar Perturbations 145
4.5.1 The Relativistic Poisson Equation 146
4.5.2 The Equation for the Anisotropic Stress 147
4.5.3 The Equation for the Velocity 148
4.5.4 The Equation for the Pressure Perturbation 148
4.6 Einstein Equations for Tensor Perturbations 149
4.7 Einstein Equations for Vector Perturbations 152
References 154

5 **Perturbed Boltzmann Equations** 157
5.1 General Form of the Perturbed Boltzmann Equation 157
5.1.1 On the Photon and Neutrino Perturbed
Distributions 159
5.2 Force Term 160
5.2.1 Scalar Perturbations 161
5.2.2 Tensor Perturbations 162
5.2.3 Vector Perturbations 163
5.3 The Perturbed Boltzmann Equation for CDM 164
5.4 The Perturbed Boltzmann Equation for Massless Neutrinos 167
5.4.1 Scalar Perturbations 168
5.5 The Perturbed Boltzmann Equation for Photons 172
5.5.1 Computing the Collisional Term Neglecting
Polarisation 173
5.5.2 Full Boltzmann Equation Including Polarisation 176

		5.5.3	Scalar Perturbations	179
		5.5.4	Tensor Perturbations	182
		5.5.5	Vector Perturbations	184
	5.6	Boltzmann Equation for Baryons		186
	References			191

6 Initial Conditions 193
	6.1	Initial Conditions		193
	6.2	Evolution Equations in the $k\eta \ll 1$ Limit		195
		6.2.1	Multipoles in the $k\eta \ll 1$ Limit	197
		6.2.2	CDM and Baryons Velocity Equations	199
		6.2.3	The $k\eta \ll 1$ Limit of the Einstein Equations	199
	6.3	The Adiabatic Primordial Mode		201
		6.3.1	Why "Adiabatic"?	203
	6.4	The Neutrino Density Isocurvature Primordial Mode		205
	6.5	The CDM and Baryons Isocurvature Primordial Modes		206
	6.6	The Neutrino Velocity Isocurvature Primordial Mode		207
	6.7	Planck Constraints on Isocurvature Modes		209
	References			210

7 Stochastic Properties of Cosmological Perturbations 211
	7.1	Stochastic Cosmological Perturbations and Power Spectrum		212
	7.2	Random Fields		213
	7.3	Power Spectrum and Gaussian Random Fields		217
		7.3.1	Definition of a Gaussian Random Field	218
		7.3.2	Estimator of the Power Spectrum and Cosmic Variance	219
	7.4	Non-Gaussian Perturbations		221
	7.5	Matter Power Spectrum, Transfer Function and Stochastic Initial Conditions		222
	7.6	CMB Power Spectra		224
		7.6.1	Cosmic Variance of Angular Power Spectra	228
	7.7	Power Spectrum for Tensor Perturbations		230
	7.8	Ergodic Theorem		231
	References			234

8 Inflation 235
	8.1	The Flatness Problem		235
	8.2	The Horizon Problem		238
	8.3	Single Scalar Field Slow-Roll Inflation		240
		8.3.1	More Slow-Roll Parameters	243
		8.3.2	Reheating	246
	8.4	Production of Gravitational Waves During Inflation		248
	8.5	Production of Scalar Perturbations During Inflation		256

8.6   Spectral Indices 262
8.7   Observational Results 264
8.8   Examples of Models of Inflation 265
      8.8.1   General Power Law Potential 266
      8.8.2   The Starobinsky Model 267
References 270

9   Evolution of Perturbations 273
    9.1   Evolution on Super-Horizon Scales 275
          9.1.1   Evolution Through Radiation-Matter Equality ... 276
          9.1.2   Evolution in the Λ-Dominated Epoch 280
          9.1.3   Evolution Through Matter-DE Equality 281
    9.2   The Matter-Dominated Epoch 284
          9.2.1   Baryons Falling into the CDM Potential Wells ... 287
    9.3   The Radiation-Dominated Epoch 290
    9.4   Deep Inside The Horizon 294
    9.5   Matching and CDM Transfer Function 296
    9.6   The Transfer Function for Tensor Perturbations 302
          9.6.1   Radiation-Dominated Epoch 303
          9.6.2   Matter-Dominated Epoch 304
          9.6.3   Deep Inside The Horizon 305
    References 308

10   Anisotropies in the Cosmic Microwave Background 309
     10.1   Free-Streaming 310
     10.2   Anisotropies on Large Scales 315
     10.3   Tight-Coupling and Acoustic Oscillations 317
            10.3.1   The Acoustic Peaks for $R = 0$ 320
            10.3.2   Baryon Loading 325
     10.4   Diffusion Damping 329
     10.5   Line-of-Sight Integration 334
     10.6   Finite Thickness Effect and Reionization 339
     10.7   Cosmological Parameters Determination 342
     10.8   Tensor Contribution to the CMB TT Correlation ... 348
     10.9   Polarisation 354
            10.9.1   Scalar Perturbations Contribution to Polarisation 354
            10.9.2   Tensor Perturbations Contribution to Polarisation 357
     References 363

11   Miscellanea 365
     11.1   Bayesian Analysis Using Type Ia Supernovae Data ... 365
     11.2   Doing Statistics in the Sky 369
            11.2.1   Top Hat and Gaussian Filters 371
            11.2.2   Sampling and Shot Noise 373

         11.2.3   Correlation Function 375

         11.2.4   Bias 377

    References 378

12 Appendices 379

    12.1   Thermal Distributions 379

         12.1.1   Derivation of the Maxwell–Boltzmann Distribution . . . 379

         12.1.2   Derivation of the Fermi–Dirac Distribution 380

         12.1.3   Derivation of the Bose–Einstein Distribution 382

    12.2   Derivation of the Poisson Distribution 382

    12.3   Helmholtz Theorem 384

    12.4   Conservation of $\mathcal{R}$ on Large Scales and for Adiabatic
        Perturbations 385

    12.5   Spherical Harmonics 387

         12.5.1   Spin-Weighted Spherical Harmonics 394

    12.6   Method of Green's Functions 396

    12.7   Polarisation 398

         12.7.1   Electromagnetic Waves 398

         12.7.2   Polarisation Ellipse and Stokes Parameters 400

    12.8   Thomson Scattering 406

    References 412

Index 413

Notation

Latin indices (e.g. i, j, k) run over the three spatial coordinates, and assume values 1, 2, 3. In a Cartesian coordinate system, we shall use $x^1 \equiv x$, $x^2 \equiv y$ and $x^3 \equiv z$.

Greek indices (e.g. μ, ν, ρ) run over the four spacetime coordinates, and assume values 0, 1, 2, 3, with x^0 being the time coordinate.

Repeated high and low indices are summed unless otherwise stated, i.e. $x^\mu y_\mu \equiv \sum_{\mu=0}^4 x^\mu y_\mu$ or $x^i y_i \equiv \sum_{i=1}^3 x^i y_i$. Repeated spatial indices are also summed, whether one is high and the other is low or not, i.e. $x_i y_i \equiv \sum_{i=0}^3 x_i y_i$.

The signature employed for the metric is $(-, +, +, +)$.

Spatial 3-vectors are indicated in boldface, e.g. $\mathbf{v}$.

The unit vector corresponding to any 3-vector $\mathbf{v}$ is denoted with a hat, i.e. $\hat{v} \equiv \mathbf{v}/|\mathbf{v}|$, where $|\mathbf{v}|$ is the modulus of the 3-vector $\mathbf{v}$ defined as $|\mathbf{v}|^2 = \delta_{ij} v^i v^j$ with δ_{ij} being the usual Kronecker delta.

A dot over any quantity denotes the derivation with respect to the cosmic time, denoted with t, of that quantity, whereas a prime denotes derivation with respect to the conformal time, denoted with η.

The operator ∇^2 is the usual Laplacian operator in the Euclidian space, i.e. $\nabla^2 \equiv \delta^{ij} \partial_i \partial_j$, where ∂_μ is used as a shorthand for the partial derivative with respect to the coordinate x^μ.

Except on vector and tensor, a 0 subscript means that a time-dependent quantity is evaluated today, i.e. at $t = t_0$ or $\eta = \eta_0$ where t_0 and η_0 represent the age of the universe in the cosmic time or in the conformal time.

The subscripts b, c, γ and ν put on matter quantities such as density and pressure refer to baryons, cold dark matter, photons and neutrinos, respectively. Subscripts m and r refer to matter and radiation, in general.

With k is denoted the comoving wavenumber.

From Chap. 4 on natural units $\hbar = c = 1$ shall be employed.

When employing the spherical coordinate system, the azimuthal coordinate is denoted with ϕ. A scalar field is denoted with φ.

Chapter 1
Cosmology

And that inverted Bowl we call the Sky,
Whereunder crawling coop't we live and die,
Lift not thy hands to It for help—for It,
Rolls impotently on as Thou or I

Omar Khayyám, Rubáiyát

In this Chapter we present an overview of cosmology, addressing its most important aspects and presenting some observational experiments and open problems.

1.1 The Expanding Universe and its Content

The starting point of our study of cosmology is the extraordinary evidence that we live in an expanding universe. This was a landmark discovery made in the XX century, usually attributed to Edwin Hubble (1929), but certainly resulting from the joint efforts of astronomers such as Vesto Slipher (1917) and cosmologists such as George Lemaître (1927). We do not enter here the debate about who is deserving more credit for the discovery of the expansion of the universe. The interested reader might want to read e.g. Way and Nussbaumer (2011); van den Bergh (2011).

Hubble discovered that the farther a galaxy is the faster it recedes from us. See Fig. 1.1. This is the famous **Hubble's law**:

$$v = H_0 r \; , \tag{1.1}$$

© Springer International Publishing AG, part of Springer Nature 2018
O. Piattella, *Lecture Notes in Cosmology*, UNITEXT for Physics,
https://doi.org/10.1007/978-3-319-95570-4_1

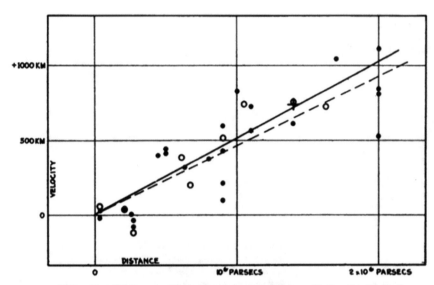

Velocity-Distance Relation among Extra-Galactic Nebulae.

Fig. 1.1 Figure 1 of Hubble's original paper (Hubble 1929)

where v is the recessional velocity, r is the distance and H_0 (usually pronounced "H-naught") is a constant named after Hubble. The value of H_0 determined by Hubble himself was:

$$H_0 = 500 \, \text{km s}^{-1} \, \text{Mpc}^{-1} \, , \tag{1.2}$$

with huge error, as can be understood from Fig. 1.1 by observing how much the data points are scattered. A more precise estimate was made by Sandage (1958) in 1958:

$$H_0 = 75 \, \text{km s}^{-1} \, \text{Mpc}^{-1} \, . \tag{1.3}$$

A recent measurement done by the BOSS collaboration (Grieb et al. 2017) gives

$$H_0 = 67.6^{+0.7}_{-0.6} \, \text{km s}^{-1} \, \text{Mpc}^{-1} \, . \tag{1.4}$$

Roughly speaking, this number means that for each Mpc away a source recedes 67.6 km/s faster. At a certain radius, the receding velocity attains the velocity of light and therefore we are unable to see farther objects. This radius is called **Hubble radius**.

The Hubble constant can be measured also with fair precision by using the time-delay among variable signals coming from lensed distant sources (Bonvin et al. 2017) and via gravitational waves (Abbott et al. 2017b).

1.1.1 Olbers's Paradox

The expansion of the universe could have been predicted a century before Hubble by solving Olbers's paradox (Olbers 1826). See e.g. Dennis Sciama's book (Sciama 2012) for a very nice account of the paradox, which is also called **the dark night sky paradox** and goes as follows: if the universe is static, infinitely large and old and with an infinite number of stars distributed uniformly, then the night sky should be bright.

Let us try to understand why. First of all, the stars are distributed uniformly, which means that their number density say n is a constant. Consider a spherical shell of thickness dR and radius R centred on the Earth. The number of stars inside this spherical shell is:

$$dN = 4\pi n R^2 dR .$$ (1.5)

The total luminosity of this spherical shell is dN multiplied by the luminosity say L of a single star, and we assume L to be the same for all the stars. Only a fraction f of the radiation produced by the star reaches Earth, but this fraction is the same for all the stars (because we assume them to be identical). Therefore, the total luminosity of a spherical shell is $dNfL$. By the inverse-square law, the total flux received on Earth is:

$$dF_{\text{tot}} = \frac{dNfL}{4\pi R^2} = nfLdR .$$ (1.6)

It does not depend on R and thus it diverges when integrated over R from zero to infinity. This means not only that the night sky should not be dark but also infinitely bright!

We can solve the problem of having an infinitely bright night sky by considering the fact that stars are not points and do eclipse each other, so that we do not really see all of them. Suppose that each star shows us a surface dA. Therefore, if a star lies at a distance R, we receive from it the flux $dF = dA\mathcal{L}/(4\pi R^2)$, where $\mathcal{L}$ is the luminosity per unit area. But $dA/R^2 = d\Omega$ is the solid angle spanned by the star in the sky. Therefore:

$$dF = \frac{\mathcal{L}}{4\pi}d\Omega .$$ (1.7)

Once again, this does not depend on R! When we integrate it over the whole solid angle, we obtain that $F_{\text{tot}} = \mathcal{L}$, i.e. the whole sky is as luminous as a star! In other words: it is true that the farther a star is the fainter it appears, but we can pack more of them in the same patch of sky.

In order to solve Olbers's paradox, we can drop one or more of the initial assumptions. For example:

- The universe is not eternal so the light of some stars has not yet arrived to us. This is plausible, but even so we could expect a bright night sky and also to see some new star to pop out from time to time, without being a transient phenomenon such as a supernova explosion. There is no record of this.

- Maybe there is not an infinite number of stars. But we have showed that taking into account their dimension we do not need an infinite number and yet the paradox still exists.
- Are stars distributed not uniformly? Even so, we would expect still a bright night sky, even if not uniformly bright.

At the end, we must do something in order for the light of some stars not to reach us. A possibility is to drop the staticity assumption. The farther a spherical shell is, the faster it recedes from us (this is Hubble's law). In this way, beyond a certain distance (the Hubble radius) light from stars cannot reach us and the paradox is solved.

1.1.2   The Accelerated Expansion of the Universe and Dark Energy

The discovery of the type Ia supernova 1997ff (Williams et al. 1996) marked the beginning of a new era in cosmology and physics. The analysis of the emission of this type of supernovae led to the discovery that our universe is in a state of **accelerated expansion**. See e.g. Perlmutter et al. (1999); Riess et al. (1998).

This is somewhat problematic because gravity as we know it should attract matter, thereby causing the expansion to decelerate. So, what does cause the acceleration in the expansion? A possibility is that there exists a new form of matter, or rather energy, which acts as anti-gravity. This is widely known today as **Dark Energy** (DE) and its nature is still a mystery to us. The most simple and successful candidate for DE is the cosmological constant Λ.

1.1.3   Dark Matter

DE is not the only dark part of our universe. Many observations of different nature and from different sources at different distance scales point out the existence of another dark component, called Dark Matter (DM). In particular, these observations are:

• **The dynamics of galaxies in clusters**. The pioneering applications of the virial theorem to the Coma cluster by Zwicky (1933) resulted in a virial mass 500 times the observed one (which can be estimated by the light emission).

• **Rotation curves of spiral galaxies**. This is the famous problem of the flattish velocity curves of stars in the outer parts of spiral galaxies. See e.g. Sofue et al. (1999); Sofue and Rubin (2001).

The surprising fact of these flattish curves is that there is no visible matter to justify them and one would then expect a Keplerian fall $V \sim 1/\sqrt{R}$, where R is the distance from the galactic centre. In order to derive the Keplerian fall, simply assume a circular orbit and use Newtonian gravity (which seems to be fine for galaxies). Then, the centrifugal force is canceled by the gravitational attraction of the galaxy as follows:

$$\frac{V^2}{R} = \frac{GM(R)}{R^2} , \qquad (1.8)$$

where we assume also a spherical distribution of matter in the galaxy and therefore use Gauss' theorem. Inside the bulge of the galaxy, the visible mass goes as $M \propto R^3$, and therefore $V \propto R$, which represents the initial part of the velocity curve. However, outside the bulge of the galaxy, the mass becomes a constant, thus the Keplerian fall $V \propto 1/\sqrt{R}$ follows.

• **The brightness in X-rays of galaxy clusters**. This depends on the gravitational potential well of the cluster, which is deduced to be much deeper than the one that would be generated by visible matter only. See e.g. Weinberg (2008).

• **The formation of structures in the universe**. In other words, the fact that the density contrast of the usual matter, called baryonic in cosmology, δ_b is today highly non-linear, i.e. $\delta_b \gg 1$. As we shall see in Chap. 9, relativistic cosmology predicts that δ_b grows by a factor of 10^3 between recombination and today and observations of CMB show that the value of δ_b at recombination is $\delta_b \sim 10^{-5}$. This means that today $\delta_b \sim 10^{-2}$ which is in stark contrast with the huge number of structures that we observe in our universe. So, there must be something else which catalyses structure formation.

• **The structure of the Cosmic Microwave Background peaks**. The temperature-temperature correlation spectrum in the CMB sky is characterised by the so-called **acoustic peaks**. The absence of DM would not allow to reproduce the structure shown in Fig. 1.2. We shall study this structure in detail in Chap. 10.

• **Weak Lensing**. The bending of light is a method for measuring the mass of the lens, and it is a classical test of General Relativity (GR) (Weinberg 1972). When the background source is distorted by the foreground lens, one has the so-called weak lensing. Analysing the emission of the distorted source allows to map the gravitational potential of the lens and therefore its matter distribution. See e.g. Dodelson (2017) for a recent textbook reference on gravitational lensing. Weak gravitational lensing is a powerful tool for the study of the geometry of the universe and its observation is one of the primary targets of forthcoming surveys such as the European Space Agency (ESA) satellite *Euclid* and the *Large Synoptic Survey Telescope* (LSST).

A remarkable combination of X-ray and weak lensing observational techniques made the **Bullet Cluster** famous (Clowe et al. 2006). Indeed X-ray maps show the result of a merging between the hot gases of two galaxy clusters which gravitational lensing maps reveal to be lagging behind their respective centres of mass. Therefore, most parts of the clusters simply went through one another, leaving behind a smaller fraction of hot gas. This is considered a direct empirical proof of the existence of DM forming a massive halo and a gravitational potential well in which gas and galaxies lie.

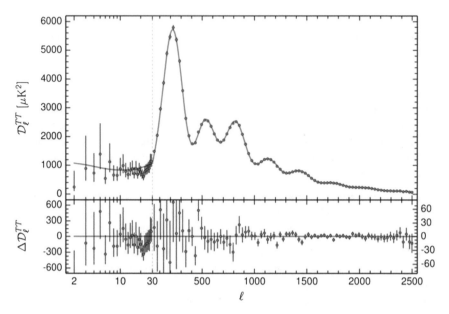

Fig. 1.2 CMB TT spectrum. Figure taken from Ade et al. (2016a). The red solid line is the best fit ΛCDM model

Popular candidates to the role of DM are particles beyond the standard model (Silk et al. 2010). Among these, the most famous are the *Sterile Neutrino* (Dodelson and Widrow 1994), the *Axion*, which is related to the process of violation of CP symmetry (Peccei and Quinn 1977), and *Weakly Interacting Massive Particles* (WIMP) which count among them the lightest supersymmetric neutral stable particle, the *Neutralino*. See Bertone and Hooper (2016) for a historical account of DM and (Profumo 2017) for a textbook on particle DM.

The observational evidences of DM that we have seen earlier do not only point to the existence of DM but also on the necessity of it being **cold**, i.e. with negligible pressure or, equivalently, with a small velocity (much less than that of light) of its particles (admitting that DM is made of particles, which is the common understanding to which we adhere in these notes). Hence **Cold Dark Matter** (CDM) shall be our DM paradigm.

As we shall see in Chap. 3 if DM was in thermal equilibrium with the rest of the known particles in the primordial plasma, i.e. it was **thermally produced**, then its being cold amounts to say that its particles have a sufficiently large mass, e.g. about 100 GeV for WIMP's. Decreasing the mass, we have DM candidates characterised by increasing velocity dispersions and thus with different impacts on the process of structure formation. Typically one refers to **Warm Dark Matter** (WDM) as a thermally produced DM with mass of the order of some keV and **Hot Dark Matter** (HDM) as thermally produced particles with small masses, e.g. of the order of the eV, or even massless. In fact neutrinos can be considered as a HDM candidate. Note that the axion has a mass of about 10^{-5} eV but nonetheless is CDM because it was not thermally produced, i.e. it never was in thermal equilibrium with the primordial plasma.

The combined observational successes of Λ and CDM form the so-called Λ**CDM** model, which is the standard model of cosmology.

1.2 Cosmological Observations

We dedicate this section to the most important cosmological observations which are ongoing or ended recently, or are planned.

1.2.1 The Cosmic Microwave Background

The cosmic microwave background (CMB) radiation provides a window onto the early universe, revealing its composition and structure. It is a relic, thermal radiation from a hot dense phase in the early evolution of our universe which has now been cooled by the cosmic expansion to just three degrees above absolute zero. Its existence had been predicted in the 1940s by Alpher and Gamow (1948) and its discovery by Penzias and Wilson at Bell Labs in New Jersey, announced in 1965 (Penzias and Wilson 1965) was convincing evidence for most astronomers that the cosmos we see today emerged from a **Hot Big Bang** more than 10 billion years ago.

Since its discovery, many experiments have been performed to observe the CMB radiation at different frequencies, directions and polarisations, mostly with ground- and balloon-based detectors. These have established the remarkable uniformity of the CMB radiation, at a temperature of 2.7 Kelvin in all directions, with a small ± 3.3 mK dipole due to the Doppler shift from our local motion (at 1 million kilometres per hour) with respect to this cosmic background.

However, the study of the CMB has been transformed over the last twenty years by three pivotal satellite experiments. The first of these was the *Cosmic Background Explorer* (CoBE, https://lambda.gsfc.nasa.gov/product/cobe/), launched by NASA in 1990. In 1992 CoBE reported the detection of statistically significant temperature anisotropies in the CMB, at the level of $\pm 30 \mu$K on 10 degree scales (Smoot et al. 1992) and it confirmed the black body spectrum with an astonishing precision, with deviations less than 50 parts per million (Smoot et al. 1992). CoBE was succeeded by the *Wilkinson Microwave Anisotropy Probe* (WMAP, https://map.gsfc.nasa.gov/) satellite, launched by NASA in 2001, which produced full sky maps in five frequencies (from 23 to 94 GHz) mapping the temperature anisotropies to sub-degree scales and determining the CMB polarisation on large angular scales for the first time.

The *Planck* satellite (http://sci.esa.int/planck/), launched by ESA in 2009, sets the current state of the art with nine separate frequency channels, measuring temperature fluctuations to a millionth of a degree at an angular resolution down to 5 arc-minutes.

Planck's mission ended in 2013 and the full-mission data were released in 2015 in Adam et al. (2016) and in many companion papers. A fourth generation of full-sky, microwave-band satellite recently proposed to ESA within Cosmic Vision 2015-2025 is the *Cosmic Origins Explorer* (COrE, http://www.core-mission.org/) (Bouchet et al. 2011).

At the moment, a great effort is being devoted to the detection of the B-mode of CMB polarization because it is the one related to the primordial gravitational waves background, as we shall see in Chap. 10. Located near the South Pole, BICEP3 (https://www.cfa.harvard.edu/CMB/bicep3/) and the *Keck Array* are telescopes devoted to this purpose.

Among the non-satellite CMB experiments we must mention the *Balloon Observations Of Millimetric Extragalactic Radiation ANd Geophysics* (BOOMERanG) which was a balloon-based mission which flew in 1998 and in 2003 and measured CMB anisotropies with great precision (higher than CoBE). From these data the Boomerang collaboration first determined that the universe is spatially flat (de Bernardis et al. 2000).

1.2.2 Redshift Surveys

Redshift surveys are observations of certain patches of sky at certain wavelengths with the aim of determining mainly the angular positions (declination and right ascension) redshifts and spectra of galaxies.

The *Sloan Digital Sky Survey* (SDSS, http://www.sdss.org/) is a massive spectroscopic redshift survey which is ongoing since the year 2000 and it is now in its stage IV with 14 data releases available. It is ground-based and uses a telescope located in New Mexico (USA). The SDSS-IV is formed by three sub-experiment:

- The *Extended Baryon Oscillation Spectroscopic Survey* (eBOSS), focusing on redshifts $0.6 < z < 2.5$ and on the Baryon Acoustic Oscillations (BAO) phenomenon[1];
- The *Apache Point Observatory Galaxy Evolution Experiment* (APOGEE-2) is dedicated to the study of our Milky Way;
- The *Mapping Nearby Galaxies at Apache Point Observatory* (MaNGA) study instead nearby galaxies by measuring their spectrum along their extension and not only at the centre.

The V generation of the SDSS will start in 2020, consisting of three surveys: the *Milky Way Mapper*, the *Black Hole Mapper* and the *Local Volume Mapper*. See Kollmeier et al. (2017).

---

[1] We shall not address BAO extensively in these notes, but only mention them in Chap. 10. Together with weak lensing, BAO are another powerful observable upon which present and future missions are planned.

The *Dark Energy Survey* (DES, https://www.darkenergysurvey.org/) measures redshifts photometrically using a telescope situated in Chile and looking for Type Ia supernovae, BAO and weak lensing signals.

Planned surveys are the already mentioned satellite *Euclid* (http://sci.esa.int/euclid/), whose launch is due possibly in 2021 and the telescope LSST, which is being built in Chile and whose first light is due in 2019. We also cite the NASA satellite *Wide Field Infrared Survey Telescope* (WFIRST, https://www.nasa.gov/wfirst) and the *Javalambre Physics of the accelerating universe Astronomical Survey* (J-PAS, http://j-pas.org/). The main cosmological goals of these experiments relay on the detection of weak lensing, BAO and type Ia supernovae signals with high precision.

1.2.3 Gravitational Waves Observatories

The recent direct detection of gravitational waves (GW) by the *LIGO-Virgo collaboration* (https://www.ligo.org/) (Abbott et al. 2016a, b) has opened a new observational window on the universe. In particular, GW are relevant in cosmology because they could be a relic from inflation containing invaluable informations on the very early universe. As already mentioned, they are being searched via the detection of the B-mode polarisation of the CMB.[2]

There are now three functioning ground-based GW observatories: *LIGO* (Hanford and Livingstone, USA) and *Virgo* (near Pisa, Italy). *KAGRA*, in Japan, is under construction and another one in India, *INDIGO*, is planned. The space-based *LISA* GW observer is still in a preliminary phase (*LISA* pathfinder).

1.2.4 Neutrino Observation

Neutrinos are relevant in cosmology, as we shall see throughout these notes, because they should form a cosmological background as CMB photons do. The great problem is that it is incredibly difficult to detect them and even more if they have low energy, as we expect to be the case for neutrinos in the cosmological background.

The most important neutrino observatory is *IceCube* (http://icecube.wisc.edu/), operating since 2005 (its construction was completed in 2010) and located near the South Pole. It detects neutrinos indirectly, via their emission of Cherenkov light.

---

[2]The events detected by the LIGO-Virgo collaboration originated from merging of black holes or neutron stars. Thus are not part of the primordial GW background.

1.2.5   Dark Matter Searches

The search for DM particles counts on many observatories and the *Large Hadron Collider* (LHC). See Gaskins (2016); Liu et al. (2017) for the status of indirect and direct DM searches which, unfortunately, have not been successful until now.

1.3   Redshift

Redshift is a fundamental observable of cosmology. Its definition is the following:

$$z = \frac{\lambda_{\text{obs}}}{\lambda_{\text{em}}} - 1 \qquad (1.9)$$

It is always positive, i.e. observed radiation is redder than the emitted one, because the universe is in expansion. For the closest sources, such as Andromeda, it is negative, i.e. the observed radiation is bluer than the emitted one, because the Hubble flow is overcome by the peculiar motion due to local gravitational effects.

For the moment we can think of the redshift as a Doppler effect due to the relative motion of the sources. In Chap. 2 we will relate it to spacetime geometry with GR.

Redshift is measured in two ways: spectroscopically or photometrically. For the former one needs to do spectroscopy, i.e. detecting known emission or absorption lines from a source and comparing their wavelengths with the ones measured in a laboratory on Earth. Hence one uses Eq. (1.9) and thus calculate z.

Photometric redshifts are calculated by assuming certain spectral features for the sources and measuring their relative brightness in certain wavebands, using filters.

A simple example is the following. Sun's spectrum is almost a blackbody one with temperature of about 6000 K and then, by using Wien's displacement law, it has a peak emission at a wavelength of 500 nm. Therefore, if a star similar to the Sun had a peak emission of say 600 nm, then using Eq. (1.9) one would calculate $z = 0.2$.

The reason for using photometry instead of spectroscopy is that it is less time-consuming and allows to obtain redshifts of very far sources, for which it is difficult to do spectroscopy. On the other hand, photometric redshifts are less precise.

1.4   Open Problems in Cosmology

The fundamental issue in cosmology is to understand what are DM and DE. The effort of answering this question makes cosmology, particle physics and quantum field theory (QFT) to merge. The ways adopted in order to tackle these problems are essentially the search for particles beyond the standard model and the investigation of new theories of gravity, which in most of the cases are extensions of GR.

1.4.1 Cosmological Constant and Dark Energy

Pure geometrical Λ and vacuum energy have the same dynamical behaviour in GR. Estimating the latter via QFT calculations and comparing the result with the observed value leads to the famous **fine-tuning** problem of the cosmological constant. See e.g. Weinberg (1989). This roughly goes as follows: the observed value of ρ_Λ is about 10^{-47} GeV^4 (Ade et al. 2016a). The natural scale for the vacuum energy density is the Planck scale, i.e. 10^{76} GeV^4. There are 123 orders of magnitude of difference! Even postulating a false vacuum state after the electro-weak phase transition at 10^8 GeV^4, the difference is 55 orders of magnitude. See Martin (2012) for a comprehensive account of Λ and the issues related to it.

Another problem with Λ is the so-called **cosmic coincidence** (Zlatev et al. 1999). This problem stems from the fact that the density of matter decreases with the inverse of the cube scale factor, whereas the energy density of the cosmological constant is, as its name indicates, constant. However, these two densities are approximately equal at the present time. This coincidence becomes all the more intriguing when we consider that if the cosmological constant had dominated the energy content of the universe earlier, galaxies would not have had time to form; on the other hand, had the cosmological constant dominated later, then the universe would still be in a decelerated phase of expansion or younger than some of its oldest structures, such as clusters of stars (Velten et al. 2014).

The cosmic coincidence problem can also be seen as a fine-tuning problem in the initial conditions of our universe. Indeed, consider the ratio ρ_Λ/ρ_m, of the cosmological constant to the matter content. This ratio goes as a^3. Suppose that we could extrapolate our classical theory (GR) up to the Planck scale, for which $a \approx 10^{-32}$. Then, at the Planck scale we have $\rho_\Lambda/\rho_m \approx 10^{-96}$. This means that, at trans-Planckian energies, possibly in the quantum universe, there must be a mechanism which establishes the ratio ρ_Λ/ρ_m with a precision of 96 significant digits! Not a digit can be missed, otherwise we would have today 10 times more cosmological constant than matter, or vice-versa, thereby being in strong disagreement with observation.

So, we find ourselves in a situation of *impasse*. On one hand, Λ is the simplest and most successful DE candidate. On the other hand it suffers from the above-mentioned issues. What do we do? Much of today research in cosmology addresses this question. Answers are looked for mostly via investigation of new theories of gravity, extensions or modifications of GR, of which DE would be a manifestation. There are so many papers addressing extended theories of gravity that it is quite difficult to choose representatives. Probably the best option is to start with a textbook, e.g. Amendola and Tsujikawa (2010).

A different approach is to accept that Λ has the value it has by chance, and it turns out to be just the right value for structures to form and for us to be here doing cosmology. This also known as **Anthropic Principle** and exists in many forms, some stronger than others. It is also possible that ours is one universe out of an infinite number of realisations, called **Multiverse**, with different values of the fundamental constants. Life as we know it then develops only in those universes where the conditions are favourable. Again, it is difficult to cite papers on these topics (which are more about metaphysics rather than physics, since there is no possibility of performing experimental tests) but a nice reading is e.g. Weinberg (1992).

1.4.2 Dark Matter and Small-Scale Anomalies

On sub-galactic scales, of about 1 kpc, the CDM paradigm displays some difficulties (Warren et al. 2006). These are called **CDM small-scales anomalies**. See e.g. Bullock and Boylan-Kolchin (2017) for a recent account. They are essentially three and stem from the results of numerical simulations of the formation of structures:

1. The *Core/Cusp* problem (Moore 1994). The CDM distribution in the centre of the halo has a cusp profile, whereas observation suggests a core one;
2. The *Missing satellites problem* (Klypin et al. 1999). Numerical simulations predict a large number of satellite structures, which are not observed;
3. The *Too big to fail* problem (Boylan-Kolchin et al. 2011). The sub-structures predicted by the simulations are too big not to be seen.

Possible solutions to these small-scale anomalies are the following:

Baryon feedback. The cross section for DM particles and the standard model particles interaction must be very small, i.e. $\sim 10^{-39}\,\mathrm{cm}^2$, but in environments of high concentration, such as in the centre of galaxies, such interactions may become important and may provide an explanation for the anomalies. The problem is that the models of baryon feedback are difficult to be simulated and they seem not to be enough to resolve the anomalies (Kirby et al. 2014).

Warm dark matter. As anticipated, WDM are particles with mass around the keV which decouple from the primordial plasma when relativistic. They are subject to free streaming that greatly cut the power of fluctuations on small scales, thereby possibly solving the anomalies of the CDM. The problem with WDM is that different observations indicate mass limits which are inconsistent among them. In particular:

- To solve the *Core/Cusp* problem is necessary a mass of $\sim 0.1\,\mathrm{keV}$ (Macciò et al. 2012).
- To solve the *Too big to fail* problem is necessary a mass of $\sim 2\,\mathrm{keV}$ (Lovell et al. 2012).
- Constraints from the Lyman-α observation require $m_{\mathrm{WDM}} > 3.3\,\mathrm{keV}$ (Viel et al. 2013).

These observational tensions disfavour WDM. In addition, it was shown in Schneider et al. (2014) that for $m_{\mathrm{WDM}} > 3.3$ keV WDM does not provide a real advantage over CDM.

Interacting Dark Matter. CDM anomalies could perhaps be understood by admitting the existence of self-interactions between dark matter particles (Spergel and

Steinhardt 2000). It has been shown that there are indications that interaction models can alleviate the *Core/Cusp* and the *Too big to fail* problems (Vogelsberger et al. 2014). Recently, Macciò et al. (2015) pointed out that a certain type of interaction and mixing between CDM and WDM particles is very satisfactory from the point of view of the resolution of the anomalies.

1.4.3 Other Problems

Understanding the nature of DE and DM is the main open question of today cosmology but here follows a small list of other open problems:

- The problem of the initial singularity, the so-called **Big Bang**. This issue is related to a quantum formulation of gravity.
- There exists a couple of more technical, but nevertheless very important, issues which are called **tensions**. These happen when observations of different phenomena provide constraints on some parameters which are different up to 68% or 95% confidence level. There is now tension between the determination of H_0 via low-redshift probes and high-redshift ones (i.e. CMB). See e.g. Marra et al. (2013); Verde et al. (2013). Moreover, there is also a tension on the determination of σ_8 (Battye et al. 2015), recently corroborated by the analysis of the first year of the DES survey data collection (Abbott et al. 2017d).
- Testing the cosmological principle and the copernican principle. See e.g. Valkenburg et al. (2014).
- The CMB anomalies (Schwarz et al. 2016). These are unexpected (in the sense of statistically relevant) features of the CMB sky.
- The Lithium problem (Coc 2016). The predicted Lithium abundance is much larger than the observed one.

References

Abbott, T.M.C., et al.: Dark energy survey year 1 results: cosmological constraints from galaxy clustering and weak lensing (2017d)

Abbott, B.P., et al.: Observation of gravitational waves from a binary black hole merger. Phys. Rev. Lett. **116**(6), 061102 (2016b)

Abbott, B.P., et al.: GW151226: observation of gravitational waves from a 22-Solar-Mass binary black hole coalescence. Phys. Rev. Lett. **116**(24), 241103 (2016a)

Abbott, B.P., et al.: A gravitational-wave standard siren measurement of the Hubble constant. Nature **551**(7678), 85–88 (2017b)

Adam, R., et al.: Planck 2015 results. I. overview of products and scientific results. Astron. Astrophys. **594**, A1 (2016)

Ade, P.A.R., et al.: Planck 2015 results. XIII. Cosmol. Parametr. Astron. Astrophys. **594**, A13 (2016a)

Alpher, R.A., Bethe, H., Gamow, G.: The origin of chemical elements. Phys. Rev. **73**, 803–804 (1948)

Amendola, L., Tsujikawa, S.: Dark Energy: Theory and Observations. Cambridge University Press, Cambridge (2010)

Battye, R.A., Charnock, T., Moss, A.: Tension between the power spectrum of density perturbations measured on large and small scales. Phys. Rev. D **91**(10), 103508 (2015)

Bertone, G., Hooper, D.: A History of Dark Matter (2016)

Bonvin, V., et al.: H0LiCOW V. New COSMOGRAIL time delays of HE 04351223: H_0 to 3.8 per cent precision from strong lensing in a flat CDM model. Mon. Not. R. Astron. Soc. **465**(4), 4914–4930 (2017)

Bouchet, F.R., et al.: COrE (Cosmic Origins Explorer) A White Paper (2011)

Boylan-Kolchin, M., Bullock, J.S., Kaplinghat, M.: Too big to fail? the puzzling darkness of massive Milky way subhaloes. Mon. Not. R. Astron. Soc. **415**, L40 (2011)

Bullock, J.S., Boylan-Kolchin, M.: Small-scale challenges to the ΛCDM paradigm. Ann. Rev. Astron. Astrophys. **55**, 343–387 (2017)

Clowe, D., Bradac, M., Gonzalez, A.H., Markevitch, M., Randall, S.W., Jones, C., Zaritsky, D.: A direct empirical proof of the existence of dark matter. Astrophys. J. **648**, L109–L113 (2006)

Coc, A.: Primordial nucleosynthesis. J. Phys. Conf. Ser. **665**(1), 012001 (2016)

de Bernardis, P., et al.: A flat universe from high resolution maps of the cosmic microwave background radiation. Nature **404**, 955–959 (2000)

Dodelson, S.: Gravitational Lensing. Cambridge University Press, UK (2017)

Dodelson, S., Widrow, L.M.: Sterile-neutrinos as dark matter. Phys. Rev. Lett. **72**, 17–20 (1994)

Gaskins, J.M.: A review of indirect searches for particle dark matter. Contemp. Phys. **57**(4), 496–525 (2016)

Grieb, J.N., et al.: The clustering of galaxies in the completed SDSS-III Baryon oscillation spectroscopic survey: cosmological implications of the fourier space wedges of the final sample. Mon. Not. R. Astron. Soc. **467**(2), 2085–2112 (2017)

Hubble, E.: A relation between distance and radial velocity among extra-galactic nebulae. Proc. Nat. Acad. Sci. **15**, 168–173 (1929)

Kirby, E.N., Bullock, J.S., Boylan-Kolchin, M., Kaplinghat, M., Cohen, J.G.: The dynamics of isolated local group galaxies. Mon. Not. R. Astron. Soc. **439**(1), 1015–1027 (2014)

Klypin, A.A., Kravtsov, A.V., Valenzuela, O., Prada, F.: Where are the missing Galactic satellites? Astrophys. J. **522**, 82–92 (1999)

Kollmeier, J.A., Zasowski, G., Rix, H.-W., Johns, M., Anderson, S.F., Drory, N., Johnson, J.A., Pogge, R.W., Bird, J.C., Blanc, G.A., Brownstein, J.R., Crane, J.D., De Lee, N.M., Klaene, M.A., Kreckel, K., MacDonald, N., Merloni, A., Ness, M.K., O'Brien, T., Sanchez-Gallego, J.R., Sayres, C.C., Shen, Y., Thakar, A.R., Tkachenko, A., Aerts, C., Blanton, M.R., Eisenstein, D.J., Holtzman, J.A., Maoz, D., Nandra, K., Rockosi, C., Weinberg, D.H., Bovy, J., Casey, A.R., Chaname, J., Clerc, N., Conroy, C., Eracleous, M., Gänsicke, B.T., Hekker, S., Horne, K., Kauffmann, J., McQuinn, K.B.W., Pellegrini, E.W., Schinnerer, E., Schlafly, E.F., Schwope, A.D., Seibert, M., Teske, J.K., van Saders, J.L.: SDSS-V: Pioneering Panoptic Spectroscopy (2017). ArXiv e-prints

Lemaître, G.: A homogeneous universe of constant mass and growing radius accounting for the radial velocity of extragalactic nebulae. Ann. Soc. Sci. Brux. Ser. I Sci. Math. Astron. Phys. **A47**, 49–59 (1927)

Liu, J., Chen, X., Ji, X.: Current status of direct dark matter detection experiments. Nature Phys. **13**(3), 212–216 (2017)

Lovell, M.R., Eke, V., Frenk, C.S., Gao, L., Jenkins, A., Theuns, T., Wang, J., White, D.M., Boyarsky, A., Ruchayskiy, O.: The haloes of bright satellite galaxies in a warm dark matter universe. Mon. Not. R. Astron. Soc. **420**, 2318–2324 (2012)

Macciò, A.V., Paduroiu, S., Anderhalden, D., Schneider, A., Moore, B.: Cores in warm dark matter haloes: a Catch 22 problem. Mon. Not. R. Astron. Soc. **424**, 1105–1112 (2012)

Macciò, A.V., Mainini, R., Penzo, C., Bonometto, S.A.: Strongly coupled dark energy cosmologies: preserving LCDM success and easing low scale problems II - cosmological simulations. Mon. Not. R. Astron. Soc. **453**, 1371–1378 (2015)

Marra, V., Amendola, L., Sawicki, I., Valkenburg, W.: Cosmic variance and the measurement of the local Hubble parameter. Phys. Rev. Lett. **110**(24), 241305 (2013)

Martin, J.: Everything you always wanted to know about the cosmological constant problem (But were afraid to ask). Comptes Rendus Physique **13**, 566–665 (2012)

Moore, B.: Evidence against dissipationless dark matter from observations of galaxy haloes. Nature **370**, 629 (1994)

Olbers, W.: Edinburgh New philso. J **1**, 141 (1826)

Peccei, R.D., Quinn, H.R.: CP conservation in the presence of instantons. Phys. Rev. Lett. **38**, 1440–1443 (1977)

Penzias, A.A., Wilson, R.W.: A measurement of excess antenna temperature at 4080- Mc/s. Astrophys. J. **142**, 419–421 (1965)

Perlmutter, S., et al.: Measurements of omega and lambda from 42 high-redshift supernovae. Astrophys. J. **517**, 565–586 (1999)

Profumo, S.: An Introduction to Particle Dark Matter. Advanced textbooks in physics, World Scientific (2017)

Riess, A.G., et al.: Observational evidence from supernovae for an accelerating universe and a cosmological constant. Astron. J. **116**, 1009–1038 (1998)

Sandage, A.: Current problems in the extragalactic distance scale. ApJ **127**, 513 (1958)

Schneider, A., Anderhalden, D., Macciò, A., Diemand, J.: Warm dark matter does not do better than cold dark matter in solving small-scale inconsistencies. Mon. Not. R. Astron. Soc. **441**, 6 (2014)

Schwarz, D.J., Copi, C.J., Huterer, D., Starkman, G.D.: CMB anomalies after planck. Class. Quantum Gravity **33**(18), 184001 (2016)

Sciama, D.W.: The Unity of the Universe. Courier Corporation (2012)

Silk, J., et al.: Particle Dark Matter: Observations. Models and searches. Cambridge University Press, Cambridge (2010)

Slipher, V.M.: Nebulae. Proc. Am. Philso. Soc. **56**, 403–409 (1917)

Smoot, G.F., et al.: Structure in the COBE differential microwave radiometer first year maps. Astrophys. J. **396**, L1–L5 (1992)

Sofue, Y., Rubin, V.: Rotation curves of spiral galaxies. Ann. Rev. Astron. Astrophys. **39**, 137–174 (2001)

Sofue, Y., Tutui, Y., Honma, M., Tomita, A., Takamiya, T., Koda, J., Takeda, Y.: Central rotation curves of spiral galaxies. Astrophys. J. **523**, 136 (1999)

Spergel, D.N., Steinhardt, P.J.: Observational evidence for selfinteracting cold dark matter. Phys. Rev. Lett. **84**, 3760–3763 (2000)

Valkenburg, W., Marra, V., Clarkson, C.: Testing the copernican principle by constraining spatial homogeneity. Mon. Not. R. Astron. Soc. **438**, L6–L10 (2014)

van den Bergh, S.: The curious case of Lemaître's equation No. 24. JRASC **105**, 151 (2011)

Velten, H.E.S., vom Marttens, R.F., Zimdahl, W.: Aspects of the cosmological coincidence problem. Eur. Phys. J. C **74**(11), 3160 (2014)

Verde, L., Protopapas, P., Jimenez, R.: Planck and the local universe: quantifying the tension. Phys. Dark Univ. **2**, 166–175 (2013)

Viel, M., Becker, G.D., Bolton, J.S., Haehnelt, M.G.: Warm dark matter as a solution to the small scale crisis: new constraints from high redshift Lyman-forest data. Phys. Rev. D **88**, 043502 (2013)

Vogelsberger, M., Zavala, J., Simpson, C., Jenkins, A.: Dwarf galaxies in CDM and SIDM with baryons: observational probes of the nature of dark matter. Mon. Not. R. Astron. Soc. **444**, 3684 (2014)

Warren, M.S., Abazajian, K., Holz, D.E., Teodoro, L.: Precision determination of the mass function of dark matter halos. Astrophys. J. **646**, 881–885 (2006)

Way, M.J., Nussbaumer, H.: Lemaître's Hubble relationship. Phys. Today **64N8**, 8 (2011)

Weinberg, S.: Dreams of a Final Theory: The Search for the Fundamental Laws of Nature (1992)

Weinberg, S.: Gravitation and Cosmology: Principles and Applications of the General Theory of Relativity. Wiley, New York (1972)

Weinberg, S.: The cosmological constant problem. Rev. Mod. Phys. **61**, 1–23 (1989)

Weinberg, S.: Cosmology. Oxford University Press, UK (2008)

Williams, R.E., Blacker, B., Dickinson, M., Dixon, W.V.D., Ferguson, H.C., Fruchter, A.S., Giavalisco, M., Gilliland, R.L., Heyer, I., Katsanis, R., Levay, Z., Lucas, R.A., McElroy, D.B., Petro, L., Postman, M., Adorf, H.-M., Hook, R.: The hubble deep field: observations, data reduction, and galaxy photometry. AJ **112**, 1335 (1996)

Zlatev, I., Wang, L.-M., Steinhardt, P.J.: Quintessence, cosmic coincidence, and the cosmological constant. Phys. Rev. Lett. **82**, 896–899 (1999)

Zwicky, F.: Die Rotverschiebung von extragalaktischen Nebeln. Helvetica Physica Acta **6**, 110–127 (1933)

Chapter 2
The Universe in Expansion

> *Oras ubicumque locaris extremas, quaeram: quid telo denique fiet?*
> *(wherever you shall set the boundaries, I will ask: what will then happen to the arrow?)*
>
> Lucretius, De Rerum Natura

We introduce in this chapter the geometric basis of cosmology and the expansion of the universe. A part from the technical treatment, historical, theological and mythological introductions to cosmology can be found in Ryden (2003) and Bonometto (2008).

2.1 Newtonian Cosmology

In order to do cosmology we need a theory of gravity, because gravity is a long-range interaction and the universe is pretty big. Electromagnetism is also a long-range interaction, but considering the lack of evidence that the universe is charged or made up of charges here and there, it seems reasonable that gravity is what we need in order to describe the universe on large scales.

Which theory of gravity do we use for describing the universe? It turns out that Newtonian physics works surprisingly well! It is also surprising that attempts of doing cosmology with Newtonian gravity are well posterior to relativistic cosmology itself.

In particular, the first work on Newtonian cosmology can be dated back to Milne and McCrea in the 1930s (McCrea and Milne 1934; Milne 1934). These were models of pure dust, while pressure was introduced later by McCrea (1951) and Harrison (1965). More recently the issue of pressure corrections in Newtonian cosmology has been tackled again in Lima et al. (1997), Fabris and Velten (2012), Hwang and Noh (2013) and Baqui et al. (2016).

© Springer International Publishing AG, part of Springer Nature 2018
O. Piattella, *Lecture Notes in Cosmology*, UNITEXT for Physics,
https://doi.org/10.1007/978-3-319-95570-4_2

Newtonian cosmology works as follows. Imagine a sphere of dust of radius r. This radius is time-dependent because the configuration is not stable since there is no pressure, thus $r = r(t)$. We assume homogeneity of the sphere during the evolution, i.e. its density depends only on the time:

$$\rho(t) = \frac{3M}{4\pi r(t)^3} \, , \tag{2.1}$$

where M is the mass of the dust sphere and is constant. Now, imagine a small test particle of mass m on the surface of the sphere. By Newton's gravitation law and Gauss's theorem one has:

$$F = -\frac{GMm}{r(t)^2} \quad \Rightarrow \quad \ddot{r} = -\frac{4\pi G}{3}\rho r \, , \tag{2.2}$$

where we have used Eq. (2.1) and the dot denotes derivation with respect to the time t. This is the same acceleration equation that we shall find later using GR, cf. Eq. (2.51).

Exercise 2.1 Integrate Eq. (2.2) and show that:

$$\frac{\dot{r}^2}{r^2} = \frac{8\pi G}{3}\rho - \frac{K}{r^2} \, , \tag{2.3}$$

where K is an integration constant.

We shall also see that Eq. (2.3) is the same as Friedmann equation in GR, cf. (2.50). The integration constant K can be interpreted as the total energy of the particle. Indeed, we can rewrite Eq. (2.3) as follows:

$$E \equiv -\frac{mK}{2} = \frac{m}{2}\dot{r}^2 - \frac{GMm}{r} \, , \tag{2.4}$$

which is the expression of the total energy of a particle of mass m in the gravitational field of the mass M.

2.2   Relativistic Cosmology

In GR we have geometry and matter related by Einstein equations:

$$G_{\mu\nu} = \frac{8\pi G}{c^4}T_{\mu\nu} \, , \tag{2.5}$$

where $G_{\mu\nu}$ is the Einstein tensor, computed from the metric, and $T_{\mu\nu}$ is the energy-momentum or stress-energy tensor, and describes the matter content.

In cosmology, which is the metric which describes the universe and what is the matter content? It turns out that both questions are very difficult to answer and, indeed, there are no still clear answers, as we stressed in Chap. 1.

2.2.1 Friedmann–Lemaître–Robertson–Walker Metric

The metric used to describe the universe on large scales is the Friedmann–Lemaître–Robertson–Walker (FLRW) metric. This is based on the assumption of very high symmetry for the universe, called the **cosmological principle**, which is minimally stated as follows: the universe is isotropic and homogeneous, i.e. there is no preferred direction or preferred position.

A more formal definition can be found in Weinberg (1972, p. 412) and is based on the following two requirements:

1. The hypersurfaces with constant cosmic standard time are maximally symmetric subspaces of the whole of the spacetime;
2. The global metric and all the cosmic tensors such as the stress-energy one $T_{\mu\nu}$ are form-invariant with respect to the isometries of those subspaces.

We shall come back in a moment to maximally symmetric spaces. Roughly speaking, the second requirement above means that the matter quantities can depend only on the time.

The cosmological principle seems to be compatible with observations at very large scales. According to Wu et al. (1999): *on a scale of about 100 h^{-1} Mpc the rms density fluctuations are at the level of ~10% and on scales larger than 300 h^{-1} Mpc the distribution of both mass and luminous sources safely satisfies the cosmological principle of isotropy and homogeneity.*

In a recent work Sarkar and Pandey (2016) find that the quasar distribution is homogeneous on scales larger than 250 h^{-1} Mpc. Moreover, numerical relativity seems to indicate that the average evolution of a generic metric on large scale is compatible with that of FLRW metric (Giblin et al. 2016).

According to the cosmological principle, the constant-time spatial hypersurfaces are maximally symmetric.[1] A maximally symmetric space is completely characterised by one number only, i.e. its scalar curvature, which is also a constant. See Weinberg (1972, Chap. 13).

Let R be this constant scalar curvature. The Riemann tensor of a maximally symmetric D-dimensional space is written as:

$$R_{\mu\nu\rho\sigma} = \frac{R}{D(D-1)}(g_{\mu\rho}g_{\nu\sigma} - g_{\mu\sigma}g_{\nu\rho}) \,. \tag{2.6}$$

Contracting with $g^{\mu\rho}$ we get for the Ricci tensor:

---

[1] This means that they possess 6 Killing vectors, i.e. there are six transformations which leave the spatial metric invariant (Weinberg 1972).

$$R_{\nu\sigma} = \frac{R}{D} g_{\nu\sigma} \ , \qquad (2.7)$$

and then R is the scalar curvature, as we stated, since $g^{\nu\sigma} g_{\nu\sigma} = D$. Since any given number can be negative, positive or zero, we have three possible maximally symmetric spaces. Now, focusing on the 3-dimensional spatial case:

1. $ds_3^2 = |d\mathbf{x}|^2 \equiv \delta_{ij} dx^i dx^j$, i.e. the Euclidean space. The scalar curvature is zero, i.e. the space is flat. This metric is invariant under 3-translations and 3-rotations.
2. $ds_3^2 = |d\mathbf{x}|^2 + dz^2$, with the constraint $z^2 + |\mathbf{x}|^2 = a^2$. This is a 3-sphere of radius a embedded in a 4-dimensional Euclidean space. It is invariant under the six 4-dimensional rotations.
3. $ds_3^2 = |d\mathbf{x}|^2 - dz^2$, with the constraint $z^2 - |\mathbf{x}|^2 = a^2$. This is a 3-hypersphere, or a hyperboloid, in a 4-dimensional pseudo-Euclidean space. It is invariant under the six 4-dimensional pseudo-rotations (i.e. Lorentz transformations).

Exercise 2.2 Why are there six independent 4-dimensional rotations in the 4-dimensional Euclidean space? How many are there in a D-dimensional Euclidean space?

Let us write in a compact form the above metrics as follows:

$$ds_3^2 = |d\mathbf{x}|^2 \pm dz^2 \ , \qquad z^2 \pm |\mathbf{x}|^2 = a^2 \ . \qquad (2.8)$$

Differentiating $z^2 \pm |\mathbf{x}|^2 = a^2$, one gets:

$$z dz = \mp \mathbf{x} \cdot d\mathbf{x} \ . \qquad (2.9)$$

Now put this back into ds_3^2:

$$ds_3^2 = |d\mathbf{x}|^2 \pm \frac{(\mathbf{x} \cdot d\mathbf{x})^2}{a^2 \mp |\mathbf{x}|^2} \ . \qquad (2.10)$$

In a more compact form:

$$ds_3^2 = |d\mathbf{x}|^2 + K \frac{(\mathbf{x} \cdot d\mathbf{x})^2}{a^2 - K|\mathbf{x}|^2} \ , \qquad (2.11)$$

with $K = 0$ for the Euclidean case, $K = 1$ for the spherical case and $K = -1$ for the hyperbolic case. The components of the spatial metric in Eq. (2.11) can be immediately read off and are:

$$g_{ij}^{(3)} = \delta_{ij} + K \frac{x_i x_j}{a^2 - K|\mathbf{x}|^2} \ . \qquad (2.12)$$

Exercise 2.3 Write down metric (2.11) in spherical coordinates. Use the fact that $|d\mathbf{x}|^2 = dr^2 + r^2 d\Omega^2$, where

$$d\Omega^2 \equiv d\theta^2 + \sin^2\theta d\phi^2 , \tag{2.13}$$

and use:

$$\mathbf{x} \cdot d\mathbf{x} = \frac{1}{2}d|\mathbf{x}|^2 = \frac{1}{2}d(r^2) = rdr . \tag{2.14}$$

Show that the result is:

$$ds_3^2 = \frac{a^2 dr^2}{a^2 - Kr^2} + r^2 d\Omega^2 \tag{2.15}$$

Calculate the scalar curvature $R^{(3)}$ for metric (2.15). Show that $R_{ij}^{(3)} = 2Kg_{ij}^{(3)}/a^2$ and thus $R^{(3)} = 6K/a^2$.

If we normalise $r \to r/a$ in metric (2.15), we can write:

$$ds_3^2 = a^2 \left(\frac{dr^2}{1 - Kr^2} + r^2 d\Omega^2 \right) , \tag{2.16}$$

and letting a to be a function of time, we finally get the FLRW metric:

$$ds^2 = -c^2 dt^2 + a^2(t) \left(\frac{dr^2}{1 - Kr^2} + r^2 d\Omega^2 \right) \tag{2.17}$$

The time coordinate used here is called **cosmic time**, whereas the spatial coordinates are called **comoving coordinates**. For each t the spatial slices are maximally symmetric; $a(t)$ is called **scale factor**, since it tells us how the distance between two points scales with time.

The FLRW metric was first worked out by Friedmann (1922, 1924) and then derived on the basis of isotropy and homogeneity by Robertson (1935, 1936) and Walker (1937). Lemaître's work (Lemaitre 1931) had been also essential to develop it.[2]

A further comment concerning FLRW metric (2.17) is in order here. The dimension of distance is being carried by the scale factor a itself, since we rescaled the radius $r \to r/a$. Indeed, as we computed earlier, the spatial curvature is $R^{(3)} = 6K/a(t)^2$, also time-varying, and it is a real, dimensional number as it should be.

---

[2]See also Lemaître (1997) for a recent republication and translation of Lemaître's 1933 paper.

2.2.2 The Conformal Time

A very useful form of rewriting FLRW metric (2.17) is via the **conformal time** η:

$$a d\eta = dt \quad \Rightarrow \quad \eta - \eta_i = \int_{t_i}^{t} \frac{dt'}{a(t')} \, . \tag{2.18}$$

As we shall see later, but as we already can guess from the above integration, $c(\eta - \eta_i)$ represents the comoving distance travelled by a photon between the times η_i and η, or t_i and t. The conformal time allows to rewrite FLRW metric (2.17) as follows:

$$ds^2 = a(\eta)^2 \left(-c^2 d\eta^2 + \frac{dr^2}{1 - Kr^2} + r^2 d\Omega^2 \right) \tag{2.19}$$

i.e. the scale factor has become a conformal factor (hence the name for η). Recalling the earlier discussion about dimensionality, if a has dimensions then $c\eta$ is dimensionless. On the other hand, if a is dimensionless, then η is indeed a time.

Note also that metric (2.19) for $K = 0$ is Minkowski metric multiplied by a conformal factor.

2.2.3 FLRW Metric Written with Proper Radius

A third useful way to write FLRW metric (2.17) is using the **proper radius**, which is defined as follows:

$$\mathcal{D}(t) \equiv a(t)r \, . \tag{2.20}$$

We shall discuss in more detail the proper radius, or proper distance, in Sect. 2.5.

Exercise 2.4 Using $\mathcal{D}$ instead of r, show that the FLRW metric (2.17) becomes:

$$ds^2 = -c^2 dt^2 \left(1 - H^2 \frac{\mathcal{D}^2/c^2}{1 - K\mathcal{D}^2/a^2} \right) - \frac{2H\mathcal{D}dt d\mathcal{D}}{1 - K\mathcal{D}^2/a^2}$$
$$+ \frac{d\mathcal{D}^2}{1 - K\mathcal{D}^2/a^2} + \mathcal{D}^2 d\Omega^2 \, , \tag{2.21}$$

where

$$H \equiv \frac{\dot{a}}{a} \tag{2.22}$$

is the **Hubble parameter**. The dot denotes derivation with respect to the cosmic time.

2.2.4 Light-Cone Structure of the FLRW Space

Let us consider the $K = 0$ case, for simplicity. Moreover, consider also $d\Omega = 0$. In this case, the radial coordinate is also the distance. Then, putting $ds^2 = 0$ in the FLRW metric gives the following light-cone structures.

Cosmic Time-Comoving Distance

From the FLRW metric (2.17), the condition $ds^2 = 0$ gives us:

$$\frac{cdt}{dr} = \pm a(t) . \tag{2.23}$$

We put our observer at $r = 0$ and $t = t_0$. The plus sign in the above equation then describes an outgoing photon, i.e. the future light-cone, whereas the negative sign describes an incoming photon, i.e. the past-light cone, which is much more interesting to us. So, let us keep the negative sign and discuss the shape of the light-cone.

Assume that $a(0) = 0$. Therefore, the slope of the past light-cone starts as $-a(t_0)$, which we can normalise as -1, i.e. locally the past light-cone is identical to the one in Minkowski space. However, a goes to zero, so the light-cone becomes flat, encompassing more radii than it would for Minkowski space. See Fig. 2.1. We can show this analytically by taking the second derivative of Eq. (2.23) with the minus sign:

$$\frac{c^2 d^2 t}{dr^2} = -\dot{a}\frac{cdt}{dr} = a\dot{a} . \tag{2.24}$$

Being $a > 0$ and $\dot{a} > 0$ (we consider just the case of an expanding universe), the function $t(r)$ is convex (i.e. it is "bent upwards").

Conformal Time-Comoving Distance

For the FLRW metric (2.19), the condition ds^2 gives:

$$\frac{cd\eta}{dr} = \pm 1 . \tag{2.25}$$

The latter is exactly the same light-cone structure of Minkowski space. Indeed, Friedmann metric written in conformal time and for $K = 0$ is Minkowski metric multiplied by a conformal factor $a(\eta)$. See Fig. 2.2.

Cosmic Time-Proper Distance

In order to find the light-cone structure for the FLRW metric (2.21) with $K = 0$, we need to solve the following equation:

$$-\frac{c^2 dt^2}{d\mathcal{D}^2}\left(1 - \frac{H^2\mathcal{D}^2}{c^2}\right) - \frac{2H\mathcal{D}}{c}\frac{cdt}{d\mathcal{D}} + 1 = 0 . \tag{2.26}$$

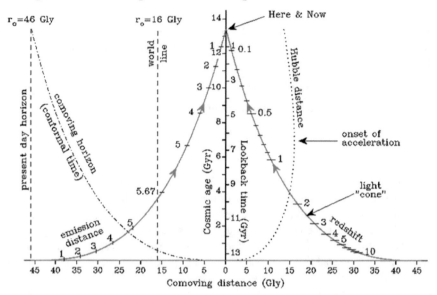

Fig. 2.1 Space-time diagram and light-cone structure for the FLRW metric (2.17). Credit: Prof. Mark Whittle, University of Virginia

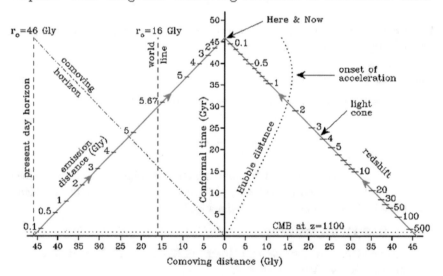

Fig. 2.2 Space-time diagram and light-cone structure for the FLRW metric (2.19). Credit: Prof. Mark Whittle, University of Virginia

Exercise 2.5 Solve Eq. (2.26) algebraically for $cdt/d\mathcal{D}$ and show that:

$$\frac{cdt}{d\mathcal{D}} = \left(\frac{H\mathcal{D}}{c} \pm 1\right)^{-1} . \tag{2.27}$$

For $t = t_0$ we have $H(t_0) > 0$ and $\mathcal{D} = 0$. Therefore, from Eq. (2.27) we have that $(cdt/d\mathcal{D})(t_0) = \pm 1$ and thus we must choose the minus sign in order to describe the past light-cone. Going back in time, $H\mathcal{D}$ grows, until

$$\frac{H\mathcal{D}}{c} = 1 , \tag{2.28}$$

for which $cdt/d\mathcal{D}$ diverges. This means that no signal can come from beyond this distance $\mathcal{D} = c/H$, which is the **Hubble radius** that we met in Chap. 1. See Fig. 2.3. The lower part of this figure is explained as follows. First of all $H\mathcal{D}$ becomes larger than 1, and this explains the change of sign of the slope of the light cone. Then, $H \rightarrow \infty$ for $a \rightarrow 0$ (if we assume a model with Big Bang) and therefore $cdt/d\mathcal{D} \rightarrow 0$. This is why the light-cone flattens close to $t = 0$ in Fig. 2.3.

2.2.5 Christoffel Symbols and Geodesics

Exercise 2.6 Assume $K = 0$ in metric (2.17), rewrite it in Cartesian coordinates and calculate the Christoffel symbols. Show that:

$$\Gamma^0_{00} = 0 , \quad \Gamma^0_{0i} = 0 , \quad \Gamma^0_{ij} = \frac{a\dot{a}}{c}\delta_{ij} , \quad \Gamma^i_{0j} = \frac{H}{c}\delta^i{}_j . \tag{2.29}$$

We now use these in the geodesic equation:

$$\frac{dP^\mu}{d\lambda} + \Gamma^\mu_{\nu\rho}P^\nu P^\rho = 0 , \tag{2.30}$$

where $P^\mu \equiv dx^\mu/d\lambda$ is the four-momentum and λ is an affine parameter. For a particle of mass m, one has $\lambda = \tau/m$, where τ is the proper time. The norm of the four-momentum is:

$$P^2 \equiv g_{\mu\nu}P^\mu P^\nu = -\frac{E^2}{c^2} + p^2 = -m^2c^2 , \tag{2.31}$$

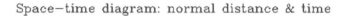

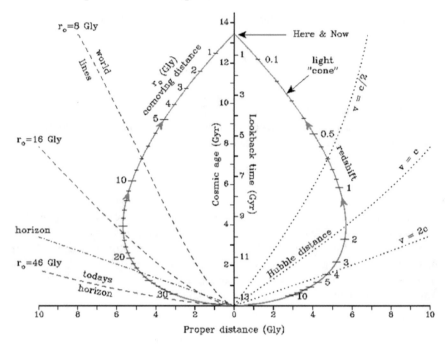

Fig. 2.3 Space-time diagram and light-cone structure for the FLRW metric (2.21). Credit: Prof. Mark Whittle, University of Virginia

where we have defined the energy and the **physical momentum** (or **proper momentum**):

$$\frac{E^2}{c^2} \equiv -g_{00}(P^0)^2 \,, \qquad p^2 \equiv g_{ij} P^i P^j \,, \tag{2.32}$$

and the last equality of Eq. (2.31), which applies only to massive particles, comes from:

$$\frac{ds^2}{d\lambda^2} = \frac{m^2 ds^2}{d\tau^2} = -m^2 c^2 \,, \tag{2.33}$$

since, by definition, $ds^2 = -c^2 d\tau^2$. We have recovered above the well-known dispersion relation of special relativity. The metric $g_{\mu\nu}$ used above is, in principle, general. But, of course, we now specialise it to the FLRW one.

For a photon, $m = 0$ and $E = pc$. The time-component of the geodesic equation is the following:

$$\frac{dP^0}{d\lambda} + \frac{a\dot{a}}{c}\delta_{ij} P^i P^j = 0 \,. \tag{2.34}$$

Introducing the proper momentum as defined in Eq. (2.32), one gets:

$$c\frac{dp}{d\lambda} + Hp^2 = 0 \, . \tag{2.35}$$

Exercise 2.7 Solve Eq. (2.35) and show that $p = E/c \propto 1/a$, i.e. the energy of the photon is proportional to the inverse scale factor.

Therefore, we can write:

$$\frac{E_{obs}}{E_{em}} = \frac{a_{em}}{a_{obs}} \, . \tag{2.36}$$

On the other hand the photon energy is $E = hf$, with f its frequency. Therefore:

$$\frac{a_{em}}{a_{obs}} = \frac{E_{obs}}{E_{em}} = \frac{f_{obs}}{f_{em}} = \frac{\lambda_{em}}{\lambda_{obs}} = \frac{1}{1+z} \, . \tag{2.37}$$

This is the relation between the redshift and the scale factor. We have connected observation with theory. Usually, $a_{obs} = 1$ and the above relation is simply written as $1 + z = 1/a$.

What does happen, on the other hand, to the energy of a massive particle? The time-geodesic equation for massive particles is identical to the one for photons, but the dispersion relation is different, i.e. $E^2 = m^2c^4 + p^2c^2$. Therefore:

$$E = \sqrt{m^2c^4 + \frac{p_i^2 a_i^2 c^2}{a^2}} \, , \tag{2.38}$$

where p_i is some initial proper momentum, at the time t_i and $a_i = a(t_i)$. For $m = 0$ we recover the result already obtained for photons. For massive particles the above relation can be approximated as follows:

$$E = mc^2\left(1 + \frac{p_i^2 a_i^2}{2a^2 m^2 c^2} + \cdots\right) \, , \qquad (mc \gg p) \, , \tag{2.39}$$

i.e. performing the expansion for small momenta which is usually done in special relativity. The second contribution between parenthesis is the classical kinetic energy of the particle, whose average is proportional to $k_B T$. Therefore:

$$T \propto a^{-1} \, , \quad \text{for relativistic particles,} \tag{2.40}$$
$$T \propto a^{-2} \, , \quad \text{for non-relativistic particles.} \tag{2.41}$$

We shall recover the above result also using the Boltzmann equation.

Exercise 2.8 Show that $p \propto 1/a$ by using not the time-component geodesic equation but the spatial one:

$$\frac{dP^i}{d\lambda} + 2\Gamma^i_{0j} P^0 P^j = 0 \,. \tag{2.42}$$

Why is there a factor two in this equation?

2.3 Friedmann Equations

Given FLRW metric, Friedmann equations can be straightforwardly computed from the Einstein equations:

$$G_{\mu\nu} + \Lambda g_{\mu\nu} = R_{\mu\nu} - \frac{1}{2} g_{\mu\nu} R + \Lambda g_{\mu\nu} = \frac{8\pi G}{c^4} T_{\mu\nu} \,, \tag{2.43}$$

where Λ is the cosmological constant.

Exercise 2.9 Calculate from FLRW metric (2.17) the components of the Ricci tensor. Show that:

$$R_{00} = -\frac{3}{c^2} \frac{\ddot{a}}{a} \,, \qquad R_{0i} = 0 \,, \qquad R_{ij} = \frac{1}{c^2} g_{ij} \left(2H^2 + \frac{\ddot{a}}{a} + 2\frac{Kc^2}{a^2} \right) \,, \tag{2.44}$$

and show that the scalar curvature is:

$$R = \frac{6}{c^2} \left(\frac{\ddot{a}}{a} + H^2 + \frac{Kc^2}{a^2} \right) \,. \tag{2.45}$$

Finally, compute the Einstein equations:

$$\boxed{ H^2 + \frac{Kc^2}{a^2} = \frac{8\pi G}{3c^2} T_{00} + \frac{\Lambda c^2}{3} } \tag{2.46}$$

$$\boxed{ g_{ij} \left(H^2 + 2\frac{\ddot{a}}{a} + \frac{Kc^2}{a^2} - \Lambda c^2 \right) = -\frac{8\pi G}{c^2} T_{ij} } \tag{2.47}$$

These are called **Friedmann equations** or **Friedmann equation** and **acceleration equation** or **Friedmann equation** and **Raychaudhuri equation**.

Which stress-energy tensor $T_{\mu\nu}$ do we use in Eqs. (2.46) and (2.47)? Having fixed the metric to be the FLRW one, we have some strong constraints:

- First of all: $G_{0i} = 0$ implies that $T_{0i} = 0$, i.e. there cannot be a flux of energy in any direction because it would violate isotropy;
- Second, since $G_{ij} \propto g_{ij}$, then $T_{ij} \propto g_{ij}$.
- Finally, since $G_{\mu\nu}$ depends only on t, then it must be so also for $T_{\mu\nu}$.

Therefore, let us stipulate that

$$T_{00} = \rho(t)c^2 = \varepsilon(t) , \qquad T_{0i} = 0 , \qquad T_{ij} = g_{ij}P(t) , \qquad (2.48)$$

where $\rho(t)$ is the rest mass density, $\varepsilon(t)$ is the energy density and $P(t)$ is the pressure. In tensorial notation we can write the following general form for the stress-energy tensor:

$$T_{\mu\nu} = \left(\rho + \frac{P}{c^2}\right) u_\mu u_\nu + P g_{\mu\nu} \qquad (2.49)$$

where u_μ is the four-velocity of the fluid element. In this form of Eq. (2.49), the stress-energy tensor does not contain either viscosity or energy transport terms. Matter described by (2.49) is known as **perfect fluid**. For more detail about the latter see Schutz (1985) whereas for more detail about viscosity, heat fluxes and the imperfect fluids see e.g. Weinberg (1972) and Maartens (1996).

Combine Eqs. (2.46), (2.47) and (2.48). The Friedmann equation becomes:

$$H^2 = \frac{8\pi G}{3}\rho + \frac{\Lambda c^2}{3} - \frac{Kc^2}{a^2} \qquad (2.50)$$

while the acceleration equation is the following:

$$\frac{\ddot{a}}{a} = -\frac{4\pi G}{3}\left(\rho + \frac{3P}{c^2}\right) + \frac{\Lambda c^2}{3} \qquad (2.51)$$

Exercise 2.10 Write Eqs. (2.50) and (2.51) using the conformal time introduced in Eq. (2.18). Show that the Friedmann equation becomes:

$$\mathcal{H}^2 = \frac{8\pi G}{3}\rho a^2 + \frac{\Lambda c^2 a^2}{3} - Kc^2 \qquad (2.52)$$

and that the acceleration equation becomes:

$$\frac{a''}{a} = \frac{4\pi G}{3}\left(\rho - \frac{3P}{c^2}\right)a^2 + \frac{2\Lambda c^2 a^2}{3} - Kc^2 , \qquad (2.53)$$

where the prime denotes derivation with respect to the conformal time η and

$$\boxed{\mathcal{H} \equiv \frac{a'}{a}} \tag{2.54}$$

is the **conformal Hubble factor**.

In the Friedmann and acceleration equations, ρ and P are the total density and pressure. Hence, they can be written as sums of the contributions of the individual components:

$$\rho \equiv \sum_x \rho_x , \qquad P \equiv \sum_x P_x . \tag{2.55}$$

The contribution from the cosmological constant can be considered either geometrically or as a matter component with the following density and pressure:

$$\rho_\Lambda \equiv \frac{\Lambda c^2}{8\pi G} , \qquad P_\Lambda \equiv -\rho_\Lambda c^2 . \tag{2.56}$$

The scale factor a is, by definition, positive, but its derivative can be negative. This would represent a contracting universe. Note that the left hand side of the Friedmann equation (2.50) is non-negative. Therefore, $\dot{a}$ can vanish only if $K > 0$, i.e. for a spatially closed universe. This implies that, if $K \leqslant 0$ and if there exists an instant for which $\dot{a} > 0$, then the universe will expand forever.

2.3.1   The Hubble Constant and the Deceleration Parameter

When the Hubble parameter H is evaluated at the present time t_0, it becomes a number: the Hubble constant H_0 which we already met in Chap. 1 in the Hubble's law (1.1). Its value is

$$H_0 = 67.74 \pm 0.46 \text{ km s}^{-1} \text{ Mpc}^{-1} , \tag{2.57}$$

at the 68% confidence level, as reported by the Planck group (Ade et al. 2016). Usually H_0 is conveniently written as

$$H_0 = 100 \, h \text{ km s}^{-1} \text{ Mpc}^{-1} . \tag{2.58}$$

The unit of measure of the Hubble constant is an inverse time:

$$H_0 = 3.24 \, h \times 10^{-18} \text{ s}^{-1} , \tag{2.59}$$

whose inverse gives the order of magnitude of the age of the universe:

$$\frac{1}{H_0} = 3.09 \, h^{-1} \times 10^{17} \, s = 9.78 \, h^{-1} \, Gyr , \qquad (2.60)$$

and multiplied by c gives the order of magnitude of the size of the visible universe, i.e. the Hubble radius that we have already seen in Eq. (2.28) but evaluated at the present time $t = t_0$:

$$\frac{c}{H_0} = 9.27 \, h^{-1} \times 10^{25} \, m = 3.00 \, h^{-1} \, Gpc . \qquad (2.61)$$

But what does "present time" t_0 mean? Time flows, therefore t_0 cannot be a constant! That is true, but if we compare a time span of 100 years (the span of some human lives) to the age of the universe (about 14 billion years), we see that the ratio is about 10^{-8}. Since this is pretty small, we can consider t_0 to be a constant, also referred to as the age of the universe.[3] We can calculate it as follows:

$$t_0 = \int_0^{t_0} dt = \int_0^1 \frac{da}{\dot{a}} = \int_0^1 \frac{da}{H(a)a} = \int_0^\infty \frac{dz}{H(z)(1+z)} . \qquad (2.62)$$

Exercise 2.11 Prove the last equality of Eq. (2.62).

The integration limits of Eq. (2.62) deserve some explanation. We assumed that $a(t = 0) = 0$, i.e. the Big Bang. This condition is not always true, since there are models of the universe, e.g. the de Sitter universe, for which a vanishes only when $t \to -\infty$. The other assumption is that $a(t_0) = 1$. This is a pure normalisation, done for convenience, which is allowed by the fact that the dynamics is invariant if we multiply the scale factor by a constant.

Recall that, in cosmology, when a quantity has subscript 0, it usually means that it is evaluated at $t = t_0$.

The Deceleration Parameter

Let us focus now on Eq. (2.51). It contains $\ddot{a}$, so it describes how the expansion of the universe is accelerating. The key-point is that if the right hand side of Eq. (2.51) is positive, i.e. $\rho + 3P/c^2 < 0$, then $\ddot{a} > 0$. There exists a parameter, named **deceleration parameter**, with which to measure the entity of the acceleration. It is defined as follows:

$$\boxed{q \equiv -\frac{\ddot{a}a}{\dot{a}^2}} \qquad (2.63)$$

---

[3]Pretty much the same happens with the redshift. A certain source has redshift z which, actually, is not a constant but varies slowly. This is called **redshift drift** and it was first considered by Sandage (1962) and McVittie (1962). Applications of the redshift drift phenomenon to gravitational lensing are proposed in Piattella and Giani (2017).

In Riess et al. (1998) and Perlmutter et al. (1999) analysis based on type Ia supernovae observation have shown that $q_0 < 0$, i.e. the deceleration parameter is negative and therefore the universe is in a state of accelerated expansion. We perform a similar but simplified analysis in Sect. 11.1 in order to illustrate how data in cosmology are analysed.

2.3.2   Critical Density and Density Parameters

Let us now rewrite Eq. (2.50) incorporating Λ in the total density ρ:

$$H^2 = \frac{8\pi G\rho}{3} - \frac{Kc^2}{a^2} \,. \tag{2.64}$$

The value of the total ρ such that $K = 0$ is called **critical energy density** and has the following form:

$$\boxed{\rho_{\mathrm{cr}} \equiv \frac{3H^2}{8\pi G}} \tag{2.65}$$

Its present value (Ade et al. 2016) is:

$$\boxed{\rho_{\mathrm{cr},0} = 1.878 \, h^2 \times 10^{-29} \ \mathrm{g \ cm}^{-3}} \tag{2.66}$$

It turns out that ρ_0 is very close to $\rho_{\mathrm{cr},0}$, so that our universe is spatially flat. Such an extreme fine-tuning in K is a really surprising coincidence, known as the **flatness problem**. A possible solution is provided by the inflationary theory which we shall see in detail in Chap. 8.

Instead of densities, it is very common and useful to employ the density parameter Ω, which is defined as

$$\boxed{\Omega \equiv \frac{\rho}{\rho_{\mathrm{cr}}} = \frac{8\pi G\rho}{3H^2}} \tag{2.67}$$

i.e. the energy density normalised to the critical one. We can then rewrite Friedmann equation (2.50) as follows:

$$1 = \Omega - \frac{Kc^2}{H^2 a^2} \,. \tag{2.68}$$

Defining

$$\Omega_K \equiv -\frac{Kc^2}{H^2 a^2} \,, \tag{2.69}$$

i.e. associating the energy density

$$\rho_K \equiv -\frac{3Kc^2}{8\pi Ga^2} , \tag{2.70}$$

to the spatial curvature, we can recast Eq. (2.68) in the following simple form:

$$1 = \Omega + \Omega_K . \tag{2.71}$$

Therefore, the sum of all the density parameters, *the curvature one included*, is always equal to unity. In particular, if it turns out that $\Omega \simeq 1$, this implies that $\Omega_K \simeq 0$, i.e. the universe is spatially flat. From the latest Planck data (Ade et al. 2016) we know that:

$$\boxed{\Omega_{K0} = 0.0008^{+0.0040}_{-0.0039}} \tag{2.72}$$

at the 95% confidence level.

It is more widespread in the literature the normalisation of ρ to the *present-time* critical density, i.e.

$$\boxed{\Omega \equiv \frac{\rho}{\rho_{cr,0}} = \frac{8\pi G\rho}{3H_0^2}} \tag{2.73}$$

because it leaves more evident the dependence on a of each material component. With this definition of Ω, Friedmann equation (2.50) is written as:

$$\frac{H^2}{H_0^2} = \sum_x \Omega_{x0} f_x(a) + \frac{\Omega_{K0}}{a^2} , \tag{2.74}$$

where $f_x(a)$ is a function which gives the a-dependence of the material component x and $f_x(a_0 = 1) = 1$. Consistently:

$$\boxed{\sum_x \Omega_{x0} + \Omega_{K0} = 1} \tag{2.75}$$

also known as **closure relation**. We shall use the definition $\Omega_x \equiv \rho_x/\rho_{cr,0}$ throughout these notes.

2.3.3 The Energy Conservation Equation

The energy conservation equation

$$\boxed{\nabla_\nu T^{\mu\nu} = 0} \tag{2.76}$$

is encapsulated in GR through the Bianchi identities. Therefore, it is not independent from the Friedmann equations (2.50) and (2.51). For the FLRW metric and a perfect fluid, it has a particularly simple form:

$$\dot{\rho} + 3H \left(\rho + \frac{P}{c^2} \right) = 0 \tag{2.77}$$

This is the $\mu = 0$ component of $\nabla_\nu T^{\mu\nu} = 0$ and it is also known from fluid dynamics as **continuity equation**.

Exercise 2.12 Derive the continuity equation (2.77) by combining Friedmann and acceleration equations (2.50) and (2.51). Derive it in a second way by explicitly calculating the four-divergence of the energy-momentum tensor.

The continuity equation can be analytically solved if we assume an equation of state of the form $P = w\rho c^2$, with w constant. The general solution is:

$$\rho = \rho_0 a^{-3(1+w)} \qquad (w = \text{constant}) , \tag{2.78}$$

where $\rho_0 \equiv \rho(a_0 = 1)$.

Exercise 2.13 Prove the above result of Eq. (2.78).

There are three particular values of w which play a major role in cosmology:

Cold matter: $w = 0$, i.e. $P = 0$, for which $\rho = \rho_0 a^{-3}$. As we have discussed in Chap. 1, the adjective cold refers to the fact that particles making up this kind of matter have a kinetic energy much smaller than the mass energy, i.e. they are non-relativistic. If they are thermally produced, i.e. if they were in thermal equilibrium with the primordial plasma, they have a mass much larger than the temperature of the thermal bath. We shall see this characteristic in more detail in Chap. 3.

Cold matter is also called **dust** and it encompasses all the non-relativistic known elementary particles, which are overall dubbed **baryons** in the jargon of cosmology. If they exist, unknown non-relativistic particles are called **cold dark matter** (CDM).

Hot matter: $w = 1/3$, i.e. $P = \rho/3$, for which $\rho = \rho_0 a^{-4}$. The adjective hot refers to the fact that particles making up this kind of matter are relativistic.

For this reason they are known, in the jargon of cosmology, as **radiation** and they encompass not only the relativistic known elementary particles, but possibly the unknown ones (i.e. **hot dark matter**). The primordial neutrino background belonged to this class, but since neutrino seems to have a mass of approximately 0.1 eV, it is now cold. We shall see why in Chap. 3.

Vacuum energy: $w = -1$, i.e. $P = -\rho c^2$ and ρ is a constant. It behaves as the cosmological constant and provides the best (and the simplest) description that we have for dark energy, though plagued by the serious issues that we have presented in Chap. 1.

2.3.4 The ΛCDM Model

The most successful cosmological model is called ΛCDM and is made up of Λ, CDM, baryons and radiation (photons and massless neutrinos). The Friedmann equation for the ΛCDM model is the following:

$$\frac{H^2}{H_0^2} = \Omega_\Lambda + \frac{\Omega_{c0}}{a^3} + \frac{\Omega_{b0}}{a^3} + \frac{\Omega_{r0}}{a^4} + \frac{\Omega_{K0}}{a^2} . \tag{2.79}$$

We already saw in Eq. (2.72) the value of the spatial curvature contribution. From Ade et al. (2016) here are the other ones:

$$\boxed{\Omega_\Lambda = 0.6911 \pm 0.0062 , \qquad \Omega_{m0} = 0.3089 \pm 0.0062} \tag{2.80}$$

at 68% confidence level, where $\Omega_{m0} = \Omega_{c0} + \Omega_{b0}$, i.e. it includes the contributions from both CDM and baryons, since they have the same dynamics (i.e. they are both cold). It is however possible to disentangle them and one observes:

$$\boxed{\Omega_{b0}h^2 = 0.02230 \pm 0.00014 , \qquad \Omega_{c0}h^2 = 0.1188 \pm 0.0010} \tag{2.81}$$

also at 68% confidence level. The radiation content, i.e. photons plus neutrinos, can be easily calculated from the temperature of the CMB, as we shall see in Chap. 3. It turns out that:

$$\boxed{\Omega_{\gamma0}h^2 \approx 2.47 \times 10^{-5} , \qquad \Omega_{\nu0}h^2 \approx 1.68 \times 10^{-5}} \tag{2.82}$$

Since $h = 0.68$, and recalling the closure relation of Eq. (2.75), we can conclude that today 69% of our universe is made of cosmological constant, 26% of CDM and 5% of baryons. Radiation and spatial curvature are negligible. That is, the situation is pretty obscure, in all senses.

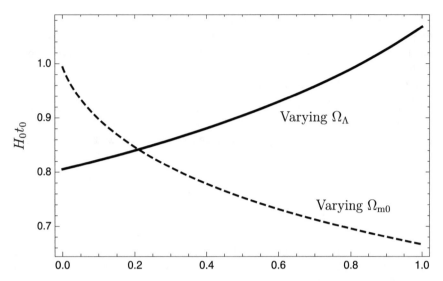

Fig. 2.4 Dimensionless age of the universe $H_0 t_0$ as function of Ω_Λ (keeping fixed the matter and radiation content) and as function of Ω_{m0} (keeping fixed the radiation content and with no Λ)

Let us now calculate the age of the universe for the ΛCDM model. Using Eq. (2.62), we get:

$$t_0 = \frac{1}{H_0} \int_0^1 da \frac{a}{\sqrt{\Omega_\Lambda a^4 + \Omega_{m0} a + \Omega_{r0} + \Omega_{K0} a^2}} . \tag{2.83}$$

Using the numbers shown insofar, we get upon numerical integration:

$$\boxed{t_0 = \frac{0.95}{H_0} = 13.73 \text{ Gyr}} \tag{2.84}$$

The value reported by Ade et al. (2016) is 13.799 ± 0.021 at 68% confidence level. Note how $H_0 t_0 \approx 1$. This fact has been dubbed **synchronicity problem** by Avelino and Kirshner (2016). In Fig. 2.4 we plot the dimensionless age of the universe $H_0 t_0$ for models with or without Λ and in Fig. 2.5 we plot the evolution of $H_0 t$ as function of a in order to show indeed how $H_0 t_0 \approx 1$ is quite a peculiar instant of the history of the universe.

As one can see, in presence of Λ the dimensionless age of the universe reaches values larger than unity. This, mathematically, is due to the a^4 factor multiplying Ω_Λ in Eq. (2.83). Note that we can obtain the observed value $H_0 t_0 \approx 0.95$ also in absence of a cosmological constant and for a curvature-dominated universe, i.e. $\Omega_{K0} \approx 0.97$.

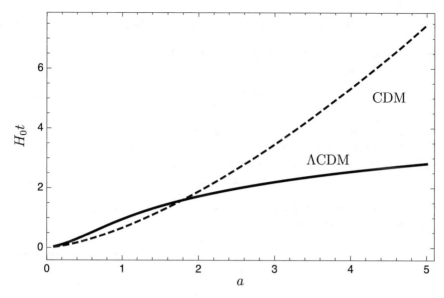

Fig. 2.5 Dimensionless age of the universe $H_0 t$ as function of a for the ΛCDM model (solid line) and in a model made only of CDM (dashed line)

2.4 Solutions of the Friedmann Equations

The Friedmann equations can be solved exactly for many cases of interest.

2.4.1 The Einstein Static Universe

As the first application of his theory to cosmology, Einstein was looking for a static universe, since at his time there was not yet compelling evidence of the contrary. Therefore, we must set $\dot{a} = \ddot{a} = 0$. Since ρ is positive, we must have $K = 1$, therefore the Einstein Static Universe (ESU) is a closed universe. Its radius is:

$$\frac{8\pi G}{3}\rho = \frac{c^2}{a^2} \quad \Rightarrow \quad a = \sqrt{\frac{3c^2}{8\pi G\rho}} . \tag{2.85}$$

From the acceleration equation we get that

$$\rho + 3P/c^2 = 0 , \tag{2.86}$$

therefore we cannot have simply ordinary matter because we need a negative pressure. Here enters the cosmological constant Λ. We assume that $\rho = \rho_m + \rho_\Lambda$, so that

$$\rho + 3P/c^2 = 0 , \quad \Rightarrow \quad \rho_m + \rho_\Lambda - 3\rho_\Lambda = 0 , \tag{2.87}$$

and therefore $\rho_m = 2\rho_\Lambda$. The radius can thus be written as

$$a = \frac{c}{\sqrt{4\pi G \rho_m}} = \frac{1}{\sqrt{\Lambda}} . \tag{2.88}$$

Until here all seems to be fine. But it is not. The problem is indeed the condition $\rho_m = 2\rho_\Lambda$, which makes the ESU unstable. In fact, if this condition is broken, say $\rho_m/\rho_\Lambda = 2 + \epsilon$, then the universe expands or collapses, depending on the sign of ϵ.

Exercise 2.14 Prove that the ESU is unstable. Hint: use $\rho_m/\rho_\Lambda = 2 + \epsilon$ in the Friedmann and acceleration equations.

2.4.2 The de Sitter Universe

For $\rho = 0$, the Friedmann equation (2.50) becomes:

$$H^2 = \frac{\Lambda c^2}{3} - \frac{K c^2}{a^2} . \tag{2.89}$$

When spatial curvature is taken into account, it is more convenient to solve the acceleration equation (2.51) rather than Friedmann equation. Indeed:

$$\ddot{a} = \frac{\Lambda c^2}{3} a , \tag{2.90}$$

is straightforwardly integrated:

$$a(t) = C_1 \exp\left(\sqrt{\frac{\Lambda}{3}} ct\right) + C_2 \exp\left(-\sqrt{\frac{\Lambda}{3}} ct\right) , \tag{2.91}$$

where C_1 and C_2 are two integration constants. One of these is constrained by Friedmann equation (2.89). Calculating $\dot{a}^2$ and a^2 from Eq. (2.91) we get:

$$\dot{a}^2 = \frac{\Lambda c^2}{3}\left[C_1^2 \exp\left(2\sqrt{\frac{\Lambda}{3}} ct\right) + C_2^2 \exp\left(-2\sqrt{\frac{\Lambda}{3}} ct\right) - 2C_1 C_2\right] , \tag{2.92}$$

$$a^2 = C_1^2 \exp\left(2\sqrt{\frac{\Lambda}{3}} ct\right) + C_2^2 \exp\left(-2\sqrt{\frac{\Lambda}{3}} ct\right) + 2C_1 C_2 . \tag{2.93}$$

Combining them, one finds:

$$\dot{a}^2 = \frac{\Lambda c^2}{3}(a^2 - 4C_1 C_2) \,. \tag{2.94}$$

Using Eq. (2.89), we can put a constraint on the product of the two integration constants:

$$\frac{4\Lambda}{3}C_1 C_2 = K \,, \tag{2.95}$$

so that we have freedom to fix just one of them. Assuming that $C_1 \neq 0$, we can write the general solution as:

$$a(t) = C_1 \exp\left(\sqrt{\frac{\Lambda}{3}}ct\right) + \frac{3K}{4\Lambda C_1} \exp\left(-\sqrt{\frac{\Lambda}{3}}ct\right) \,. \tag{2.96}$$

When $K = \pm 1$ we can set C_1 such that we can write the solution as follows:

$$a(t) = \begin{cases} \sqrt{3/\Lambda} \sinh\left(\sqrt{\Lambda/3}ct\right) \,, & \text{for } K = -1 \,, \\ a_0 \exp\left(\sqrt{\Lambda/3}ct\right) \,, & \text{for } K = 0 \,, \\ \sqrt{3/\Lambda} \cosh\left(\sqrt{\Lambda/3}ct\right) \,, & \text{for } K = 1 \,. \end{cases} \tag{2.97}$$

where a_0 is some initial $t = 0$ scale factor. Note that there is no a_0 for the solutions with spatial curvature because we fixed the integration constant in order to have the hyperbolic sine and cosine.

Exercise 2.15 From the Einstein equations (2.43) show that $R = 4\Lambda$ for the de Sitter universe. Verify that the above solutions (2.96) and (2.97) satisfy this relation by substituting them into the expression in Eq. (2.45) for the Ricci scalar.

The de Sitter universe (de Sitter 1917, 1918a, b, c) is eternal with no Big-Bang (i.e. when $a = 0$) for $K = 1$. Here we have rather a bounce at the minimum value a_0 for the scale factor. For $K = 0$ there is a Big-Bang at $t = -\infty$. For $K = -1$ we might have negative scale factors, which we however neglect and consider the evolution as starting only at $t = 0$, for which there is another Big-Bang. Note that the Big-Bang's we are mentioning here are not singularities. There is no physical singularity in the de Sitter space, since it is maximally symmetric.

The deceleration parameter is the following

$$q = -\frac{\ddot{a}a}{\dot{a}^2} = -\frac{\Lambda c^2}{3}\frac{a^2}{\dot{a}^2} = -\left(1 - \frac{3K}{\Lambda a^2}\right)^{-1} \,, \tag{2.98}$$

i.e. always negative.

2.4.3   Radiation-Dominated Universe

For $\rho = \rho_0 a^{-4}$, $K = 0$ and $\Lambda = 0$, the solution of Eq. (2.50) is:

$$a = \sqrt{\frac{t}{t_0}} \, , \tag{2.99}$$

The deceleration parameter is $q_0 = 1$ and the age of the universe is:

$$t_0 = \frac{1}{2H_0} \, . \tag{2.100}$$

Exercise 2.16   Prove the results of Eqs. (2.99) and (2.100).

It is quite complicated to analytically solve Friedmann equation (2.50) for a radiation-dominated universe when $K \neq 0$. On the other hand, solving the acceleration equation (2.53) is much easier. For $\rho c^2 = 3P$, Eq. (2.53) becomes:

$$a'' + Kc^2 a = 0 \, , \tag{2.101}$$

whose general solution is:

$$a(\eta) = \begin{cases} C_1 \exp(c\eta) + C_2 \exp(-c\eta) \, , & \text{for } K = -1 \, , \\ C_3 + C_4 \eta \, , & \text{for } K = 0 \, , \\ C_5 \sin(c\eta) + C_6 \cos(c\eta) \, , & \text{for } K = 1 \, . \end{cases} \tag{2.102}$$

First of all, we can choose $a(0) = 0$. Thus, the general solution (2.102) becomes:

$$a(\eta) = \begin{cases} 2C_1 \sinh(c\eta) \, , & \text{for } K = -1 \, , \\ C_4 \eta \, , & \text{for } K = 0 \, , \\ C_5 \sin(c\eta) \, , & \text{for } K = 1 \, . \end{cases} \tag{2.103}$$

Second, these solutions are subject to the constraint of Friedmann equation (2.52), written of course in the radiation-dominated case:

$$a'^2 = \frac{8\pi G}{3} \rho a^4 - Kc^2 a^2 \, , \tag{2.104}$$

where notice that $\rho a^4 = \rho_0$, i.e. a constant. When $a = 0$, i.e. $\eta = 0$, then

$$a'^2(\eta = 0) = \frac{8\pi G}{3} \rho_0 \equiv a_m^2 c^2 \, , \tag{2.105}$$

and the solutions (2.103) become:

$$a(\eta) = a_m \begin{cases} \sinh(c\eta), & \text{for } K = -1, \\ c\eta, & \text{for } K = 0, \\ \sin(c\eta), & \text{for } K = 1. \end{cases} \quad (2.106)$$

If we want to recover the cosmic time from the above solutions, we need to solve the following integration:

$$\int_0^\eta a(\eta')d\eta' = t. \quad (2.107)$$

Using Eq. (2.106), one obtains:

$$ct = a_m \begin{cases} \cosh(c\eta) - 1, & \text{for } K = -1, \\ (c\eta)^2/2, & \text{for } K = 0, \\ 1 - \cos(c\eta), & \text{for } K = 1. \end{cases} \quad (2.108)$$

Inverting these relations allows you to find $\eta = \eta(t)$, which once substituted in Eq. (2.106) allows to find $a = a(t)$.

Exercise 2.17 Using the solutions (2.108), find the explicit form of $a(t)$. Show that $ct = a_m \eta^2/2$ leads to Eq. (2.99).

2.4.4 Cold Matter-Dominated Universe

For $\rho = \rho_0 a^{-3}$, $K = 0$ and $\Lambda = 0$, the solution of Friedmann equation (2.50) is straightforwardly obtained:

$$a = \left(\frac{t}{t_0}\right)^{2/3}. \quad (2.109)$$

The deceleration parameter is $q_0 = 1/2$ and the age of the universe is:

$$t_0 = \frac{2}{3H_0} = 6.52 \, h^{-1} \, \text{Gyr}. \quad (2.110)$$

This model of universe is also known as the **Einstein-de Sitter universe**. A part the fact that it does not predict any accelerated expansion, there are also problems with the age of the universe given in Eq. (2.110): it is smaller than the one of some globular clusters (Velten et al. 2014).

Exercise 2.18 Prove the results of Eqs. (2.109) and (2.110).

Worse than the radiation-dominated case, it is impossible to analytically solve Friedmann equation (2.50) for a dust-dominated universe when $K \neq 0$. But, as in the radiation-dominated case, it is possible to find an exact solution for $a(\eta)$. Let's write Eq. (2.53) for the dust-dominated case:

$$a'' = \frac{4\pi G}{3}\rho a^3 - Kc^2 a .$$
(2.111)

Note that $\rho a^3 = \rho_0 = $ constant. The general solution is therefore the general solution of Eq. (2.101) plus a particular solution of Eq. (2.111), that is:

$$a(\eta) = \begin{cases} C_1 \sinh(c\eta) + C_2 \cosh(c\eta) - \frac{4\pi G}{3c^2}\rho_0 , & \text{for } K = -1 , \\ C_3 + C_4\eta + \frac{2\pi G}{3}\rho_0\eta^2 , & \text{for } K = 0 , \\ C_5 \sin(c\eta) + C_6 \cos(c\eta) + \frac{4\pi G}{3c^2}\rho_0 , & \text{for } K = 1 . \end{cases}$$
(2.112)

Exercise 2.19 Using the condition $a(0) = 0$ and employing Friedmann equation for a dust-dominated universe, i.e.

$$a'^2 + Kc^2a^2 = \frac{8\pi G}{3}\rho a^4 ,$$
(2.113)

as constraint, show that Eq. (2.112) can be cast as:

$$a(\eta) = \frac{4\pi G}{3c^2}\rho_0 \begin{cases} \cosh(c\eta) - 1 , & \text{for } K = -1 , \\ (c\eta)^2/2 , & \text{for } K = 0 , \\ 1 - \cos(c\eta) , & \text{for } K = 1 . \end{cases}$$
(2.114)

Recovering the cosmic time from Eq. (2.114) one has:

$$ct = \frac{4\pi G}{3c^2}\rho_0 \begin{cases} \sinh(c\eta) - c\eta , & \text{for } K = -1 , \\ (c\eta)^3/6 , & \text{for } K = 0 , \\ c\eta - \sin(c\eta) , & \text{for } K = 1 . \end{cases}$$
(2.115)

Unfortunately, the above relations for $K = \pm 1$ cannot be explicitly inverted in order to give $\eta(t)$ and then $a(t)$.

2.4.5 Radiation Plus Dust Universe

The mixture of radiation plus matter is a cosmological model closer to reality and with which we can describe the evolution of our universe on a larger timespan than the single component-dominated cases. Consider the total density:

$$\rho = \rho_m + \rho_r = \frac{\rho_{eq}}{2} \frac{a_{eq}^3}{a^3} + \frac{\rho_{eq}}{2} \frac{a_{eq}^4}{a^4} , \tag{2.116}$$

where a_{eq} is the equivalence scale factor, i.e. the scale factor evaluated at the time at which dust and radiation densities were equal. At this time, we dub the total density as ρ_{eq}. Now write down the acceleration equation (2.53) for the dust plus radiation model:

$$a'' = \frac{4\pi G}{3} \rho_m a^3 - Kc^2 a . \tag{2.117}$$

It is identical to the dust-dominated case, viz. Eq. (2.111)! Indeed, the fact that radiation is also present will enter when we set the constraint from Friedmann equation, which is the following:

$$a'^2 + Kc^2 a^2 = \frac{4\pi G \rho_{eq}}{3} \left(a_{eq}^3 a + a_{eq}^4 \right) . \tag{2.118}$$

Solving Eq. (2.117) with the condition $a(0) = 0$ leads to the following solutions:

$$a(\eta) = \frac{2\pi G \rho_{eq} a_{eq}^3}{3c^2} \begin{cases} C_1 \sinh(c\eta) + \cosh(c\eta) - 1 , & \text{for } K = -1 , \\ C_2 \eta + (c\eta)^2/2 , & \text{for } K = 0 , \\ C_3 \sin(c\eta) + 1 - \cos(c\eta) , & \text{for } K = 1 . \end{cases} \tag{2.119}$$

Now use the constraint from Friedmann equation, i.e.

$$a'^2(\eta = 0) = \frac{4\pi G \rho_{eq}}{3} a_{eq}^4 , \tag{2.120}$$

and find that:

$$C_1 = C_3 = c\sqrt{\frac{3}{\pi G \rho_{eq} a_{eq}^2}} \equiv c\tilde{\eta} , \qquad C_2 = c^2 \tilde{\eta} , \tag{2.121}$$

so that:

$$a(\eta) = \frac{2a_{eq}}{c^2 \tilde{\eta}^2} \begin{cases} c\tilde{\eta} \sinh(c\eta) + \cosh(c\eta) - 1 , & \text{for } K = -1 , \\ c^2 \tilde{\eta}\eta + (c\eta)^2/2 , & \text{for } K = 0 , \\ c\tilde{\eta} \sin(c\eta) + 1 - \cos(c\eta) , & \text{for } K = 1 . \end{cases} \tag{2.122}$$

In particular, the solution for $K = 0$ is:

$$a(\eta) = a_{eq} \left(2\frac{\eta}{\tilde{\eta}} + \frac{\eta^2}{\tilde{\eta}^2} \right) . \tag{2.123}$$

Exercise 2.20 Show that the conformal time at equivalence η_{eq} and $\tilde{\eta}$ are related by:

$$\eta_{eq} = (\sqrt{2} - 1)\tilde{\eta} . \tag{2.124}$$

Exercise 2.21 Solve Friedmann equation for the ΛCDM model, neglecting radiation and spatial curvature:

$$\frac{H^2}{H_0^2} = \frac{\Omega_{m0}}{a^3} + \Omega_\Lambda . \tag{2.125}$$

Show that:

$$a(t) = \left[\frac{\Omega_{m0}}{\Omega_\Lambda} \sinh^2 \left(\frac{3}{2}\sqrt{\Omega_\Lambda} H_0 t \right) \right]^{1/3} . \tag{2.126}$$

Exercise 2.22 Solve the Friedmann equation for the curvature-dominated universe:

$$H^2 = -\frac{Kc^2}{a^2} . \tag{2.127}$$

This is the Milne model (Milne 1935). Clearly, only $K = -1$ is allowed.

Show then that $a = ct$. Substitute this solution into the expression for the Ricci scalar (2.45). Show that $R = 0$.

Write down explicitly the FLRW metric with $a = ct$ and show that it is Minkowski metric written in a coordinate systems different from the usual.

The above last result is not completely surprising, since Milne model has no matter (empty universe) and no cosmological constant. The spatial hypersurfaces are already maximally symmetric because of the cosmological principle and the absence of matter add even more symmetry to the spacetime.

2.5 Distances in Cosmology

We present and discuss in this section the various notions of distance that are employed in cosmology. See e.g. Hogg (1999) for a reference on the subject.

2.5.1 Comoving Distance and Proper Distance

We have already encountered comoving coordinates in the FLRW metric (2.17) and the proper radius $\mathcal{D}(t) \equiv a(t)r$ in the FLRW metric (2.21). We must be clearer about the difference between the radial coordinate and the distance. They are equal only when $d\Omega = 0$. The comoving square infinitesimal distance is indeed, from FLRW metric (2.17) the following:

$$d\chi^2 = \frac{dr^2}{1 - Kr^2} + r^2 d\Omega^2 , \tag{2.128}$$

i.e. it has indeed a radial part, but also has a transversal part. So, if χ is the comoving distance between two points, the proper distance at a certain time t is $d(\chi, t) = a(t)\chi$.

The comoving distance is a notion of distance which does not include the expansion of the universe and thus does not depend on time.

The proper distance is the distance that would be measured instantaneously by rulers. For example, imagine to extend a ruler between GN-z11 (the farthest known galaxy, $z = 11.09$) and us. Our reading at the time t would be the proper distance at that time.

Suppose that $d\Omega = 0$. Then the comoving distance to an object with radial coordinate r is the following:

$$\chi = \int_0^r \frac{dr'}{\sqrt{1 - Kr'^2}} = \begin{cases} \arcsin r , & \text{for } K = 1 , \\ r , & \text{for } K = 0 , \\ \text{arcsinh } r , & \text{for } K = -1 . \end{cases} \tag{2.129}$$

Deriving d with respect to the time one gets:

$$\dot{d} = \dot{a}\chi = \frac{\dot{a}}{a}d = Hd , \tag{2.130}$$

which recovers the Hubble's law for $t = t_0$.

2.5.2 The Lookback Time

Imagine a photon emitted by a galaxy at a time t_{em} and detected at the time t_0 on Earth. A very basic notion of distance is $c(t_0 - t_{em})$, i.e. it is the light-travel distance, based on the fact that light always travels with speed c. The quantity $t_0 - t_{em}$ is called **lookback time** and suggestively reminds the fact that when we observe some source in the sky we are actually looking into the past, because of the finiteness of c.

From the FLRW metric, by putting $ds^2 = 0$, we can relate the lookback time with the comoving distance as follows:

$$cdt = a(t)d\chi . \tag{2.131}$$

This seems quite similar to the proper distance, but careful: the proper distance is defined as $a\chi$ and evidently $ad\chi \neq d(a\chi)$. The lookback time is the photon time of flight and thus it includes cumulatively the expansion of the universe. On the other hand, the proper distance is the distance considered between two simultaneous events and therefore the expansion of the universe is not taken into account cumulatively.

Since we observe redshifts, is there a way to calculate the lookback time from z? In principle yes: one solves Friedmann equation, finds $a(t)$, inverts this function in order to find $t = t(a)$, uses $1 + z = 1/a$ and finally gets a relation $t = t(z)$. For example, for the flat Einstein-de Sitter universe, using Eqs. (2.109) and (2.110) one gets:

$$1 + z = \left(\frac{2}{3H_0t}\right)^{2/3} \quad \Rightarrow \quad t = \frac{2}{3H_0(1 + z)^{3/2}} . \tag{2.132}$$

This approach is model-dependent because in order to solve the Friedmann equation we must know it and this is possible only if we know, or *model*, the energy content of the universe. Hence the model-dependence.

A model-independent way of relating lookback time and redshift is **cosmography**, a word which means "measuring the universe". In practice, cosmography consists in a Taylor expansion of the scale factor about its today value:

$$a(t) = a(t_0) + \left.\frac{da}{dt}\right|_{t_0} (t - t_0) + \frac{1}{2}\left.\frac{d^2a}{dt^2}\right|_{t_0} (t - t_0)^2 + \cdots \tag{2.133}$$

where we stop at the second order, for simplicity. This can be written as

$$a(t) = a(t_0)\left[1 + H_0(t - t_0) - \frac{1}{2}q_0H_0^2(t - t_0)^2 + \cdots\right], \tag{2.134}$$

i.e. the first coefficient of the expansion is the Hubble constant, whereas the second one is proportional to the deceleration parameter. The third is usually called *jerk* and the fourth *snap*. All these parameters are evaluated at t_0 in the above expansion.

Exercise 2.23 For $a(t_0) = 1$ and introducing the redshift show that:

$$z \sim H_0(t_0 - t) + \frac{1}{2}(q_0 + 2)H_0^2(t_0 - t)^2 . \tag{2.135}$$

Here is a direct, model-independent relation between the redshift and the lookback time $(t_0 - t)$.

2.5.3 Distances and Horizons

For a photon, not unexpectedly,

$$dχ = \frac{cdt}{a(t)} = cd\eta \,, \tag{2.136}$$

i.e. the comoving distance is equal to the conformal time, which we introduced in Eq. (2.18). We might say that the comoving distance is a *lookback conformal time*.

By integrating $cdt/a(t)$ from t_{em} to t_0 we get the comoving distance from the source to us, or the conformal time spent by the photon travelling from the source to us:

$$χ = \int_{t_{em}}^{t_0} \frac{cdt'}{a(t')} = \int_a^1 \frac{cda'}{H(a')a'^2} \,. \tag{2.137}$$

For the dust-dominated case one has $H = H_0/a^{3/2}$ and the comoving distance as a function of the scale factor and of the redshift is:

$$χ(a) = \frac{c}{H_0} \int_a^1 \frac{da'}{\sqrt{a'}} = \frac{2c}{H_0}\left(1 - \sqrt{a}\right), \quad χ(z) = \frac{2c}{H_0}\left(1 - \frac{1}{\sqrt{1+z}}\right). \tag{2.138}$$

When $z \to 0$, $χ \sim cz/H_0$. Comparing with Eq. (2.135) one sees that, at the first order in the redshift, the lookback time distance is equivalent to the comoving one.

Exercise 2.24 Calculate the comoving distance as a function of the scale factor and of the redshift for a radiation-dominated universe and for the de Sitter universe.

When the lower integration limit in Eq. (2.137) is $a = 0$, i.e. the Big Bang, one defines the **comoving horizon** $χ_p$ (also known as **particle horizon** or **cosmological horizon**). This is the conformal time spent from the Big Bang until the cosmic time t or scale factor a. It is also the maximum comoving distance travelled by a photon (hence the name particle horizon) since the Big Bang and so it is the comoving size of the visible universe.

In the dust-dominated case, using Eq. (2.138), with $a = 0$ or $z = \infty$ one obtains:

$$χ_p = c\eta_0 = \frac{2c}{H_0} \,. \tag{2.139}$$

Note that this is not the age of the universe given in Eq. (2.110), but three times its value.

When the upper integration limit of Eq. (2.137) is infinite, one defines the **event horizon**:

$$χ_e(t) \equiv c \int_t^\infty \frac{dt'}{a(t')} = c \int_a^\infty \frac{da'}{H(a')a'^2} \,, \tag{2.140}$$

which of course makes sense only if the universe does not collapse. This represents the maximum distance travelled by a photon from a time t. If it diverges, then no event horizon exists and therefore eventually all the events in the universe will be causally connected. This happens, for example, in the dust-dominated case:

$$\chi_e = \frac{c}{H_0} \int_a^\infty \frac{da'}{\sqrt{a'}} = \infty . \tag{2.141}$$

But, in the de Sitter universe we have

$$\chi_e = \frac{c}{H_0} \int_a^\infty \frac{da'}{a'^2} = \frac{c}{H_0 a} . \tag{2.142}$$

The proper event horizon for the de Sitter universe is a constant:

$$a\chi_e = \frac{c}{H_0} . \tag{2.143}$$

2.5.4   The Luminosity Distance

The luminosity distance is a very important notion of distance for observation. It is based on the knowledge of the intrinsic luminosity L of a source, which is therefore called **standard candle**. Type Ia supernovae are standard candles, for example. Then, measuring the flux F of that source and dividing L by F, one obtains the square luminosity distance:

$$d_L^2 \propto \frac{L}{F} . \tag{2.144}$$

Now, imagine a source at a certain redshift z with intrinsic luminosity $L = dE/dt$. The observed flux is given by the following formula:

$$F = \frac{dE_0}{dt_0 A_0} , \tag{2.145}$$

where A_0 is the area of the surface on which the radiation is spread:

$$A_0 = 4\pi a_0^2 \chi^2 , \tag{2.146}$$

i.e. over a sphere with the proper distance as the radius. We must use the proper distance, because this is the instantaneous distance between source and observer at the time of detection. Note that χ is the comoving distance between the source and us.

We do not observe the same photon energy as the one emitted, because photons suffer from the cosmological redshift, thus:

$$\frac{dE}{dE_0} = \frac{a_0}{a} \ . \tag{2.147}$$

Finally, the time interval used at the source is also different from the one used at the observer location:

$$\frac{dt}{dt_0} = \frac{a}{a_0} \ . \tag{2.148}$$

We can easily show this by using FLRW metric with $ds^2 = 0$, i.e. $cdt = a(t)d\chi$. Consider the same $d\chi$ at the source and at the observer's location. Thus, $cdt = a(t)d\chi$ and $cdt_0 = a(t_0)d\chi$ and the above result follows.

Putting all the contributions together, we get

$$F = \frac{dE_0}{dt_0 A_0} = \frac{a^2 dE}{a_0^2 dt 4\pi a_0^2 \chi^2} = \frac{dE}{dt 4\pi a_0^2 \chi^2 (1+z)^2} \ . \tag{2.149}$$

Hence, the luminosity distance is defined as:

$$\boxed{d_{\rm L} \equiv a_0(1+z)\chi} \tag{2.150}$$

From this formula and the observed redshifts of type Ia supernovae we can determine if the universe is in an accelerated expansion, in a model-independent way. In order to do this, we first need to know how to expand χ in series of powers of the redshift.

Using the definition (2.137) and the expansion (2.135), we get:

$$\chi = \int_t^{t_0} \frac{cdt'}{a(t')} = \frac{c(t_0 - t)}{a_0} + \frac{cH_0}{2a_0}(t_0 - t)^2 + \cdots \tag{2.151}$$

where we stop at the second order only. This is the expansion of the comoving distance with respect to the lookback time. We must invert the power series of Eq. (2.135) in order to find the expansion of the lookback time with respect to the redshift. This can be done, for example, by assuming the following ansatz:

$$H_0(t_0 - t) = \alpha + \beta z + \gamma z^2 + \cdots \tag{2.152}$$

and substitute it into Eq. (2.135), keeping at most terms $\mathcal{O}(z^2)$.

Exercise 2.25 Show that:

$$\alpha = 0 \ , \qquad \beta = 1 \ , \qquad \gamma = -\frac{1}{2}(q_0 + 2) \ , \tag{2.153}$$

and thus

$$H_0(t_0 - t) = z - \frac{1}{2}(q_0 + 2)z^2 + \cdots \tag{2.154}$$

Substituting the expansion of Eq. (2.154) back into Eq. (2.150), one gets

$$d_L = \frac{c}{H_0} \left[z + \frac{1}{2}(1 - q_0)z^2 + \cdots \right] , \tag{2.155}$$

where note again that at the lowest order the luminosity distance is cz/H_0, identical to the comoving distance and to the lookback time distance.

Since d_L and z are measured, one can fit the data with this quadratic function and determine q_0, thereby establishing if the universe expansion is accelerated or not. Note that H_0 is an overall multiplicative factor, thus does not determine the shape of the function $d_L(z)$.

In the case of a dust-dominated universe, using Eq. (2.138), the luminosity distance has the following expression:

$$d_L = \frac{2c}{H_0} \left(1 + z - \sqrt{1+z} \right) . \tag{2.156}$$

For small z, this distance can be expanded in powers of the redshift as:

$$d_L = \frac{c}{H_0} \left(z + \frac{1}{4}z^2 + \cdots \right) , \tag{2.157}$$

which, when compared with Eq. (2.150), provides $q_0 = 1/2$, as expected.

2.5.5   Angular Diameter Distance

The angular diameter distance is based on the knowledge of proper sizes. Objects with a known proper size are called **standard rulers**. Suppose a standard ruler of transversal proper size ds (small) to be at a redshift z and comoving distance χ. Moreover, this object has an angular dimension $d\phi$, also small. See Fig. 2.6 for reference.

At a fixed time t, we can write the FLRW metric as:

$$ds^2 = a(t)^2 d\chi^2 . \tag{2.158}$$

Fig. 2.6 Defining the angular diameter distance

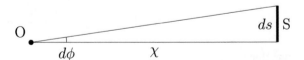

Since the object is small and we are at the origin of the reference frame, the comoving distance χ is also the radial distance. Therefore, the transversal distance is:

$$ds = a(t)\chi d\phi .\tag{2.159}$$

Dividing the proper dimension of the object by its angular size provides us with the angular diameter distance:

$$d_A = a(t)\chi .\tag{2.160}$$

For the case of a dust-dominated universe, one has:

$$d_A = \frac{2c}{H_0}\left[\frac{1}{1+z} - \frac{1}{(1+z)^{3/2}}\right] .\tag{2.161}$$

In the limit of small z, we find $d_A \sim cz/H_0$. All the distances that we defined insofar coincide at the first order expansion in z.

Note the relation:

$$d_L = (1+z)^2 d_A ,\tag{2.162}$$

known as **Etherington's distance duality** (Etherington 1933).

In gravitational lensing applications it is often necessary to know the angular-diameter distance between two sources at different redshifts (i.e. the angular-diameter distance between the lens and the background source). In order to compute this, let us refer to Fig. 2.7.

The problem is to determine the angular-diameter distance between L and S, say $d_A(LS)$. Is this the difference between the angular-diameter distances $d_A(S) - d_A(L)$? We now show that this is not the case. Simple trigonometry is sufficient to establish that:

$$ds = a(t_S)\chi_S d\phi_S = a(t_S)\chi_{LS} d\phi_L ,\tag{2.163}$$

And for comoving distances we do have that $\chi_{LS} = \chi_S - \chi_L$. Therefore, we have

$$\boxed{d_A(LS) = a(t_S)\chi_{LS} = a(t_S)(\chi_S - \chi_L)}\tag{2.164}$$

Fig. 2.7 The angular diameter distance between two different redshifts

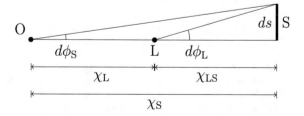

which is the relation we were looking for, and it is different from the difference between the angular diameter distances:

$$d_{\mathrm{A}}(\mathrm{S}) - d_{\mathrm{A}}(\mathrm{L}) = a(t_{\mathrm{S}})\chi_{\mathrm{S}} - a(t_{\mathrm{L}})\chi_{\mathrm{L}} \ . \tag{2.165}$$

References

Ade, P.A.R., et al.: Planck 2015 results. XIII. Cosmological parameters. Astron. Astrophys. **594**, A13 (2016)

Avelino, A., Kirshner, R.P.: The dimensionless age of the universe: a riddle for our time. Astrophys. J. **828**(1), 35 (2016)

Baqui, P.O., Fabris, J.C., Piattella, O.F.: Cosmology and stellar equilibrium using Newtonian hydrodynamics with general relativistic pressure. JCAP **1604**(04), 034 (2016)

Bonometto, S.: Cosmologia & Cosmologie. Zanichelli (2008)

de Sitter, W.: On the relativity of inertia. Remarks concerning Einstein's latest hypothesis. Koninklijke Nederlandsche Akademie van Wetenschappen Proceedings **19**, 1217–1225 (1917)

de Sitter, W.: Einstein's theory of gravitation and its astronomical consequences. Third paper. Mon. Not. R. Astron. Soc. **78**, 3–28 (1918a)

de Sitter, W.: Further remarks on the solutions of the field-equations of the Einstein's theory of gravitation. Koninklijke Nederlandsche Akademie van Wetenschappen Proceedings **20**, 1309–1312 (1918b)

de Sitter, W.: On the curvature of space. Koninklijke Nederlandsche Akademie van Wetenschappen Proceedings **20**, 229–243 (1918c)

Etherington, I.M.H.: On the definition of distance in general relativity. Philos. Mag. **15** (1933)

Fabris, J.C., Velten, H.: Neo-Newtonian cosmology: an intermediate step towards general relativity. RBEF **4302** (2012)

Friedmann, A.: Ueber die Kruemmung des Raumes. Z. Phys. **10**, 377–386 (1922)

Friedmann, A.: Ueber die Moeglichkeit einer Welt mit konstanter negativer Kruemmung des Raumes. Z. Phys. **21**, 326–332 (1924)

Giblin, J.T., Mertens, J.B., Starkman, G.D.: Observable deviations from homogeneity in an inhomogeneous universe. Astrophys. J. **833**(2), 247 (2016)

Harrison, E.R.: Cosmology without general relativity. Ann. Phys. **35**, 437–446 (1965)

Hogg, D.W.: Distance measures in cosmology (1999)

Hwang, J.-C., Noh, H.: Newtonian hydrodynamics with general relativistic pressure. JCAP **1310**, 054 (2013)

Lemaitre, G.: The expanding universe. Mon. Not. R. Astron. Soc. **91**, 490–501 (1931)

Lemaître, G.: The expanding universe. Gen. Rel. Gravit. **29**, 641–680 (1997)

Lima, J.A.S., Zanchin, V., Brandenberger, R.H.: On the Newtonian cosmology equations with pressure. Mon. Not. R. Astron. Soc. **291**, L1–L4 (1997)

Maartens, R.: Causal thermodynamics in relativity (1996)

McCrea, W.H., Milne, E.A.: Newtonian Universes and the curvature of space. Q. J. Math. **5** (1934)

McCrea, W.H.: Relativity theory and the creation of matter. Proc. R. Soc. Lond. Ser. A **206**, 562–575 (1951)

McVittie, G.C.: Appendix to the change of redshift and apparent luminosity of galaxies due to the deceleration of selected expanding universes. Astrophys. J. **136**, 334 (1962)

Milne, E.A.: A Newtonian expanding Universe. Q. J. Math. **5** (1934)

Milne, E.A.: Relativity, gravitation and world-structure. The Clarendon Press, Oxford (1935)

Perlmutter, S., et al.: Measurements of omega and lambda from 42 high-redshift supernovae. Astrophys. J. **517**, 565–586 (1999)

Piattella, O.F., Giani, L.: Redshift drift of gravitational lensing. Phys. Rev. D **95**(10), 101301 (2017)

Riess, A.G., et al.: Observational evidence from supernovae for an accelerating universe and a cosmological constant. Astron. J. **116**, 1009–1038 (1998)

Robertson, H.P.: Kinematics and world-structure. Astrophys. J. **82**, 284 (1935)

Robertson, H.P.: Kinematics and world-structure III. Astrophys. J. **83**, 257 (1936)

Ryden, B.: Introduction to Cosmology, 244 p. Addison-Wesley, San Francisco (2003)

Sandage, A.: The change of redshift and apparent luminosity of galaxies due to the deceleration of selected expanding universes. Astrophys. J. **136**, 319 (1962)

Sarkar, S., Pandey, B.: An information theory based search for homogeneity on the largest accessible scale. Mon. Not. R. Astron. Soc. **463**(1), L12–L16 (2016)

Schutz, B.F.: A First Course In General Relativity. Cambridge University Press, Cambridge (1985)

Velten, H.E.S., vom Marttens, R.F., Zimdahl, W.: Aspects of the cosmological coincidence problem. Eur. Phys. J. C **74**(11), 3160 (2014)

Walker, A.G.: On milne's theory of world-structure. Proc. Lond. Math. Soc. **2**(1), 90–127 (1937)

Weinberg, S.: Gravitation and Cosmology: Principles and Applications of the General Theory of Relativity. Wiley, New York (1972)

Wu, K.K.S., Lahav, O., Rees, M.J.: The large-scale smoothness of the universe. Nature **397**, 225–230 (1999). (**19** (1998))

Chapter 3
Thermal History

L'umanità non sopporta il pensiero che il mondo sia nato per caso, per sbaglio. Solo perché quattro atomi scriteriati si sono tamponati sull'autostrada bagnata
(Humanity cannot bear the thought that the world was born by accident, by mistake. Just because four mindless atoms crashed on the wet highway)

Umberto Eco, Il Pendolo di Foucault

In this Chapter we discuss the application of Boltzmann equation in cosmology. In particular, we address Big Bang Nucleosynthesis (BBN), recombination of protons and electrons in neutral hydrogen atoms and the relic abundance of CDM. Our main references are Dodelson (2003), Kolb and Turner (1990) and Daniel Baumann's lecture notes (Chap. 3).[1] See also Bernstein (1988).

3.1 Thermal Equilibrium and Boltzmann Equation

We have encountered in the previous chapter the continuity equation (2.77):

$$\dot{\varepsilon} + 3H(\varepsilon + P) = 0 \, , \tag{3.1}$$

where recall that the dot represents derivation with respect to the cosmic time. Surprisingly, it is possible to obtain the continuity equation by using the first and second law of thermodynamics, i.e.

$$TdS = PdV + dU \, , \tag{3.2}$$

---

[1] http://www.damtp.cam.ac.uk/user/db275/Cosmology/Lectures.pdf.

© Springer International Publishing AG, part of Springer Nature 2018
O. Piattella, *Lecture Notes in Cosmology*, UNITEXT for Physics,
https://doi.org/10.1007/978-3-319-95570-4_3

where T is the temperature, S is the entropy, V is the volume and U is the internal energy of the cosmic fluid. Now, assuming adiabaticity, i.e. $dS = 0$, and writing $U = \varepsilon V$, one gets

$$PdV + d(\varepsilon V) = 0 \quad \Rightarrow \quad Vd\varepsilon + (\varepsilon + P)dV = 0 . \tag{3.3}$$

Exercise 3.1 Since the volume is proportional to the cube of the scale factor, i.e. $V \propto a^3$, show that Eq. (3.3) leads to the the continuity equation (3.1).

In using Eq. (3.2), we have made a very strong assumption: the evolution of the universe is an adiabatic reversible transformation, i.e. at each instant the universe is in an equilibrium state.

In some instances we can trust this assumption and it gives the correct continuity equation. In particular, we will see that Eq. (3.1) can be obtained from Boltzmann equation assuming no interactions among particles or assuming a very high rate of interactions so that thermal equilibrium is reached. The latter instance can be mathematically represented as follows:

$$\boxed{\Gamma \gg H} \tag{3.4}$$

i.e. **the interaction rate is much larger than the Hubble rate**, where the interaction rate is defined as follows:

$$\boxed{\Gamma \equiv n\sigma v_{\mathrm{rel}}} \tag{3.5}$$

where n is the particle number density of projectiles, v_{rel} is the relative velocity between projectile and targets and σ is the cross section.

Equation (3.4) can also be rephrased as the fact that the mean-free-path is much smaller than the Hubble radius. In this situation, particles interact so frequently that they do not even care about the cosmological expansion and any fluctuation in their energy density is rapidly smoothed out, thus recovering thermal equilibrium.

It is important to make distinction between **kinetic equilibrium** and **chemical equilibrium**. When $\Gamma \gg H$ refers to a process of the type:

$$1 + 2 \leftrightarrow 3 + 4 , \tag{3.6}$$

i.e. we have four different particle species which transform into each other in a balanced way, then we have **chemical equilibrium**. This can be also reformulated as:

$$\boxed{\mu_1 + \mu_2 = \mu_3 + \mu_4} \tag{3.7}$$

where the μ's are the chemical potentials.

On the other hand, when $\Gamma \gg H$ refers to a reaction such as a scattering:

$$1 + 2 \leftrightarrow 1 + 2 \,, \tag{3.8}$$

then we have **kinetic equilibrium**. Kinetic or chemical equilibrium (or both) imply thermal equilibrium. In general, it is possible for a species to break chemical equilibrium and still remain in kinetic, therefore thermal, equilibrium with the rest of the cosmic plasma through scattering processes.[2]

Note that Γ is different for different fundamental interactions and for different particle species masses. Therefore, the above condition (3.4) is valid for all the known (and perhaps unknown) particles in the very early universe but is broken at different times for different species. This is the essence of the **thermal history of the universe**.

So, at the very beginning (we are talking about tiny fractions of seconds after the Big Bang) all the particles were in thermal equilibrium in a primordial soup, the **primordial plasma**. When for a species, the condition $\Gamma \sim H$ is reached, it **decouples** from the primordial plasma. If it does this by breaking the chemical equilibrium, then it is said to **freeze out** and attains some fixed abundance.[3]

When we want to explicitly calculate the residual abundance of some species, we have to track its evolution until $\Gamma \sim H$. In this instance, equilibrium thermodynamics fails and we are compelled to use Boltzmann equation. For example, we shall use Boltzmann equation when analysing the formation of light elements during BBN, the recombination of protons and electrons in neutral hydrogen atoms and the relic abundance of CDM.

The fundamental interactions which characterise the above-mentioned processes compel some particles to react and transform into others and vice-versa, such as in Eq. (3.6). When $\Gamma \gg H$ these reactions take place with equal probability in both directions, hence the $\leftrightarrow$ symbol, but when $\Gamma \sim H$ eventually one direction is preferred over the other. This is the characteristic of irreversibility which demands the use of Boltzmann equation.

3.2 Short Summary of Thermal History

We present in this section a brief scheme of the main events characterising the thermal history of the universe. A minimal knowledge of the standard model of particle physics is required.

Planck scale, inflation and Grand Unified Theory. Planck scale is usually considered as an upper threshold in the energy, being the Planck mass is $M_{Pl}c^2 = 10^{19}\,\text{GeV}$, or as a lower threshold in the time, 10^{-43} s, to which we can extend our classical the-

---

[2]This occurs in some DM particle models. For example a $m = 100\,\text{GeV}$ WIMP chemically decouples at $5\,\text{GeV}$ and kinetically decouples at $25\,\text{MeV}$. See e.g. Profumo et al. (2006).

[3]It attains a fixed abundance if it is a stable particle, of course. If not it disappears.

ory of gravity. Beyond that threshold, the common understanding is that we should incorporate quantum effects.

Inflation is a very important piece of the current description of the primordial universe. We shall dedicate to it Chap. 8 so, for now, it is enough to say that it occurs at an energy scale of the order 10^{16} GeV, which is the same of the Grand Unified Theory (GUT), i.e. a model in which electromagnetic, weak and strong interactions are unified.

Baryogenesis and Leptogenesis. Baryogenesis is the creation, via some still unclear mechanism, of a positive baryon number. In other words, the creation of a primordial quark-antiquark asymmetry by virtue of which protons and neutrons are much more common than anti-protons and anti-neutrons. In order to maintain the neutrality of the universe, we need thus also Leptogenesis, i.e. a mechanism which produces a non-vanishing lepton number in the form of an excess of electrons over positrons. There is no evidence for a similar asymmetry in neutrinos and anti-neutrinos,[4] therefore in the universe energy budget, after the annihilation epochs, we shall take into account both of them while neglecting anti-protons, anti-neutrons and positrons.

We know that antimatter exists but we also know that matter is the most abundant in the universe. If they were produced exactly in the same quantity, they should have annihilated almost completely when in equilibrium in the primordial plasma, leaving just photons. Instead, we have a small but nonvanishing baryon-to-photon ratio $\eta_b = 5.5 \times 10^{-10}$.

Decoupling of the Top quark. The Top quark is the most massive of all observed fundamental particles, $m_{Top}c^2 = 173$ GeV, so it is probably the first to decouple from the primordial plasma. For this reason it is considered here in this list.

Electroweak phase transition. At a thermal energy of about 100 GeV, which corresponds to 10^{-12} s after the Big Bang, the electromagnetic and weak forces start to behave distinctly. This happens because the vector bosons $W^\pm$ and Z^0 gain their masses, of roughly 80 and 90 GeV respectively, through the Higgs mechanism and the weak interaction "weakens", since it is now ruled by the Fermi constant:

$$\frac{G_F}{(\hbar c)^3} = 1.17 \times 10^{-5}\, \text{GeV}^{-2} \ . \tag{3.9}$$

QCD phase transition. Below 150 MeV quarks pass from their asymptotic freedom to bound states form by two (mesons) or three (baryons) of them. The above energy corresponds roughly to 20 μs after the Big Bang.

DM freeze-out. We do not know if DM is made up of particles and, if so, which ones. For the neutralino case, the freeze-out takes place at about 25 MeV.

Neutrino decoupling. Neutrinos maintain thermal equilibrium with the primordial plasma through interactions such as

---

[4]It is still unclear whether neutrino is a Majorana fermion, i.e. a fermion which is its own anti-particle, or a Dirac fermion, i.e. a fermion which is distinct from its anti-particle.

$$p + e^- \leftrightarrow n + \nu , \qquad p + \bar{\nu} \leftrightarrow n + e^+ , \qquad n \leftrightarrow p + e^- + \bar{\nu} , \qquad (3.10)$$

down to a thermal energy of 1 MeV, which corresponds roughly to 1 s after the Big Bang. Below this energy threshold they decouple. We can calculate roughly the 1 MeV scale of decoupling in the following way. Using Eq. (3.5) with $v_{rel} = c$, since neutrinos are relativistic, one gets:

$$\frac{\Gamma}{H} = \frac{n_\nu \sigma c}{H} \approx \frac{G_F^2}{(\hbar c)^6} M_{Pl} c^2 (k_B T)^3 \approx \left(\frac{k_B T}{1 \, \text{MeV}} \right)^3 , \qquad (3.11)$$

where we have used some results that we shall prove later, such as that the particle number density n_ν goes as T^3 and $H \propto T^2/M_{Pl}$. When $k_B T \sim 1$ MeV then decoupling occurs.

We have just used the **effective field theory** of the weak interaction in assuming $\sigma \propto G_F^2 T^2$. An effective field theory is a low energy approximation of the full theory. Indeed, the cross section $\sigma \propto G_F^2 T^2$ diverges at high energies and this is unphysical. In the case of the weak interaction, we have chosen to work at an energy scale much smaller than 80 GeV, which is the mass of the boson vector $W^\pm$ (which mediates the weak interaction). In this approximation, it is as if the boson vectors $W^\pm$ and Z^0 had infinite masses and therefore the range of the weak interaction is zero. In other words, we have considered the interactions of Eq. (3.10) as if they occurred in a point. The result 1 MeV $\ll$ 80 GeV is consistent with the effective field theory approximation and thus is reliable.

Another approximation that we have used is the neutrino masslessness. Even considering a small mass $m_\nu c^2 \lesssim 0.1$ eV, the above calculation is solid, since $m_\nu c^2 \ll k_B T \sim 1$ MeV. The condition $mc^2 \ll k_B T$ for a generic particle of mass m in thermal equilibrium guarantees that such particle is relativistic, as we shall demonstrate later.

Other kinds of interactions involving neutrinos are those of annihilation, such as:

$$e^- + e^+ \leftrightarrow \nu_e + \bar{\nu}_e . \qquad (3.12)$$

and these also are no more efficient on energies scales below 1 MeV.

Electron-positron annihilation. The interaction

$$e^- + e^+ \leftrightarrow \gamma + \gamma , \qquad (3.13)$$

is balanced for energies higher than the mass of electrons and positrons $m_e c^2 = 511$ keV. When the temperature of the thermal bath drops below this value pair production is no longer possible and annihilation takes over. Just for completeness, the cross section for pair production is Berestetskii et al. (1982):

$$\sigma_{\gamma\gamma} = \frac{3}{16} \sigma_T (1 - v^2) \left[(3 - v^4) \ln \left(\frac{1+v}{1-v} \right) - 2v(2 - v^2) \right] , \qquad (3.14)$$

where the parameter v is defined as follows:

$$v \equiv \sqrt{1 - (m_e c^2)^2/(\hbar^2 \omega_1 \omega_2)} \; ; \tag{3.15}$$

the **Thomson cross section** is

$$\sigma_T = \frac{8\pi}{3} \left(\frac{\alpha \hbar c}{m_e c^2} \right)^2 \approx 66.52 \, \text{fm}^2 \, , \tag{3.16}$$

with $\alpha = 1/137$ being the **fine structure constant**. Finally, ω_1 and ω_2 are the frequencies of the two photons. Being in a thermal bath, we have that $\hbar \omega_1 \approx \hbar \omega_2 \approx k_B T$. From Eq. (3.15) it is then clear that the condition

$$\hbar^2 \omega_1 \omega_2 \approx (k_B T)^2 > (m_e c^2)^2 \, , \tag{3.17}$$

must be satisfied in order to produce pairs. Therefore, when the temperature of the thermal bath drops below the value of the electron mass, annihilation becomes the only relevant process. Thanks to leptogenesis, positrons disappear and only electrons are left.

Big Bang Nucleosynthesis (BBN). We shall discuss BBN in great detail in Sect. 3.9. It occurs at about 0.1 MeV (some three minutes after the Big Bang) as deuterium and Helium form.

Recombination and photon decoupling. Proton and electrons form neutral hydrogen at about 0.3 eV. Having no more free electrons with which to scatter, photons decouple and can be seen today as CMB. We shall study this process in detail in Sect. 3.10.

3.3   The Distribution Function

Before applying Boltzmann equation to cosmology, we have to introduce the main character of our story: the **distribution function** f. This is a function $f = f(\mathbf{x}, \mathbf{p}, t)$ of the position, of the proper momentum and of the time, i.e. it is a function which takes its values in the **phase space**. It can be thought of as a probability density, i.e.

$$f(t, \mathbf{x}, \mathbf{p}) \frac{d^3 \mathbf{x} d^3 \mathbf{p}}{\mathcal{V}} \, , \tag{3.18}$$

is the probability of finding a particle at the time t in a small volume $d^3 \mathbf{x} d^3 \mathbf{p}$ of the phase space centred in $(\mathbf{x}, \mathbf{p})$, and $\mathcal{V}$ is some suitable normalisation.

Because of Heisenberg uncertainty principle of quantum mechanics no particle can be localised in the phase space in a point $(\mathbf{x}, \mathbf{p})$, but at most in a small volume $\mathcal{V} = h^3$ about that point, where h is Planck constant. Therefore, the probability density is the following:

$$dP(t, \mathbf{x}, \mathbf{p}) = f(t, \mathbf{x}, \mathbf{p})\frac{d^3x d^3\mathbf{p}}{h^3} = f(t, \mathbf{x}, \mathbf{p})\frac{d^3x d^3\mathbf{p}}{(2\pi\hbar)^3} . \tag{3.19}$$

Note the dimensions: h has dimensions of energy times time, or momentum times space. Hence f is dimensionless.

3.3.1 Volume of the Fundamental Cell

Why the fundamental cell has volume precisely equal to h^3? We will see that this value is very important in order to calculate correctly the abundances of the universe components. That is, if it were $2h^3$ instead of h^3 we would estimate half of the present abundance of photons.

Consider a quantum particle confined in a cube of side L. The eigenfunctions $u_{\mathbf{p}}(\mathbf{x})$ of the momentum operator $\hat{\mathbf{p}}$ are determined by the equation:

$$\hat{\mathbf{p}} u_{\mathbf{p}}(\mathbf{x}) = \mathbf{p} u_{\mathbf{p}}(\mathbf{x}) \quad \Rightarrow \quad -i\hbar\nabla u_{\mathbf{p}}(\mathbf{x}) = \mathbf{p} u_{\mathbf{p}}(\mathbf{x}) . \tag{3.20}$$

Assuming variable separation,

$$u_{\mathbf{p}}(\mathbf{x}) = u_x(x)u_y(y)u_z(z) , \tag{3.21}$$

Eq. (3.20) is easily solved:

$$u_{\mathbf{p}}(\mathbf{x}) = \frac{1}{L^{3/2}} \exp\left(\frac{i\mathbf{p} \cdot \mathbf{x}}{\hbar}\right) , \tag{3.22}$$

where the factor $1/L^{3/2}$ comes from the normalisation of the eigenfunction, which has integrated square modulus equal to 1 in the box.

Exercise 3.2 Prove the result (3.22).

Since we have used variable separation, from now on focus only on the x dimension, for simplicity. Since the particle is restricted to be in the box, we must impose periodic boundary conditions in Eq. (3.22):

$$u_x(0) = u_x(L) . \tag{3.23}$$

These conditions imply that the momentum is quantised, i.e.

$$\frac{p_x L}{\hbar} = 2\pi n_x , \tag{3.24}$$

where $n_x \in \mathbb{Z}$.

Exercise 3.3 Prove the result (3.24).

The phase space occupied by a single state is thus:

$$\Delta(p_x L) = 2\pi \hbar = h , \tag{3.25}$$

and recovering the three dimensions we get the expected result

$$V = \Delta(p_x L)\Delta(p_y L)\Delta(p_z L) = h^3 . \tag{3.26}$$

3.3.2   Integrals of the Distribution Function

Integrating the distribution function with respect to the momentum, one gets the particle number density:

$$n(t, \mathbf{x}) \equiv g_s \int \frac{d^3\mathbf{p}}{(2\pi\hbar)^3} f(t, \mathbf{x}, \mathbf{p}) \tag{3.27}$$

where g_s is the degeneracy of the species, e.g. the number of spin states. If we weigh the energy $E(p) = \sqrt{p^2 c^2 + m^2 c^4}$ with the distribution function, we get the energy density

$$\varepsilon(t, \mathbf{x}) = g_s \int \frac{d^3\mathbf{p}}{(2\pi\hbar)^3} f(t, \mathbf{x}, \mathbf{p}) \sqrt{p^2 c^2 + m^2 c^4} \tag{3.28}$$

where note that we are identifying $p \equiv |\mathbf{p}|$. The pressure is defined as follows:

$$P(t, \mathbf{x}) = g_s \int \frac{d^3\mathbf{p}}{(2\pi\hbar)^3} f(t, \mathbf{x}, \mathbf{p}) \frac{p^2 c^2}{3E(p)} \tag{3.29}$$

For photons one has $E(p) = pc$ and therefore combining Eqs. (3.28) and (3.29) one gets the familiar result $P = \varepsilon/3$.

In general, the energy-momentum tensor written in terms of the distribution function has the following form:

$$T^\mu{}_\nu(t, \mathbf{x}) = g_s \int \frac{dP_1 dP_2 dP_3}{(2\pi\hbar)^3} \frac{1}{\sqrt{-g}} \frac{cP^\mu P_\nu}{P^0} f(t, \mathbf{x}, \mathbf{p}) \tag{3.30}$$

where $P_\mu \equiv dx_\mu/d\lambda$ is the comoving momentum and g is the determinant of $g_{\mu\nu}$. Note that the combination $dP_1 dP_2 dP_3/(\sqrt{-g}P^0)$ which appears in the integrand is actually covariant. We shall prove this later in Sect. 3.8, but we give now a simple proof within special relativity. Indeed, for $g = -1$, the volume element reduces to $d^3\mathbf{P}/E$, which is invariant under Lorentz transformations. Let us show this, focussing on a single spatial dimension, i.e. the one along which the boost takes place and along which the particle travels. Lorentz transformations between two inertial reference frames are:

$$E' = \gamma(E - \beta pc) , \qquad p'c = \gamma(pc - \beta E) , \tag{3.31}$$

with $\beta \equiv V/c$, being V the boost velocity. Combining the differential form of the second, with the first, we get:

$$\frac{dp'c}{E'} = \frac{dpc - \beta dE}{E - \beta pc} . \tag{3.32}$$

Exercise 3.4 Differentiate the dispersion relation $E^2 = p^2c^2 + m^2c^4$, use it into (3.32) and show that

$$\frac{dp'}{E'} = \frac{dp}{E} , \tag{3.33}$$

i.e. this combination is invariant.

We can rewrite Eq. (3.30) in terms of the proper momentum instead of the comoving one. Using the FLRW metric with $K = 0$, for simplicity, the dispersion relation (2.31) and Eq. (2.32) we can write

$$g_{00}(P^0)^2 = -E^2/c^2 \quad \Rightarrow \quad \sqrt{-g_{00}}P^0 = E/c , \tag{3.34}$$

so that Eq. (3.30) becomes:

$$\boxed{T^\mu{}_\nu(t, \mathbf{x}) = g_s \int \frac{dP_1 dP_2 dP_3}{(2\pi\hbar)^3} \frac{1}{a^3} \frac{c^2 P^\mu P_\nu}{E} f(t, \mathbf{x}, \mathbf{p})} \tag{3.35}$$

Exercise 3.5 Using the definition (2.32) of the proper momentum, i.e.

$$p^2 = g_{ij}P^i P^j = g^{ij}P_i P_j = \frac{1}{a^2}\delta^{ij}P_i P_j , \tag{3.36}$$

show that:

$$\boxed{P_i = ap_i = ap\hat{p}_i} \qquad \boxed{P^i = \frac{p^i}{a} = \frac{p\hat{p}^i}{a}} \tag{3.37}$$

where $\hat{p}_i$ is the *direction* of the proper or comoving momentum, satisfying:

$$\delta^{ij}\hat{p}_i\hat{p}_j = 1 \tag{3.38}$$

Note the following:

$$ap_i = P_i = g_{ij}P^j = a^2\delta_{ij}\frac{p^j}{a} = a\delta_{ij}p^j . \tag{3.39}$$

This means that

$$p_i = \delta_{ij}p^j \tag{3.40}$$

i.e. it is as if the proper momentum were a 3-vector in the Euclidean space.

Therefore, Eq. (3.35) becomes:

$$T^{\mu}{}_{\nu}(t, \mathbf{x}) = g_s \int \frac{d^3\mathbf{p}}{(2\pi\hbar)^3}\frac{c^2 P^{\mu}P_{\nu}}{E}f(t, \mathbf{x}, \mathbf{p}) \tag{3.41}$$

For $\mu = \nu = 0$:

$$T^0{}_0(t, \mathbf{x}) = g_s \int \frac{d^3\mathbf{p}}{(2\pi\hbar)^3}cP_0f(t, \mathbf{x}, \mathbf{p}) = g_s \int \frac{d^3\mathbf{p}}{(2\pi\hbar)^3}E(p)f(t, \mathbf{x}, \mathbf{p}) . \tag{3.42}$$

For $\mu = 0$ and $\nu = i$:

$$T^0{}_i(t, \mathbf{x}) = g_s \int \frac{d^3\mathbf{p}}{(2\pi\hbar)^3}cP_i f(t, \mathbf{x}, \mathbf{p}) = g_s \int \frac{d^3\mathbf{p}}{(2\pi\hbar)^3}apc\hat{p}_i f(t, \mathbf{x}, \mathbf{p}) . \tag{3.43}$$

Finally, for $\mu = i$ and $\nu = j$:

$$T^i{}_j(t, \mathbf{x}) = g_s \int \frac{d^3\mathbf{p}}{(2\pi\hbar)^3}\frac{c^2 p^2 \hat{p}^i\hat{p}_j}{E}f(t, \mathbf{x}, \mathbf{p}) . \tag{3.44}$$

Taking the spatial trace and dividing by 3 we get the pressure, consistently with Eq. (3.29).

3.4　The Entropy Density

For many of the forthcoming purposes the hypothesis of thermal equilibrium is suitable and very useful. As we have stated at the beginning of this chapter, it is justified in those instances in which the interaction rate among particles is much higher than the expansion rate. In these cases, one can use equilibrium thermodynamics and a very useful quantity is the entropy density:

$$s \equiv \frac{S}{V} , \tag{3.45}$$

because, as we will show in a moment, sa^3 is conserved.

In thermal equilibrium, we can cast the thermodynamical relation (3.2) in the following form:

$$TdS = Vd\varepsilon + (\varepsilon + P)dV = V\frac{d\varepsilon}{dT}dT + (\varepsilon + P)dV , \tag{3.46}$$

because the energy density (and also the pressure) only depends on the temperature T. The integrability condition applied to Eq. (3.46) yields to:

$$\frac{\partial^2 S}{\partial T \partial V} = \frac{\partial^2 S}{\partial V \partial T} \quad \Rightarrow \quad T\frac{dP}{dT} = \varepsilon + P . \tag{3.47}$$

Bosons and fermions in thermal equilibrium are distributed according to the Bose–Einstein and Fermi-Dirac distributions:

$$\boxed{f_{\text{BE}} = \frac{1}{\exp\left(\frac{E-\mu}{k_B T}\right) - 1} , \quad f_{\text{FD}} = \frac{1}{\exp\left(\frac{E-\mu}{k_B T}\right) + 1}} \tag{3.48}$$

We derive these distributions in Chap. 12.

We now prove Eq. (3.47) in another way, assuming a distribution function of the type $f = f(E/T)$. We first must know how to calculate dP/dT.

Call $E/T = x$ and $f' \equiv df/dx$. Then:

$$df = f'dx = \frac{f'}{T}dE - \frac{f'E}{T^2}dT . \tag{3.49}$$

Comparing this with

$$df = \frac{\partial f}{\partial E}dE + \frac{\partial f}{\partial T}dT , \tag{3.50}$$

we can establish that

$$\frac{\partial f}{\partial T} = -\frac{E}{T}\frac{\partial f}{\partial E} . \tag{3.51}$$

Now we use this result into:

$$\frac{dP}{dT} = g_s \int \frac{d^3\mathbf{p}}{(2\pi\hbar)^3} \frac{\partial f}{\partial T} \frac{p^2 c^2}{3E} , \tag{3.52}$$

and obtain

$$\frac{dP}{dT} = -g_s \int \frac{d^3\mathbf{p}}{(2\pi\hbar)^3} \frac{E}{T} \frac{\partial f}{\partial E} \frac{p^2c^2}{3E} . \tag{3.53}$$

Now we introduce spherical coordinates in the proper momentum space:

$$d^3\mathbf{p} = p^2 dp d^2\hat{p} , \tag{3.54}$$

and use the dispersion relation $E^2 = p^2c^2 + m^2c^4$ in order to write

$$\frac{\partial f}{\partial E} = \frac{\partial f}{\partial p} \frac{dp}{dE} = \frac{\partial f}{\partial p} \frac{E}{pc^2} . \tag{3.55}$$

Equation (3.53) thus becomes:

$$\frac{dP}{dT} = -g_s \int \frac{p^2 dp d^2\hat{p}}{(2\pi\hbar)^3} \frac{E}{T} \frac{\partial f}{\partial p} \frac{p}{3} . \tag{3.56}$$

Integrating by parts:

$$\frac{dP}{dT} = -g_s \frac{4\pi}{(2\pi\hbar)^3} \frac{Ep^3}{3T} f \Big|_0^\infty + g_s \int \frac{dp d^2\hat{p}}{(2\pi\hbar)^3} \frac{1}{3T} f \left(3p^2 E + p^3 \frac{pc^2}{E} \right) , \tag{3.57}$$

where we have used again $dE/dp = pc^2/E$. The first contribution vanishes for $p \to \infty$, since $f \to 0$, i.e. there are no particles with infinite momentum. Recovering the momentum volume $d^3\mathbf{p}$, we get:

$$\frac{dP}{dT} = g_s \int \frac{d^3\mathbf{p}}{(2\pi\hbar)^3} \frac{1}{T} f \left(E + \frac{p^2c^2}{3E} \right) \quad \Rightarrow \quad \boxed{\frac{dP}{dT} = \frac{\varepsilon + P}{T}} \tag{3.58}$$

as we wanted to show.

We now prove that the temperature derivative of the pressure is the entropy density. Substituting Eq. (3.47) into Eq. (3.46), i.e.

$$TdS = d(\varepsilon V) + PdV = d[(\varepsilon + P)V] - VdP , \tag{3.59}$$

one gets

$$dS = \frac{1}{T} d[(\varepsilon + P)V] - V \frac{\varepsilon + P}{T^2} dT = d \left[\frac{(\varepsilon + P)V}{T} \right] , \tag{3.60}$$

i.e. up to an additive constant:

$$\boxed{s \equiv \frac{S}{V} = \frac{\varepsilon + P}{T}} . \tag{3.61}$$

Taking into account the chemical potential μ, the thermodynamical relation (3.2) becomes

$$TdS = d(\varepsilon V) + PdV - \mu d(nV) \,, \tag{3.62}$$

and the entropy density is redefined as

$$\boxed{s \equiv \frac{S}{V} = \frac{\varepsilon + P - \mu n}{T}} \tag{3.63}$$

Exercise 3.6 Using the continuity equation and/or Eq. (3.2) show that sa^3 is a constant:

$$\boxed{\frac{d}{dt}(sa^3) = 0} \tag{3.64}$$

i.e. the entropy density is proportional to $1/a^3$.

3.5 Photons

In these notes we use "photons" as synonym of CMB, though this is not correct since there exist photons whose origin is not cosmological, e.g. those produced in our Sun as well as in other stars or emitted by hot interstellar gas. These say non-cosmological photons contribute at least one order of magnitude less than CMB photons (Camarena and Marra 2016) so our sloppiness is partially justified.

Assuming a vanishing chemical potential since $\mu/(k_B T) < 9 \times 10^{-5}$, as reported by Fixsen et al. (1996), the Bose–Einstein distribution for photons becomes:

$$f_\gamma = \frac{1}{\exp\left(\frac{E}{k_B T}\right) - 1} = \frac{1}{\exp\left(\frac{pc}{k_B T}\right) - 1} \,, \tag{3.65}$$

where we have used the dispersion relation $E = pc$. Taking into account the chemical potential is important in order to study distortions in the CMB spectrum, which is a very recent and promising research field (Chluba and Sunyaev 2012; Chluba 2014).

Let us calculate the photon energy density:

$$\varepsilon_\gamma = 2 \int \frac{d^3\mathbf{p}}{(2\pi\hbar)^3} \frac{pc}{\exp\left(\frac{pc}{k_B T}\right) - 1} \,, \tag{3.66}$$

where the factor 2 represents the two states of polarisation of the photon. The angular part can be readily integrated out, giving a factor 4π. We are left then with:

$$\varepsilon_\gamma = \frac{c}{\pi^2 \hbar^3} \int_0^\infty dp \frac{p^3}{\exp\left(\frac{pc}{k_B T}\right) - 1} . \tag{3.67}$$

Let us do the substitution $x \equiv pc/(k_B T)$. We obtain:

$$\varepsilon_\gamma = \frac{c}{\pi^2 \hbar^3} \left(\frac{k_B T}{c}\right)^4 \int_0^\infty dx \frac{x^3}{e^x - 1} . \tag{3.68}$$

The integration is proportional to the Riemann ζ function, which has the following integral representation:

$$\zeta(s) = \frac{1}{\Gamma(s)} \int_0^\infty dx \frac{x^{s-1}}{e^x - 1} = \frac{1}{(1 - 2^{1-s})\Gamma(s)} \int_0^\infty dx \frac{x^{s-1}}{e^x + 1} , \tag{3.69}$$

where $\Gamma(s)$ is Euler gamma function.

In alternative, a possible way to perform the integration is the following. Let I_- be:

$$I_- \equiv \int_0^\infty dx \frac{x^3}{e^x - 1} = \int_0^\infty dx \frac{e^{-x} x^3}{1 - e^{-x}} . \tag{3.70}$$

Now use the geometric series in order to write

$$\frac{1}{1 - e^{-x}} = \sum_{n=0}^\infty e^{-nx} , \tag{3.71}$$

and substitute this in the integral I_-:

$$I_- = \sum_{n=0}^\infty \int_0^\infty dx \, e^{-(n+1)x} x^3 . \tag{3.72}$$

Exercise 3.7 Integrate the above equation three times by part and show that:

$$I_- = 6 \sum_{n=1}^\infty \frac{1}{n^3} \int_0^\infty dx \, e^{-nx} . \tag{3.73}$$

Integrate again and find:

$$I_- = 6 \sum_{n=1}^\infty \frac{1}{n^4} \equiv 6\zeta(4) = \frac{\pi^4}{15} , \tag{3.74}$$

in agreement with Eq. (3.69).

Therefore, the photon energy density is the following:

$$\varepsilon_\gamma = \frac{\pi^2}{15\hbar^3 c^3}(k_B T)^4 . \tag{3.75}$$

This is the Stefan–Boltzmann law, of the black-body radiation. From the continuity equation we know that $\varepsilon_\gamma = \varepsilon_{\gamma 0}/a^4$ so that we can infer that

$$T = \frac{T_0}{a}, \qquad T_0 = 2.725\,\text{K} \tag{3.76}$$

i.e. the temperature of the photons decreases with the inverse scale factor. This is a result that we will prove also using the Boltzmann equation (indeed the continuity equation that provides $\varepsilon_\gamma = \varepsilon_{\gamma 0}/a^4$ is a way of writing the Boltzmann equation).

In the above Eq. (3.76), the value of T_0 is the measured one of the CMB. Knowing this value, we can estimate the photon energy content today (and thus at any times):

$$\Omega_{\gamma 0} = \frac{\varepsilon_{\gamma 0}}{\varepsilon_{cr0}} = \frac{8\pi^3 G}{45\hbar^3 c^3 H_0^2 c^2}(k_B T_0)^4 . \tag{3.77}$$

Exercise 3.8 Using $H_0 = 100\,h\,\text{km s}^{-1}\,\text{Mpc}^{-1}$ and the known constants of nature show that:

$$\Omega_{\gamma 0} h^2 = 2.47 \times 10^{-5} . \tag{3.78}$$

The photon number density is calculated as follows:

$$n_\gamma = \frac{1}{\pi^2 \hbar^3} \int_0^\infty dp \frac{p^2}{\exp\left(\frac{pc}{k_B T}\right) - 1} . \tag{3.79}$$

Making the usual substitution $x \equiv pc/(k_B T)$ and using Eq. (3.69) we get:

$$n_\gamma = \frac{(k_B T)^3}{\pi^2 \hbar^3 c^3} \int_0^\infty dx \frac{x^2}{e^x - 1} \Rightarrow n_\gamma = \frac{2\zeta(3)}{\pi^2 \hbar^3 c^3}(k_B T)^3 \tag{3.80}$$

Exercise 3.9 Calculate the photon number density today. Show that it is $n_{\gamma 0} = 411\,\text{cm}^{-3}$.

3.6 Neutrinos

The same comment made at the beginning of the previous section also applies here: with "neutrinos" we mean the cosmological, or primordial, ones and not those produced e.g. in supernovae explosions.

The massless neutrino energy density also scales as $\varepsilon_\nu = \varepsilon_{\nu 0}/a^4$, as the photons energy density, but since neutrinos are fermions we need now to employ the Fermi-Dirac distribution.

Exercise 3.10 Show that the energy-density of a massless fermion species is given by the following integral:

$$\varepsilon = \frac{c}{2\pi^2 \hbar^3} \left(\frac{k_B T}{c} \right)^4 g_s \int_0^\infty dx \frac{x^3}{e^x + 1} . \tag{3.81}$$

Since neutrinos are spin $1/2$ fermions, then $g_\nu = 2$, where g_ν is the neutrino g_s. On the other hand, neutrinos are particles which interact only via weak interaction and this violates parity. In other words, only left-handed neutrinos can be detected. **Right-handed neutrinos**, if they exist, would interact only via gravity and via the **seesaw mechanism** with the left-handed neutrino (Gell-Mann et al. 1979). Right-handed neutrinos are also called **sterile neutrinos** and are advocated as possible candidates for DM (Dodelson and Widrow 1994).

Let I_+ be the integral in Eq. (3.81):

$$I_+ \equiv \int_0^\infty dx \frac{x^3}{e^x + 1} . \tag{3.82}$$

Using Eq. (3.69), we have:

$$I_+ = (1 - 2^{-3})I_- = \frac{7}{8}I_- , \tag{3.83}$$

i.e. the difference between bosons and fermions energy densities per spin state is simply a factor $7/8$. Taking into account that $I_- = \pi^4/15$, the neutrino energy density (3.81) is:

$$\boxed{\varepsilon_\nu = \frac{7}{8} N_\nu g_\nu \frac{\pi^2}{30\hbar^3 c^3} (k_B T_\nu)^4} \tag{3.84}$$

where N_ν is the number of neutrino families. Of course, an equivalent expression holds true for antineutrinos.

3.6.1 Temperature of the Massless Neutrino Thermal Bath

As we have seen in Eq. (3.64), for a species in thermal equilibrium its entropy density s is proportional to $1/a^3$. Moreover, if that species is relativistic then its temperature T scales as $1/a$. Therefore, for a relativistic species in thermal equilibrium $s \propto T^3$.

Since photons and neutrinos do not interact, it is reasonable to ask whether the temperatures of the photon thermal bath and of the neutrino thermal bath are the same. We cannot observe the neutrino thermal bath today, but we can indeed predict different temperatures.

When the temperature of the photon thermal bath was sufficiently high, positron-electron annihilation and positron-electron pair production were balanced reactions:

$$e^+ + e^- \leftrightarrow \gamma + \gamma . \tag{3.85}$$

In order to produce e^+-e^- pairs, the photons must have temperature of the order of 1 MeV at least.

Exercise 3.11 Knowing that today, i.e. for $z = 0$, the photon thermal bath has temperature $T_0 = 2.725$ K, estimate the redshift at which $k_B T = 1$ MeV.

Therefore, when the temperature of the photon thermal bath drops below that value, the above reactions are unbalanced and more photons are thus injected in the thermal bath. For this reason we expect the photon temperature to drop more slowly than $1/a$ and then to be different from the neutrino temperature. We now quantify this difference using the conservation of sa^3.

For a relativistic bosonic species (such as the photon), the entropy density is:

$$s_{\text{boson}} = \frac{\varepsilon + P}{T} = \frac{4\varepsilon}{3T} = g_s \frac{2\pi^2 k_B^4}{45\hbar^3 c^3} T^3 \tag{3.86}$$

whereas for a relativistic fermion species (such as the neutrino), the entropy density is:

$$s_{\text{fermion}} = \frac{7}{8} g_s \frac{2\pi^2 k_B^4}{45\hbar^3 c^3} T^3 \tag{3.87}$$

Therefore, at a certain scale factor a_1 earlier than e^--e^+ annihilation, the entropy density is:

$$s(a_1) = \frac{2\pi^2 k_B^4}{45\hbar^3 c^3} T_1^3 \left[2 + \frac{7}{8}(2+2) + \frac{7}{8} N_\nu (g_\nu + g_\nu) \right] , \tag{3.88}$$

where we have left explicit all the degrees of freedom, i.e. 2 for the photons, 2 for the electrons, 2 for the positrons, g_ν for the neutrinos and g_ν for the antineutrinos.

Moreover, we have assumed the same temperature T_1 for photons and neutrinos because they came from the original thermal bath (the Big Bang).

At a certain scale factor a_2 after the annihilation the entropy density is:

$$s(a_2) = \frac{2\pi^2 k_B^4}{45\hbar^3 c^3}\left(2T_\gamma^3 + \frac{7}{4}N_\nu g_\nu T_\nu^3\right), \qquad (3.89)$$

where we have now made distinction between the two temperatures. Equating

$$s(a_1)a_1^3 = s(a_2)a_2^3, \qquad (3.90)$$

we obtain

$$(a_1 T_1)^3\left[2 + \frac{7}{4}\left(N_\nu g_\nu + 2\right)\right] = (a_2 T_\nu)^3\left[2\left(\frac{T_\gamma}{T_\nu}\right)^3 + \frac{7}{4}N_\nu g_\nu\right]. \qquad (3.91)$$

Since $(a_1 T_1)^3 = (a_2 T_\nu)^3$, because the neutrino temperature did not change its $\propto 1/a$ behaviour, we are left with:

$$\boxed{\frac{T_\nu}{T_\gamma} = \left(\frac{4}{11}\right)^{1/3} \approx 0.714} \qquad (3.92)$$

Therefore, since the CMB temperature today is of $T_{\gamma 0} = 2.725$ K, we expect a thermal neutrino background of temperature $T_{\nu 0} \approx 1.945$ K. Remarkably, the result of Eq. (3.92) does not depend on the neutrino g_ν.

Let us open a brief parenthesis in order to justify the procedure that we have used in order to determine the result in Eq. (3.92).

We saw that for a single species in thermal equilibrium sa^3 is conserved because of the continuity equation. However, during electron-positron annihilation the continuity equation *does not* hold true neither for photons nor for electron and positron, i.e.

$$\dot\varepsilon_\gamma + 4H\varepsilon_\gamma = +\Gamma_{\text{ann}}, \qquad (3.93)$$
$$\dot\varepsilon_e + 3H(\varepsilon_e + P_e) = -\Gamma_{\text{ann}}, \qquad (3.94)$$

where Γ_{ann} is the $e^- \text{-} e^+$ annihilation rate. Therefore, we cannot use the constancy of sa^3 for each of these species separately. However, we can and did use it for the *total*, since the sum of the two above equations gives:

$$\dot\varepsilon_{\text{tot}} + 3H(\varepsilon_{\text{tot}} + P_{\text{tot}}) = 0. \qquad (3.95)$$

In other words, the continuity equation always applies if one suitably extends the set of species, ultimately because of Bianchi identities.

Using Eq. (3.92), the neutrino and antineutrino energy densities can be thus related to the photon energy density as follows:

$$\varepsilon_\nu = \varepsilon_{\bar{\nu}} = \frac{7}{8}\frac{N_\nu g_\nu}{2}\left(\frac{4}{11}\right)^{4/3}\varepsilon_\gamma \qquad (3.96)$$

where we have left unspecified the value of the neutrino degeneracy g_ν and the number of neutrino families N_ν.

The total radiation energy content can thus be written as

$$\varepsilon_r \equiv \varepsilon_\gamma + \varepsilon_\nu + \varepsilon_{\bar{\nu}} = \varepsilon_\gamma\left[1 + \frac{7}{8}N_\nu g_\nu\left(\frac{4}{11}\right)^{4/3}\right]. \qquad (3.97)$$

The Planck collaboration (Ade et al. 2016a) has put the constraint

$$N_{\text{eff}} \equiv N_\nu g_\nu = 3.04 \pm 0.33 \qquad (3.98)$$

at 95% CL. Therefore, three neutrino families ($N_\nu = 3$) and one spin state for each neutrino ($g_\nu = 1$) are values which work fine.

Calculating the neutrino + antineutrino number density today is straightforward:

$$n_\nu = \frac{N_\nu g_\nu}{\pi^2\hbar^3}\int_0^\infty dp\,\frac{p^2}{\exp\left(\frac{pc}{k_B T_\nu}\right) + 1}, \qquad (3.99)$$

where now the subscript ν indicates both neutrinos and antineutrinos. Making the usual substitution $x \equiv pc/(k_B T_\nu)$, we get:

$$n_\nu = \frac{N_\nu g_\nu (k_B T_\nu)^3}{\pi^2\hbar^3 c^3}\int_0^\infty dx\,\frac{x^2}{e^x + 1} = \frac{N_\nu g_\nu (k_B T_\nu)^3}{\pi^2\hbar^3 c^3}\frac{3\zeta(3)\Gamma(3)}{2} = \frac{3\zeta(3)N_\nu g_\nu(k_B T_\nu)^3}{2\pi^2\hbar^3 c^3}. \qquad (3.100)$$

Exercise 3.12 Taking into account (3.92), show that

$$n_\nu = \frac{3}{11}N_\nu g_\nu n_\gamma \qquad (3.101)$$

3.6.2   Massive Neutrinos

The 2015 Nobel Prize in Physics has been awarded to Takaaki Kajita and Arthur B. McDonald *for the discovery of neutrino oscillations, which shows that neutrinos have mass* (quoting from the Nobel Prize website). Indeed, neutrino flavours oscillate among the leptonic families (electron, muon and tau), e.g. an electronic neutrino can turn into a muonic one and a tau one, depending on its energy and on how far it travels.

The most stringent constraints on neutrino mass do not come from particle accelerators but cosmology: $\sum m_\nu < 0.194$ eV at 95% CL from the Planck collaboration (Ade et al. 2016a). The neutrino mass is thus very small and this does not change relevantly the early history of the universe whereas it has some impact at late-times for structure formation (Lesgourgues and Pastor 2006).

Using Eq. (3.28) and the FD distribution, the massive neutrino energy density can be calculated as follows:

$$\varepsilon_\nu = \int \frac{d^3\mathbf{p}}{(2\pi\hbar)^3} \frac{\sqrt{p^2c^2 + m_\nu^2 c^4}}{\exp\left[\sqrt{p^2c^2 + m_\nu^2 c^4}/(k_B T)\right] + 1} , \tag{3.102}$$

where we are assuming $g_s = 1$.

Actually, Eq. (3.102) is valid for any fermion (or boson, if we have the -1 at the denominator), provided we can neglect the chemical potential. For this reason, we drop the subscript ν and consider a generic species of mass m, fermion or boson, in order to make more general statements.

Rewrite Eq. (3.102) as follows:

$$\varepsilon = \frac{mc^2}{2\pi^2\hbar^3} \int_0^\infty dp\, p^2 \frac{\sqrt{p^2/(m^2c^2) + 1}}{\exp\left[A\sqrt{p^2/(m^2c^2) + 1}\right] \pm 1} , \qquad A \equiv \frac{mc^2}{k_B T} . \tag{3.103}$$

Exercise 3.13 Calling $x \equiv A\sqrt{p^2/(m^2c^2) + 1}$, show that the above integration (3.103) becomes:

$$\varepsilon = \frac{m^4c^5}{2\pi^2\hbar^3 A^4} \int_A^\infty dx\, \frac{x^2\sqrt{x^2 - A^2}}{e^x \pm 1} = \frac{(k_B T)^4}{2\pi^2\hbar^3 c^3} \int_A^\infty dx\, \frac{x^2\sqrt{x^2 - A^2}}{e^x \pm 1} . \tag{3.104}$$

Note the lower integration limit.

Relativistic and Non-relativistic Regimes of Particles in Thermal Equilibrium

Unfortunately, the above integral in Eq. (3.104) cannot be solved analytically. When $A \ll 1$, i.e. the thermal energy is much larger than the mass energy, the integral in Eq. (3.104) can be expanded as follows:

$$\int_A^\infty dx \, \frac{x^2 \sqrt{x^2 - A^2}}{e^x - 1} = \frac{\pi^4}{15} - \frac{\pi^2 A^2}{12} + \mathcal{O}(A^3) , \tag{3.105}$$

$$\int_A^\infty dx \, \frac{x^2 \sqrt{x^2 - A^2}}{e^x + 1} = \frac{7\pi^4}{120} - \frac{\pi^2 A^2}{24} + \mathcal{O}(A^3) . \tag{3.106}$$

The zero-order terms recovers the result of Eq. (3.75), for photons, and of Eq. (3.84), obtained for massless neutrinos. In general, we can state that **particles with mass m in thermal equilibrium at temperature T behave as relativistic particles when** $mc^2 \ll k_B T$.

The opposite limit $mc^2 \gg k_B T$ is much trickier to investigate analytically, so we shall do it numerically.

Exercise 3.14 Calling $x \equiv A\sqrt{p^2/(m^2 c^2) + 1}$, show that the number density can be written as:

$$n = \frac{m^3 c^3}{2\pi^2 \hbar^3 A^3} \int_A^\infty dx \, \frac{x\sqrt{x^2 - A^2}}{e^x \pm 1} . \tag{3.107}$$

The ratio $\varepsilon/(nmc^2)$ can be written as:

$$\gamma \equiv \frac{\varepsilon}{(nmc^2)} = \frac{1}{A} \frac{\int_A^\infty dx \, \frac{x^2\sqrt{x^2 - A^2}}{e^x \pm 1}}{\int_A^\infty dx \, \frac{x\sqrt{x^2 - A^2}}{e^x \pm 1}} , \tag{3.108}$$

where γ is indeed some sort of averaged Lorentz factor since it is the ratio of the energy density to the mass energy density. The behaviour of γ as function of A is shown in Fig. 3.1.

From Fig. 3.1 we can infer that $\varepsilon/(nmc^2) \approx 1$ when $mc^2 \gg k_B T$ and therefore the particle becomes non-relativistic since all its energy is mass energy. In general, we can state that **particles with mass m in thermal equilibrium at temperature T behave as non-relativistic particles when $mc^2 \gg k_B T$**. The transition relativistic $\rightarrow$ non-relativistic takes place for $k_B T \approx 10mc^2$.

Now, suppose that a single family of neutrinos and antineutrinos became non-relativistic only recently, what would be their energy density and density parameter today? Being non-relativistic, we can write their energy density as follows:

$$\varepsilon_{m_\nu} = \rho_\nu c^2 = n_\nu m_\nu c^2 , \tag{3.109}$$

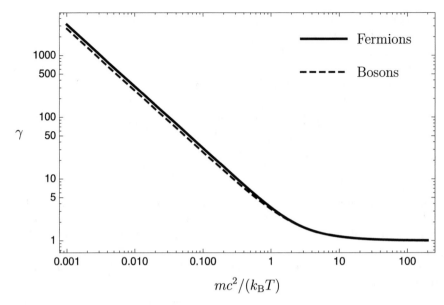

Fig. 3.1 Plot of γ as function of A. The solid line is for fermions whereas the dashed one for bosons

and thus the density parameter today is:

$$\Omega_{m_\nu 0} = \frac{8\pi G n_{\nu 0} m_\nu}{3H_0^2} .$$

(3.110)

Exercise 3.15 Using Eq. (3.101) and the result for $n_{\gamma 0}$, prove that Eq. (3.110) can be written as:

$$\boxed{\Omega_{m_\nu 0} = \frac{1}{94h^2} g_\nu \frac{m_\nu c^2}{\text{eV}} \text{eV}}$$

(3.111)

3.6.3   Matter-Radiation Equality

The epoch, or instant, at which the energy density of matter (i.e. baryons plus CDM) equals the energy density of radiation (i.e. photons plus neutrinos) is particularly important from the point of view of the evolution of perturbations, as we shall see in Chap. 9. From Eq. (3.97) we have that:

$$\boxed{\Omega_{r0} = \Omega_{\gamma 0} \left[1 + \frac{7}{8} N_{\text{eff}} \left(\frac{4}{11} \right)^{4/3} \right]} \tag{3.112}$$

In order to calculate the scale factor a_{eq} of the equivalence we only need to solve the following equation:

$$\frac{\Omega_{r0}}{a_{\text{eq}}^4} = \frac{\Omega_{m0}}{a_{\text{eq}}^3}, \tag{3.113}$$

which gives

$$a_{\text{eq}} = \frac{\Omega_{r0}}{\Omega_{m0}} = \frac{\Omega_{\gamma 0}}{\Omega_{m0}} \left[1 + \frac{7}{8} N_{\text{eff}} \left(\frac{4}{11} \right)^{4/3} \right]. \tag{3.114}$$

Using Eq. (3.78) and $N_{\text{eff}} = 3$ one obtains:

$$\boxed{a_{\text{eq}} = \frac{4.15 \times 10^{-5}}{\Omega_{m0} h^2}} \qquad \Rightarrow \qquad \boxed{1 + z_{\text{eq}} = 2.4 \times 10^4 \, \Omega_{m0} h^2} \tag{3.115}$$

What does happen to the equivalence redshift z_{eq} if one of the neutrino species has mass $m_\nu \neq 0$?

If that neutrino species becomes non-relativistic *after* the equivalence epoch, then the above calculation still holds true. Therefore, let us assume that the neutrino species becomes non-relativistic, thereby counting as matter, before the equivalence. Since there is more matter and less radiation we expect the equivalence to take place earlier, i.e.

$$a_{\text{eq}} = \frac{3.59 \times 10^{-5}}{(\Omega_{m0} + \Omega_{m_\nu 0}) h^2} \quad 1 + z_{\text{eq}} = 2.79 \times 10^4 \, (\Omega_{m0} + \Omega_{m_\nu 0}) h^2 \,. \tag{3.116}$$

Since $T = T_0(1 + z)$, the photon temperature at equivalence is:

$$T_{\gamma,\text{eq}} = 2.79 \times T_0 \cdot 10^4 \, (\Omega_{m0} + \Omega_{m_\nu 0}) h^2 = 7.60 \times 10^4 \, (\Omega_{m0} + \Omega_{m_\nu 0}) h^2 \, \text{K} \,. \tag{3.117}$$

Using Eq. (3.92), the temperature of neutrinos at equivalence is:

$$T_{\nu,\text{eq}} = 5.43 \times 10^4 \, (\Omega_{m0} + \Omega_{m_\nu 0}) h^2 \, \text{K} \,. \tag{3.118}$$

The neutrino mass energy $m_\nu c^2$ has to be larger than $k_B T_{\nu,\text{eq}}$ in order for the above calculation to be consistent. This yields:

$$m_\nu c^2 > 5.43 \, k_B \times 10^4 \, (\Omega_{m0} + \Omega_{m_\nu 0}) h^2 \, \text{K} = 4.68 \, (\Omega_{m0} + \Omega_{m_\nu 0}) h^2 \, \text{eV} \,. \tag{3.119}$$

Using $\Omega_{m0} h^2 = 0.14$ and Eq. (3.111) we get $m_\nu c^2 > 0.69 \, \text{eV}$, which is incompatible with the constraint $\sum m_\nu < 0.194$ found by the Planck collaboration.

3.7 Boltzmann Equation

Here is the main character of this Chapter: Boltzmann equation. It is very simple to write:

$$\frac{df}{dt} = C[f] ,\qquad (3.120)$$

but nonetheless very meaningful, as we shall appreciate. Here f is the one-particle distribution function and $C[f]$ is the collisional term, i.e. a functional of f describing the interactions among the particles constituting the system under investigation. The one-particle distribution is a function of time t, of the particle position $\mathbf{x}$ and of the particle momentum $\mathbf{p}$. In turn, also $\mathbf{x}$ and $\mathbf{p}$ are functions of time, because of the particle motion. Therefore, the total time derivative can be written as:

$$\frac{df}{dt} = \frac{\partial f}{\partial t} + \frac{d\mathbf{x}}{dt} \cdot \nabla_{\mathbf{x}} f + \frac{d\mathbf{p}}{dt} \cdot \nabla_{\mathbf{p}} f = \frac{\partial f}{\partial t} + \mathbf{v} \cdot \nabla_{\mathbf{x}} f + \mathbf{F} \cdot \nabla_{\mathbf{p}} f \equiv \hat{L}(f) ,\quad (3.121)$$

where $\mathbf{v}$ is the particle velocity and $\mathbf{F}$ is the force acting on the particle. The operator $\hat{L}$ acting on f is similar to the convective derivative used in fluid dynamics and is also called **Liouville operator**.

If interactions are absent, then

$$\frac{df}{dt} = 0 ,\qquad (3.122)$$

is the collisionless Boltzmann equation, or Vlasov equation. It represents mathematically the fact that the number of particles in a phase space volume element does not change with the time. Note that we are starting here with the non relativistic version of the Boltzmann equation. We shall see it later in the general relativistic case and cosmology.

The collisionless Boltzmann equation is a direct consequence of Liouville theorem:

$$\frac{d\rho(t, \mathbf{x}_i, \mathbf{p}_i)}{dt} = 0 ,\qquad i = 1, \ldots, N \qquad (3.123)$$

where $\rho(\mathbf{x}_i, \mathbf{p}_i, t)$ is the N-particle distribution function, i.e.

$$\rho(t, \mathbf{x}_i, \mathbf{p}_i) d^N \mathbf{x} d^N \mathbf{p} ,\qquad (3.124)$$

is the probability of finding our system of N particles in a small volume $d^N \mathbf{x} d^N \mathbf{p}$ of the phase space centred in $(\mathbf{x}_i, \mathbf{p}_i)$.

If the particles are not interacting, then the probability of finding N particles in some configuration is the product of the single probabilities. That is, the positions in the phase space of the individual particles are independent events. Therefore:

$$\rho \propto f^N , \qquad \frac{d\rho}{dt} = Nf^{N-1}\frac{df}{dt} , \tag{3.125}$$

and using Liouville theorem (3.123) one obtains the collisionless Boltzmann equation (3.122).

3.7.1 Proof of Liouville Theorem

In order to complete this brief introduction to the Boltzmann equation, we prove Liouville theorem. As first step, expand the time derivative of ρ:

$$\frac{d\rho(t, \mathbf{x}_i, \mathbf{p}_i)}{dt} = \frac{\partial \rho}{\partial t} + \sum_{i=1}^{N} \left(\frac{d\mathbf{x}_i}{dt}\nabla_{\mathbf{x}_i}\rho + \frac{d\mathbf{p}_i}{dt}\nabla_{\mathbf{p}_i}\rho \right) . \tag{3.126}$$

This can be written as

$$\frac{d\rho(t, \mathbf{x}_i, \mathbf{p}_i)}{dt} = \frac{\partial \rho}{\partial t} + \sum_{i=1}^{N} \left[\nabla_{\mathbf{x}_i}(\rho\dot{\mathbf{x}}_i) + \nabla_{\mathbf{p}_i}(\rho\dot{\mathbf{p}}_i) \right] , \tag{3.127}$$

since the extra term

$$\nabla_{\mathbf{x}_i}(\dot{\mathbf{x}}_i) + \nabla_{\mathbf{p}_i}(\dot{\mathbf{p}}_i) = \nabla_{\mathbf{p}_i}\nabla_{\mathbf{x}_i}(H) - \nabla_{\mathbf{x}_i}\nabla_{\mathbf{p}_i}(H) = 0 , \tag{3.128}$$

is vanishing because of Hamilton equations of motion. Moreover, Eq. (3.127) can be cast as:

$$\frac{d\rho(t, \mathbf{x}_i, \mathbf{p}_i)}{dt} = \frac{\partial \rho}{\partial t} + \nabla_{\mathbf{y}} \cdot (\rho\dot{\mathbf{y}}) , \tag{3.129}$$

where we have indicated as $\mathbf{y}$ the generic variable of the phase space, i.e. $\mathbf{y} = \{\mathbf{x}_i, \mathbf{p}_i\}$ for $i = 1, \dots, N$. Equation (3.129) is a continuity equation. Therefore, if the particle number N is conserved, then Eq. (3.123) must hold.

3.7.2 Example: The One-Dimensional Harmonic Oscillator

Consider the one-dimensional harmonic oscillator:

$$H = \frac{p^2}{2m} + \frac{kx^2}{2} . \tag{3.130}$$

The Boltzmann equation is the following:

$$\frac{\partial f}{\partial t} + \frac{\partial f}{\partial x}\frac{dx}{dt} + \frac{\partial f}{\partial p}\frac{dp}{dt} = 0 , \tag{3.131}$$

where

$$\frac{dx}{dt} = \frac{\partial H}{\partial p} = \frac{p}{m} , \qquad \frac{dp}{dt} = -\frac{\partial H}{\partial x} = -kx . \tag{3.132}$$

The general strategy for solving Boltzmann equation is to use the method of the characteristics. A characteristic is a curve in the space formed by the phase space plus the time axis and it is parametrised by its arc length s such that

$$\frac{df(s)}{ds} = 0 , \tag{3.133}$$

i.e. f is constant along the characteristic. So, given an initial value s_0, one has $f = f(s_0)$.

In our example of the harmonic oscillator, the characteristic is a curve in the 3-dimensional space (t, x, p), described by $t(s), x(s)$ and $p(s)$. Using the chain rule, we get:

$$\frac{df}{ds} = \frac{\partial f}{\partial t}\frac{dt}{ds} + \frac{\partial f}{\partial x}\frac{dx}{ds} + \frac{\partial f}{\partial p}\frac{dp}{ds} = 0 . \tag{3.134}$$

Comparing with (3.131), we get the following system:

$$\begin{cases} \frac{dt}{ds} = 1 \\ \frac{dx}{ds} = \frac{p}{m} \\ \frac{dp}{ds} = -kx \end{cases} \Rightarrow \begin{cases} t = s - s_0 \\ p = m\frac{dx}{ds} \\ \frac{d^2x}{ds^2} + \frac{k}{m}x = 0 \end{cases} . \tag{3.135}$$

We solve the last equation supposing that $x(s_0) = x_0$. Therefore:

$$x(s) = x_0 \cos\left[\sqrt{\frac{k}{m}}(s - s_0)\right] , \tag{3.136}$$

and

$$p(s) = m\frac{dx(s)}{ds} = -x_0\sqrt{km}\sin\left[\sqrt{\frac{k}{m}}(s - s_0)\right] \equiv p_0 \sin\left[\sqrt{\frac{k}{m}}(s - s_0)\right] . \tag{3.137}$$

The distribution function is thus a function of:

$$f = f(s_0) = f\left[t(s_0), x(s_0), p(s_0)\right] , \tag{3.138}$$

that is

$$f = f(s_0) = f[x(s_0)] = f(x_0) , \tag{3.139}$$

i.e. it is simply a function of the initial position x_0. The latter can be related to the position and momentum at any time in the following way:

$$\frac{x^2}{x_0^2} + \frac{p^2}{p_0^2} = 1 , \tag{3.140}$$

which turns out to be:

$$\frac{x^2}{x_0^2} + \frac{p^2}{x_0^2 km} = 1 \quad \Rightarrow \quad x_0^2 = \frac{2E}{k} , \tag{3.141}$$

where E is the energy of the harmonic oscillator. Therefore, $f = f(E)$, i.e. the distribution function is a function of the energy. The precise functional form depends on the initial condition that we give on f.

3.7.3 Boltzmann Equation in General Relativity and Cosmology

In GR the distribution function must be expressed covariantly as $f = f(x^\mu, P^\mu)$, and the total derivative of f cannot be taken with respect to the time because this would violate the general covariance of the theory. The total derivative of f is taken with respect to an affine parameter λ, as follows:

$$\frac{df}{d\lambda} = \frac{\partial f}{\partial x^\mu} \frac{dx^\mu}{d\lambda} + \frac{\partial f}{\partial P^\mu} \frac{dP^\mu}{d\lambda} . \tag{3.142}$$

The geometry enters through the derivative of the four-momentum, which can be expressed via the geodesic equation:

$$\frac{dP^\mu}{d\lambda} + \Gamma^\mu_{\nu\rho} P^\nu P^\rho = 0 , \tag{3.143}$$

so that

$$\frac{df}{d\lambda} = P^\mu \frac{\partial f}{\partial x^\mu} - \Gamma^\mu_{\nu\rho} P^\nu P^\rho \frac{\partial f}{\partial P^\mu} \equiv \hat{L}_{\text{rel}}(f) , \tag{3.144}$$

where we have defined the relativistic Liouville operator $\hat{L}_{\text{rel}}$. It might seem that in the relativistic case we have gained one variable, i.e. P^0, but this is not so because P^0 is related to the spatial momentum P^i via the relation $g_{\mu\nu} P^\mu P^\nu = -m^2 c^2$. For this reason, we can reformulate the Liouville operator as follows:

$$\frac{df}{d\lambda} = P^\mu \frac{\partial f}{\partial x^\mu} - \Gamma^i_{\nu\rho} P^\nu P^\rho \frac{\partial f}{\partial P^i} , \tag{3.145}$$

i.e. by considering $f = f(x^\mu, P^i)$.

Collisionless Boltzmann Equation in Relativistic Cosmology

When we couple Eq. (3.145) with FLRW metric, we must take into account that f cannot depend on the position x^i, because of homogeneity and isotropy. The collisionless Boltzmann equation thus becomes:

$$P^0 \frac{\partial f}{\partial t} - \Gamma^i_{\nu\rho} P^\nu P^\rho \frac{\partial f}{\partial P^i} = 0 \; . \tag{3.146}$$

Exercise 3.16 Considering the spatially flat case $K = 0$, show that the above equation can be cast as follows:

$$\frac{\partial f}{\partial t} - 2HP^i \frac{\partial f}{\partial P^i} = 0 \; . \tag{3.147}$$

Again, because of isotropy, f cannot depend on the direction of P^i, but only on its modulus $P^2 = \delta_{ij} P^i P^j$.

Exercise 3.17 Show for a generic function $f = f(x^2)$ that:

$$x^i \frac{\partial f}{\partial x^i} = x \frac{\partial f}{\partial x} \; , \tag{3.148}$$

with $x^2 \equiv \delta_{ij} x^i x^j$.

Therefore, we can write

$$\frac{\partial f}{\partial t} - 2HP \frac{\partial f}{\partial P} = 0 \; . \tag{3.149}$$

Exercise 3.18 Show that the solution of the above equation is a generic function:

$$f = f(a^2 P) = f(ap) \; , \tag{3.150}$$

where in the second equality we have used the definition of the proper momentum.

Show that the Boltzmann equation written using the proper momentum has the following form:

$$\boxed{\frac{\partial f}{\partial t} - Hp \frac{\partial f}{\partial p} = 0} \tag{3.151}$$

Moments of the Collisionless Boltzmann Equation

Taking moments of the Boltzmann equation means to integrate it in the momentum space, weighed with powers of the proper momentum. This method is due to Grad (Grad 1958). For example, the moment zero of Eq. (3.151) is the following:

$$\int \frac{d^3\mathbf{p}}{(2\pi\hbar)^3} \left(\frac{\partial f}{\partial t} - Hp\frac{\partial f}{\partial p} \right) = 0 \quad \text{(Moment zero)} . \tag{3.152}$$

Exercise 3.19 Using the definition of the particle number density (3.27), show that Eq. (3.152) becomes:

$$\boxed{\dot{n} + 3Hn = 0 \quad \Rightarrow \quad \frac{1}{a^3}\frac{d(na^3)}{dt} = 0} \tag{3.153}$$

The particle number na^3 is conserved. This is an expected result since we have considered a collisionless Boltzmann equation, i.e. absence of interactions and thus no source of creation or destruction of particles.

Weighing Eq. (3.151) with the energy and integrating in the momentum space we get:

$$\dot{\varepsilon} - H \int \frac{d^3\mathbf{p}}{(2\pi\hbar)^3} pE(p)\frac{\partial f}{\partial p} = 0 . \tag{3.154}$$

Exercise 3.20 Integrate by parts the above equation and show that it becomes:

$$\dot{\varepsilon} + 3H (\varepsilon + P) = 0 , \tag{3.155}$$

i.e. the continuity equation.

Weighing Boltzmann equation (3.151) with $\hat{p}^i$ will always result in an identity, because of isotropy. We can show this as follows:

$$\int \frac{d^3\mathbf{p}}{(2\pi\hbar)^3}\hat{p}^i \left(\frac{\partial f}{\partial t} - Hp\frac{\partial f}{\partial p} \right) = \int d^2\hat{p}\, \hat{p}^i \int \frac{dp\, p^2}{(2\pi\hbar)^3} \left(\frac{\partial f}{\partial t} - Hp\frac{\partial f}{\partial p} \right) = 0 . \tag{3.156}$$

Since $f = f(ap)$ then only the integration in p contains f and the angular integration is just a multiplicative factor.

Exercise 3.21 Show that:

$$\int d^2\hat{p}\, \hat{p}^i = 0 \,, \qquad (3.157)$$

and thus Eq. (3.156) is an identity.

3.8 Boltzmann Equation with a Collisional Term

For the forthcoming applications we shall need to investigate interactions of the following type:

$$1 + 2 \leftrightarrow 3 + 4 \,, \qquad (3.158)$$

among generic species that we dub 1, 2, 3 and 4. This reaction can describe scattering or annihilation and will be suitable for discussing BBN, recombination and calculating the expected relic abundance of CDM.

Let us take the particle 1 as reference and let us focus on its Boltzmann equation:

$$\frac{df_1}{d\lambda} = C[f] \,. \qquad (3.159)$$

The collisional term is the same for all the particles, and has dimension of an inverse time, i.e. it is a scattering rate. Using the Liouville operator of Eq. (3.146), we can cast the above equation as

$$\frac{\partial f_1}{\partial t} - Hp_1\frac{\partial f_1}{\partial p_1} = \frac{1}{P_1^0}C[f] \,, \qquad (3.160)$$

where we have passed from the affine parameter λ to the time t, and for this reason P_1^0 appears. Taking the moment zero of the above equation, and using the result of Eq. (3.153), we get:

$$\frac{1}{a^3}\frac{d(n_1a^3)}{dt} = \int \frac{d^3\mathbf{p}_1}{(2\pi\hbar)^3 P_1^0}C[f] \,. \qquad (3.161)$$

Introducing the general expression of the right hand side (Kolb and Turner 1990), the above equation becomes:

$$\frac{1}{a^3}\frac{d(n_1 a^3)}{dt} = \int \frac{d^3\mathbf{p}_1}{(2\pi\hbar)^3 2E_1} \int \frac{d^3\mathbf{p}_2}{(2\pi\hbar)^3 2E_2} \int \frac{d^3\mathbf{p}_3}{(2\pi\hbar)^3 2E_3} \int \frac{d^3\mathbf{p}_4}{(2\pi\hbar)^3 2E_4}$$
$$(2\pi)^4 \delta^{(3)}(\mathbf{p}_1 + \mathbf{p}_2 - \mathbf{p}_3 - \mathbf{p}_4)\delta(E_1 + E_2 - E_3 - E_4)|\mathcal{M}|^2$$
$$\left[f_3 f_4 (1 \pm f_1)(1 \pm f_2) - f_1 f_2 (1 \pm f_3)(1 \pm f_4) \right] ,$$

$$(3.162)$$

where $E_i = \sqrt{p_i^2 + m_i^2}$. We have to provide several comments.

• Since we took the zero moment of Eq. (3.159), the left hand side is the time derivative of the number of particles of the species 1. See also Eq. (3.153).

• We have incorporated the particles degeneracies g_s in the distribution functions.

• The integrals are over the particles momenta. On the other hand, the total four-momentum must be conserved, hence the Dirac deltas in the second line. Note the E_1 contribution in the first integration. You may think that it comes from the P_1^0 factor in Eq. (3.161), but this would not explain the factor 2. Indeed, one must rather look at the right hand side of Eq. (3.162) as a *definition* of the momentum integrated $C[f]$.

• The fundamental physics of the interaction is represented by the amplitude $|\mathcal{M}|^2$. We have also assumed symmetric interaction, i.e. for fixed particles four-momenta the amplitude probability for $1 + 2 \to 3 + 4$ is the same as for $1 + 2 \leftarrow 3 + 4$.

• In the last line, we have a balance: the more particle 3 and 4 we have, the more they react and produce particles 1 and 2. And vice-versa. Since our reference particle is the 1, we have the combination $f_3 f_4 - f_1 f_2$.

• Finally, the contributions of the type $1 + f$ and $1 - f$ are called Bose enhancement and Pauli blocking, respectively. They represent the fact that it is easier to produce a boson rather than a fermion because, due to Pauli exclusion principle, there are more states available to the former than to the latter.

• The volume element $d^3\mathbf{p}/2E$ is covariant. It comes from the fact that we have enforced $E^2 = p^2 c^2 + m^2 c^4$ for each particle. We now prove it. When we want to enforce the dispersion relation $E^2 = p^2 c^2 + m^2 c^4$ we use a Dirac delta in the four-momentum space:

$$\int d^3\mathbf{p} \int_0^\infty dE\, \delta(\mathcal{P}^2 + m^2 c^2) = \int d^3\mathbf{p} \int_{-\infty}^\infty dE\, \delta(E^2 - p^2 c^2 - m^2 c^4)\, \theta(E) ,$$

$$(3.163)$$

where the square modulus of the four-momentum is $\mathcal{P}^2 = -E^2/c^2 + p^2$. The Heaviside function $\theta(E)$ serves to choose only the positive values of the energy. The above equation is covariant. Using the known relation:

$$\delta[F(x)] = \sum_i \frac{\delta(x - x_i)}{|F'(x_i)|} ,$$

$$(3.164)$$

where the x_i's are the roots of the generic function $F(x)$, one gets:

$$\delta(E^2 - p^2c^2 - m^2c^4) = \frac{\delta(E - \sqrt{p^2c^2 + m^2c^4})}{2E_+} + \frac{\delta(E + \sqrt{p^2c^2 + m^2c^4})}{2E_-} ,$$

$$(3.165)$$

where $E_{\pm} = \pm\sqrt{p^2c^2 + m^2c^4}$.

The second term of Eq. (3.165) integrated with $\theta(E)$ in Eq. (3.163) vanishes and we are left with:

$$\int d^3\mathbf{p} \int_{-\infty}^{\infty} dE \, \delta(E^2 - p^2c^2 - m^2c^4) \, \theta(E) = \int \frac{d^3\mathbf{p}}{2E_+} ,$$

$$(3.166)$$

which is what we wanted to prove.

We now focus our attention to Eq. (3.162) and make some assumptions in order to simplify it. In particular, we shall always assume thermal equilibrium and sufficiently small temperatures such that:

$$\boxed{E - \mu \gg k_B T}$$

$$(3.167)$$

in order to use the FD and BE distributions simplified as follows:

$$f \approx e^{-E/(k_B T)} e^{\mu/(k_B T)} ,$$

$$(3.168)$$

i.e. forgetting the ± 1 (using thus the Maxwell-Boltzmann distribution).

Exercise 3.22 Show that using Eq. (3.168), the last line of Eq. (3.162) can be simplified as:

$$f_3 f_4 (1 \pm f_1)(1 \pm f_2) - f_1 f_2 (1 \pm f_3)(1 \pm f_4) \approx$$
$$e^{-(E_1+E_2)/(k_B T)} \left[e^{(\mu_3+\mu_4)/(k_B T)} - e^{(\mu_1+\mu_2)/(k_B T)} \right] .$$

$$(3.169)$$

In particular, we can neglect the Bose enhancement and Pauli blocking terms. One has to use energy conservation $E_3 + E_4 = E_1 + E_2$ in order to obtain the above equation.

With the approximation given in Eq. (3.168), the particle number density can be simplified as follows:

$$n = g_s \int \frac{d^3\mathbf{p}}{(2\pi\hbar)^3} f \approx g_s e^{\mu/(k_B T)} \int \frac{d^3\mathbf{p}}{(2\pi\hbar)^3} e^{-E/(k_B T)} ,$$

$$(3.170)$$

where we could extract the chemical potential from the integral since it does not depend on the particle momentum but rather on the temperature of the thermal bath. For $\mu = 0$ we define the **equilibrium number density** as:

$$n^{(0)} = g_s \int \frac{d^3\mathbf{p}}{(2\pi\hbar)^3} e^{-E/(k_B T)} \qquad (3.171)$$

3.8.1 Detailed Calculation of the Equilibrium Number Density

Now we show how to calculate the equilibrium number density in the regimes $mc^2 \gg k_B T$ and $mc^2 \ll k_B T$, which we proved earlier to be regimes in which the particles are non-relativistic and relativistic, respectively. At the same time, we justify the approximation of Eq. (3.168). Let us start from:

$$n^{(0)} = g_s \int \frac{d^3\mathbf{p}}{(2\pi\hbar)^3} \frac{1}{e^{E/(k_B T)} \pm 1} = \frac{g_s}{2\pi^2\hbar^3} \int_0^\infty dp \frac{p^2}{e^{\sqrt{p^2c^2+m^2c^4}/(k_B T)} \pm 1} , \qquad (3.172)$$

and split the integration, let us dub it I, into two contributions, one from zero up to mc and the other from mc to infinity, i.e.

$$I = \int_0^{mc} dp \frac{p^2}{e^{\sqrt{p^2c^2+m^2c^4}/(k_B T)} \pm 1} + \int_{mc}^\infty dp \frac{p^2}{e^{\sqrt{p^2c^2+m^2c^4}/(k_B T)} \pm 1} . \qquad (3.173)$$

The Non-relativistic Case $mc^2 \gg k_B T$

Now let us start considering the non-relativistic case $mc^2 \gg k_B T$. We shall use different approximations within the two integrals in which we split I:

$$I = \int_0^{mc} dp \frac{p^2}{e^{mc^2\sqrt{p^2/(m^2c^2)+1}/(k_B T)} \pm 1} + \int_{mc}^\infty dp \frac{p^2}{e^{pc\sqrt{1+m^2c^2/p^2}/(k_B T)} \pm 1} . \qquad (3.174)$$

Now we use the $mc^2 \gg k_B T$ condition. In the first integral $p < mc$ because of the integration limits, so we can expand the square root with respect to $p/(mc)$. The second integral is negligible with respect to the first one because, since $pc > mc^2 \gg k_B T$, the integrand is always exponentially small. Thus we are left with:

$$I = e^{-mc^2/(k_B T)} \int_0^{mc} dp \, p^2 e^{-p^2/(2mk_B T)} , \qquad (3.175)$$

where we kept just the dominant term for the first integral. This is the same integral that we would have obtained had we started from Eq. (3.171).

After changing variable, we get:

$$I = e^{-mc^2/(k_BT)} (2mk_BT)^{3/2} \int_0^{mc/\sqrt{2mk_BT}} dx\, x^2 e^{-x^2} , \qquad (3.176)$$

and the result is:

$$I = e^{-mc^2/(k_BT)} (2mk_BT)^{3/2} \left[-\sqrt{\frac{mc^2}{8mk_BT}} e^{-mc^2/(2k_BT)} + \frac{\sqrt{\pi}}{4} \mathrm{Erf}\left(\sqrt{\frac{mc^2}{2mk_BT}} \right) \right] . \qquad (3.177)$$

The dominant contribution of the above expression comes from the Erf function, which can be approximated to 1 in the $mc^2 \gg k_BT$ limit, and thus:

$$I = e^{-mc^2/(k_BT)} (2mk_BT)^{3/2} \frac{\sqrt{\pi}}{4} . \qquad (3.178)$$

Finally, we can express the equilibrium number density for any species in thermal equilibrium and $mc^2 \gg k_BT$ as:

$$\boxed{ n^{(0)} = g_s \left(\frac{mk_BT}{2\pi\hbar^2} \right)^{3/2} e^{-mc^2/(k_BT)} , \qquad \text{for } mc^2 \gg k_BT . } \qquad (3.179)$$

The Relativistic Case $mc^2 \ll k_BT$

Now we adopt the same technique in the other relevant limit, the one of relativistic particles, for which $mc^2 \ll k_BT$. We split the integral I in the same way as before:

$$I = \int_0^{mc} dp \frac{p^2}{e^{mc^2\sqrt{p^2/(m^2c^2)+1}/(k_BT)} \pm 1} + \int_{mc}^{\infty} dp \frac{p^2}{e^{pc\sqrt{1+m^2c^2/p^2}/(k_BT)} \pm 1} , \qquad (3.180)$$

but now the $mc^2 \ll k_BT$ condition allows us to make the following approximations:

$$I = \int_0^{mc} dp \frac{p^2}{1 + \frac{mc^2}{k_BT}\sqrt{p^2/(m^2c^2)+1} \pm 1} + \int_{mc}^{\infty} dp \frac{p^2}{e^{pc(1+m^2c^2/2p^2)/(k_BT)} \pm 1} . \qquad (3.181)$$

Now it is the first integral that is subdominant with respect to the second one, and I can be written as:

$$I = \int_{mc}^{\infty} dp\, p^2 e^{-pc/(k_BT)} = \frac{(k_BT)^3}{c^3} \int_{mc^2/(k_BT)}^{\infty} dx\, x^2 e^{-x} . \qquad (3.182)$$

The solution of the integration is:

$$I = \frac{(k_B T)^3}{c^3} e^{-mc^2/(k_B T)} \left[2 + \frac{mc^2}{k_B T} \left(2 + \frac{mc^2}{k_B T} \right) \right] . \tag{3.183}$$

In the $mc^2 \ll k_B T$ regime, this becomes:

$$I = \frac{2(k_B T)^3}{c^3} . \tag{3.184}$$

Finally, we can express the equilibrium number density for any species in thermal equilibrium and $mc^2 \ll k_B T$ as:

$$\boxed{n^{(0)} = g_s \frac{(k_B T)^3}{\pi^2 \hbar^3 c^3} , \qquad \text{for } mc^2 \ll k_B T .} \tag{3.185}$$

3.8.2 Saha Equation

Expressing the contributions $e^{\mu_i/(k_B T)}$ as ratios $n_i/n_i^{(0)}$, we can recast the collisional Boltzmann's equation (3.162) as follows:

$$\boxed{\frac{1}{a^3} \frac{d(n_1 a^3)}{dt} = n_1^{(0)} n_2^{(0)} \langle \sigma v \rangle \left(\frac{n_3 n_4}{n_3^{(0)} n_4^{(0)}} - \frac{n_1 n_2}{n_1^{(0)} n_2^{(0)}} \right)} \tag{3.186}$$

where we have defined the **thermally averaged cross section** as follows:

$$\langle \sigma v \rangle \equiv \frac{1}{n_1^{(0)} n_2^{(0)}} \int \frac{d^3 \mathbf{p}_1}{(2\pi\hbar)^3 2E_1} \int \frac{d^3 \mathbf{p}_2}{(2\pi\hbar)^3 2E_2} \int \frac{d^3 \mathbf{p}_3}{(2\pi\hbar)^3 2E_3} \int \frac{d^3 \mathbf{p}_4}{(2\pi\hbar)^3 2E_4}$$
$$(2\pi)^4 \delta^{(3)}(\mathbf{p}_1 + \mathbf{p}_2 - \mathbf{p}_3 - \mathbf{p}_4) |\mathcal{M}|^2 e^{-(E_1 + E_2)/(k_B T)} . \tag{3.187}$$

Observe the following very important point about Eq. (3.186). When there are no interactions, i.e. when $\langle \sigma v \rangle = 0$, we recover the collisionless Boltzmann equation that we have discussed earlier. On the other hand, when the interaction rate is extremely high, i.e. it is much larger than than the Hubble rate:

$$n_1^{(0)} n_2^{(0)} \langle \sigma v \rangle \gg \frac{1}{a^3} \frac{d(n_1 a^3)}{dt} \sim n_1 H , \tag{3.188}$$

then in order for Eq. (3.186) to hold true, we must have:

$$\boxed{\frac{n_3 n_4}{n_3^{(0)} n_4^{(0)}} = \frac{n_1 n_2}{n_1^{(0)} n_2^{(0)}}} \tag{3.189}$$

This equation can be written as Eq. (3.7) and thus represents **chemical equilibrium**, as we saw at the beginning of this chapter. Equation (3.189) is also known as **Saha equation**.

So we see that when $\Gamma \gg H$ then Saha equation (3.189) must hold and Boltzmann equation (3.186) becomes:

$$\frac{1}{a^3} \frac{d(n_1 a^3)}{dt} = 0 \,, \tag{3.190}$$

which is the same Boltzmann equation that we would have if interactions were absent! This similarity between high interaction rate and no interaction at all is intriguing and responsible of the very low degree of CMB polarisation, as we shall see in Chap. 10.

3.9 Big-Bang Nucleosynthesis

The BBN is the formation of the primordial light elements, mainly helium. It took place at a temperature (photon temperature) of about 0.1 MeV, which corresponds to a redshift $z \approx 10^9$.

In order to investigate the BBN, we need to know the characters of the story. At temperatures say larger than 1 MeV the primordial plasma was formed by photons, electrons, positrons, neutrinos, antineutrinos, protons and neutrons. We have already seen how photons, electrons, positrons, neutrinos, antineutrinos interact among each other, so now we focus on protons and neutrons. Their interactions relevant to BBN are:

$$n \leftrightarrow p + e^- + \bar{\nu}_e \quad \text{(beta decay)} \,, \tag{3.191}$$

$$p + e^- \leftrightarrow \nu_e + n \quad \text{(electron capture)} \,, \tag{3.192}$$

$$p + \bar{\nu}_e \leftrightarrow e^+ + n \quad \text{(inverse beta decay)} \,. \tag{3.193}$$

As we have seen, at a temperature of about 1 MeV neutrinos decouple. Therefore, the β-decay reaction in Eq. (3.191) (from left to right) takes over and the number of neutrons starts to diminish. On the other hand, they can also be captured by protons and form deuterium nuclei. The BBN is essentially a competition in capturing neutrons before they decay.

3.9.1 The Baryon-to-Photon Ratio

A very important number for BBN and cosmology is the **baryon-to-photon ratio** η_b, which we have already encountered at the beginning of this chapter. It is defined as follows:

$$\eta_b \equiv \frac{n_b}{n_\gamma} = 5.5 \times 10^{-10} \left(\frac{\Omega_{b0}h^2}{0.020} \right) \tag{3.194}$$

i.e. as the ratio between the number of baryons and the number of photons. We have defined it via number densities, which individually are time-dependent quantities but whose ratio is fixed since both scales as $1/a^3$. The above numbers in Eq. (3.194) can be found as follows:

$$\eta_b = \frac{n_b}{n_\gamma} = \frac{\varepsilon_{b0}}{m_b c^2 n_{\gamma 0}}, \tag{3.195}$$

where we have assumed the baryons to be nonrelativistic, which is indeed the case at the temperatures we are dealing with ($k_B T \sim \text{MeV}$) since the proton mass is of the order of 1 GeV.

Using Eqs. (3.75) and (3.80), we can write

$$\frac{n_{\gamma 0}}{\varepsilon_{\gamma 0}} = \frac{30\zeta(3)}{\pi^2 k_B T_0}. \tag{3.196}$$

Therefore,

$$\eta_b = \frac{\pi^4 k_B T_0 \varepsilon_{b0}}{30\zeta(3)m_b c^2 \varepsilon_{\gamma 0}} = \frac{\pi^4 k_B T_0 \Omega_{b0}}{30\zeta(3)m_b c^2 \Omega_{\gamma 0}}, \tag{3.197}$$

where in the last equality we have multiplied and divided by the present critical energy density in order for the density parameters to appear.

Exercise 3.23 Show that Eq. (3.197) leads to Eq. (3.194). Use the mass of the proton as m_b since the baryon energy density is indeed dominated by the mass energy density of the protons.

The fact that there is a billion photon for each proton and electron is very important for the following reason. Even if the temperature of the thermal bath is lower than the binding energy of deuterium, i.e. 2.2 MeV, there are still many photons with energy higher than 2.2 MeV which are able to break newly formed deuterium nuclei. This is also known as **deuterium bottleneck**.

As Kolb and Turner comment in their book (Kolb and Turner 1990, page 92), it is not deuterium's fault if BBN takes place at temperature much smaller than 2.2 MeV. Rather, the very high entropy of the universe, i.e. the smallness of η_b, is the culprit.

3.9.2 The Deuterium Bottleneck

The deuterium bottleneck is the situation in which newly formed deuterium nuclei are destroyed by photons. With no deuterium available, BBN cannot take place.

Let us start considering the reaction:

$$p + n \leftrightarrow D + \gamma , \tag{3.198}$$

at chemical equilibrium. Using Saha equation (3.189), we have

$$\frac{n_D n_\gamma}{n_D^{(0)} n_\gamma^{(0)}} = \frac{n_p n_n}{n_p^{(0)} n_n^{(0)}} . \tag{3.199}$$

Neglecting the photon chemical potential, i.e. $n_\gamma = n_\gamma^{(0)}$, and using Eq. (3.179) we obtain:

$$\frac{n_D}{n_p n_n} = \frac{n_D^{(0)}}{n_p^{(0)} n_n^{(0)}} = \frac{g_D}{g_p g_n} \left(\frac{2\pi \hbar^2 m_D}{m_p m_n k_B T} \right)^{3/2} e^{-(m_D - m_p - m_n)c^2/(k_B T)} . \tag{3.200}$$

The deuterium has spin 1, whereas protons and neutrons have spin $1/2$. Therefore,

$$\frac{n_D}{n_p n_n} = \frac{3}{4} \left(\frac{2\pi \hbar^2 m_D}{m_p m_n k_B T} \right)^{3/2} e^{B_D/(k_B T)} , \tag{3.201}$$

where $B_D = 2.22 \, \text{MeV}$ is the deuterium binding energy. We can write the masses ratio as follows:

$$\frac{m_D}{m_p m_n} = \frac{m_p + m_n - B_D/c^2}{m_p m_n} = \frac{1}{m_n} + \frac{1}{m_p} - \frac{B_D}{m_p m_n c^2} . \tag{3.202}$$

Introducing the neutron-proton mass difference (Wilczek 2015):

$$Q \equiv (m_n - m_p)c^2 = 1.239 \, \text{MeV} , \tag{3.203}$$

we can write

$$\frac{m_D}{m_p m_n} = \frac{1}{m_p(1 + Q/m_p c^2)} + \frac{1}{m_p} - \frac{B_D}{m_p^2(1 + Q/m_p c^2)c^2} . \tag{3.204}$$

Since $Q/m_p c^2 \sim B_D/m_p c^2 \sim 10^{-3}$, we approximate:

$$\frac{m_D}{m_p m_n} \approx \frac{2}{m_p} . \tag{3.205}$$

Moreover, being $n_n = n_p = n_b$, we can write:

$$\frac{n_D}{n_b} = \frac{3}{4}n_b \left(\frac{4\pi\hbar^2}{m_p k_B T}\right)^{3/2} e^{B_D/(k_B T)} , \tag{3.206}$$

i.e. we obtain a deuterium-to-baryon ratio which defines how many baryons exist that are deuterium nuclei. Dealing with the n_b on the right hand side as follows:

$$n_b = \eta_b n_\gamma = \eta_b n_\gamma^{(0)} = 2\eta_b \frac{(k_B T)^3}{\pi^2 \hbar^3 c^3} , \tag{3.207}$$

where we have used Eq. (3.185) for the photon number density, we get

$$\frac{n_D}{n_b} = \frac{12}{\sqrt{\pi}}\eta_b \left(\frac{k_B T}{m_p c^2}\right)^{3/2} e^{B_D/(k_B T)} . \tag{3.208}$$

Even when $k_B T \sim B_D$ the relative abundance is very small, because of the prefactor η_b. This is the deuterium bottleneck.

Typically, the temperature T_{BBN} at which the BBN starts is the one at which the bottleneck is overcome. This is because, as numerical calculations show, once deuterium is formed it rapidly combines into Helium.

So, we define T_{BBN} as the one for which $n_D = n_b$:

$$\log\left(\frac{12}{\sqrt{\pi}}\eta_b\right) + \frac{3}{2}\log\left(\frac{k_B T_{BBN}}{m_p c^2}\right) = -\frac{B_D}{k_B T_{BBN}} . \tag{3.209}$$

Numerically solving this equation, one finds

$$\boxed{k_B T_{BBN} \approx 0.07\,\text{MeV}} \tag{3.210}$$

Note that we can use Saha equation only until chemical equilibrium holds true. When the reaction $p + n \leftrightarrow D + \gamma$ unbalances and deuterium is formed, we must use the full Boltzmann equation (3.186). However, it can be shown numerically that the two equations provide compatible results up to the moment in which the equilibrium is broken. Therefore, Saha equation is an useful tool for estimating *when* the equilibrium is broken but, of course, if one needs precise results one should solve the full Boltzmann equation.

3.9.3 Neutron Abundance

After the deuterium bottleneck is overcome, BBN takes place in the following chain reactions:

$$p + n \rightarrow D + \gamma , \tag{3.211}$$

$$D + D \rightarrow {}^{3}\text{He} + n , \tag{3.212}$$

$$D + {}^{3}\text{He} \rightarrow {}^{4}\text{He} + p . \tag{3.213}$$

In the following, we shall assume that the three above reactions take place instantaneously and the neutrons are captured in Helium nuclei. This is not what occurred, of course, but it turns out to be a good approximation which allows us to perform easy calculations.

Note that Lithium ^{3}Li is also produced, but in tiny fraction (one billionth of the hydrogen abundance). However, measuring its abundance in the universe is a very important independent measure of Ω_{b0}.

The main prediction of BBN is on the abundance of ^{4}He because this is the element which is mostly formed, due to both its high binding energy per nucleon, which is about 7 MeV, see Fig. 3.2, but also to the fact that there is not much time for forming heavier nuclei since the thermal bath is rapidly cooling and η_b is so small.

Our objective is thus to determine the neutron abundance at T_{BBN}. This is done considering two reactions. The electron capture:

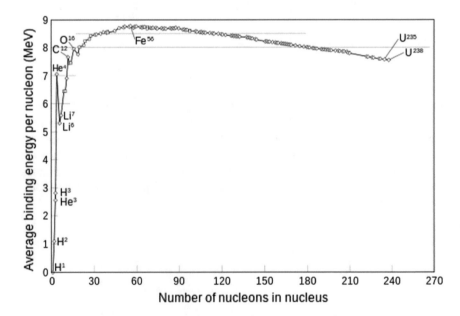

Fig. 3.2 Binding energy per nucleon. Figure taken from https://en.wikipedia.org/wiki/Helium-4

$$p + e^- \leftrightarrow n + \nu , \tag{3.214}$$

and the β-decay:

$$n \leftrightarrow p + e^- + \bar{\nu} . \tag{3.215}$$

The β-decay will provide just an exponential suppression on the abundance predicted by the the electron capture. Therefore, we focus on the latter.

For temperature $k_B T \gg 1$ MeV, protons and neutrons are in chemical equilibrium:

$$\frac{n_p}{n_n} = \frac{n_p^{(0)}}{n_n^{(0)}} = \left(\frac{m_p}{m_n}\right)^{3/2} e^{(m_n - m_p)c^2/(k_B T)} \sim e^{Q/(k_B T)} , \tag{3.216}$$

where $Q = 1.239$ MeV, see Eq. (3.203). When $k_B T \gg Q$, the mass difference between protons and neutrons is irrelevant, and therefore they are in chemical equilibrium. When $k_B T$ drops below Q nature starts to favor protons because they are energetically more "economic" and neutrons start to disappear. However, as we mentioned earlier, out of equilibrium we cannot use Saha equation but have to solve the full Boltzmann equation:

$$\frac{1}{a^3} \frac{d(n_n a^3)}{dt} = n_n^{(0)} n_l^{(0)} \langle \sigma v \rangle \left(\frac{n_p n_l}{n_p^{(0)} n_l^{(0)}} - \frac{n_n n_l}{n_n^{(0)} n_l^{(0)}}\right) , \tag{3.217}$$

where we have denoted with the subscript l the leptons, either electron or neutrino, involved in the electron capture process. We assume their chemical potentials to be zero and simplify Eq. (3.217) as follows:

$$\frac{1}{a^3} \frac{d(n_n a^3)}{dt} = n_l^{(0)} \langle \sigma v \rangle \left(\frac{n_p n_n^{(0)}}{n_p^{(0)}} - n_n\right) . \tag{3.218}$$

Let us define the neutron abundance and the scattering rate as follows:

$$X_n \equiv \frac{n_n}{n_n + n_p} , \qquad n_l^{(0)} \langle \sigma v \rangle \equiv \lambda_{np} . \tag{3.219}$$

Exercise 3.24 Show that Eq. (3.218) can be cast, using Eq. (3.219), as follows:

$$\frac{dX_n}{dt} = \lambda_{np} \left[(1 - X_n) e^{-Q/(k_B T)} - X_n\right] . \tag{3.220}$$

In order to solve (numerically) Eq. (3.220), we introduce the variable:

$$x \equiv \frac{Q}{k_B T} . \tag{3.221}$$

Exercise 3.25 Show that the time derivative of $x \equiv Q/(k_B T)$ can be cast as follows:

$$\frac{dx}{dt} = Hx = x\sqrt{\frac{8\pi G\varepsilon}{3c^2}} . \tag{3.222}$$

Since we are deep in the radiation-dominated era, we can write the energy density of Eq. (3.222) as follows:

$$\boxed{\varepsilon = \frac{\pi^2(k_B T)^4}{30(\hbar c)^3} g_*} \tag{3.223}$$

where **the effective number of relativistic degrees of freedom** is

$$g_* \equiv \sum_{i=\text{bosons}} g_i + \frac{7}{8} \sum_{i=\text{fermions}} g_i , \tag{3.224}$$

where recall the $7/8$ factor coming from Eq. (3.83). The effective number of relativistic degrees of freedom is actually a function of the temperature, since it decreases when a certain species becomes non-relativistic. For temperature larger than $1\,\text{MeV}$ g_* is roughly a constant and its value is:

$$g_* = 2 + \frac{7}{8}(3 + 3 + 2 + 2) = 10.75 , \tag{3.225}$$

where we have considered two degrees of freedom coming from the photons, $3 + 3$ coming from neutrinos and anti-neutrinos and $2 + 2$ coming from electrons and positrons. We have considered just a single spin state for each neutrino and anti-neutrino.

Exercise 3.26 Show that the Hubble parameter can thus be cast in the following form:

$$H^2 = \frac{8\pi G}{3c^2} \frac{\pi^2(k_B T)^4}{30(\hbar c)^3} g_* = \frac{4\pi^3 G g_* Q^4}{45c^2(\hbar c)^3} \frac{1}{x^4} . \tag{3.226}$$

Calculate:

$$H(x = 1) = 1.13\,\text{s}^{-1} . \tag{3.227}$$

Show that Eq. (3.220) can be cast as:

$$\frac{dX_n}{dx} = \frac{x\lambda_{np}}{H(x = 1)} \left[e^{-x} - X_n(1 + e^{-x}) \right] . \tag{3.228}$$

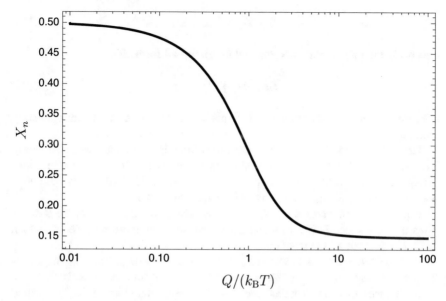

Fig. 3.3 Evolution of X_n from Eq. (3.228)

We only need a last piece of information, i.e. the interaction rate λ_{np}:

$$\lambda_{np} = \frac{255}{\tau_n x^5}(12 + 6x + x^2) , \tag{3.229}$$

where

$$\tau_n = 886.7\,\text{s} , \tag{3.230}$$

is the neutron lifetime. The above scattering rate can be found in Bernstein (1988).

We can now solve numerically Eq. (3.228) together with Eq. (3.229). The initial condition on X_n is of course $X_n(x \to 0) = 1/2$, as we can see from the Saha equation (3.216). We plot the evolution of X_n in Fig. 3.3.

As we can appreciate from Fig. 3.3, $X_n \to 0.15$ as $x \to \infty$. This number is not a very good approximation of the residual abundance of neutrons at T_{BBN} because there are other relevant processes which contribute to deplete or enhance the number of neutrons. Namely:

$$n \to p + e^- + \bar{\nu} , \tag{3.231}$$
$$n + p \to D + \gamma , \tag{3.232}$$

i.e. the β-decay and the neutron-proton capture. These processes lower the number of free neutrons. The Helium-3 formation:

$$D + D \rightarrow {}^3\text{He} + n \,, \qquad (3.233)$$

puts back into play another neutron and the Helium-4 formation:

$$^3\text{He} + D \rightarrow {}^4\text{He} + p \,, \qquad (3.234)$$

reinserts into play another proton, which helps in capturing neutrons, thus lowering their number.

The right way to calculate the abundances of the light elements produced during BBN is to consider all the coupled Boltzmann equations for all the relevant reactions taking place. This is, of course, done numerically and the standard code is Wagoner's one (Wagoner 1973) (there has been refinements since then).

We now show that correcting $X_n = 0.15$ by taking into account only the β-decay gives a result which is in surprising agreement with the more reliable one which takes into account all the reactions.

What we have to do is to weigh $X_n = 0.15$ with $\exp(-t_{\text{BBN}}/\tau_n),$[5] where t_{BBN} is the time corresponding to $k_B T_{\text{BBN}} = 0.07\,\text{MeV}$, i.e. the time at which BBN starts. Moreover, we suppose that at this time all the free neutrons are immediately captured and produce Helium-4. Thereby, estimating X_n gives a direct estimation of $X_{^4\text{He}}$.

At $k_B T_{\text{BBN}} = 0.07\,\text{MeV}$, electrons and positrons have already annihilated. Therefore, the effective relativistic degrees of freedom are:

$$g_* = 2 + \frac{7}{8}\left(\frac{4}{11}\right)^{4/3} 6 \approx 3.36 \,, \qquad (3.235)$$

where we have taken into account the temperature difference between photons and neutrinos.

Since $T \propto 1/a$ and in the radiation-dominated epoch $a \propto \sqrt{t}$, we can relate time and temperature as follows:

$$\frac{1}{4t^2} = H^2 = \frac{4\pi^3 G (k_B T)^4}{45 c^2 (\hbar c)^3} g_* \,. \qquad (3.236)$$

Exercise 3.27 Show from Eq. (3.236) that:

$$t = 271 \left(\frac{0.07\,\text{MeV}}{k_B T}\right)^2 \text{s} \,. \qquad (3.237)$$

Therefore, the expected abundance of neutrons at $k_B T_{\text{BBN}} = 0.07\,\text{MeV}$ is:

$$X_n(T_{\text{BBN}}) = 0.15 \cdot e^{-271/886.7} \approx 0.11 \,. \qquad (3.238)$$

---

[5]This exponential weight comes from Poisson distribution, which governs stochastic processes such as the β-decay. We derive it in Chap. 12.

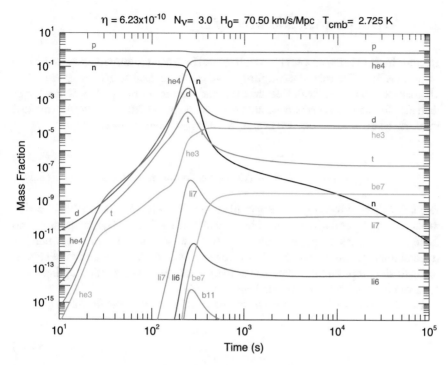

Fig. 3.4 Time-evolution of the mass fraction of various elements. Figure taken from http://cococubed.asu.edu/images/net_bigbang/bigbang_time_2010.pdf

Assuming that all the neutrons end up in Helium-4 nuclei, we have the prediction:

$$Y_P \equiv 4X_{^4\text{He}} \equiv \frac{4n_{^4\text{He}}}{n_b} = 2X_n(T_{\text{BBN}}) = 0.22 \ . \tag{3.239}$$

The factor 4 comes from the fact that Y_P is a mass fraction and each Helium-4 nucleus contains 4 baryons. We have also assumed $m_b = m_p = m_n$.

The more accurate numerical result is Kolb and Turner (1990):

$$Y_P = 0.2262 + 0.0135 \log \left(\frac{\eta_b}{10^{-10}} \right) \ , \tag{3.240}$$

which is in very good agreement with our "back-of-the-envelope" calculation. In Fig. 3.4 the time-evolution plots of the mass fractions of various elements are displayed.

3.10 Recombination and Decoupling

Recombination is the process by which neutral hydrogen is formed via combination of protons and electrons. **Decoupling** is generally refered to be the epoch when photons stop to interact with free electrons and their mean free path becomes larger than the Hubble radius and we are able to detect them as CMB coming from the **last scattering surface**. For the two events, the relevant interactions are:

$$p + e^- \leftrightarrow H + \gamma \,, \tag{3.241}$$

$$e^- + \gamma \leftrightarrow e^- + \gamma \quad \text{(Compton/Thomson scattering)} \,. \tag{3.242}$$

Recombination and decoupling temporally occur close to each other for the following reason. At sufficiently low temperatures, which we will calculate, photons are no more able to break hydrogen atoms and so these start to form in larger number (recombination). Being captured in hydrogen atoms, the number of free electrons dramatically drops and the Thomson scattering rate goes to zero (decoupling). The seminal paper on recombination is Peebles (1968).

In order to determine the epoch of recombination, let us use again Saha equation:

$$\frac{n_e n_p}{n_H} = \frac{n_e^{(0)} n_p^{(0)}}{n_H^{(0)}} \,. \tag{3.243}$$

Let us assume neutrality of the universe, i.e. $n_e = n_p$ and define the free electron fraction:

$$X_e \equiv \frac{n_e}{n_e + n_H} = \frac{n_p}{n_p + n_H} \,, \tag{3.244}$$

Exercise 3.28 Considering that the degeneracy of the hydrogen atom, in the state $1s$, is $g_{1s} = 4$ (it has two hyperfine states, one of spin 0 and the other of spin 1), show that Saha equation can be written, using Eq. (3.179), as:

$$\frac{X_e^2}{1 - X_e} = \frac{1}{n_e + n_H} \left(\frac{m_e m_p k_B T}{2 m_H \pi \hbar^2} \right)^{3/2} e^{-(m_e + m_p - m_H) c^2 / (k_B T)} \,. \tag{3.245}$$

Consider the contribution at the denominator of the right hand side as:

$$n_e + n_H = n_b = \eta_b n_\gamma = \eta_b \frac{2\zeta(3)}{\pi^2 \hbar^3 c^3} (k_B T)^3 \approx 10^{-9} \frac{(k_B T)^3}{\hbar^3 c^3} \,. \tag{3.246}$$

Look at the first equality as follows: the total electron number is made up of those which are free plus those which have already been captured. Moreover, the total electron number density is the same as the baryon number density because electrons *are* baryons (in the jargon of cosmology).

We again neglect the mass difference elsewhere than at the exponential, and write:

$$\frac{X_e^2}{1 - X_e} \approx 10^9 \left(\frac{m_e c^2}{2\pi k_B T}\right)^{3/2} \exp\left(-\frac{13.6\,\text{eV}}{k_B T}\right), \tag{3.247}$$

where we used

$$\varepsilon_0 \equiv (m_e + m_p - m_H)c^2 = 13.6\,\text{eV}, \tag{3.248}$$

i.e. the ionisation energy of the hydrogen atom.

The high photon-to-baryon number delays recombination as well as it delayed BBN. Indeed, when $k_B T = 13.6\,\text{eV}$, we get from Eq. (3.247):

$$\frac{X_e^2}{1 - X_e} \approx 10^{15}, \tag{3.249}$$

From which one gets that $X_e \approx 1$. This means that even when the energy of the thermal bath drops below the ionisation energy of the hydrogen atom, still no hydrogen is formed and the electrons remain free. This, again, happens because there are still many photons with energy much higher than $13.6\,\text{eV}$.

As we already mentioned, Saha equation works until chemical equilibrium is maintained. In Table 3.1 we show numerical calculations of X_e from Eq. (3.247) in order to have a hint about the time of recombination.

From Table 3.1 we see that the free electron fraction falls abruptly at about $k_B T \approx 0.30\,\text{eV}$.

Exercise 3.29 Calculate at which redshift corresponds the energy $k_B T = 0.3\,\text{eV}$ of the photon thermal bath.

In Fig. 3.5 we numerically solve Saha equation (3.247) and use both $k_B T$ and the redshift as variables.

Table 3.1 Free electron fraction at different photon temperatures

$k_B T$ [eV]	X_e
0.5	1
0.38	0.995
0.36	0.970
0.34	0.819
0.32	0.434
0.30	0.137
0.29	0.067
0.25	0.001

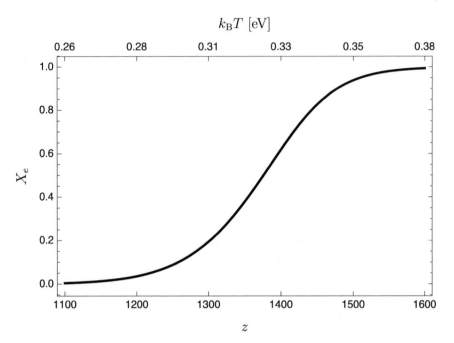

Fig. 3.5 Numerical solution of the Saha equation (3.247)

In order to accurately calculate X_e, we need to use the full Boltzmann equation (3.186), which for recombination becomes:

$$\frac{1}{a^3}\frac{d(n_e a^3)}{dt} = n_e^{(0)} n_p^{(0)} \langle \sigma v \rangle \left(\frac{n_H n_\gamma}{n_H^{(0)} n_\gamma^{(0)}} - \frac{n_e n_p}{n_e^{(0)} n_p^{(0)}} \right) . \tag{3.250}$$

We assume $n_\gamma = n_\gamma^{(0)}$ and $n_e = n_p$ again. Therefore,

$$\frac{1}{a^3}\frac{d(n_e a^3)}{dt} = \langle \sigma v \rangle \left[n_H \left(\frac{m_e k_B T}{2\pi \hbar^2} \right)^{3/2} e^{-\varepsilon_0/(k_B T)} - n_e^2 \right] . \tag{3.251}$$

Introducing now the free electron fraction X_e defined in Eq. (3.244), we get:

$$\frac{dX_e}{dt} = \langle \sigma v \rangle \left[(1 - X_e) \left(\frac{m_e k_B T}{2\pi \hbar^2} \right)^{3/2} e^{-\varepsilon_0/(k_B T)} - X_e^2 n_b \right] . \tag{3.252}$$

As we did for BBN in Eq. (3.207), we can replace n_b with:

$$n_b = 2\eta_b \frac{(k_B T)^3}{\pi^2 \hbar^3 c^3} . \tag{3.253}$$

Now we need the fundamental physics of the capture process. It is given by:

$$\langle \sigma v \rangle \equiv \alpha^{(2)} = 9.78\, \alpha^2 \frac{\hbar^2}{m_e^2 c} \left(\frac{\varepsilon_0}{k_B T} \right)^{1/2} \log\left(\frac{\varepsilon_0}{k_B T} \right) , \tag{3.254}$$

where $\alpha = 1/137$ is the fine structure constant. The superscript (2) serves to indicate that the best way to form hydrogen is not via the capture of an electron in the $1s$ state, because this generates a 13.6 eV photon which ionises another newly formed H.

The efficient way to form hydrogen is to form it in a excited state. When it relaxes to the ground state, the photons emitted have not enough energy to ionise other hydrogen atoms.

For example, an electron captured in the $n = 2$ state generates a 3.4 eV photon. Subsequently, when the electron falls in the ground state, the hydrogen releases another 10.2 eV photon. Neither of the two photons has sufficient energy for ionising another hydrogen atom in the ground state.

We now solve numerically Eq. (3.252) together with Eqs. (3.207) and (3.254). We use the redshift as independent variable:

$$\frac{dX_e}{dt} = \frac{dX_e}{dz}\frac{dz}{dt} = -\frac{dX_e}{dz} H(1+z) , \tag{3.255}$$

and the following Hubble parameter:

$$\frac{H^2}{H_0^2} = \Omega_{m0}(1+z)^3 + \Omega_{r0}(1+z)^4 + \Omega_\Lambda , \tag{3.256}$$

where for $z \approx 1100$ is the matter contribution the dominant one.

Therefore, we rewrite Eq. (3.252) as follows:

$$\frac{dX_e}{dz} = -\frac{\langle \sigma v \rangle}{H(1+z)} \left[(1 - X_e) \left(\frac{m_e k_B T}{2\pi \hbar^2} \right)^{3/2} e^{-\varepsilon_0/(k_B T)} - X_e^2 n_b \right] , \tag{3.257}$$

also taking into account that the photon temperature T scales as:

$$T = T_0(1+z) , \tag{3.258}$$

with $T_0 = 2.725$ K.

In Fig. 3.6 we show the numerical solution of the Boltzmann equation (3.257), compared with the solution of the Saha equation (3.247). Note how the two solutions are compatible for high redshifts, but that of the Boltzmann equation predicts a residual free electron fraction of about $X_e \approx 10^{-3}$.

At the same time of recombination the decoupling of photons from electron takes place. As we anticipated, this happens because very few free electrons remain after

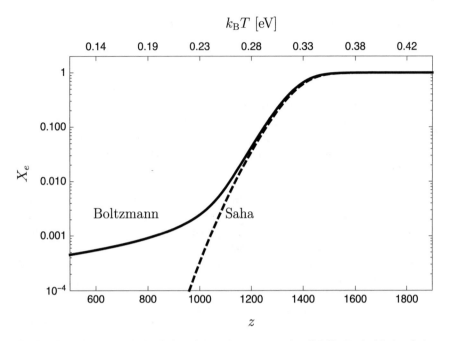

Fig. 3.6 Solid line: numerical solution of the Boltzmann equation (3.257). Dashed line: solution of the Saha equation (3.247)

hydrogen formation and therefore photons are free to propagate undisturbed and seen by us as the CMB.

3.10.1 Decoupling

As we have already mentioned at the beginning of this chapter, roughly speaking in the expanding universe any kind of reaction stops occurring when its interaction rate Γ becomes of the order of H. In the case of photons and electrons, the relevant process is Thomson scattering, for which:

$$\Gamma_{\mathrm{T}} = n_e \sigma_{\mathrm{T}} c = X_e n_{\mathrm{b}} \sigma_{\mathrm{T}} c \ , \tag{3.259}$$

where n_e is the free-electron number density, which we have written as $X_e n_{\mathrm{b}}$ because we have neglected the Helium abundance. The baryon number density can be expressed as

$$n_{\mathrm{b}} = \frac{\rho_{\mathrm{b}}}{m_{\mathrm{b}}} = \frac{3 H_0^2 \Omega_{\mathrm{b}0}}{8 \pi G m_p a^3} \ , \tag{3.260}$$

where we have identified $m_b = m_p$ since it is the proton mass that dominates the baryon energy density.

Exercise 3.30 Show that:

$$\frac{\Gamma_T}{H} = \frac{n_e \sigma_T c}{H} = 0.0692\, h \frac{X_e \Omega_{b0} H_0}{H a^3}\,.\tag{3.261}$$

As for H, we consider a matter plus radiation universe, for which:

$$H^2 = H_0^2 \frac{\Omega_{m0}}{a^3}\left(1 + \frac{a_{eq}}{a}\right)\,.\tag{3.262}$$

Exercise 3.31 Using the above Hubble parameter show that:

$$\frac{\Gamma_T}{H} = 113 X_e \left(\frac{\Omega_{b0}h^2}{0.02}\right)\left(\frac{0.15}{\Omega_{m0}h^2}\right)^{1/2}\left(\frac{1+z}{1000}\right)^{3/2}\left(1 + \frac{1+z}{3600}\frac{0.15}{\Omega_{m0}h^2}\right)^{-1/2}\,.\tag{3.263}$$

The decoupling redshift z_{dec} is defined to be that for which $\Gamma_T = H$. Note that in the above equation X_e is a function of the redshift, which we have plotted in Fig. 3.6. On the other hand, X_e drops abruptly during recombination, so that the factor 113 is easily overcome. For this reason recombination and decoupling take place at roughly the same time (Fig. 3.7).

Now imagine that no recombination takes place, i.e. $X_e = 1$. The above equation then gives the decoupling redshift:

$$1 + z_{dec} = 43 \left(\frac{0.02}{\Omega_{b0}h^2}\right)^{2/3}\left(\frac{\Omega_{m0}h^2}{0.15}\right)^{1/3}\,.\tag{3.264}$$

This is the freeze-out redshift of the electrons, i.e. eventually photons and electrons do not interact anymore because they are too diluted by the cosmological expansion. This would happen for a redshift 42. This number is important because of the following. Well after decoupling, ultraviolet light emitted by stars and gas is able to ionise again hydrogen atoms. This phase is called **reionisation**. If the latter ocurred for redshifts smaller than 42, the newly freed electron would not interact with photons because they are too much diluted by the cosmological expansion. Indeed, reionization takes place for $z_{reion} \approx 10$, so that the CMB spectrum is poorly affected, as we shall see in Chap. 10.

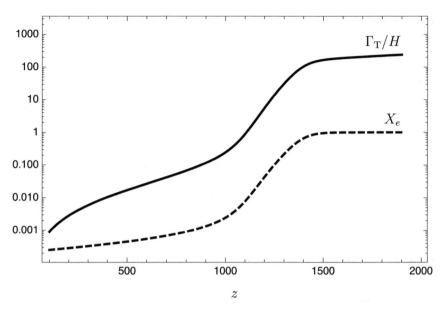

Fig. 3.7 Solid line: numerical solution of the ratio Γ_T/H of Eq. (3.263). Dashed line: numerical solution of Eq. (3.257) for X_e

3.11   Thermal Relics

In this section we investigate in some detail the freeze-out and relic abundance of CDM. In general, with **thermal relic** one refers to the abundance of a certain species left over from the annihilation suffered in thermal bath and, after its decoupling, from the dilution caused by the expansion of the universe. To this purpose, consider the following process:

$$X + \bar{X} \leftrightarrow l + \bar{l} \,, \tag{3.265}$$

where X represents the DM particle and l a lepton. Due to the expansion of the universe, at a certain point the annihilation of DM is no more efficient and thus its abundance is fixed. This is precisely what we want to calculate. We could then put constraints on the cross-section of the above process and on the mass of the DM particle by measuring the abundance of DM necessary today in order to be in agreement with the cosmological observations.

We are assuming here that DM is made up of massive particles which were in thermal equilibrium with the rest of the standard model particles in the primordial universe, hence the name **thermal relics**. Moreover, since we are considering CDM, or more in general particle which decouple from the primordial plasma when non-relativistic, ours is a calculation of **cold relics abundance**.

The Boltzmann equation for the above process has the following form:

$$\frac{1}{a^3}\frac{d(n_X a^3)}{dt} = \langle \sigma v \rangle \left(n_X^{(0)2} - n_X^2 \right) , \tag{3.266}$$

where we have assumed $n_l^{(0)} = n_l$, because we are in thermal bath with very high temperature (e.g. larger than $100\,\text{GeV}$ for WIMPs).

Now we take advantage of the scaling $T \propto 1/a$ and define the following dimensionless quantities:

$$Y \equiv \frac{n_X (\hbar c)^3}{(k_B T)^3} , \qquad Y_{eq} \equiv \frac{n_X^{(0)} (\hbar c)^3}{(k_B T)^3} , \qquad x \equiv \frac{mc^2}{k_B T} , \qquad \lambda \equiv \frac{(mc^2)^3 \langle \sigma v \rangle}{H(m)(\hbar c)^3} , \tag{3.267}$$

where $H(m) = H(x = 1)$ is the Hubble parameter corresponding to a thermal energy $k_B T = mc^2$, with m being the mass of the DM particle. Note that, being deep in the radiation-dominated era, the Hubble parameter scales as follows:

$$H = \frac{H(x = 1)}{x^2} . \tag{3.268}$$

The Boltzmann equation thus becomes:

$$\frac{(k_B T)^3}{(\hbar c)^3}\frac{dY}{dx}\frac{H(x = 1)}{x} = \langle \sigma v \rangle \frac{(k_B T)^6}{(\hbar c)^6} \left(Y_{eq}^2 - Y^2 \right) , \tag{3.269}$$

and finally

$$\boxed{\frac{dY}{dx} = -\frac{\lambda}{x^2} \left(Y^2 - Y_{eq}^2 \right)} \tag{3.270}$$

In this case Saha equation is simply $Y = Y_{eq}$ and from Eq. (3.107) we know that $Y_{eq} \to 0$ for $x \to \infty$, because of the dilution due to the cosmological expansion. On the other hand, we also expect, as we saw for recombination, that Y attains an asymptotic value which we call Y_∞ and with which we shall calculate the present abundance of DM. The departure between the Saha equation solution and the Boltzmann equation solution is the **freeze-out** and, as we know, it takes place approximately when $\Gamma \sim H$.

Using Eq. (3.107) in order to express Y_{eq}, we can numerically solve Eq. (3.270). In order to do this, we set a small initial value $x = x_i$ such that $Y(x_i) = Y_{eq}(x_i)$ and fix λ to be a constant. In Figs. 3.8 and 3.9 we choose $x_i = 0.01$ and $\lambda = 1, 10, 100$ for a fermionic DM species.

From Fig. 3.8 one can appreciate that Y attains a residual abundance and that, as expected, the latter is smaller the larger λ is. This happens because for larger values of λ the interaction is more efficient.

In Fig. 3.9 we show the behaviour of the relative difference $Y/Y_{eq} - 1$ in order to understand when the freeze-out approximately takes place. Of course, the freeze-out

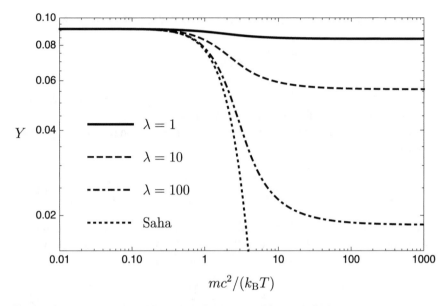

Fig. 3.8 Numerical solution of Eq. (3.270) for the case of fermionic DM

is not a specific instant, but depends on a criterion that we choose. For example, from inspection of Fig. 3.9, we see that $Y/Y_{eq} - 1$ at $x \approx 10$ starts to increase with more steepness and therefore we might establish that $x_f \approx 10$. Moreover, this value is very weakly dependent on λ.

Now, what is the difference between fermionic and bosonic DM? In Fig. 3.10 we plot Y in the bosonic DM (solid lines) and fermionic DM (dashed lines) for $\lambda = 10, 100$. As one can see, there are two differences: (*i*) When x is small the baryonic abundance is larger than the fermionic one (because of the ± 1 at the denominator of the distribution function) of a factor 4/3; (*ii*) for large x, the larger λ is, the smaller becomes the difference between the fermionic and bosonic relic abundances.

We now relate the relic abundance to the DM annihilation cross-section. For $x \gg 1$ we know that Y_{eq} is vanishing, and the Boltzmann equation can be written as:

$$\frac{dY}{dx} \approx -\frac{\lambda}{x^2} Y^2 , \tag{3.271}$$

whose solution is:

$$\frac{1}{Y_\infty} - \frac{1}{Y_f} \approx \frac{\lambda}{x_f} . \tag{3.272}$$

We have again considered here a constant λ, for simplicity. As we can see from Fig. 3.8, $Y_f - Y_\infty > 0$ and the difference gets larger the larger λ is. Moreover, $x_f \approx 10$, so that we can simplify

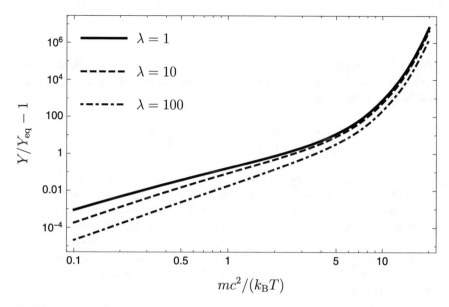

Fig. 3.9 Relative difference $Y/Y_{eq} - 1$

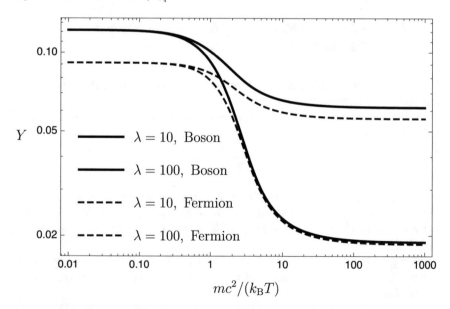

Fig. 3.10 Comparison of the numerical solutions of Eq. (3.270) for the cases of bosonic DM (solid lines) and fermionic DM (dashed lines) and $\lambda = 10$ (top two lines) and $\lambda = 100$ (bottom two lines, almost superposed)

$$Y_\infty \approx \frac{10}{\lambda} . \tag{3.273}$$

When Y has attained Y_∞, the abundance of particle is fixed and their number density starts to be diluted as $n_X \propto a^{-3}$. Therefore, we can can write down the present-time density as follows:

$$\rho_{X0} = n_1 m \frac{a_1^3}{a_0^3} = m Y_\infty \frac{(k_B T_1)^3}{(\hbar c)^3} \frac{a_1^3}{a_0^3} = \frac{10m}{\lambda} \frac{(k_B T_0)^3}{(\hbar c)^3} \left(\frac{a_1^3 T_1^3}{a_0^3 T_0^3} \right) . \tag{3.274}$$

We have introduced here the photon temperature (the only one we can measure). The ratio between parenthesis is not just equal to 1. We have seen an example of this discrepancy when we calculated the photon-neutrinos temperatures ratio. The reason is that not all along the cosmological evolution T decays as the inverse scale factor. When there are processes such as electron-positron annihilation, more photons are injected in the thermal bath and the temperature scales in a milder way than $1/a$.

It is now time to tackle more seriously the issue of the effective numbers of relativistic degrees of freedom.

3.11.1 The Effective Numbers of Relativistic Degrees of Freedom

Let T be the photon temperature, which we always use as reference since it is the only one we can measure, from CMB.

In Eq. (3.223), we wrote the energy density of all the relativistic species in the following way:

$$\varepsilon = \frac{\pi^2 (k_B T)^4}{30(\hbar c)^3} g_* , \tag{3.275}$$

where, actually $g_* = g_*(T)$ because when $k_B T$ drops below the mc^2 of a species, this becomes non-relativistic and is removed from the above equation. Thus g_* varies, but very rapidly close to the thresholds of the mass energies. Far from those, g_* it is practically constant.

In Fig. 3.11 we plot g_* calculated from Eqs. (3.103) and (3.223) for the species of Table 3.2. The residual non-vanishing value of $g_*(T)$ for low temperatures is due to the massless species, i.e. photons and neutrinos.

Note that this plot is just an illustrative example, because it employs Eq. (3.103) during the entire evolution, i.e. it assumes thermal equilibrium all the time, and, moreover, the QCD phase transition at 200 MeV and the difference between photon and neutrino temperature after electron-positron annihilation at 0.5 MeV have been put "by hand". The correct calculation should employ the energy density obtained from the solutions of the Boltzmann equations of the various species, which correctly track the evolutions when $\Gamma \sim H$, i.e. out of equilibrium.

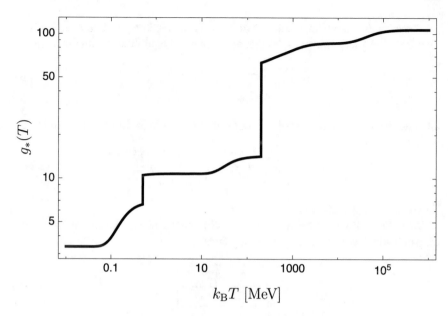

Fig. 3.11 Evolution of g_* calculated from Eqs. (3.103) and (3.223) for the species of Table 3.2. Note the QCD phase transition at 200 MeV and the difference between photon and neutrino temperature after electron-positron annihilation at 0.5 MeV

Table 3.2 The standard model particles, with their mass, spin and degeneracies g_s

Particle		Mass	Spin	g_s
Quarks	$t, \bar{t}$	173 GeV	$\frac{1}{2}$	$2 \cdot 2 \cdot 3 = 12$
	$b, \bar{b}$	4 GeV		
	$c, \bar{c}$	1 GeV		
	$s, \bar{s}$	100 MeV		
	$d, \bar{d}$	5 MeV		
	$u, \bar{u}$	2 MeV		
Gluons	g_i	0 GeV	1	$8 \cdot 2 = 16$
Leptons	$\tau^{\pm}$	1777 MeV	$\frac{1}{2}$	$2 \cdot 2 = 4$
	$\mu^{\pm}$	106 MeV		
	$e^{\pm}$	511 keV		
	$\nu_{\tau}, \bar{\nu}_{\tau}$	<0.6 eV	$\frac{1}{2}$	$2 \cdot 1 = 2$
	$\nu_{\mu}, \bar{\nu}_{\mu}$	<0.6 eV		
	$\nu_e, \bar{\nu}_e$	<0.6 eV		
Gauge Bosons	W^+	80 GeV	1	3
	W^-	80 GeV		
	Z^0	91 GeV		
	γ	0		2
Higgs Bosons	H^0	125 GeV	0	1

The effective numbers of relativistic degrees of freedom gets two contributions. One from the relativistic particles that are in thermal equilibrium with the photons:

$$g_*^{\text{therm}} \equiv \sum_{i=\text{bosons}} g_i + \frac{7}{8} \sum_{i=\text{fermions}} g_i \, , \tag{3.276}$$

and another from the relativistic particles that are no more in thermal equilibrium with the photons:

$$g_*^{\text{dec}} \equiv \sum_{i=\text{bosons}} g_i \frac{T_i^4}{T^4} + \frac{7}{8} \sum_{i=\text{fermions}} g_i \frac{T_i^4}{T^4} \, . \tag{3.277}$$

The latter are basically neutrinos only.

Why do we insist on relativistic species? Because, as we showed, these are the only one which contribute to the entropy density s:

$$\boxed{s = \frac{2\pi^2 k_B^4 T^3}{45(\hbar c)^3} g_{*S}} \tag{3.278}$$

where g_{*S} is the effective number of degrees of freedom for the entropy. Now, also to g_{*S} contribute species which are in thermal equilibrium with the photons and for which:

$$g_{*S}^{\text{therm}} = g_*^{\text{therm}} = \sum_{i=\text{bosons}} g_i + \frac{7}{8} \sum_{i=\text{fermions}} g_i \, , \tag{3.279}$$

and species that are no more in thermal equilibrium with the photons and for which:

$$g_{*S}^{\text{dec}} = \sum_{i=\text{bosons}} g_i \frac{T_i^3}{T^3} + \frac{7}{8} \sum_{i=\text{fermions}} g_i \frac{T_i^3}{T^3} \neq g_*^{\text{dec}} \, . \tag{3.280}$$

Because of the last inequality, due to the different scaling of s and ε with the temperature, we expect $g_* \neq g_{*S}$.

Now, sa^3 is a very useful quantity since it is conserved, as we showed earlier. When a species becomes non-relativistic, its contribution to s is exponentially suppressed as $\exp(-mc^2/k_B T)$. Therefore, the non-relativistic species passes its entropy to rest of the thermal bath such that sa^3 does not change.

What is the value of g_* and of g_{*S}? We need to recover our knowledge of particle physics in Table 3.2.

Summing up all the contributions we have from bosons and fermions:

$$g_{\text{bosons}} = 28 \, , \qquad g_{\text{fermions}} = 90 \, , \tag{3.281}$$

so that

$$g_* = 28 + \frac{7}{8} \cdot 90 = 106.75 \, . \tag{3.282}$$

When the temperature drops below one species mass, this becomes non-relativistic and then its g_s does not contribute anymore to the above sum. Note that before neutrino decoupling $g_* = g_{*S}$.

3.11.2 Relic Abundance of DM and the WIMP Miracle

On the basis of the discussion of the previous subsection, we can therefore write Eq. (3.274) as follows:

$$\Omega_{X0} = \frac{\rho_{X0}}{\rho_{\mathrm{cr},0}} = \frac{10m}{\lambda \rho_{\mathrm{cr},0}} \frac{(k_B T_0)^3}{(\hbar c)^3} \frac{g_{*S}(T_0)}{g_{*S}(m)} , \tag{3.283}$$

where $g_{*S}(m)$ is the number of effective degrees of freedom in entropy for a thermal energy equal to the DM particle mass. This must be certainly larger than $1\,\mathrm{MeV}$, roughly when neutrino decoupling takes place, therefore $g_* = g_{*S}$. Moreover,

$$g_{*S}(T_0) = 2 + \frac{7}{8} \cdot 6 \cdot \left(\frac{T_\nu}{T_0}\right)^3 = 2 + \frac{7}{8} \cdot 6 \cdot \frac{4}{11} = 3.91 , \tag{3.284}$$

and so we can rewrite Eq. (3.283) as follows:

$$\Omega_{X0} = \frac{8\pi G H(m) x_f (k_B T_0)^3}{3H_0^2 m^2 c^6 \langle \sigma v \rangle} \frac{g_{*S}(T_0)}{g_*(m)} , \tag{3.285}$$

where we have also made λ explicit. The Hubble parameter $H(m)$ can be written as:

$$H^2(m) = \frac{8\pi G}{3c^2} g_*(m) \frac{\pi^2}{30} \frac{(mc^2)^4}{(\hbar c)^3} , \tag{3.286}$$

so that we have finally:

$$\Omega_{X0} h^2 \approx 0.331 \frac{x_f}{\sqrt{g_*(m)}} \frac{10^{-37}\,\mathrm{cm}^2}{\langle \sigma v/c \rangle} \approx 0.331 \frac{x_f}{\sqrt{g_*(m)}} \frac{2.57 \times 10^{-10}\,\mathrm{GeV}^{-2}}{\langle \sigma v/c \rangle} . \tag{3.287}$$

As we saw, reasonable values are $x_f \approx 10$ and $g_* \approx 100$. A cross-section of the order of $G_F^2 \sim 10^{-10}\,\mathrm{GeV}^{-2}$ gives the right order of magnitude of the present abundance of CDM. This coincidence is known as **WIMP miracle**.[6]

Relic Abundance of Baryons

The very same result of Eq. (3.287) can be used for the annihilation of baryons:

---

[6]One can think of different cross sections and thus find CDM candidates lighter than WIMPs. See e.g. Profumo (2017). This possibility is also called **WIMPless miracle**.

$$b + \bar{b} \leftrightarrow \gamma + \gamma . \tag{3.288}$$

Let us consider only nucleons, i.e. protons and neutrons, since their mass is the dominant one for the baryonic energy density. The annihilation cross-section is of the order of:

$$\langle \sigma v / c \rangle \approx \frac{(\hbar c)^2}{(m_\pi c^2)^2} , \tag{3.289}$$

where $m_\pi c^2 \approx 140 \, \text{MeV}$ is the mass of the meson π, which can be thought as the mediator of the strong interaction among nucleons. Substituting into Eq. (3.287) we get:

$$\Omega_{b0} h^2 \approx 10^{-11} , \tag{3.290}$$

i.e. a value many orders of magnitude below the observed one and that thus constitutes a compelling argument for the necessity of baryogenesis.

Focusing on electrons and positrons, the annihilation cross section is of the order

$$\langle \sigma v / c \rangle \approx \frac{\alpha^2 (\hbar c)^2}{(m_e c^2)^2} \approx 204 \, \text{GeV}^{-2} , \tag{3.291}$$

and from Eq. (3.287) we get:

$$\Omega_{e0} h^2 \approx 10^{-12} . \tag{3.292}$$

References

Ade, P.A.R., et al.: Planck 2015 results. XIII. Cosmological parameters. Astron. Astrophys. **594**, A13 (2016a)

Berestetskii, V.B., Lifshitz, E.M., Pitaevskii, L.P.: *Quantum Electrodynamics Volume 4 of Course of Theoretical Physics*. Pergamon Press, Oxford (1982)

Bernstein, J.: Kinetic Theory in the Expanding Universe. Cambridge University Press, Cambridge (1988)

Camarena, D., Marra, V.: Cosmological constraints on the radiation released during structure formation. Eur. Phys. J. C **76**(11), 644 (2016)

Chluba, J.: Science with CMB spectral distortions. In *Proceedings, 49th Rencontres de Moriond on Cosmology: La Thuile, Italy, March 15–22, 2014* (pp. 327–334). (2014)

Chluba, J., Sunyaev, R.A.: The evolution of CMB spectral distortions in the early Universe. Mon. Not. Roy. Astron. Soc. **419**, 1294–1314 (2012)

Dodelson, S.: Modern Cosmology. Academic Press, Amsterdam (2003)

Dodelson, S., Widrow, L.M.: Sterile-neutrinos as dark matter. Phys. Rev. Lett. **72**, 17–20 (1994)

Fixsen, D.J., Cheng, E.S., Gales, J.M., Mather, J.C., Shafer, R.A., Wright, E.L.: The cosmic microwave background spectrum from the full COBE FIRAS data set. Astrophys. J. **473**, 576 (1996)

Gell-Mann, M., Ramond, P., Slansky, R.: Complex Spinors and Unified Theories. Conf. Proc. **C790927**, 315–321 (1979)

Grad, H.: Principles of the kinetic theory of gases. Thermodynamik der Gase/Thermodynamics of Gases, pp. 205–294. Springer, Berlin (1958)

Kolb, E.W., Turner, M.S.: The early Universe. Front. Phys. **69**, 1–547 (1990)

Lesgourgues, J., Pastor, S.: Massive neutrinos and cosmology. Phys. Rept. **429**, 307–379 (2006)

Peebles, P.J.E.: Recombination of the primeval plasma. Astrophys. J. **153**, 1 (1968)

Profumo, S.: An Introduction to Particle Dark Matter. Advanced textbooks in physics. World Scientific, Singapore (2017)

Profumo, S., Sigurdson, K., Kamionkowski, M.: What mass are the smallest protohalos? Phys. Rev. Lett. **97**, 031301 (2006)

Wagoner, R.V.: Big bang nucleosynthesis revisited. Astrophys. J. **179**, 343–360 (1973)

Wilczek, F.: Particle physics: a weighty mass difference. Nature **520**, 303–304 (2015)

Chapter 4
Cosmological Perturbations

Без труда не вынешь рыбку из пруда
(Without effort, you can't pull a fish out of the pond)
Russian proverb

As we have seen in the previous Chapters, the assumption of homogeneous and isotropic universe is very useful and productive, but it is reliable only on very large scales (above 200 Mpc). Its shortcomings become evident when we start to investigate how structures, such as galaxies and their clusters, form, since these are huge deviations from the cosmological principle.

In this Chapter we address **small** deviations from the cosmological principle, considering perturbations in the FLRW metric. This is the starting point of the incredibly difficult task of understanding how structures form in an expanding universe, which ultimately needs powerful machines and numerical simulations.

The material for this Chapter is mainly drawn from the textbooks (Dodelson 2003; Mukhanov 2005; Weinberg 2008) and from the papers/reviews (Bardeen 1980; Kodama and Sasaki 1984; Mukhanov et al. 1992; Ma and Bertschinger 1995).

We assume hereafter $K = 0$, both for simplicity and because we have seen that there is strong observational evidence of a spatially flat universe. From this Chapter on, we also start to adopt natural $\hbar = c = 1$ units.

4.1 From the Perturbations of the FLRW Metric to the Linearised Einstein Tensor

Let $\bar{g}_{\mu\nu}$ be the FLRW metric, and write it using the conformal time:

$$\bar{g}_{\mu\nu} = a^2(\eta)(-d\eta^2 + \delta_{ij}dx^i dx^j) . \tag{4.1}$$

© Springer International Publishing AG, part of Springer Nature 2018
O. Piattella, *Lecture Notes in Cosmology*, UNITEXT for Physics,
https://doi.org/10.1007/978-3-319-95570-4_4

This metric describes the **background spacetime**, or manifold. However, the background spacetime is fictitious, in the sense that we are now considering deviations from homogeneity and isotropy and therefore the actual **physical spacetime** is a different manifold described by the metric $g_{\mu\nu}$. Defining the difference:

$$\delta g_{\mu\nu}(x) = g_{\mu\nu}(x) - \bar{g}_{\mu\nu}(x) \,, \tag{4.2}$$

at a certain spacetime coordinate x is an ill-posed statement because $g_{\mu\nu}$ and $\bar{g}_{\mu\nu}$ are tensors defined on different manifolds and x is a coordinate defined through different charts. Even if we embed the two manifolds in a single one, still the difference between two tensors evaluated at different points is an ill-defined operation. Therefore, in order to make Eq. (4.2) meaningful, we need an extra ingredient: a map which identifies points of the background manifold with those of the physical manifold. This map is called **gauge**, is arbitrary and allows us to use a fixed coordinate system (a chart) in the background manifold also for the points in the physical manifold. In other words, we shall still use conformal or cosmic time plus comoving spatial coordinates even when describing perturbative quantities. This property leads to the so-called **problem of the gauge**, which we will briefly discuss later. For more details on these topics, see Stewart (1990) and Malik and Matravers (2013).

Metric $g_{\mu\nu}$ has in general 10 independent components that, in a generic gauge, we write down in the following form:

$$g_{\mu\nu} = a^2(\eta) \left\{ \begin{array}{cc} -[1 + 2\psi(\eta, \mathbf{x})] & w_i(\eta, \mathbf{x}) \\ w_i(\eta, \mathbf{x}) & \delta_{ij}[1 + 2\phi(\eta, \mathbf{x})] + \chi_{ij}(\eta, \mathbf{x}) \end{array} \right\} , \qquad \delta^{ij}\chi_{ij} = 0 \,, \tag{4.3}$$

where ψ, ϕ, w_i, χ_{ij} $(i, j = 1, 2, 3)$ are functions of the background spacetime coordinates x^μ, in our case conformal time and comoving spatial coordinates.[1] From now on we omit their explicit functional dependence wherever possible, in order to keep a lighter notation.

As we mentioned above, the liberty of choosing a gauge allows us to fix a coordinate system in the background manifold. The latter shall be one in which homogeneity and isotropy are manifest, of course. Therefore, we could use any time parametrisation, though we will employ conformal time the most because it has the important physical meaning of the comoving particle horizon, and as for the spatial coordinates we shall always choose the comoving ones for which at any given and fixed time the background spatial metric is Euclidean, i.e. $\delta_{ij}dx^i dx^j$. For this reason, the perturbations in Eq. (4.3) can be regarded as usual 3-vectors, defined from the rotation group SO(3).

---

[1]The reason why ψ and ϕ are multiplied by 2 in the perturbed metric is just for pure future convenience of calculation.

Let us elaborate more on this point. Consider the coordinate transformation:

$$\frac{\partial x^\mu}{\partial x^{\nu'}} = \begin{pmatrix} 1 & 0 \\ 0 & R^i{}_j \end{pmatrix} , \tag{4.4}$$

where $R^i{}_j$ is a rotation. By definition, a rotation is characterised by $R^T R = I$, i.e. the transposed matrix is also the inverse, hence $\delta_{kl} R^k{}_i R^l{}_j = \delta_{ij}$. Applying this transformation to metric (4.3) we get:

$$g'_{00} = g_{00} , \quad g'_{0i} = R^k{}_i g_{0k} , \quad g'_{ij} = R^k{}_i R^l{}_j g_{kl} , \tag{4.5}$$

and hence, recalling that $R^k{}_i R^l{}_j \delta_{kl} = \delta_{ij}$, we finally find:

$$\psi' = \psi , \quad w'_i = R^k{}_i w_k , \quad \phi' = \phi , \quad \chi'_{ij} = R^k{}_i R^l{}_j \chi_{kl} . \tag{4.6}$$

Therefore, ψ and ϕ are two 3-scalars, w_i ($i = 1, 2, 3$) is a 3-vector, χ_{ij} is a 3-tensor ($i, j = 1, 2, 3$) and the indices of w_i and χ_{ij} are raised and lowered by δ_{ij}. Note that the 6 components of χ_{ij} are not independent because $\delta^{ij} \chi_{ij} = 0$. In other words, χ_{ij} is traceless and we have already put in evidence the spatial trace of the metric through ϕ. One does this also because the spatial intrinsic curvature depends only on ϕ, not on χ_{ij}.[2]

Now, if the gauge chosen in Eq. (4.3) is such that:

$$|\bar{g}_{\mu\nu}| \gg |\delta g_{\mu\nu}| , \tag{4.7}$$

then we are dealing with **perturbations**. They are considered small, or linear, or at first-order if we neglect powers with exponent larger than one in the quantities themselves and in their derivatives. Not only, also combinations among different perturbations are neglected. For example, ψ^2, ϕw_i, $w^i \chi_{ij}$, $\phi' \phi$ and so on are all second order perturbations, and therefore negligible. Let us see how this works when computing the Christoffel symbols for metric (4.3). Substituting the decomposition (4.2) into the definition of Christoffel symbol we have:

$$\Gamma^\mu_{\nu\rho} = \frac{1}{2} g^{\mu\sigma} \left(g_{\sigma\nu,\rho} + g_{\sigma\rho,\nu} - g_{\nu\rho,\sigma} \right) = \frac{1}{2} \bar{g}^{\mu\sigma} \left(\bar{g}_{\sigma\nu,\rho} + \bar{g}_{\sigma\rho,\nu} - \bar{g}_{\nu\rho,\sigma} \right)$$
$$\frac{1}{2} \bar{g}^{\mu\sigma} \left(\delta g_{\sigma\nu,\rho} + \delta g_{\sigma\rho,\nu} - \delta g_{\nu\rho,\sigma} \right) + \frac{1}{2} \delta g^{\mu\sigma} \left(\bar{g}_{\sigma\nu,\rho} + \bar{g}_{\sigma\rho,\nu} - \bar{g}_{\nu\rho,\sigma} \right) , \tag{4.8}$$

where the comma denotes the usual partial derivative. Note that we have assumed the same decomposition of Eq. (4.2) also for the covariant components of the metric and neglected terms such as $\delta g^{\mu\sigma} \delta g_{\sigma\nu,\rho}$, i.e. perturbative quantities multiplied by their

---

[2]One can guess this by simply noting that the spatial curvature is a 3-scalar and one cannot form any 3-scalar at first-order from χ_{ij} since it is traceless.

derivatives. It is important to realise that $\delta g^{\mu\sigma}$ is not simply $\delta g_{\mu\sigma}$ with indices raised by $\bar{g}^{\mu\sigma}$.

Exercise 4.1 Since

$$g^{\mu\rho} g_{\rho\nu} = \delta^\mu{}_\nu , \qquad \bar{g}^{\mu\rho} \bar{g}_{\rho\nu} = \delta^\mu{}_\nu , \qquad (4.9)$$

because both are metrics, using Eq. (4.2) show that

$$\boxed{\delta g^{\mu\nu} = -\bar{g}^{\mu\rho} \delta g_{\rho\sigma} \bar{g}^{\nu\sigma}} \qquad (4.10)$$

In particular, in our scenario of cosmological perturbations, we shall write the total metric, using the conformal time, as follows:

$$g_{\mu\nu} = a^2 (\eta_{\mu\nu} + h_{\mu\nu}) . \qquad (4.11)$$

Hence, the perturbed contravariant metric is the following:

$$\delta g^{00} = -\frac{1}{a^2} h_{00} \quad \delta g^{0i} = \frac{1}{a^2} \delta^{il} h_{0l} = \frac{1}{a^2} h_{0i} , \quad \delta g^{ij} = -\frac{1}{a^2} \delta^{il} h_{lm} \delta^{mj} = -\frac{1}{a^2} h_{ij} ,$$
$$(4.12)$$

where we have used our hypothesis that the indices of h_{ij} are raised by δ^{ij} and the property $h^{ij} = h_{ij}$.

Do not be confused by the fact that $\delta g_{\mu\nu}$ is the perturbed covariant metric but $\delta g^{\mu\nu}$ *is not* the contravariant perturbed metric.

One does raise the indices of the contravariant perturbed metric with the background one, but a minus sign must be taken into account. This fact is not dissimilar from considering the Taylor expansion

$$\frac{1}{1+x} = 1 - x + \mathcal{O}(x^2) . \qquad (4.13)$$

4.1.1 The Perturbed Christoffel Symbols

It is clear from Eq. (4.8) that we can decompose the affine connection as follows:

$$\Gamma^\mu_{\nu\rho} = \bar{\Gamma}^\mu_{\nu\rho} + \delta\Gamma^\mu_{\nu\rho} , \qquad (4.14)$$

where the barred one is computed from the background metric only.

Exercise 4.2 Show that

$$\delta\Gamma^\mu_{\nu\rho} = \frac{1}{2}\bar{g}^{\mu\sigma}\left(\delta g_{\sigma\nu,\rho} + \delta g_{\sigma\rho,\nu} - \delta g_{\nu\rho,\sigma} - 2\delta g_{\sigma\alpha}\bar{\Gamma}^\alpha_{\nu\rho}\right) \tag{4.15}$$

The background Christoffel symbols were calculated already in Chap. 2 but for the FLRW metric written in the cosmic time.

Exercise 4.3 Show that the only non-vanishing background Christoffel symbols are:

$$\bar{\Gamma}^0_{00} = \frac{a'}{a}, \quad \bar{\Gamma}^0_{ij} = \frac{a'}{a}\delta_{ij}, \quad \bar{\Gamma}^i_{0j} = \frac{a'}{a}\delta^i{}_j, \tag{4.16}$$

where the prime denotes derivation with respect to the conformal time. One can make the calculation directly from the FLRW metric written in the conformal time or use the results already found in Eq. (2.29) for the FLRW metric written in the cosmic time and use the transformation relation for the Christoffel symbol:

$$\bar{\Gamma}'^\mu_{\nu\rho} = \bar{\Gamma}^\alpha_{\beta\gamma}\frac{\partial x^\beta}{\partial x'^\nu}\frac{\partial x^\gamma}{\partial x'^\rho}\frac{\partial x'^\mu}{\partial x^\alpha} + \frac{\partial x'^\mu}{\partial x^\sigma}\frac{\partial^2 x^\sigma}{\partial x'^\nu\partial x'^\rho}, \tag{4.17}$$

where the primed coordinates are in the conformal time and hence:

$$\frac{\partial x'^0}{\partial x^0} = \frac{1}{a}, \quad \frac{\partial x'^l}{\partial x^m} = \delta^l{}_m, \tag{4.18}$$

being the other cases vanishing. It is a good and reassuring exercise to do in both ways and check that the result is the same.

A moment of reflection. First of all, we shall use mostly the conformal time throughout these notes since, as we saw earlier, it represents the comoving particle horizon and it will allow us to clearly distinguish the evolution of super-horizon (hence causally disconnected) scales from sub-horizon ones.

On the other hand, the most economic way to compute the linearised Einstein equations is using the cosmic time and stopping to the calculation of the Ricci tensor by considering the Einstein equations in the form

$$R_{\mu\nu} = 8\pi G\left(T_{\mu\nu} - \frac{1}{2}g_{\mu\nu}T\right), \tag{4.19}$$

as done e.g. in Weinberg (2008). It is the most economic way because in the cosmic time we have only 2 non-vanishing Christoffel symbols, whereas in the conformal time we have three, and because we are spared to compute the perturbed Ricci scalar. Once done this calculation, it is straightforward to come back to the conformal time.

In the following we shall anyway go on using the conformal time, because it is a good workout. Compare the results found here with those in Weinberg (2008) Chap. 5 using the tensorial properties:

$$R_{\mu\nu} = \tilde{R}_{\rho\sigma} \frac{\partial \tilde{x}^\rho}{\partial x^\mu} \frac{\partial \tilde{x}^\sigma}{\partial x^\nu} , \quad h_{\mu\nu} = \tilde{h}_{\rho\sigma} \frac{\partial \tilde{x}^\rho}{\partial x^\mu} \frac{\partial \tilde{x}^\sigma}{\partial x^\nu} , \tag{4.20}$$

where the quantities with tilde are in the cosmic time. Since the change from cosmic to conformal time does not affect the spatial coordinates, we have that:

$$R_{00} = \tilde{R}_{00} a^2 , \quad R_{0i} = \tilde{R}_{0i} a , \quad R_{ij} = \tilde{R}_{ij} , \tag{4.21}$$

and similarly for $h_{\mu\nu}$. With this map, we can check the results that we are going to find here with those in Weinberg (2008). Mind that in Weinberg (2008) the Ricci tensor is defined with the opposite sign with respect to ours here.

Exercise 4.4 We are now in the position of writing the perturbed Christoffel symbols. We do so without leaving $h_{\mu\nu}$ explicit from Eq. (4.3). Find that:

$$\delta\Gamma^0_{00} = -\frac{1}{2} h'_{00} , \quad \delta\Gamma^0_{i0} = -\frac{1}{2} \left(h_{00,i} - 2\mathcal{H} h_{0i} \right) , \tag{4.22}$$

$$\delta\Gamma^i_{00} = h'_{i0} + \mathcal{H} h_{i0} - \frac{1}{2} h_{00,i} , \tag{4.23}$$

$$\delta\Gamma^0_{ij} = -\frac{1}{2} \left(h_{0i,j} + h_{0j,i} - h'_{ij} - 2\mathcal{H} h_{ij} - 2\mathcal{H} \delta_{ij} h_{00} \right) , \tag{4.24}$$

$$\delta\Gamma^i_{j0} = \frac{1}{2} h'_{ij} + \frac{1}{2} \left(h_{i0,j} - h_{0j,i} \right) , \tag{4.25}$$

$$\delta\Gamma^i_{jk} = \frac{1}{2} \left(h_{ij,k} + h_{ik,j} - h_{jk,i} - 2\mathcal{H} \delta_{jk} h_{i0} \right) . \tag{4.26}$$

The prime denotes derivative with respect to the conformal time. The indices might seem unbalanced, but we have used the fact that h_{i0} and h_{ij} are 3-tensors with respect to the metric δ_{ij} and hence, for example, $h^i_0 = h_{i0}$. Recall that

$$\mathcal{H} \equiv \frac{a'}{a} , \tag{4.27}$$

i.e. $\mathcal{H}$ is the Hubble factor written in conformal time.

4.1.2 The Perturbed Ricci Tensor and Einstein Tensor

With this result we are going to compute the components of the perturbed Ricci tensor. Recall that this is defined as:

$$R_{\mu\nu} = \Gamma^{\rho}_{\mu\nu,\rho} - \Gamma^{\rho}_{\mu\rho,\nu} + \Gamma^{\rho}_{\mu\nu}\Gamma^{\sigma}_{\rho\sigma} - \Gamma^{\rho}_{\mu\sigma}\Gamma^{\sigma}_{\nu\rho} , \tag{4.28}$$

and hence, when substituting Eq. (4.14), we get

$$R_{\mu\nu} = \bar{\Gamma}^{\rho}_{\mu\nu,\rho} - \bar{\Gamma}^{\rho}_{\mu\rho,\nu} + \bar{\Gamma}^{\rho}_{\mu\nu}\bar{\Gamma}^{\sigma}_{\rho\sigma} - \bar{\Gamma}^{\rho}_{\mu\sigma}\bar{\Gamma}^{\sigma}_{\nu\rho}$$
$$+\delta\Gamma^{\rho}_{\mu\nu,\rho} - \delta\Gamma^{\rho}_{\mu\rho,\nu} + \bar{\Gamma}^{\rho}_{\mu\nu}\delta\Gamma^{\sigma}_{\rho\sigma} + \delta\Gamma^{\rho}_{\mu\nu}\bar{\Gamma}^{\sigma}_{\rho\sigma} - \bar{\Gamma}^{\rho}_{\mu\sigma}\delta\Gamma^{\sigma}_{\nu\rho} - \delta\Gamma^{\rho}_{\mu\sigma}\bar{\Gamma}^{\sigma}_{\nu\rho} , \tag{4.29}$$

by neglecting second order terms in the connection. It is clear that we can expand also the Ricci tensor as:

$$R_{\mu\nu} = \bar{R}_{\mu\nu} + \delta R_{\mu\nu} , \tag{4.30}$$

and we compute now its perturbed components.

Exercise 4.5 After opening up all the sums, show that:

$$\delta R_{00} = \delta\Gamma^{l}_{00,l} - \delta\Gamma^{l}_{0l,0} - \mathcal{H}\delta\Gamma^{l}_{0l} + 3\mathcal{H}\delta\Gamma^{0}_{00} , \tag{4.31}$$

and substituting the previous results, one gets:

$$\delta R_{00} = -\frac{1}{2}\nabla^2 h_{00} - \frac{3}{2}\mathcal{H}h'_{00} + h'_{k0,k} + \mathcal{H}h_{k0,k} - \frac{1}{2}\left(h''_{kk} + \mathcal{H}h'_{kk}\right) . \tag{4.32}$$

Here we have defined $\delta^{ij}\partial_i\partial_j \equiv \nabla^2$ as the Laplacian in comoving coordinates. Note that $h_{kk} = \delta^{lm}h_{lm}$, i.e. in the present instance repeated indices are summed even if both covariant or contravariant.

Exercise 4.6 Of course, repeat the previous exercise also for the other components. Find that:

$$\delta R_{0i} = -\mathcal{H}h_{00,i} - \frac{1}{2}\left(\nabla^2 h_{0i} - h_{k0,ik}\right) + \left(\frac{a''}{a} + \mathcal{H}^2\right)h_{0i} - \frac{1}{2}\left(h'_{kk,i} - h'_{ki,k}\right) , \tag{4.33}$$

and

$$\delta R_{ij} = \frac{1}{2}h_{00,ij} + \frac{\mathcal{H}}{2}h'_{00}\delta_{ij} + \left(\mathcal{H}^2 + \frac{a''}{a}\right)h_{00}\delta_{ij}$$

$$-\frac{1}{2}\left(\nabla^2 h_{ij} - h_{ki,kj} - h_{kj,ki} + h_{kk,ij}\right) + \frac{1}{2}h''_{ij} + \mathcal{H}h'_{ij} + \left(\mathcal{H}^2 + \frac{a''}{a}\right)h_{ij}$$

$$+\frac{\mathcal{H}}{2}h'_{kk}\delta_{ij} - \mathcal{H}h_{k0,k}\delta_{ij} - \frac{1}{2}(h'_{0i,j} + h'_{0j,i}) - \mathcal{H}(h_{0i,j} + h_{0j,i}) \,. \quad (4.34)$$

In the same way we decomposed metric (4.2), we decompose the Einstein tensor. We shall work with mixed indices:

$$G^{\mu}{}_{\nu} = g^{\mu\rho}R_{\rho\nu} - \frac{1}{2}\delta^{\mu}{}_{\nu}R = \bar{g}^{\mu\rho}\bar{R}_{\rho\nu} - \frac{1}{2}\delta^{\mu}{}_{\nu}\bar{R} + \bar{g}^{\mu\rho}\delta R_{\rho\nu} + \delta g^{\mu\rho}\bar{R}_{\rho\nu} - \frac{1}{2}\delta^{\mu}{}_{\nu}\delta R \,,$$
$$(4.35)$$

where $\bar{G}^{\mu}{}_{\nu} = \bar{R}^{\mu}{}_{\nu} - \frac{1}{2}\delta^{\mu}{}_{\nu}\bar{R}$ is the background Einstein tensor and depends purely from the background metric $\bar{g}_{\mu\nu}$ whereas

$$\delta G^{\mu}{}_{\nu} = \bar{g}^{\mu\rho}\delta R_{\rho\nu} + \delta g^{\mu\rho}\bar{R}_{\rho\nu} - \frac{1}{2}\delta^{\mu}{}_{\nu}\delta R \,, \quad (4.36)$$

is the linearly perturbed Einstein tensor, which depends from both $\bar{g}_{\mu\nu}$ and $h_{\mu\nu}$.

Exercise 4.7 Compute the perturbed Ricci scalar:

$$\delta R = g^{\mu\nu}R_{\mu\nu} = \bar{g}^{\mu\nu}\delta R_{\mu\nu} + \delta g^{\mu\nu}\bar{R}_{\mu\nu} \,. \quad (4.37)$$

Expand the above expression and use formula (4.10) in order to find:

$$\delta R = -\frac{1}{a^2}\delta R_{00} + \frac{1}{a^2}\delta^{ij}\delta R_{ij} - a^2 h_{\rho\sigma}\bar{g}^{\rho\mu}\bar{g}^{\sigma\nu}\bar{R}_{\mu\nu} \,, \quad (4.38)$$

and then, recalling that the background Ricci tensor is:

$$\bar{R}_{00} = 3\left(\mathcal{H}^2 - \frac{a''}{a}\right) \,, \quad \bar{R}_{ij} = \delta_{ij}\left(\mathcal{H}^2 + \frac{a''}{a}\right) \,, \quad (4.39)$$

one can write:

$$\delta R = -\frac{1}{a^2}\delta R_{00} - \frac{3}{a^2}h_{00}\left(\mathcal{H}^2 - \frac{a''}{a}\right) + \frac{1}{a^2}\delta^{ij}\delta R_{ij} - \frac{1}{a^2}h_{kk}\left(\mathcal{H}^2 + \frac{a''}{a}\right) \,,$$
$$(4.40)$$

Substituting the formulae for the perturbed Ricci tensor, one finds:

$$a^2 \delta R = \nabla^2 h_{00} + 3\mathcal{H}h'_{00} + 6\frac{a''}{a}h_{00} - 2h'_{k0,k} - 6\mathcal{H}h_{k0,k}$$
$$+ h''_{kk} + 3\mathcal{H}h'_{kk} - \nabla^2 h_{kk} + h_{kl,kl} . \tag{4.41}$$

The second line represents $a^2 \delta R^{(3)}$, i.e. the intrinsic spatial perturbed curvature scalar.

Now, let us calculate the mixed components of the perturbed Einstein tensor.

Exercise 4.8 Show that:

$$\boxed{2a^2 \delta G^0{}_0 = -6\mathcal{H}^2 h_{00} + 4\mathcal{H}h_{k0,k} - 2\mathcal{H}h'_{kk} + \nabla^2 h_{kk} - h_{kl,kl}} \tag{4.42}$$

and

$$\boxed{2a^2 \delta G^0{}_i = 2\mathcal{H}h_{00,i} + \nabla^2 h_{0i} - h_{k0,ki} + h'_{kk,i} - h'_{ki,k}} \tag{4.43}$$

and

$$2a^2 \delta G^i{}_j = \left[-4\frac{a''}{a}h_{00} - 2\mathcal{H}h'_{00} - \nabla^2 h_{00} + 2\mathcal{H}^2 h_{00} - 2\mathcal{H}h'_{kk} + \nabla^2 h_{kk} \right.$$
$$\left. - h_{kl,kl} + 2h'_{k0,k} + 4\mathcal{H}h_{k0,k} - h''_{kk} \right] \delta^i{}_j + h_{00,ij} - \nabla^2 h_{ij} + h_{ki,kj}$$
$$+ h_{kj,ki} - h_{kk,ij} + h''_{ij} + 2\mathcal{H}h'_{ij} - (h'_{0i,j} + h'_{0j,i}) - 2\mathcal{H}(h_{0i,j} + h_{0j,i}). \tag{4.44}$$

Now we turn to the right hand side of the Einstein equations, i.e. the energy-momentum tensor.

4.2 Perturbation of the Energy-Momentum Tensor

In the following we shall use mostly the energy-momentum tensor defined through the distribution function and its perturbation. However, let us see how to perturb the background, perfect fluid energy-momentum tensor. This was introduced as:

$$\bar{T}_{\mu\nu} = (\bar{\rho} + \bar{P})\bar{u}_\mu \bar{u}_\nu + \bar{P}\bar{g}_{\mu\nu} , \tag{4.45}$$

as the tensor describing a fluid with no dissipation, i.e. constant entropy along the flow.

Exercise 4.9 Compute $\bar{u}^\nu \nabla_\nu \bar{u}^\mu$, demand its vanishing and check via the second law of thermodynamics that this corresponds to a constant entropy.

Let us rewrite $\bar{T}_{\mu\nu}$ as follows:

$$\bar{T}_{\mu\nu} = \bar{\rho}\bar{u}_\mu \bar{u}_\nu + \bar{P}\bar{\theta}_{\mu\nu} \,, \qquad \bar{\theta}_{\mu\nu} \equiv \bar{g}_{\mu\nu} + \bar{u}_\mu \bar{u}_\nu \,. \tag{4.46}$$

The tensor $\bar{\theta}_{\mu\nu}$ acts as a projector on the hypersurface orthogonal to the four-velocity.

Exercise 4.10 Show that $\bar{\theta}_{\mu\nu}\bar{u}^\mu = 0$.

This is an example of 3+1 decomposition, which is particularly useful when we study a fluid flow.

Exercise 4.11 Show that:

$$\bar{\rho} = \bar{T}_{\mu\nu}\bar{u}^\mu \bar{u}^\nu \,, \quad 3\bar{P} = \bar{T}_{\mu\nu}\bar{\theta}^{\mu\nu} \,, \quad T \equiv \bar{g}^{\mu\nu}\bar{T}_{\mu\nu} = -\rho + 3P \,, \tag{4.47}$$

i.e. the fluid density is the projection of the energy-momentum tensor along the 4-velocity of the fluid element and the pressure is the projection of the energy-momentum tensor on the 3-hypersurface orthogonal to the four-velocity.

The most general energy-momentum tensor, which also includes the possibility of dissipation, can be then written by straightforwardly generalising Eq. (4.46), i.e.

$$\boxed{T_{\mu\nu} = \rho u_\mu u_\nu + q_\mu u_\nu + q_\nu u_\mu + (P + \pi)\theta_{\mu\nu} + \pi_{\mu\nu}} \tag{4.48}$$

where q_μ is the **heat transfer** contribution, satisfying $q_\mu u^\mu = 0$ and thus contributing with 3 independent components; $\pi_{\mu\nu}$ is the **anisotropic stress**, it is traceless and satisfies $\pi_{\mu\nu}u^\mu = 0$, hence providing 5 independent components. The trace of $\pi_{\mu\nu}$ is π, it is called **bulk viscosity** and has been put in evidence together with the pressure. For more detail about dissipative processes in cosmology and the above decomposition of the energy-momentum tensor, see the review (Maartens 1996). The anisotropic stress $\pi_{\mu\nu}$ is not necessarily related to viscosity, but can exist for relativistic species such as photons and neutrinos because of the quadrupole moments of their distributions, as we shall see later. On the other hand, heat fluxes and bulk viscosity are related to dissipative processes, and we neglect them in these notes starting from the next section.

Now we consider the energy-momentum tensor of Eq. (4.48) as made up of a background contribution plus a linear perturbation, i.e. we expand

$$\rho = \bar{\rho} + \delta\rho(\eta, \mathbf{x}) \,, \quad P = \bar{P} + \delta P(\eta, \mathbf{x}) \,, \quad u^{\mu} = \bar{u}^{\mu} + \delta u^{\mu}(\eta, \mathbf{x}) \,, \quad (4.49)$$

which are the physical density, pressure and four-velocity, which, remember, depend on the background quantities because of our choice of a gauge. The barred quantities depend only on η, since they are defined on the FLRW background. Heat fluxes, bulk viscosity and anisotropic stresses are purely perturbed quantities.[3] Therefore, we have:

$$T_{\mu\nu} = \bar{T}_{\mu\nu} + \delta\rho\bar{u}_{\mu}\bar{u}_{\nu} + \bar{\rho}\delta u_{\mu}\bar{u}_{\nu} + \bar{\rho}\bar{u}_{\mu}\delta u_{\nu} + q_{\mu}\bar{u}_{\nu} + q_{\nu}\bar{u}_{\mu} + \bar{\theta}_{\mu\nu}(\delta P + \pi) + \bar{P}\delta\theta_{\mu\nu} + \pi_{\mu\nu} \,,$$
$$(4.50)$$

where one can straightforwardly identify the perturbed energy-momentum tensor and where

$$\delta\theta_{\mu\nu} = \delta g_{\mu\nu} + \delta u_{\mu}\bar{u}_{\nu} + \bar{u}_{\mu}\delta u_{\nu} \,. \qquad (4.51)$$

Now, recall that the background four-velocity satisfies the normalisation:

$$\bar{g}_{\mu\nu}\bar{u}^{\mu}\bar{u}^{\nu} = -1 \,. \qquad (4.52)$$

Let us choose a frame in which to make explicit the components of the energy-momentum tensor. Of course, we use comoving coordinates, for which one has $\bar{u}^i = 0$. From Eq. (4.52) we have then

$$a^2(\bar{u}^0)^2 = 1 \,, \qquad (4.53)$$

and we choose the positive solution $\bar{u}^0 = 1/a$, which implies $u_0 = -a$.[4] When we choose these coordinates, the relations $q_{\mu}u^{\mu} = \pi_{\mu\nu}u^{\mu} = 0$ imply that $q_0 = \pi_{\mu 0} = 0$, i.e. the heat flux and the anisotropic stress have only spatial components.

Exercise 4.12 Calculate the components of the energy-momentum tensor (4.50). Show that:

$$T_{00} = \bar{\rho}(1 + \delta)a^2 - 2a(\bar{\rho} + \bar{P})\delta u_0 + \bar{P}\delta g_{00} \,, \qquad (4.54)$$
$$T_{0i} = -a(\bar{\rho} + \bar{P})\delta u_i - aq_i + \bar{P}\delta g_{0i} \,, \qquad (4.55)$$
$$T_{ij} = (\bar{P} + \delta P + \pi)a^2\delta_{ij} + \bar{P}\delta g_{ij} + \pi_{ij} \,, \qquad (4.56)$$

where we have introduced one of the main characters of these notes, the **density contrast**:

$$\boxed{\delta \equiv \frac{\delta\rho}{\bar{\rho}}} \qquad (4.57)$$

---

[3]Bulk viscosity is compatible with the cosmological principle and can be contemplated also at background level. It plays a central role in the so-called **bulk viscous cosmology**, see Zimdahl (1996).

[4]It amounts to choose that the conformal time and the fluid element proper time flow in the same direction.

The density contrast is very important because describes how structure formation begins. Note how the presence of perturbations in the four-velocity gives rise to mixed time-space components in the energy-momentum tensor, i.e. the breaking of homogeneity and isotropy allows for extra fluxes beyond the Hubble one.

Note that the total four-velocity also satisfies a normalisation relation, with respect to the total metric, i.e.

$$g_{\mu\nu}u^\mu u^\nu = -1 \ . \tag{4.58}$$

If we expand this relation up to the first order we find:

$$\bar{g}_{\mu\nu}\bar{u}^\mu\bar{u}^\nu + \delta g_{\mu\nu}\bar{u}^\mu\bar{u}^\nu + 2\bar{g}_{\mu\nu}\delta u^\mu\bar{u}^\nu = -1 \ . \tag{4.59}$$

Using Eq. (4.52) and $\bar{u}^i = 0$, we find that:

$$\delta g_{00} + 2\bar{g}_{00}a\delta u^0 = 0 \ , \tag{4.60}$$

and thus we can relate the metric perturbation h_{00} to δu^0 as follows:

$$\boxed{\delta u^0 = \frac{h_{00}}{2a}} \tag{4.61}$$

Care is needed when we want to compute the covariant components of the perturbed four-velocity δu_μ. These are not simply $\delta u_\mu = \bar{g}_{\mu\nu}\delta u^\nu$. It is the same care we had to apply when considering the relation between $\delta g^{\mu\nu}$ and $h_{\mu\nu}$. So, let us define:

$$\boxed{\delta u_i = av_i} \tag{4.62}$$

with v_i components of a 3-vector, i.e. its index is raised by δ^{ij} so that $v^i = v_i$. Let us compute now the components δu^i. We must start from the covariant expression for the total four velocity, i.e.

$$\bar{u}_\mu + \delta u_\mu = u_\mu = g_{\mu\nu}u^\nu = g_{\mu\nu}(\bar{u}^\nu + \delta u^\nu) \ , \tag{4.63}$$

and expanding up to first order, we get

$$\bar{u}_\mu + \delta u_\mu = \bar{g}_{\mu\nu}\bar{u}^\nu + \bar{g}_{\mu\nu}\delta u^\nu + \delta g_{m u\nu}\bar{u}^\nu \ . \tag{4.64}$$

Equating order by order we obtain $\bar{u}_\mu = \bar{g}_{\mu\nu}\bar{u}^\nu$, as expected, and

$$\boxed{\delta u_\mu = \bar{g}_{\mu\nu}\delta u^\nu + \delta g_{\mu\nu}\bar{u}^\nu} \tag{4.65}$$

so that

$$\boxed{\delta u^\mu = \bar{g}^{\mu\nu}\delta u_\nu - \bar{g}^{\mu\rho}\delta g_{\rho\nu}\bar{u}^\nu} \tag{4.66}$$

So, there is an extra term $\delta g_{\mu\nu}\bar{u}^\nu$. This comes from the fact that $\bar{g}_{\mu\nu}$ raises or lowers indices for the background quantities only whereas $g_{\mu\nu}$ raises or lowers indexes for the full quantities only, and δu^μ is neither. From Eqs. (4.65) and (4.66) we have that

$$\boxed{\delta u_0 = \frac{h_{00}a}{2}, \qquad a\delta u^i = v_i - 1h_{i0}} \tag{4.67}$$

Exercise 4.13 Rewrite the components of the energy-momentum tensor as:

$$T_{00} = \bar{\rho}(1+\delta)a^2 - \bar{\rho}a^2 h_{00}, \tag{4.68}$$
$$T_{0i} = -a^2(\bar{\rho}+\bar{P})v_i - aq_i + \bar{P}a^2 h_{0i}, \tag{4.69}$$
$$T_{ij} = (\bar{P}+\delta P+\pi)a^2\delta_{ij} + \bar{P}a^2 h_{ij} + \pi_{ij}, \tag{4.70}$$

In order to calculate the mixed components we use the standard relation:

$$T^\mu{}_\nu = g^{\mu\rho}T_{\rho\nu} = \bar{g}^{\mu\rho}\bar{T}_{\rho\nu} + \bar{g}^{\mu\rho}\delta T_{\rho\nu} + \delta g^{\mu\rho}\bar{T}_{\rho\nu}. \tag{4.71}$$

Exercise 4.14 Compute the mixed components of the energy-momentum tensor. Show that:

$$T^0{}_0 = -\bar{\rho}(1+\delta), \tag{4.72}$$
$$T^0{}_i = \left(\bar{\rho}+\bar{P}\right)v_i + a^{-1}q_i, \tag{4.73}$$
$$T^i{}_0 = -\left(\bar{\rho}+\bar{P}\right)(v_i - h_{0i}) - a^{-1}q_i, \tag{4.74}$$
$$T^i{}_j = \delta^i{}_j(\bar{P}+\delta P+\pi) + \pi^i{}_j, \tag{4.75}$$

where we have stipulated that $\delta^{il}\pi_{lj}a^{-2} = \pi^i{}_j$.

In the following we shall mainly use, especially for photons and neutrinos, the energy-momentum tensor computed from kinetic theory, i.e. Equation (3.41):

$$T^\mu{}_\nu = \int \frac{d^3\mathbf{p}}{(2\pi)^3}\frac{p^\mu p_\nu}{p^0}f, \tag{4.76}$$

in which the distribution function is also perturbed:

$$\boxed{f = \bar{f} + \mathcal{F}} \qquad (4.77)$$

thus allowing to define a perturbed energy-momentum tensor as follows:

$$\delta T^{\mu}{}_{\nu} = \int \frac{d^3\mathbf{p}}{(2\pi)^3} \frac{p^{\mu} p_{\nu}}{p^0} \mathcal{F} , \qquad (4.78)$$

The momentum used here is the proper one which, recall, has the metric embedded in its definition so that:

$$p^i = p_i , \quad p^2 = \delta_{ij} p^i p^j , \quad E = p^0 , \quad p_0 = -E . \qquad (4.79)$$

These relations provide directly the mass-shell one $E^2 = p^2 + m^2$ from $g_{\mu\nu} P^{\mu} P^{\nu} = -m^2$. Note that we are assuming $g_{0i} = 0$, also at perturbative level thanks to gauge freedom, otherwise the relations above would be incorrect since the spatial metric would not be g_{ij} but $g_{ij} - g_{0i} g_{0j}/g_{00}$. See Landau and Lifschits (1975) for a nice explanation of this fact.

Then, from the above definition of perturbed energy-momentum, we have:

$$\delta T^0{}_0 = - \int \frac{d^3\mathbf{p}}{(2\pi)^3} E(p)\mathcal{F} = -\delta\rho , \qquad (4.80)$$

which makes sense, and is consistent with Eq. (4.72), because we are weighting the particle energy with the perturbed distribution function. The mixed components are:

$$\delta T^0{}_i = \int \frac{d^3\mathbf{p}}{(2\pi)^3} p_i \mathcal{F} = (\bar{\rho} + \bar{P})v_i , \quad \delta T^i{}_0 = - \int \frac{d^3\mathbf{p}}{(2\pi)^3} p^i \mathcal{F} = -(\bar{\rho} + \bar{P})v_i , \qquad (4.81)$$

compatible with Eqs. (4.73) and (4.74) since we are assuming $h_{0i} = 0$ (and neglecting already the heat fluxes). Note the multiplication by $\bar{\rho} + \bar{P}$. Physically the integral gives the perturbed spatial momentum density, which can be decomposed in the velocity flow times the inertial mass density, which is $\bar{\rho} + \bar{P}$ in GR.

Finally:

$$\delta T^i{}_j = \int \frac{d^3\mathbf{p}}{(2\pi)^3} \frac{p^i p_j}{E} \mathcal{F} = \delta^i{}_j \delta P + \pi^i{}_j , \qquad (4.82)$$

from which we can define the perturbed pressure as the trace part:

$$\delta P = \frac{1}{3}\delta^{ij}\delta T_{ij} = \int \frac{d^3\mathbf{p}}{(2\pi)^3} \frac{p^2}{3E} \mathcal{F} , \qquad (4.83)$$

and the anisotropic stress as the traceless part:

$$\pi^i{}_j = \delta T^i{}_j - \frac{1}{3}\delta^i{}_j\delta^{lm}\delta T_{lm} = \int \frac{d^3\mathbf{p}}{(2\pi)^3}\frac{1}{E}\left(p^i p_j - \frac{1}{3}\delta^i{}_j p^2\right)\mathcal{F}. \tag{4.84}$$

The evolution equations for the perturbations are given by the **linearised Einstein equations**:

$$\delta G^\mu{}_\nu = 8\pi G \delta T^\mu{}_\nu. \tag{4.85}$$

Unfortunately, Eq. (4.85) is not sufficient to completely describe the behaviour of both matter and metric quantities if the fluid components are more than one. See e.g. Gorini et al. (2008). We shall make use of the perturbed Boltzmann equations for each component of our cosmological model and we shall derive them in Chap. 5.

4.3 The Problem of the Gauge and Gauge Transformations

As we have mentioned in the previous section, a gauge is a map between the points of the physical manifold and those of the background one which allows us to define the difference between tensors defined on the two manifolds. Suppose we change gauge from a $\mathcal{G}$ to a $\hat{\mathcal{G}}$. Metric (4.3) then becomes:

$$g_{\mu\nu} = a^2(\eta)\begin{Bmatrix} -[1 + 2\hat{\psi}(\eta, \mathbf{x})] & \hat{w}_i(\eta, \mathbf{x}) \\ \hat{w}_i(\eta, \mathbf{x}) & \delta_{ij}[1 + 2\hat{\phi}(\eta, \mathbf{x})] + \hat{\chi}_{ij}(\eta, \mathbf{x}) \end{Bmatrix}, \qquad \delta^{ij}\hat{\chi}_{ij} = 0. \tag{4.86}$$

We have new hatted functions representing perturbations depending again on the background coordinates, which we have fixed.

A similar map can be defined also on the single background manifold, so in absence of perturbations, and it can be deceiving, in the sense that quantities similar to perturbations might appear even if we are in the background manifold. Let us rephrase this. The problem when considering fluctuations in GR is that we cannot be sure, by only looking at a metric, that there are real fluctuations about a known background or it is the metric which is written in a not very appropriate coordinate system. For example, consider the following time transformation for the FLRW metric:

$$d\eta = g(t, \mathbf{x})dt. \tag{4.87}$$

Then, the FLRW metric becomes:

$$ds^2 = -a(t, \mathbf{x})^2 g(t, \mathbf{x})^2 dt^2 + a(t, \mathbf{x})^2 \delta_{ij}dx^i dx^j, \tag{4.88}$$

and recalling back $t \to \eta$ we get:

$$ds^2 = -a(\eta, \mathbf{x})^2 g(\eta, \mathbf{x})^2 d\eta^2 + a(\eta, \mathbf{x})^2 \delta_{ij}dx^i dx^j. \tag{4.89}$$

Now the metric coefficients depend on the position so we may think that homogeneity and isotropy are lost, but it was just a coordinate transformation. This *is not* the problem of the gauge, but is the general covariance typical of GR. As we show above, it is not even necessarily related to perturbations but it is just a coordinate transformation masking the metric in the form that we are used to see it. Computing relativistic invariants or Killing vectors allows to determine if the above is the FLRW metric or not.

The problem of the gauge is the dependence of the perturbations on the gauge. In the following we will see how a gauge transformation manifests itself through a change of coordinates and deal with the problem of the gauge by introducing **gauge-invariant variables**. Loosely speaking, we will see that the gauge is the *functional dependence* of the perturbative quantities on the coordinates, whereas the background quantities maintain their functional form.

For a more complete treatment of the gauge problem, see for example Stewart (1990), Mukhanov et al. (1992) and Malik and Matravers (2013).

4.3.1 Coordinates and Gauge Transformations

In order to understand how a gauge transformation changes the functional dependence of the perturbative quantities we express the change from a gauge $\mathcal{G}$ to a gauge $\hat{\mathcal{G}}$ as the following infinitesimal coordinate transformation:

$$x^\mu \to \hat{x}^\mu = x^\mu + \xi^\mu(x) , \tag{4.90}$$

where x^μ are the background coordinates and ξ^μ is a generic vector field, **the gauge generator**, which must be $|\xi^\mu| \ll 1$ in order to preserve the smallness of the perturbation. From a geometric point of view, we fix a point on the background manifold and by changing gauge we change the corresponding point on the physical manifold. This point has coordinates different from the first one and given by Eq. (4.90). The gauge generator can be seen then as a vector field on the physical manifold and the Lie derivative of the metric along ξ tells us how the perturbative quantities change their functional form.

Under a coordinate transformation, the metric tensor $g_{\mu\nu}$, as well as any other tensor of the same rank, transforms in the following way:

$$g_{\mu\nu}(x) = \frac{\partial \hat{x}^\rho}{\partial x^\mu} \frac{\partial \hat{x}^\sigma}{\partial x^\mu} \hat{g}_{\rho\sigma}(\hat{x}) , \tag{4.91}$$

which, using Eq. (4.90), can be cast as follows:

$$g_{\mu\nu}(x) = \left(\delta^\rho{}_\mu + \partial_\mu \xi^\rho\right)\left(\delta^\sigma{}_\nu + \partial_\nu \xi^\sigma\right)\hat{g}_{\rho\sigma}(\hat{x}) . \tag{4.92}$$

Writing down Eq. (4.92) up to first order, one obtains:

$$g_{\mu\nu}(x) = \hat{g}_{\mu\nu}(x) + \partial_\alpha \hat{g}_{\mu\nu}(x)\xi^\alpha + \partial_\mu \xi^\rho \hat{g}_{\rho\nu}(x) + \partial_\nu \xi^\rho \hat{g}_{\rho\mu}(x) , \tag{4.93}$$

where we have also expanded $\hat{g}_{\rho\sigma}(\hat{x})$ about x using Eq. (4.90).

Exercise 4.15 Show that Eq. (4.93) can be cast in the following form:

$$g_{\mu\nu}(x) = \hat{g}_{\mu\nu}(x) + \partial_\alpha g_{\mu\nu}(x)\xi^\alpha + \partial_\mu \xi^\rho g_{\rho\nu}(x) + \partial_\nu \xi^\rho g_{\rho\mu}(x) , \tag{4.94}$$

i.e. prove that we can remove the hat from the metric when it is multiplied by ξ. Then, cast the above equation as follows:

$$g_{\mu\nu}(x) = \hat{g}_{\mu\nu}(x) + \nabla_\nu \xi_\mu + \nabla_\mu \xi_\nu . \tag{4.95}$$

This equation shows how the functional form of the metric components, i.e. the gauge, changes upon a coordinate transformation.

If Eq. (4.90) is an isometry, i.e. $g_{\mu\nu}(x) = \hat{g}_{\mu\nu}(x)$, one gets the **Killing equation**:

$$\nabla_\nu \xi_\mu + \nabla_\mu \xi_\nu = 0 . \tag{4.96}$$

Now use the perturbed FLRW metric (4.3) in Eq. (4.93).

Exercise 4.16 Show that at the order zero:

$$\boxed{\hat{a}(\eta) = a(\eta)} \tag{4.97}$$

i.e. the scale factor maintains its functional form, confirming the property of the gauge, which maintains the functional form of the background quantities. Then, show that at first-order the following transformation relations hold:

$$\boxed{\hat{\psi} = \psi - \mathcal{H}\xi^0 - \xi^{0\prime}} \qquad \boxed{\hat{w}_i = w_i - \zeta_i' + \partial_i \xi^0} \tag{4.98}$$

$$\boxed{\hat{\phi} = \phi - \mathcal{H}\xi^0 - \frac{1}{3}\partial_l \xi^l} \qquad \boxed{\hat{\chi}_{ij} = \chi_{ij} - \partial_j \zeta_i - \partial_i \zeta_j + \frac{2}{3}\delta_{ij}\partial_l \xi^l} \tag{4.99}$$

where $\zeta_i \equiv \delta_{il}\xi^l$. We have introduced ζ_i in order not to make confusion with the spatial part of ξ_μ, which is $\xi_i = a^2 \delta_{il}\xi^l = a^2 \zeta_i$.

In the very same fashion we adopted for the metric, we can also find the transformation rules for the components of the energy-momentum tensor. That is, through the same steps that we have just used for the metric, we can write:

$$\hat{T}_{\mu\nu}(x) = T_{\mu\nu}(x) - \partial_\alpha T_{\mu\nu}(x)\xi^\alpha - \partial_\mu \xi^\rho T_{\rho\nu}(x) - \partial_\nu \xi^\rho T_{\rho\mu}(x) \ . \qquad (4.100)$$

Exercise 4.17 Use the above transformation with $T_{\mu\nu}$ given by Eqs. (4.68)–(4.70). Find that at zeroth order one has:

$$\boxed{\hat{\bar{\rho}}(\eta) = \bar{\rho}(\eta)} \qquad \boxed{\hat{\bar{P}}(\eta) = \bar{P}(\eta)} \qquad (4.101)$$

i.e. the background density and pressure maintain their functional forms. Then, show that at first order the following transformation relations hold:

$$\boxed{\hat{\delta\rho} = \delta\rho - \bar{\rho}'\xi^0} \qquad \boxed{\hat{v}_i = v_i + \partial_i \xi^0} \qquad \boxed{\hat{q}_i = q_i} \qquad (4.102)$$

$$\boxed{\hat{\pi} = \pi} \qquad \boxed{\hat{\delta P} = \delta P - \bar{P}'\xi^0} \qquad \boxed{\hat{\pi}_{ij} = \pi_{ij}} \qquad (4.103)$$

Hint. In order to find these relations you have to use those for the metric quantities. Moreover, one obtains $\hat{q}_i = q_i$ noticing that $\bar{\rho} + \bar{P}$ is arbitrary. One the other hand, we do not have a mathematical way to separate the transformations for δP and π. We do that by giving the physical argument by which π is related to dissipative processes whereas δP is not.

In general, perturbations of quantities which are vanishing or constant in the background are automatically gauge-invariant and one can see this explicitly above for the heat flux, the bulk viscosity and the anisotropic stress. This property is formalised in the **Stewart-Walker lemma** (Stewart and Walker 1974). It stimulates for example the use of the perturbed Weyl tensor (which is vanishing in FLRW metric) and the quasi-Maxwellian equations (Hawking 1966; Jordan et al. 2009).

4.3.2 The Scalar-Vector-Tensor Decomposition

The scalar-vector-tensor (SVT) decomposition was introduced by Lifshitz in 1946 (Lifshitz 1946), who was the first to address cosmological perturbations. See also Lifshitz and Khalatnikov (1963) and Ma and Bertschinger (1995), for a particularly detailed account. It consists of the following procedure. We have already seen that the perturbed metric can be written in terms of two scalars ψ and ϕ, a 3-vector w_i and a 3-tensor χ_{ij}. However, we can "squeeze out" two more scalars from w_i and χ_{ij} and one more vector from χ_{ij}.

Helmholtz theorem (see Sect. 12.3 for a brief reminder) states that, under certain conditions of regularity, any spatial vector w_i can be uniquely decomposed in its longitudinal part plus its orthogonal contribution:

$$w_i = w_i^\parallel + w_i^\perp , \qquad (4.104)$$

which are respectively irrotational and solenoidal (divergenceless), namely:

$$\epsilon^{ijk} \partial_j w_k^\parallel = 0 , \qquad \partial^k w_k^\perp = 0 , \qquad (4.105)$$

where ϵ^{ijk} is the Levi-Civita symbol.[5] By Stokes theorem, the irrotational part can be written as the gradient of a scalar say w so that, finally, we can write w_i as follows:

$$\boxed{w_i = \partial_i w + S_i} \qquad (4.106)$$

where we have defined $S_i \equiv w_i^\perp$, because it is simpler to write. Therefore, w is the **scalar** part of w_i and S_i is the **vector** part of w_i. Usually, when in cosmology one talks about a **vector perturbation** one is referring to S_i, i.e. to a vector which cannot be written as a gradient of a scalar.

Similarly to the vector case, any spatial rank-2 tensor say χ_{ij} can be decomposed in its longitudinal part $\chi_{ij}^\parallel$ plus its orthogonal part $\chi_{ij}^\perp$ plus the transverse contribution χ_{ij}^T:

$$\chi_{ij} = \chi_{ij}^\parallel + \chi_{ij}^\perp + \chi_{ij}^T , \qquad (4.107)$$

defined as follows:

$$\epsilon^{ijk} \partial^l \partial_j \chi_{lk}^\parallel = 0 , \qquad \partial^i \partial^j \chi_{ij}^\perp = 0 , \qquad \partial^i \chi_{ij}^T = 0 . \qquad (4.108)$$

Basically, one builds a vector by taking the divergence of χ_{ij} and then applies Helmholtz theorem to it. This implies that the longitudinal and the orthogonal parts can be further decomposed in the same spirit of Eq. (4.105) in the following way:

$$\chi_{ij}^\parallel = \left(\partial_i \partial_j - \frac{1}{3} \delta_{ij} \nabla^2 \right) 2\mu , \qquad \chi_{ij}^\perp = \partial_j A_i + \partial_i A_j , \qquad \partial^i A_i = 0 , \qquad (4.109)$$

where μ is a scalar, A_i is a divergenceless vector and recall that $\nabla^2 \equiv \delta^{lm} \partial_l \partial_m$. We can thus write χ_{ij} in the following form:

$$\boxed{\chi_{ij} = \left(\partial_i \partial_j - \frac{1}{3} \delta_{ij} \nabla^2 \right) 2\mu + \partial_j A_i + \partial_i A_j + \chi_{ij}^T} \qquad (4.110)$$

---

[5] Reminder: $\epsilon^{123} = 1$ and it changes sign upon any odd permutation of its indices. It follows that $\epsilon^{ijk} = 0$ if two or more indices are equal.

The transverse part χ_{ij}^T cannot be decomposed in any scalar or divergenceless vector. Therefore, it constitutes a **tensor perturbation**.

The SVT decomposition is a fundamental tool for the investigation of first order perturbations because the three classes do not mix up and therefore they can be independently analysed. The absence of mixing is due to the fact that any kind of interaction term among the three categories would be of second order and therefore negligible.

Let us see how each class of perturbations transforms. Apply Helmholtz theorem also to the spatial part of ξ^μ as follows:

$$\xi^0 \equiv \alpha , \qquad \zeta_i = \partial_i \beta + \epsilon_i , \qquad (\partial^l \epsilon_l = 0) , \qquad (4.111)$$

where α and β are scalars and ϵ^i is a divergenceless vector. Now let us write the transformations found in Eqs. (4.98) and (4.99) using the SVT decomposition:

$$\hat{\psi} = \psi - \mathcal{H}\alpha - \alpha' , \qquad (4.112)$$

$$\partial_i \hat{w} + \hat{S}_i = \partial_i w + S_i - \partial_i \beta' - \epsilon_i' + \partial_i \alpha , \qquad (4.113)$$

$$\hat{\phi} = \phi - \mathcal{H}\alpha - \frac{1}{3}\nabla^2 \beta , \qquad (4.114)$$

$$\left(\partial_i \partial_j - \frac{1}{3}\delta_{ij}\nabla^2\right) 2\hat{\mu} + \partial_j \hat{A}_i + \partial_i \hat{A}_j + \hat{\chi}_{ij}^T = \left(\partial_i \partial_j - \frac{1}{3}\delta_{ij}\nabla^2\right) 2\mu$$

$$+\partial_j A_i + \partial_i A_j + \chi_{ij}^T - 2\partial_j \partial_i \beta - \partial_j \epsilon_i - \partial_i \epsilon_j + \frac{2}{3}\delta_{ij}\nabla^2 \beta . \qquad (4.115)$$

We are now in the position of writing explicitly the transformation rules for each class of perturbation.

Scalar Perturbations and Their Gauge-Invariant Combinations

By taking the divergence ∂^i of Eq. (4.113) and twice the divergence $\partial^i \partial^j$ of Eq. (4.115), we eliminate all the vector and tensor contributions and are left with the transformation equations for the scalar perturbations only:

$$\hat{\psi} = \psi - \mathcal{H}\alpha - \alpha' , \qquad (4.116)$$

$$\nabla^2 \hat{w} = \nabla^2 \left(w - \beta' + \alpha\right) , \qquad (4.117)$$

$$\hat{\phi} = \phi - \mathcal{H}\alpha - \frac{1}{3}\nabla^2 \beta , \qquad (4.118)$$

$$\nabla^2 \nabla^2 \hat{\mu} = \nabla^2 \nabla^2 (\mu - \beta) . \qquad (4.119)$$

From the above transformations we obtain:

$$\boxed{\hat{\psi} = \psi - \mathcal{H}\alpha - \alpha'} \tag{4.120}$$

$$\boxed{\hat{w} = w - \beta' + \alpha} \tag{4.121}$$

$$\boxed{\hat{\phi} = \phi - \mathcal{H}\alpha - \frac{1}{3}\nabla^2\beta} \tag{4.122}$$

$$\boxed{\hat{\mu} = \mu - \beta} \tag{4.123}$$

In principle, we should have extra functions in the second and fourth equations, coming from the integration of ∇^2 and $\nabla^2\nabla^2$, but these are **spurious gauge modes** which can be set to zero without losing of generality.

The following combinations of scalar perturbations are gauge-invariant:

$$\boxed{\Psi = \psi + \frac{1}{a}\left[(w - \mu')a\right]'} \quad \boxed{\Phi = \phi + \mathcal{H}(w - \mu') - \frac{1}{3}\nabla^2\mu} \tag{4.124}$$

They are the famous **Bardeen's potentials** (Bardeen 1980).

Exercise 4.18 Prove that the Bardeen potentials are gauge-invariant. Why are there only two of them?

The same technique that we have just used for the metric perturbations can be applied to the matter quantities in Eqs. (4.102) and (4.103).

Exercise 4.19 Applying the SVT decomposition to v_i:

$$v_i = \partial_i v + U_i, \qquad (\partial^l U_l = 0), \tag{4.125}$$

show that, for scalar perturbations, one gets:

$$\boxed{\hat{\delta\rho} = \delta\rho - \bar{\rho}'\alpha} \quad \Rightarrow \quad \boxed{\hat{\delta} = \delta + 3\mathcal{H}(1 + \bar{P}/\bar{\rho})\alpha} \tag{4.126}$$

$$\boxed{\hat{v} = v + \alpha} \tag{4.127}$$

$$\boxed{\hat{\delta P} = \delta P - \bar{P}'\alpha} \tag{4.128}$$

Note how a cosmological constant has gauge invariant perturbations.

As we did earlier for the geometric quantities, we can combine the above transformations for matter in order to obtain gauge-invariant variables. The strategy is, in general, to combine the transformations in order to eliminate α and β. The result is then manifestly gauge-invariant. Using matter quantities only, we can eliminate α

from Eqs. (4.126) and (4.128), thus obtaining the following gauge-invariant perturbation:

$$\Gamma \equiv \delta P - \frac{\bar{P}'}{\bar{\rho}'}\delta\rho \tag{4.129}$$

This is the **entropy perturbation**. The ratio $\delta P/\delta\rho$ is called **effective speed of sound**, whereas the ratio $\bar{P}'/\bar{\rho}'$ is called **adiabatic speed of sound**. When the two are equal, i.e. $\Gamma = 0$ one finds $dS = 0$, i.e. one has adiabaticity.

The other gauge-invariant combinations are:

$$\delta\rho_m^{(gi)} \equiv \bar{\rho}\Delta \equiv \delta\rho + \bar{\rho}'v \qquad \delta P_m^{(gi)} \equiv \delta P + \bar{P}'v \tag{4.130}$$

i.e. the gauge-invariant density, also called **comoving-gauge density perturbation**, and pressure perturbations. The subscript m refers to "matter", since it is also possible to build gauge-invariant perturbations of the density and pressure using metric quantities. We are borrowing this notation from Bardeen (1980).

We can now think of combining the geometric and matter transformations, a total of 7 relations, trying to eliminate α and β in order to create new gauge-invariant variables.

Indeed, we can form the so-called **comoving curvature perturbation**

$$\mathcal{R} \equiv \phi + \mathcal{H}v - \frac{1}{3}\nabla^2\mu \tag{4.131}$$

also known as **Lukash variable** (Lukash 1980), and we can form the quantity:

$$\zeta \equiv \phi + \frac{\delta\rho}{3(\bar{\rho} + \bar{P})} - \frac{1}{3}\nabla^2\mu \tag{4.132}$$

which was introduced first in Bardeen et al. (1983) but started to be exploited in Wands et al. (2000). These $\mathcal{R}$ and ζ are especially important in the framework of inflation because they are conserved on large scales and for adiabatic perturbations, as noticed in Bardeen (1980) (at least for $\mathcal{R}$), and as we shall prove in Sect. 12.4.

Again, we can form gauge-invariant density, velocity and pressure perturbations:

$$\delta\rho_g^{(gi)} \equiv \delta\rho + \bar{\rho}'\left(w - \mu'\right) \qquad v^{(gi)} \equiv v - (w - \mu') \qquad \delta P_g^{(gi)} \equiv \delta P + \bar{P}'\left(w - \mu'\right) \tag{4.133}$$

In general, having 2 gauge variables α and β and 7 transformations, we can build 5 independent scalar gauge-invariant perturbations.

Vector Perturbations and Their Gauge-Invariant Combinations

We can now eliminate the scalar contribution from Eq. (4.113) and consider the divergence ∂^j of Eq. (4.115). In this way we shall find the transformations for vector perturbations:

$$\boxed{\hat{S}_i = S_i - \epsilon_i'} \tag{4.134}$$

$$\nabla^2 \hat{A}_i = \nabla^2 (A_i - \epsilon_i) . \tag{4.135}$$

From the second equation, we can find the following transformation:

$$\boxed{\hat{A}_i = A_i - \epsilon_i} \tag{4.136}$$

It is possible to define a new gauge-invariant vector potential, which has the following form:

$$\boxed{W_i \equiv S_i - A_i'} \tag{4.137}$$

Exercise 4.20 Prove that W_i is gauge-invariant. Why is there only one gauge-invariant vector perturbation?

Using the SVT decomposition of Eq. (4.125), show that from the matter sector we just have:

$$\hat{U}_l = U_l , \tag{4.138}$$

i.e. the vector contribution of v_i is already gauge-invariant.

Tensor Perturbations

Since an infinitesimal gauge transformation, cf. Eq. (4.90), cannot be realised by any rank-2 tensor, the following result is not unexpected:

$$\boxed{\hat{\chi}_{ij}^T = \chi_{ij}^T} \tag{4.139}$$

i.e. that the transverse part of χ_{ij} is already gauge-invariant.

Summary

We have thus seen that a generic perturbation of the metric can be split in:

- 4 scalar functions;
- 2 divergenceless 3-vectors, for a total of 4 independent components (2 each);
- A transverse, traceless spatial tensor of rank 2, χ_{ij}^T. It has 2 independent components.

The total number of independent components sums up to 10, as it should be.

Exercise 4.21 Why does χ_{ij}^T have two independent components only?

The above decomposition holds true not only for the metric but for any rank-2 symmetric tensor.

4.3.3 Gauges

Thanks to gauge freedom we can set any 4 components of the metric to zero. There are two particularly useful choices: the synchronous gauge and the Newtonian gauge.

Synchronous gauge. This is realised by the choice:

$$\hat\psi = 0, \qquad \hat w_i = 0. \tag{4.140}$$

Note that $\hat w_i = 0$ means that both the scalar and the vector part of w_i are being set to zero. Using the transformations found in the previous subsection, we find:

$$\begin{cases} \psi - \mathcal{H}\alpha - \alpha' = 0 \\[2mm] w - \beta' + \alpha = 0 \\[2mm] S_i - \epsilon_i' = 0 \end{cases} \tag{4.141}$$

This must be interpreted as a system of equations for the unknowns α, β and ϵ_i, i.e. from a generic gauge we want to know which transformations we have to perform in order to go to the synchronous gauge. We have 4 equation for 4 unknowns, so we expect to determine a single solution. However, the above equations are differential and this implies the following:

$$\begin{cases} \alpha = \frac{1}{a} \int d\eta \, a\psi + f(\mathbf{x}) \\[2mm] \beta = \int d\eta \, (w + \alpha) + g(\mathbf{x}) \\[2mm] \epsilon_i = \int d\eta \, S_i + h_i(\mathbf{x}) \end{cases} \tag{4.142}$$

Since we have only time derivatives, the integrations give rise to purely space-dependent functions, which we have called f, g and h here, and which are **spurious gauge modes**.

Newtonian gauge. This is realised by the choice:

$$\hat{w} = 0 , \qquad \hat{\mu} = 0 , \qquad \hat{\chi}_{ij}^{\perp} = 0 . \tag{4.143}$$

Using the transformations found in the previous subsection, we find:

$$
\begin{cases}
w - \beta' + \alpha = 0 \\[2mm]
\mu - \beta = 0 \\[2mm]
A_i - \epsilon_i = 0
\end{cases}
. \tag{4.144}
$$

It easy to see that the second and the third equation are algebraic and determine β and ϵ_i. Substituting the solution for β into the first equation, we then find α. There is no integration to perform, therefore no spurious gauge mode appears.

An important fact that makes the conformal Newtonian gauge somewhat special is that $\hat{\psi} = \Psi$ and $\hat{\phi} = \Phi$, i.e. the metric perturbations become identical to the Bardeen potentials. These lecture notes are based on the use of the Newtonian gauge.

Transformations Between the Two Gauges

It is useful to provide the transformation rules among the metric perturbations in the two gauges, for those readers who might want to translate into the synchronous gauge the results of these notes and comparing them with the huge literature in which this gauge is employed. For the scalar case, using Eqs. (4.120)–(4.123) and assuming that the hatted quantities are the synchronous ones whereas the non-hatted perturbations are the conformal Newtonian ones, we get:

$$0 = \psi_N - \mathcal{H}\alpha - \alpha' , \tag{4.145}$$

$$0 = 0 - \beta' + \alpha , \tag{4.146}$$

$$\phi_S = \phi_N - \mathcal{H}\alpha - \frac{1}{3}\nabla^2\beta , \tag{4.147}$$

$$\mu_S = -\beta . \tag{4.148}$$

The second and the fourth equation completely specify the transformation in terms of μ_S, i.e. we have:

$$\boxed{\alpha = \beta' = -\mu_S'} \tag{4.149}$$

and the metric potentials are related by:

$$\boxed{\psi_N = -\mathcal{H}\mu_S' - \mu_S''} \qquad \boxed{\phi_N = \phi_S - \mathcal{H}\mu_S' - \frac{1}{3}\nabla^2\mu_S} \tag{4.150}$$

In the literature, see e.g. Ma and Bertschinger (1995), the Fourier transforms of ϕ_S and μ_S are usually named h and 6η, respectively.

For vector perturbations we have:

$$0 = S_N^i - \epsilon^{i'} \,, \tag{4.151}$$

$$A_S^i = 0 - \epsilon^i \,, \tag{4.152}$$

from which it is straightforward to obtain:

$$S_N^i = -A_S^{i'} \,. \tag{4.153}$$

Of course, tensor perturbations are naturally gauge-invariant and thus have the same functional form in the two gauges. This is also true for any tensor of rank equal or higher than 2.

4.4   Normal Mode Decomposition

We are now in the position of writing down explicitly the Einstein equations, fixing a gauge of our choice, which will be the Newtonian one. We expect these equations to be linear second order partial differential equations, given the perturbation scheme employed. Because of this, it is very convenient to express the perturbations as superpositions of normal modes, i.e. the eigenmodes $Q(\mathbf{k}, \mathbf{x})$ of the Laplacian operator, defined via the Helmholtz equation:

$$\nabla^2 Q(\mathbf{k}, \mathbf{x}) = -k^2 Q(\mathbf{k}, \mathbf{x}) \,. \tag{4.154}$$

For flat spatial slicing, which is considered here, this normal mode decomposition of course amounts to a Fourier transform, i.e.

$$Q(\mathbf{k}, \mathbf{x}) = e^{i\mathbf{k}\cdot\mathbf{x}} \,, \tag{4.155}$$

and a given quantity $X(\eta, \mathbf{x})$ is expressed as:

$$\tilde{X}(\eta, \mathbf{k}) = \int d^3\mathbf{x}\, X(\eta, \mathbf{x}) e^{-i\mathbf{k}\cdot\mathbf{x}} \,, \qquad X(\eta, \mathbf{x}) = \int \frac{d^3\mathbf{k}}{(2\pi)^3}\, \tilde{X}(\eta, \mathbf{k}) e^{i\mathbf{k}\cdot\mathbf{x}} \,. \tag{4.156}$$

Consider for example the metric perturbations ψ, ϕ, w_i and χ_{ij}. For ψ and ϕ we simply have that:

$$\psi(\eta, \mathbf{x}) = \int \frac{d^3\mathbf{k}}{(2\pi)^3}\, \tilde{\psi}(\eta, \mathbf{k}) e^{i\mathbf{k}\cdot\mathbf{x}} \,, \qquad \phi(\eta, \mathbf{x}) = \int \frac{d^3\mathbf{k}}{(2\pi)^3}\, \tilde{\phi}(\eta, \mathbf{k}) e^{i\mathbf{k}\cdot\mathbf{x}} \,, \tag{4.157}$$

and similar expressions also apply for $\delta\rho$ and δP.

Using its SVT decomposition in Eq. (4.106), we have for w_i the following Fourier transformation:

$$w_i(\eta, \mathbf{x}) = \int \frac{d^3\mathbf{k}}{(2\pi)^3} \, \tilde{w}_i(\eta, \mathbf{k}) e^{i\mathbf{k}\cdot\mathbf{x}} = \int \frac{d^3\mathbf{k}}{(2\pi)^3} \, [\tilde{\partial_i w}(\eta, \mathbf{k}) + \tilde{S}_i(\eta, \mathbf{k})] e^{i\mathbf{k}\cdot\mathbf{x}} \, . \tag{4.158}$$

The Fourier transform of a partial spatial derivative is

$$\tilde{\partial_i w}(\eta, \mathbf{k}) = i k_i \tilde{w}(\eta, \mathbf{k}) \, , \tag{4.159}$$

and therefore we have:

$$w_i(\eta, \mathbf{x}) = i \int \frac{d^3\mathbf{k}}{(2\pi)^3} \, k_i \tilde{w}(\eta, \mathbf{k}) e^{i\mathbf{k}\cdot\mathbf{x}} + \int \frac{d^3\mathbf{k}}{(2\pi)^3} \, \tilde{S}_i(\eta, \mathbf{k}) e^{i\mathbf{k}\cdot\mathbf{x}} \, . \tag{4.160}$$

So we see that the Fourier transforms of ψ or ϕ and w are not treated on an equal footing, because $\tilde{w}$ is multiplied by a k_i. A similar argument goes also for v_i.

Finally, doing the same for χ_{ij} in Eq. (4.110), one gets:

$$\chi_{ij}(\eta, \mathbf{x}) = \int \frac{d^3\mathbf{k}}{(2\pi)^3} \, \left(-k_i k_j + \frac{1}{3}\delta_{ij} k^2 \right) 2\tilde{\mu}(\eta, \mathbf{k}) e^{i\mathbf{k}\cdot\mathbf{x}}$$
$$+ i \int \frac{d^3\mathbf{k}}{(2\pi)^3} \, [k_j \tilde{A}_i(\eta, \mathbf{k}) + k_i \tilde{A}_j(\eta, \mathbf{k})] e^{i\mathbf{k}\cdot\mathbf{x}} + \int \frac{d^3\mathbf{k}}{(2\pi)^3} \, \tilde{\chi}_{ij}^T(\eta, \mathbf{k}) e^{i\mathbf{k}\cdot\mathbf{x}} \, , \tag{4.161}$$

with a similar expansion holding true for $\pi^i{}_j$. We see that $\tilde{\mu}$ is multiplied by a factor k^2 and $\tilde{A}_i$ is multiplied by a factor k. Therefore, in order to properly compare perturbations, *par condicio* is restored by "correcting" as follows:

$$\tilde{w} \equiv -\frac{1}{k}\tilde{B} \, , \quad \tilde{v} \equiv -\frac{1}{k}\tilde{V} \, , \quad \tilde{\mu} \equiv \frac{1}{k^2}\tilde{E} \, , \quad \tilde{A}_i \equiv -\frac{1}{2k}\tilde{F}_i \, . \tag{4.162}$$

In this way, all scalar quantities are treated on the same footing. Let us see what happens to the comoving gauge density perturbation:

$$\boxed{\tilde{\Delta} = \tilde{\delta} + 3(1 + \bar{P}/\bar{\rho})\frac{\mathcal{H}}{k}\tilde{V}} \tag{4.163}$$

Since $\mathcal{H} \propto 1/\eta$, on sub-horizon scales, i.e. for $k\eta \gg 1$, the density contrast becomes gauge-invariant. We present this fact in Fig. 4.1, where we plot the evolution of the modulus of the CDM density contrast δ_c as function of k and for $z = 0$, computed with CLASS (Lesgourgues 2011) in the synchronous (solid line) and Newtonian (dashed line) gauges.

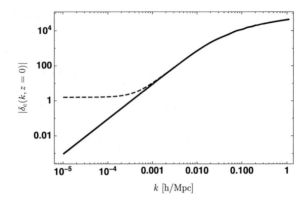

Fig. 4.1 Plot of the modulus of the CDM density contrast at $z = 0$ (today) as function of k, using CLASS. The solid line is the result obtained using the synchronous gauge whereas the dashed one is obtained using the Newtonian one. The initial conditions are adiabatic and normalised in order to have $\mathcal{R} = 1$. All the cosmological parameters have been set, as default, corresponding to the Planck best fit of the ΛCDM model (Ade et al. 2016a)

The plots in Fig. 4.1 are drawn for $z = 0$, hence the value of the Hubble parameter is the Hubble constant, $H_0 = 3 \times 10^{-4} \, h \, \text{Mpc}^{-1}$. Indeed, when $k > H_0$ the two evolutions coincide.

Now we have to understand how to extract the scalar and vector part from a full 3-vector quantity. Let us reformulate the FT of w_i as follows:

$$\tilde{w}_i(\eta, \mathbf{k}) = -i\hat{k}_i \tilde{B}(\eta, \mathbf{k}) + \tilde{S}_i(\eta, \mathbf{k}) \ . \tag{4.164}$$

We see that we can isolate the scalar part of the FT transform of a 3-vector perturbation by contracting it with $i\hat{k}^i$. Indeed:

$$i\hat{k}^i \tilde{w}_i(\eta, \mathbf{k}) = \tilde{B}(\eta, \mathbf{k}) \ , \tag{4.165}$$

because $\hat{k}^i \tilde{S}_i = 0$, since $\partial^i S_i = 0$. The vector part is therefore obtained by subtracting the scalar part:

$$\tilde{S}_i = \left(\delta^j{}_i - \hat{k}^j \hat{k}_i\right) \tilde{w}_j \ . \tag{4.166}$$

We can easily check that this formula satisfies $\hat{k}^i \tilde{S}_i = 0$.

What about a tensorial quantity? For the traceless spatial metric perturbation in Eq. (4.110), we can write:

$$\tilde{\chi}_{ij} = -\left(\hat{k}_i \hat{k}_j - \frac{1}{3}\delta_{ij}\right) 2\tilde{E} - \frac{i}{2}\hat{k}_j \tilde{F}_i - \frac{i}{2}\hat{k}_i \tilde{F}_j + \tilde{\chi}_{ij}^T \ . \tag{4.167}$$

Here the scalar contribution can be isolated by contracting with $-3\hat{k}^i \hat{k}^j/2$. In fact:

$$-\frac{3}{2}\hat{k}^i\hat{k}^j\tilde{\chi}_{ij} = 2\tilde{E} .\tag{4.168}$$

Exercise 4.22 Show that the vector contribution is obtained contracting once with $2i\hat{k}^i$ and using Eq. (4.168), i.e.

$$\tilde{F}_i = 2i\left(\delta^j{}_i - \hat{k}^j\hat{k}_i\right)\hat{k}^l\tilde{\chi}_{jl} .\tag{4.169}$$

For the tensor part, show that:

$$\tilde{\chi}^T_{ij} = \left(\delta^l{}_i - \hat{k}^l\hat{k}_i\right)\left(\delta^m{}_j - \hat{k}^m\hat{k}_j\right)\tilde{\chi}_{lm} + \frac{1}{2}(\delta_{ij} - \hat{k}_i\hat{k}_j)\hat{k}^l\hat{k}^m\tilde{\chi}_{lm} .\tag{4.170}$$

Verify that $\hat{k}^i\tilde{\chi}^T_{ij} = \hat{k}^j\tilde{\chi}^T_{ij} = 0$ and $\delta^{ij}\tilde{\chi}^T_{ij} = 0$. Recall that $\tilde{\chi}_{ij}$ is already traceless.

4.5 Einstein Equations for Scalar Perturbations

In this section we focus on scalar perturbations and employ the Newtonian gauge. Our perturbed metric is then:

$$g_{00} = -a(\eta)^2[1 + 2\Psi(\eta, \mathbf{x})] , \quad g_{0i} = 0 , \quad g_{ij} = a(\eta)^2\delta_{ij}[1 + 2\Phi(\eta, \mathbf{x})] ,\tag{4.171}$$

where we are employing the Bardeen potentials, exploiting the fact that $w = \mu = 0$. In metric (4.171) Ψ plays the role of the Newtonian potential and Φ is the spatial curvature perturbation.

Exercise 4.23 Calculate from scratch the perturbed Einstein tensor $\delta G^\mu{}_\nu$ from metric (4.171), AND do it again using also Eqs. (4.42)–(4.44). Show that:

$$a^2\delta G^0{}_0 = -6\mathcal{H}\Phi' + 6\mathcal{H}^2\Psi + 2\nabla^2\Phi ,\tag{4.172}$$

$$a^2\delta G^0{}_i = 2\partial_i(\Phi' - \mathcal{H}\Psi) ,\tag{4.173}$$

$$a^2\delta G^i{}_j = \left[-2\Phi'' - 4\mathcal{H}\Phi' + 2\mathcal{H}\Psi' + 4\frac{a''}{a}\Psi - 2\mathcal{H}^2\Psi + \nabla^2(\Phi + \Psi)\right]\delta^i{}_j$$

$$-\partial^i\partial_j(\Phi + \Psi) .\tag{4.174}$$

Note again that $\partial^i = \partial_i$ since it is the partial derivative with respect to comoving coordinates and $\nabla^2 = \delta^{lm}\partial_l\partial_m$ is the comoving Laplacian. Being comoving, it is always accompanied by a factor $1/a^2$, since together they form the physical Laplacian.

4.5.1 The Relativistic Poisson Equation

The 0–0 Einstein equation is the following:

$$-3\mathcal{H}\Phi' + 3\mathcal{H}^2\Psi + \nabla^2\Phi = 4\pi Ga^2 \delta T^0{}_0 . \tag{4.175}$$

Using the perturbed energy momentum tensor component $\delta T^0{}_0$, as we can read from Eq. (4.72), we have:

$$\boxed{3\mathcal{H}\Phi' - 3\mathcal{H}^2\Psi - \nabla^2\Phi = 4\pi Ga^2 \delta\rho_{\text{tot}}} \tag{4.176}$$

where the total perturbed density is

$$\delta\rho_{\text{tot}} = \sum_i \delta\rho_i = \sum_i \rho_i\delta_i , \tag{4.177}$$

i.e. it is the sum of the perturbed densities of all the material components that constitute our cosmological model, in the same way that we did for the background. We start here to eliminate the bar over the background quantities. We shall consider the ΛCDM as our standard cosmological model. Thus, we have to deal with 4 contributions:

1. Photons;
2. Neutrinos;
3. CDM;
4. Baryons.

Each of these has its own energy-momentum tensor and the total one, which enters the right hand side of the Einstein equations, is their sum. The cosmological constant only contributes at background level.

We have found above the **relativistic Poisson equation**. Indeed, if we consider $a = 1$, and hence $\mathcal{H} = 0$, we recover the usual Newtonian Poisson equation (or almost, since we have two potentials in GR). Written in terms of the density contrast, using Eq. (4.72), the relativistic Poisson equation is the following:

$$\boxed{3\mathcal{H}\Phi' - 3\mathcal{H}^2\Psi - \nabla^2\Phi = 4\pi Ga^2 \left(\rho_c\delta_c + \rho_b\delta_b + \rho_\gamma\delta_\gamma + \rho_\nu\delta_\nu\right)} \tag{4.178}$$

As one can see, Eq. (4.178) is a second order partial differential equation (PDE) which is *linear*, because we are doing first-order perturbation theory and thus all the perturbative variables appear with power 1. Because of this, as we anticipated, it is very convenient to introduce the Fourier transform of the latter, thus transforming Eq. (4.178) in a linear ordinary differential equation (ODE).

Hereafter, we shall constantly employ the Fourier transform, but drop the tilde above the transformed quantities as it customary in cosmology because almost always one deals directly with the Fourier modes rather than with the configuration space.

Therefore, Eq. (4.178) Fourier-transformed and written in the conformal time is the following:

$$\boxed{3\mathcal{H}\Phi' - 3\mathcal{H}^2\Psi + k^2\Phi = 4\pi Ga^2 \left(\rho_c\delta_c + \rho_b\delta_b + \rho_\gamma\delta_\gamma + \rho_\nu\delta_\nu\right)} \tag{4.179}$$

4.5.2 The Equation for the Anisotropic Stress

The next Einstein equation that we present is the traceless part of $\delta G^i{}_j$, which we know from Eq. (4.75) to be related to the anisotropic stress π_{ij}.

Exercise 4.24 From Eq. (4.174), calculate the trace and the traceless part of $\delta G^i{}_j$. Show that:

$$a^2\delta G^l{}_l = -6\Phi'' - 12\mathcal{H}\Phi' + 6\mathcal{H}\Psi' + 12\frac{a''}{a}\Psi - 6\mathcal{H}^2\Psi + 2\nabla^2(\Phi + \Psi),$$
$$\tag{4.180}$$

and then

$$a^2\delta G^i{}_j - \frac{1}{3}\delta^i{}_j a^2\delta G^l{}_l = -\left(\partial^i\partial_j - \frac{1}{3}\delta^i{}_j\nabla^2\right)(\Phi + \Psi). \tag{4.181}$$

The Fourier transform of the latter can be written in the following form:

$$a^2\delta G^i{}_j - \frac{1}{3}\delta^i{}_j a^2\delta G^l{}_l = k^2\left(\hat{k}^i\hat{k}_j - \frac{1}{3}\delta^i{}_j\right)(\Phi + \Psi), \tag{4.182}$$

where we have used $k^i = k\hat{k}^i$ and $\hat{k}^i$ is the unit vector denoting the direction of **k**. The spatial traceless Einstein equation can thus be written as:

$$\boxed{k^2\left(\hat{k}^i\hat{k}_j - \frac{1}{3}\delta^i{}_j\right)(\Phi + \Psi) = 8\pi Ga^2\pi^i{}_j} \tag{4.183}$$

since $\pi^i{}_j$ is the spatial traceless part of the energy-momentum tensor, as we know from Eq. (4.75). On the left hand side, we notice the same operator multiplying the scalar contribution of χ_{ij} in Eq. (4.167). Hence, only the scalar contribution of $\pi^i{}_j$ would contribute on the right hand side. Indeed, contracting the above equation with $\hat{k}_i\hat{k}^j$, as in Eq. (4.168), we obtain:

$$\boxed{k^2(\Phi + \Psi) = 12\pi Ga^2\hat{k}_i\hat{k}^j\pi^i{}_j} \tag{4.184}$$

Our $\hat{k}_i \hat{k}^j \pi^i{}_j$ corresponds to the $-(\rho + P)\sigma$ used in Ma and Bertschinger (1995). We leave this equation as it is for the moment. We shall see that $\hat{k}_i \hat{k}^j \pi^i{}_j$ is sourced by the quadrupole moments of the photon and neutrino distributions.

This equation tells us that $\Phi = -\Psi$, i.e. there exists only one gravitational potential, unless a quadrupole moment of the energy content distribution is present. For example, when CDM dominated the universe then $\Phi = -\Psi$ but this is not the case in the early universe, because of neutrinos. Even when CDM or DE dominates but the underlying theory of gravity is not GR one might have $\Phi \neq -\Psi$.[6]

4.5.3 The Equation for the Velocity

The $0 - i$ Einstein equation can be written using Eqs. (4.173) and (4.73) as follows:

$$\partial_i(\Phi' - \mathcal{H}\Psi) = 4\pi G a^2 (\rho + P) v_i . \tag{4.185}$$

Upon FT we get:

$$ik_i(\Phi' - \mathcal{H}\Psi) = 4\pi G a^2 (\rho + P) v_i , \tag{4.186}$$

and now we must get the scalar part of this vectorial equation, contracting by $i\hat{k}^i$, as we showed in Eq. (4.165). Hence, we get:

$$\boxed{k(-\Phi' + \mathcal{H}\Psi) = 4\pi G a^2 (\rho + P) V} \tag{4.187}$$

Comparing with the notation employed in Ma and Bertschinger (1995), one has $\theta = kV$. Note that the $(\rho + P) V$ in the above equation is the total one, hence making explicit the various contributions one has:

$$\boxed{k(-\Phi' + \mathcal{H}\Psi) = 4\pi G a^2 \left(\rho_c V_c + \rho_b V_b + \frac{4}{3}\rho_\gamma V_\gamma + \frac{4}{3}\rho_\nu V_\nu \right)} \tag{4.188}$$

where we have considered the usual equations of state for the various components, i.e. $P_c = P_b = 0$ and $P_\gamma = \rho_\gamma/3$ and $P_\nu = \rho_\nu/3$.

4.5.4 The Equation for the Pressure Perturbation

Using Eq. (4.180), we can immediately write down the last Einstein equation for scalar perturbations:

---

[6]One can probe the value of $\Phi + \Psi$ via weak lensing, which we do not address in these notes. See e.g. Dodelson (2017) for a treatise on gravitational lensing.

$$\boxed{\Phi'' + 2\mathcal{H}\Phi' - \mathcal{H}\Psi' - (2\mathcal{H}' + \mathcal{H}^2)\Psi + \frac{k^2}{3}(\Phi + \Psi) = -4\pi Ga^2\delta P} \quad (4.189)$$

Exercise 4.25 Use the previously derived transformations (4.150) from the Newtonian to the synchronous gauge and write there the Einstein equations.

4.6 Einstein Equations for Tensor Perturbations

Our perturbed FLRW metric, with tensor perturbations only, can be cast as follows:

$$g_{00} = -a^2, \qquad g_{0i} = 0, \qquad g_{ij} = a^2(\delta_{ij} + h_{ij}^T), \quad (4.190)$$

where h_{ij}^T is divergenceless and traceless.

Exercise 4.26 Start from metric (4.190) and calculate the perturbed Einstein tensor $\delta G^\mu{}_\nu$. Verify the calculations also using Eqs. (4.42)–(4.44). Show that the only non vanishing components are:

$$\boxed{2a^2\delta G^i{}_j = h_{ij}^{T''} + 2\mathcal{H}h_{ij}^{T'} - \nabla^2 h_{ij}^T} \quad (4.191)$$

Notice that the wave operator has appeared.

The calculation of the above exercise is pretty straightforward because the tensor nature of the perturbation h_{ij}^T already suggests that it cannot contribute to R_{00}, R_{0i} and the Ricci scalar R (at first-order).

The tensor part of the Einstein equations is thus:

$$\boxed{h_{ij}^{T''} + 2\mathcal{H}h_{ij}^{T'} + k^2 h_{ij}^T = 16\pi Ga^2\pi_{ij}^T} \quad (4.192)$$

where π_{ij}^T is the tensorial part of the anisotropic stress, which can be computed from the total as in Eq. (4.170):

$$\pi_{ij}^T = \left(\delta^l{}_i - \hat{k}^l\hat{k}_i\right)\left(\delta^m{}_j - \hat{k}^m\hat{k}_j\right)\pi_{lm} + \frac{1}{2}\hat{k}^l\hat{k}^m\pi_{lm}\left(\delta_{ij} - \hat{k}_i\hat{k}_j\right). \quad (4.193)$$

In Fourier space, the divergenceless condition can be written down as:

$$\hat{k}^i h_{ij}^T = 0. \quad (4.194)$$

Therefore, h_{ij}^T can be expanded with respect to a 2-dimensional basis $\{\hat{e}_1, \hat{e}_2\}$ defined on the 2-dimensional subspace orthogonal to $\hat{k}$. The basis satisfies thus the condition:

$$\gamma^{ij} e_{a,i} \hat{k}_j = 0 , \qquad \gamma^{ij} e_{a,i} e_{b,j} = \delta_{ab} , \tag{4.195}$$

where γ_{ij} is the metric on the 2-dimensional subspace and $a, b \in \{1, 2\}$. We can write then h_{ij}^T as follows:

$$h_{ij}^T(\mathbf{k}) = (e_{1,i} e_{1,j} - e_{2,i} e_{2,j})(\hat{k}) h_+(\mathbf{k}) + (e_{1,i} e_{2,j} + e_{2,i} e_{1,j})(\hat{k}) h_\times(\mathbf{k}) . \tag{4.196}$$

Note the dependence on $\hat{k}$ of the combinations of the 2-dimensional basis vectors. In fact these depend on the orientation of $\hat{k}$.

Substituting the above expansion into Eq. (4.192) and in absence of quadrupole moments, $h_{+,\times}$ satisfy then the equation:

$$\boxed{h''_{+,\times} + 2\mathcal{H} h'_{+,\times} + k^2 h_{+,\times} = 0} \tag{4.197}$$

which we will employ in order to investigate the GW production during inflation in Sect. 8.4.

If we choose a Cartesian reference frame and $\hat{k} = \hat{z}$, i.e. a propagation direction of a gravitational wave along $\hat{z}$, then a natural choice is $\hat{e}_1 = \hat{x}$ and $\hat{e}_2 = \hat{y}$ and the perturbed metric can be written as:

$$h_{ij}^T(k\hat{z}) = \begin{pmatrix} h_+ & h_\times & 0 \\ h_\times & -h_+ & 0 \\ 0 & 0 & 0 \end{pmatrix} . \tag{4.198}$$

Though this a convenient way of expressing $h_{ij}^T(k\hat{z})$, one usually prefers to use the combinations:

$$h_+ \mp i h_\times , \tag{4.199}$$

since these have helicity ± 2, see e.g. Weinberg (1972). In order to see this, we apply a rotation about $\hat{z}$ and calculate how $h_{ij}^T(k\hat{z})$ transforms.

Exercise 4.27 Apply the rotation:

$$R^i{}_j(\theta) = \begin{pmatrix} \cos\theta & -\sin\theta & 0 \\ \sin\theta & \cos\theta & 0 \\ 0 & 0 & 1 \end{pmatrix} , \tag{4.200}$$

about the $\hat{z}$ axis to h_{ij}^T, i.e. compute the components:

$$\bar{h}_{lm}^T = R^i{}_l R^j{}_m h_{ij}^T , \tag{4.201}$$

and show that:

$$\bar{h}_+ = \cos^2\theta h_+ + 2\sin\theta\cos\theta h_\times - \sin^2\theta h_+ = \cos 2\theta h_+ + \sin 2\theta h_\times \;, \quad (4.202)$$
$$\bar{h}_\times = \cos^2\theta h_\times - 2\sin\theta\cos\theta h_+ - \sin^2\theta h_\times = \cos 2\theta h_\times - \sin 2\theta h_+ \;. \quad (4.203)$$

Hence, the aforementioned combinations $h_+ \pm ih_\times$ transform as:

$$\boxed{\bar{h}_+ \pm i\bar{h}_\times = e^{\mp 2i\theta}(h_+ \pm ih_\times)} \qquad (4.204)$$

and have thus helicity ∓ 2. Sometimes the sign could be a bit confusing depending in which sense the rotation is performed. By convention, $\theta > 0$ denotes an anti-clockwise rotation so a rotation of θ about the $\hat{z}$ axis corresponds to a $-\theta$ rotation about the $-\hat{z}$ axis, which is the line of sight and therefore the relevant direction for the observer. So, the observed helicities have opposite sign with respect to the propagating ones.

So, we write the total tensor perturbation as a sum over the helicities:

$$h_{ij}^T(\eta, k\hat{z}) = \sum_{\lambda=\pm 2} e_{ij}(\hat{z}, \lambda)h(\eta, k\hat{z}, \lambda) \;, \qquad (4.205)$$

where:

$$e_{11}(\hat{z}, \pm 2) = -e_{22}(\hat{z}, \pm 2) = \mp i e_{12}(\hat{z}, \pm 2) = \mp i e_{21}(\hat{z}, \pm 2) = \frac{1}{\sqrt{2}} \;, \qquad (4.206)$$

and of course $e_{3i} = e_{i3} = 0$. Therefore, from Eq. (4.205) we get:

$$h_+(\eta, k\hat{z}) = \frac{1}{\sqrt{2}}h(\eta, k\hat{z}, +2) + \frac{1}{\sqrt{2}}h(\eta, k\hat{z}, -2) \;, \qquad (4.207)$$

$$h_\times(\eta, k\hat{z}) = \frac{i}{\sqrt{2}}h(\eta, k\hat{z}, +2) - \frac{i}{\sqrt{2}}h(\eta, k\hat{z}, -2) \;, \qquad (4.208)$$

and inverting:

$$\sqrt{2}h(\eta, k\hat{z}, +2) = h_+(\eta, k\hat{z}) - ih_\times(\eta, k\hat{z}) \;, \qquad (4.209)$$
$$\sqrt{2}h(\eta, k\hat{z}, -2) = h_+(\eta, k\hat{z}) + ih_\times(\eta, k\hat{z}) \;. \qquad (4.210)$$

For **k** in a generic direction, we have that:

$$h_{ij}^T(\eta, \mathbf{k}) = \sum_{\lambda=\pm 2} e_{ij}(\hat{k}, \lambda)h(\eta, \mathbf{k}, \lambda) \;, \qquad (4.211)$$

where the polarisation tensor is defined as:

$$e_{ij}(\hat{k}, \pm 2) = \sqrt{2} e_{\pm,i} e_{\pm,j} , \tag{4.212}$$

where the **polarisation vectors** are:

$$e_{\pm,i}(\hat{k}) \equiv \frac{(e_1 \pm i e_2)_i}{\sqrt{2}} . \tag{4.213}$$

Of course $e_{ij}(\hat{k}, \lambda)$ has the same symmetry of $h_{ij}^T(\eta, \mathbf{k})$ and thus is traceless and transverse, i.e.

$$\hat{k}^l e_{lm}(\hat{k}, \lambda) = 0 , \tag{4.214}$$

Again, the two $h(\eta, \mathbf{k}, \pm 2)$ satisfy the same Eq. (4.197) as for $h_{+,\times}$.

When we will discuss about the effect of GW on photon propagation and CMB we shall have then to deal with two directions: one is the GW direction $\hat{k}$ and the other is the photon direction of propagation, $\hat{p}$. In general a frame $\hat{k} = \hat{z}$ is chosen in order to simplify the calculations. But then, before taking the anti-Fourier transform and obtaining the physical quantities in the real space one has to remember of performing a rotation which brings $\hat{k}$ back in a generic direction. We shall see this in detail in Chap. 10.

4.7   Einstein Equations for Vector Perturbations

Finally, we address vector perturbations. The vector-perturbed FLRW metric, in the Newtonian gauge, has the following form:

$$g_{00} = -a^2 , \qquad g_{0i} = 0 , \qquad g_{ij} = a^2(\delta_{ij} + h_{ij}^V) , \tag{4.215}$$

where

$$h_{ij}^V = \partial_i A_j + \partial_j A_i , \tag{4.216}$$

and A_i is a divergenceless vector, i.e. $\partial_i A^i = 0$.

Exercise 4.28   Repeat the very same calculations performed in the tensor case but now for h_{ij}^V. Notice a very important difference: h_{ij}^V is traceless but NOT divergenceless. Compare the results with those found using Eqs. (4.42)–(4.44). Show that the non-vanishing components of the perturbed Einstein tensor are:

$$2a^2 \delta G^0{}_i = -\partial_l h_{li}^{V'} = -\nabla^2 A_i' , \tag{4.217}$$

$$2a^2 \delta G^i{}_j = h_{ij}^{V''} + 2\mathcal{H} h_{ij}^{V'} = (\partial_i A_j + \partial_j A_i)'' + 2\mathcal{H}(\partial_i A_j + \partial_j A_i)' . \tag{4.218}$$

With the Laplacian missing, the last equation has no more the wave behaviour that the corresponding tensor equation has. With no vector sources, show then that in the early, radiation-dominated universe, for which $\mathcal{H} = 1/\eta$, one has:

$$h_{ij}^V \propto 1/\eta^2 , \tag{4.219}$$

and hence vector perturbations vanish, if not sourced.

Einstein equations are thus:

$$k F_i' = -32\pi G a^2 (\rho + P) U_i , \tag{4.220}$$
$$(i\hat{k}_i F_j + i\hat{k}_j F_i)'' + 2\mathcal{H}(i\hat{k}_i F_j + i\hat{k}_j F_i)' = -32\pi G a^2 \pi_{ij}^V , \tag{4.221}$$

where U_i is the vector part of v_i, defined in Eq. (4.125), F_i is defined in Eq. (4.162) and π_{ij}^V is the vector part of the anisotropic stress, defined as from Eq. (4.169):

$$\pi_{ij}^V = 2i \left(\delta^m_{\ i} - \hat{k}^m \hat{k}_i \right) \hat{k}^l \pi_{lm} \hat{k}_j + 2i \left(\delta^m_{\ j} - \hat{k}^m \hat{k}_j \right) \hat{k}^l \pi_{lm} \hat{k}_i , \tag{4.222}$$

Of course, we have that:

$$\hat{k}^i F_i = 0 , \qquad \hat{k}^i U_i = 0 , \qquad \hat{k}^i \hat{k}^j \pi_{ij}^V = 0 . \tag{4.223}$$

Contracting the second Einstein equation with $i\hat{k}^j$ leaves us with:

$$\boxed{F_i'' + 2\mathcal{H} F_i' = 32\pi G a^2 i\hat{k}^j \pi_{ij}^V} \tag{4.224}$$

We choose, as in the tensor case, to align $\hat{k}$ to $\hat{z}$. Therefore, the divergenceless of A_i implies that

$$ik^i A_i = 0 \quad \Rightarrow \quad A_3 = 0 , \tag{4.225}$$

and from Eq. (4.216) we have:

$$h_{ij}^V = ik_i A_j + ik_j A_i = -\frac{i}{2}\hat{k}_i F_j - \frac{i}{2}\hat{k}_j F_i . \tag{4.226}$$

So, h_{ij}^V can be written as follows:

$$h_{ij}^V = \begin{pmatrix} 0 & 0 & -iF_1/2 \\ 0 & 0 & -iF_2/2 \\ -iF_1/2 & -iF_2/2 & 0 \end{pmatrix} . \tag{4.227}$$

Exercise 4.29   Applying the same rotation about $\hat{z}$ as we did for tensor perturbations. Show that:

$$\bar{F}_1 = F_1 \cos\theta - F_2 \sin\theta \,, \tag{4.228}$$

$$\bar{F}_2 = F_1 \sin\theta + F_2 \cos\theta \,. \tag{4.229}$$

Hence we have:

$$\bar{F}_1 + i\bar{F}_2 = (F_1 + iF_2)e^{i\theta} \,, \tag{4.230}$$

$$\bar{F}_1 - i\bar{F}_2 = (F_1 - iF_2)e^{-i\theta} \,, \tag{4.231}$$

and therefore the quantities:

$$F_{\pm} \equiv F_1 \pm iF_2 \,, \tag{4.232}$$

are fields with helicity ± 1.

References

Ade, P.A.R., et al.: Planck 2015 results. XIII. Cosmological parameters. Astron. Astrophys. **594**, A13 (2016a)

Bardeen, J.M.: Gauge invariant cosmological perturbations. Phys. Rev. D **22**, 1882–1905 (1980)

Bardeen, J.M., Steinhardt, P.J., Turner, M.S.: Spontaneous creation of almost scale - free density perturbations in an inflationary universe. Phys. Rev. D **28**, 679 (1983)

Dodelson, S.: Gravitational Lensing, p. 2017. Cambridge University Press, Cambridge (2017)

Dodelson, S.: Modern Cosmology. Academic Press, Netherlands (2003)

Gorini, V., Kamenshchik, A.Y., Moschella, U., Piattella, O.F., Starobinsky, A.A.: Gauge-invariant analysis of perturbations in Chaplygin gas unified models of dark matter and dark energy. JCAP **0802**, 016 (2008)

Hawking, S.W.: Perturbations of an expanding universe. Astrophys. J. **145**, 544–554 (1966)

Jordan, P., Ehlers, J., Kundt, W.: Republication of: exact solutions of the field equations of the general theory of relativity. Gen. Relativ. Gravit. **41**(9), 2191–2280 (2009)

Kodama, H., Sasaki, M.: Cosmological perturbation theory. Prog. Theor. Phys. Suppl. **78**, 1–166 (1984)

Landau, L.D., Lifschits, E.M.: The Classical Theory of Fields. Volume 2 of Course of Theoretical Physics. Pergamon Press, Oxford (1975)

Lesgourgues, J.: The Cosmic Linear Anisotropy Solving System. CLASS) I, Overview (2011)

Lifshitz, E.: On the Gravitational stability of the expanding universe. J. Phys. (USSR) **10**, 116 (1946)

Lifshitz, E.M., Khalatnikov, I.M.: Investigations in relativistic cosmology. Adv. Phys. **12**, 185–249 (1963)

Lukash, V.N.: Production of phonons in an isotropic universe. Sov. Phys. JETP **52**, 807–814 (1980). [Zh. Eksp. Teor. Fiz.79,1601(1980)]

Ma, C.-P., Bertschinger, E.: Cosmological perturbation theory in the synchronous and conformal Newtonian gauges. Astrophys. J. **455**, 7–25 (1995)

Maartens, R.: Causal thermodynamics in relativity(1996)

Malik, K.A., Matravers, D.R.: Comments on gauge-invariance in cosmology. Gen. Relativ. Gravit. **45**, 1989–2001 (2013)

Mukhanov, V.: Physical Foundations of Cosmology. Cambridge University Press, Cambridge (2005)

Mukhanov, V.F., Feldman, H.A., Brandenberger, R.H.: Theory of cosmological perturbations. Part 1. Classical perturbations. Part 2. Quantum theory of perturbations. Part 3. Extensions. Phys. Rep. **215**, 203–333 (1992)

Stewart, J.M.: Perturbations of Friedmann-Robertson-Walker cosmological models. Class. Quantum Gravity **7**, 1169–1180 (1990)

Stewart, J.M., Walker, M.: Perturbations of spacetimes in general relativity. Proc. R. Soc. Lond. **A341**, 49–74 (1974)

Wands, D., Malik, K.A., Lyth, D.H., Liddle, A.R.: A new approach to the evolution of cosmological perturbations on large scales. Phys. Rev. D **62**, 043527 (2000)

Weinberg, S.: Cosmology. Oxford University Press, Oxford (2008)

Weinberg, S.: Gravitation and Cosmology: Principles and Applications of the General Theory of Relativity. Wiley, New York (1972)

Zimdahl, W.: Bulk viscous cosmology. Phys. Rev. D **53**, 5483–5493 (1996)

Chapter 5
Perturbed Boltzmann Equations

The Boltzmann equation plays a similar role for physicists and astronomers: no one ever talks about it, but everyone is always thinking about it

Scott Dodelson, Modern Cosmology

We derive in this chapter the perturbed Boltzmann equations for photons, massless neutrinos, CDM and baryons. We shall use these in order to track the evolution of small fluctuations in these components, and couple them to Einstein equations. For deriving the hierarchy of temperature and polarisation for photons we follow mainly Ma and Bertschinger (1995), Hu and White (1997) and Tram and Lesgourgues (2013). We shall focus most of the time on the photon perturbed Boltzmann equation which is much more laborious than the others. This is because photons are massless and interact, therefore we cannot truncate the hierarchy and a collisional term must be taken into account. The latter comes from Thomson scattering, whose cross-section depends also on the polarisation, thus further complicating the treatment of photons fluctuations, which we nonetheless will bravely face.

5.1 General Form of the Perturbed Boltzmann Equation

Let us build in a general fashion the perturbed Boltzmann equation, in order to understand first and to tackle separately afterwards the terms which constitute it. As we did in Eq. (4.77), the distribution function f can be split in a background contribution plus a perturbation:

$$f(\eta, \mathbf{x}, \mathbf{p}) = \bar{f}(\eta, p) + \mathcal{F}(\eta, \mathbf{x}, \mathbf{p}) , \tag{5.1}$$

© Springer International Publishing AG, part of Springer Nature 2018
O. Piattella, *Lecture Notes in Cosmology*, UNITEXT for Physics,
https://doi.org/10.1007/978-3-319-95570-4_5

where we have left explicit the functional dependence in order to stress that, after breaking homogeneity and isotropy, the perturbed distribution function depends on 7 variables. Covariance demands the total distribution function f to be a function of the 4-position and of the 4-momentum:

$$f = f(x^\mu, P^\mu) \,, \qquad P^\mu = \frac{dx^\mu}{d\lambda} \,, \tag{5.2}$$

where λ is an affine parameter. However, the total number of independent variables is not 8 but 7, due to the mass-shell relation $g_{\mu\nu} P^\mu P^\nu = -m^2$.

We choose these seven variables to be x^μ, the proper momentum modulus p and the proper momentum direction $\hat{p}^i$. Now, we are ready to write down the Liouville operator with the above choice of coordinates in the phase space:

$$\frac{df}{d\lambda} = P^0 \frac{\partial f}{\partial \eta} + \underbrace{\frac{\partial f}{\partial x^i}}_{\text{1st order}} P^i + \frac{\partial f}{\partial p} \frac{dp}{d\lambda} + \underbrace{\frac{\partial f}{\partial \hat{p}^i} \frac{d\hat{p}^i}{d\lambda}}_{\text{2nd order}} \,, \tag{5.3}$$

where we have already put in evidence the fact that $\partial f/\partial x^i$, $\partial f/\partial \hat{p}^i$ and $d\hat{p}^i/d\lambda$ are pure first-order quantities, thereby making the last term of second order and thus negligible. Let us see in some more detail why.

The term $\partial f/\partial x^i$ is of first order because it breaks homogeneity and the latter can be broken only at first order. The same reasoning goes for $\partial f/\partial \hat{p}^i$, which breaks isotropy. Finally, $d\hat{p}^i/d\lambda$ is identically zero in the background cosmology, because there is nothing that could deviate the path of a photon in a homogeneous and isotropic space. Therefore, $d\hat{p}^i/d\lambda$ is a first-order quantity.

Dividing by P^0, we can write the first-order perturbed Boltzmann equation as follows:

$$\boxed{\frac{df}{d\eta} = \frac{\partial f}{\partial \eta} + \frac{\partial f}{\partial x^i} \frac{P^i}{P^0} + \frac{\partial f}{\partial p} \frac{dp}{d\eta} = \frac{1}{P^0} C[f]} \tag{5.4}$$

So, we have to deal with 3 terms:

1. The velocity term P^i/P^0.
2. The force term $dp/d\eta$.
3. The collisional term $C[f]$.

The velocity term is the easiest one to compute because it must be at zeroth order, being $\partial f/\partial x^i$ a first-order quantity, as we have just discussed. Therefore, as we already showed in Chap. 3:

$$\frac{P^i}{P^0} = \frac{p}{E} \hat{p}^i \,, \tag{5.5}$$

which for photons and neutrinos simply becomes $\hat{p}^i$.

The force term is made explicit through the geodesic equation and hence depends on the metric perturbations. The only collisional term that we shall consider explicitly is the one related to Thomson scattering and will be of interest just for photons and free electrons.

5.1.1 On the Photon and Neutrino Perturbed Distributions

As we know, in the Hot Big Bang model the background distribution function for photons and neutrinos are the BD and the FD ones:

$$\bar{f}_\gamma = \frac{1}{e^{p_\gamma/T_\gamma} - 1} \, , \quad \bar{f}_\nu = \frac{1}{e^{p_\nu/T_\nu} + 1} \, , \tag{5.6}$$

with $T_\gamma(\eta) \propto 1/a(\eta)$ (far from electron-positron annihilation) and $T_\nu(\eta) \propto 1/a(\eta)$, because of the thermal equilibrium and the cosmological principle. Their perturbed distribution functions can be written in the same form, but with a perturbed temperature:

$$\mathcal{T}_\gamma(\eta, \mathbf{x}, \mathbf{p}) = T_\gamma(\eta) + \delta T_\gamma(\eta, \mathbf{x}, \mathbf{p}) \, , \quad \mathcal{T}_\nu(\eta, \mathbf{x}, \mathbf{p}) = T_\nu(\eta) + \delta T_\nu(\eta, \mathbf{x}, \mathbf{p}) \, . \tag{5.7}$$

So, let us write the distribution functions as follows:

$$f^{-1} = \exp\left[\frac{p}{T + \delta T}\right] \mp 1 \, . \tag{5.8}$$

Exercise 5.1 For $\delta T \ll T$, show that:

$$\boxed{f = \bar{f} - p\frac{\partial \bar{f}}{\partial p}\frac{\delta T}{T}} \tag{5.9}$$

Hence:

$$\mathcal{F} = -p\frac{\partial \bar{f}}{\partial p}\frac{\delta T}{T} \, , \tag{5.10}$$

the perturbed distribution function for neutrinos and photons can be physically interpreted as their relative temperature fluctuations.

5.2   Force Term

We now deal with the force term, which originates from the geodesic equation. We shall calculate it first for a perturbed metric $a^2 h_{\mu\nu}$ with $h_{0i} = 0$ and then specify the result obtained for the various perturbation types: scalar, tensor and vector.

Let us write the mass-shell relation $g_{\mu\nu} P^\mu P^\nu$ as follows:

$$-a^2(1 - h_{00})(P^0)^2 + a^2(\delta_{ij} + h_{ij}) P^i P^j = -m^2 , \qquad (5.11)$$

and hence define the energy and proper momentum modulus as:

$$E^2 = a^2(1 - h_{00})(P^0)^2 , \qquad p^2 = a^2(\delta_{ij} + h_{ij}) P^i P^j . \qquad (5.12)$$

Exercise 5.2 Determine $dE/d\eta$ using the geodesic equation:

$$\frac{dP^0}{d\lambda} = -\Gamma^0_{\alpha\beta} P^\alpha P^\beta . \qquad (5.13)$$

First show that:

$$\frac{dE}{d\eta} = \mathcal{H} E - E \frac{d(h_{00}/2)}{d\eta} - \Gamma^0_{\alpha\beta} \frac{P^\alpha P^\beta}{E} a^2(1 - h_{00}) . \qquad (5.14)$$

Working out the Christoffel symbol term:

$$-2\Gamma^0_{\alpha\beta} P^\alpha P^\beta a^2(1 - h_{00}) = -2aa'(1 - h_{00})(P^0)^2 + a^2 h'_{00}(P^0)^2 + 2a^2 \partial_l h_{00} P^0 P^l$$
$$-2aa'(\delta_{ij} + h_{ij}) P^i P^j - a^2 h'_{ij} P^i P^j . \qquad (5.15)$$

Hence, using again the definitions of P^0 and of the proper momentum, get:

$$-\Gamma^0_{\alpha\beta} \frac{P^\alpha P^\beta}{E} a^2(1 - h_{00}) = -\mathcal{H} E + E \frac{h'_{00}}{2} + p\hat{p}^l \partial_l h_{00} - \mathcal{H} \frac{p^2}{E} - \frac{1}{2E} h'_{ij} p^i p^j . \qquad (5.16)$$

Collecting all the above found contributions, we have

$$\frac{dE}{d\eta} = -\mathcal{H} \frac{p^2}{E} + p\hat{p}^l \partial_l \frac{h_{00}}{2} - \frac{1}{2E} h'_{ij} p^i p^j , \qquad (5.17)$$

and using the differentiated mass-shell relation $E dE = p dp$, we can write:

$$\boxed{\frac{dp}{d\eta} = -\mathcal{H}p + E\hat{p}^l\partial_l\frac{h_{00}}{2} - \frac{p}{2}h'_{ij}\hat{p}^i\hat{p}^j} \tag{5.18}$$

As we can see, if we choose the synchronous gauge then $h_{00} = 0$ and all the perturbation types are elegantly taken into account in the last term. The Boltzmann equation thus becomes:

$$f' + \frac{p\hat{p}^i}{E}\frac{\partial f}{\partial x^i} + p\left(-\mathcal{H} + \frac{E}{p}\hat{p}^l\partial_l\frac{h_{00}}{2} - \frac{1}{2}h'_{ij}\hat{p}^i\hat{p}^j\right)\frac{\partial f}{\partial p} = \frac{a}{E}C[f], \tag{5.19}$$

where we have already used the fact that the collisional term is a first-order quantity and thus neglected its multiplication by h_{00}. Separating the distribution function in its background plus perturbed part $f = \bar{f} + \mathcal{F}$, the perturbed part of the above Boltzmann equation can be written as:

$$\boxed{\mathcal{F}' + \frac{ik\mu p}{E}\mathcal{F} - \mathcal{H}p\frac{\partial\mathcal{F}}{\partial p} + \left(ik\mu E\frac{h_{00}}{2} - \frac{p}{2}h'_{ij}\hat{p}^i\hat{p}^j\right)\frac{\partial\bar{f}}{\partial p} = \frac{a}{E}C[\mathcal{F}]} \tag{5.20}$$

where we have defined

$$\boxed{\mu \equiv \hat{k}\cdot\hat{p}} \tag{5.21}$$

This quantity plays a major role in the rest of these notes, so keep it in mind. Note that the left hand side of the perturbed Boltzmann equation has an azimuthal dependence possibly coming from $h'_{ij}\hat{p}^i\hat{p}^j$. We shall see that for scalar perturbations no azimuthal dependence appears.

5.2.1 Scalar Perturbations

Using metric (4.171), we can identify $h_{00} = -2\Psi$ and $h_{ij} = 2\Phi\delta_{ij}$. the proper momentum modulus is defined as:

$$p^2 = a^2(1 + 2\Phi)\delta_{ij}P^iP^j, \qquad p^i = a(1 + \Phi)P^i, \tag{5.22}$$

and

$$P^0 = \frac{E}{a}(1 - \Psi). \tag{5.23}$$

The velocity term up to first-order is thus the following:

$$\frac{dx^i}{d\eta} = \frac{P^i}{P^0} = \frac{p(1-\Phi)\hat{p}^i}{E(1-\Psi)} = \frac{p}{E}\hat{p}^i(1 - \Phi + \Psi), \tag{5.24}$$

but of course we have to neglect $\Psi - \Phi$ in the Liouville operator since $\partial f/\partial x^i$ is of first order.

From Eq. (5.18), we have:

$$\frac{dE}{d\eta} = -\mathcal{H}\frac{p^2}{E} - \frac{p^2}{E}\Phi' - p\hat{p}^i\partial_i\Psi , \tag{5.25}$$

and then:

$$\boxed{\frac{dp}{d\eta} = -\mathcal{H}p - p\Phi' - E\hat{p}^i\partial_i\Psi} \tag{5.26}$$

The first term on the right hand side of the above equation is of zeroth order and is responsible for the cosmological redshift. The second takes into account the time variation of the spatial curvature and the third is the gradient of the gravitational potential along the direction of the photon.

Therefore, the perturbed Boltzmann equation with scalar perturbations has the following general form:

$$\boxed{\mathcal{F}' + \frac{p}{E}\hat{p}^i\partial_i\mathcal{F} - \mathcal{H}p\frac{\partial\mathcal{F}}{\partial p} - \left(p\Phi' + E\hat{p}^i\partial_i\Psi\right)\frac{\partial\bar{f}}{\partial p} = \frac{a}{E}C[\mathcal{F}]} \tag{5.27}$$

The Fourier transform of the above Boltzmann equation can be cast as follows:

$$\boxed{\mathcal{F}' + ik\mu\frac{p}{E}\mathcal{F} - \mathcal{H}p\frac{\partial\mathcal{F}}{\partial p} - \left(p\Phi' + ik\mu E\Psi\right)\frac{\partial\bar{f}}{\partial p} = \frac{a}{E}C[\mathcal{F}]} \tag{5.28}$$

5.2.2 Tensor Perturbations

From Eq. (4.190) we have that $h_{00} = 0$ and $h_{ij} = h_{ij}^T$, transverse and traceless. From Eq. (5.18), we have:

$$\frac{dE}{d\eta} = -\mathcal{H}\frac{p^2}{E} - \frac{p^2}{2E}h_{ij}^{T'}\hat{p}^i\hat{p}^j , \qquad \frac{dp}{d\eta} = -\mathcal{H}p - \frac{p}{2}h_{ij}^{T'}\hat{p}^i\hat{p}^j , \tag{5.29}$$

and the Fourier transform of the Boltzmann equation can be cast as follows:

$$\boxed{\mathcal{F}' + ik\mu\frac{p}{E}\mathcal{F} - \mathcal{H}p\frac{\partial\mathcal{F}}{\partial p} - \frac{p}{2}h_{ij}^{T'}\hat{p}^i\hat{p}^j\frac{\partial\bar{f}}{\partial p} = \frac{a}{E}C[\mathcal{F}]} \tag{5.30}$$

If we choose $\hat{k} = \hat{z}$, from Eq. (4.198) we have $h_{11}^T = -h_{22}^T = h_+$ and $h_{12}^T = h_{21}^T = h_\times$, hence the contributions $\hat{p}^1$ and $\hat{p}^2$ shall be selected, producing thus an azimuthal dependence. The particle direction unit vector can be written in spherical coordinates as follows:

$$\hat{p} = \hat{p} = (\sqrt{1-\mu^2}\cos\phi, \sqrt{1-\mu^2}\sin\phi, \mu) \tag{5.31}$$

again, because we have fixed beforehand $\hat{k} = \hat{z}$.

Exercise 5.3 Show that:

$$h_{ij}^{T'}\hat{p}^i\hat{p}^j = h'_+(1-\mu^2)\cos 2\phi + h'_\times(1-\mu^2)\sin 2\phi =$$

$$2\sqrt{\frac{2\pi}{15}}\left[Y_2^2(\mu,\phi)(h'_+ - ih'_\times) + Y_2^{-2}(\mu,\phi)(h'_+ + ih'_\times)\right] =$$

$$4\sqrt{\frac{\pi}{15}}\left[Y_2^2(\mu,\phi)h'(\lambda=+2) + Y_2^{-2}(\mu,\phi)h'(\lambda=-2)\right] . \tag{5.32}$$

So the metric contribution in the case of tensor perturbations carries a $Y_2^{\pm2}(\mu,\phi)$ proportionality, each coupled to the respective helicity of the GW.

5.2.3 Vector Perturbations

From Eq. (4.215) we have that $h_{00} = 0$ and $h_{ij} = h_{ij}^V$, traceless and with vanishing double divergence. From Eq. (5.18), we have:

$$\frac{dE}{d\eta} = -\mathcal{H}\frac{p^2}{E} - \frac{p^2}{2E}h_{ij}^{V'}\hat{p}^i\hat{p}^j , \quad \frac{dp}{d\eta} = -\mathcal{H}p - \frac{p}{2}h_{ij}^{V'}\hat{p}^i\hat{p}^j , \tag{5.33}$$

and the Fourier transform of the Boltzmann equation can be cast in a form identical to the tensor:

$$\mathcal{F}' + ik\mu\frac{p}{E}\mathcal{F} - \mathcal{H}p\frac{\partial\mathcal{F}}{\partial p} - \frac{p}{2}h_{ij}^{V'}\hat{p}^i\hat{p}^j\frac{\partial\bar{f}}{\partial p} = \frac{a}{E}C[\mathcal{F}] \tag{5.34}$$

In Eq. (4.227) we have defined $h_{13}^V = -iF_1/2$ and $h_{23}^V = -iF_2/2$, hence again an azimuthal dependence shall appear. In particular, using Eq. (5.31):

$$h_{ij}^{V'}\hat{p}^i\hat{p}^j = -iF_1'\sqrt{1-\mu^2}\mu\cos\phi - iF_2'\sqrt{1-\mu^2}\mu\sin\phi =$$

$$-\sqrt{\frac{2\pi}{15}}\left[Y_2^1(\mu,\phi)(iF_1' + F_2') + Y_2^{-1}(\mu,\phi)(iF_1' - F_2')\right] =$$

$$-i\sqrt{\frac{2\pi}{15}}\left[Y_2^1(\mu,\phi)F_-' + Y_2^{-1}(\mu,\phi)F_+'\right] . \tag{5.35}$$

So the metric contribution in the case of tensor perturbations carries a $Y_2^{\pm 1}(\mu, \phi)$ proportionality. Again, this result holds true only for $\hat{k} = \hat{z}$.

One might ask about a possible $Y_1^{\pm 1}(\mu, \phi)$ contribution. This is absent because of our choice $h_{0i} = 0$.

5.3 The Perturbed Boltzmann Equation for CDM

We are going to consider only scalar perturbations sourcing the evolution of CDM since, as we are going to see, we shall need only two equations: one for the density contrast and the other for the fluid velocity. The former only has a scalar contribution whereas the latter does have a vector contribution, but negligible, as vector perturbations usually are.

Hence, the perturbed Boltzmann equation for CDM is Eq. (5.27) with the collisional term vanishing. We have seen that the latter may be important, at least for WIMPs, at energies of the order of tens of MeV, before kinetic decoupling. However, we are not interested in epochs so primordial.

We are going to take moments of Eq. (5.27) and show that we can neglect all of them except the first two, because CDM particles must be very massive, if thermally produced. That is, we can treat CDM in the **fluid approximation**.

Recalling the definitions Eqs. (4.80)–(4.84), multiply Eq. (5.27) with no $C[\mathcal{F}]$ by E and integrate in the momentum space:

$$\delta\rho' + (\rho + P)\partial_i v^i - \mathcal{H} \int \frac{d^3\mathbf{p}}{(2\pi)^3} \frac{\partial \mathcal{F}}{\partial p} pE - \Phi' \int \frac{d^3\mathbf{p}}{(2\pi)^3} \frac{\partial \bar{f}}{\partial p} pE = 0 . \quad (5.36)$$

The integral multiplying $\partial_i \Psi$ is vanishing because $\bar{f}$ has no angular dependence and thus when integrated with $\hat{p}^i$ the result is zero.

Exercise 5.4 Show, integrating by parts, that:

$$-\mathcal{H} \int \frac{d^3\mathbf{p}}{(2\pi)^3} \frac{\partial \mathcal{F}}{\partial p} pE = 3\mathcal{H} \int \frac{d^3\mathbf{p}}{(2\pi)^3} \mathcal{F}\left(E + \frac{p^2}{3E}\right) = 3\mathcal{H}\,(\delta\rho + \delta P) , \quad (5.37)$$

and that

$$-\Phi' \int \frac{d^3\mathbf{p}}{(2\pi)^3} \frac{\partial \bar{f}}{\partial p} pE = 3\Phi' \int \frac{d^3\mathbf{p}}{(2\pi)^3} \bar{f}\left(E + \frac{p^2}{3E}\right) = 3\Phi'(\rho + P) . \quad (5.38)$$

Thus the integrated perturbed Boltzmann equation becomes:

$$\delta\rho' + (\rho + P)\partial_i v^i + 3\mathcal{H}(\delta\rho + \delta P) + 3(\rho + P)\Phi' = 0 . \tag{5.39}$$

Exercise 5.5 Show, using $\delta\rho = \rho\delta$, that:

$$\boxed{\delta' + (1 + w)kV + 3\mathcal{H}\left(\frac{\delta P}{\delta\rho} - w\right)\delta + 3(1 + w)\Phi' = 0} \tag{5.40}$$

where we have introduced the background equation of state

$$w \equiv \frac{P}{\rho} . \tag{5.41}$$

It is necessary to use the background conservation equation $\rho' = -3\mathcal{H}(\rho + P)$ at some point during the calculation.

Equation (5.40) is valid for any kind of fluid with pressure, though for CDM we shall consider $w = 0$ and $\delta P = 0$.

Equation (5.40) is not enough for describing the behaviour of CDM because we do not know how V evolves. In order to find an evolution equation for V we take the first moment of Eq. (5.28). That is, we multiply it by $p\hat{p}^i$ and then integrate. There are four contributions which we address separately. We already consider contraction with $i\hat{k}_i$ in order to single out the scalar contribution.

1. The first term is very straightforward:

$$i\hat{k}_i \int \frac{d^3\mathbf{p}}{(2\pi)^3} \frac{\partial\mathcal{F}}{\partial\eta} p\hat{p}^i = \frac{\partial}{\partial\eta}[(\rho + P)V] = \rho(1 + w)V' - 3\mathcal{H}\rho(1 + w)^2 V + \rho w' V . \tag{5.42}$$

For CDM, we shall take $w = 0$.

2. The second one is:

$$-\hat{k}_i k_l \int \frac{d^3\mathbf{p}}{(2\pi)^3} \mathcal{F}\frac{p\hat{p}^l}{E} p\hat{p}^i = -k\hat{k}_i\hat{k}_l\delta T^{il} = -k\delta P - k\hat{k}_i\hat{k}_l\pi^{il} , \tag{5.43}$$

where we have already met $\hat{k}_i\hat{k}_l\pi^{il}$ in Eq. (4.184). It corresponds to $-(\rho + P)\sigma$ of Ma and Bertschinger (1995). For CDM we shall neglect these contributions because they are of order p^2/E^2 and hence negligible, being CDM cold.

3. The third integration is the following:

$$-i\hat{k}_i\mathcal{H} \int \frac{d^3\mathbf{p}}{(2\pi)^3} \frac{\partial\mathcal{F}}{\partial p} p^2\hat{p}^i . \tag{5.44}$$

Exercise 5.6 Integrate by parts and show that:

$$- i\hat{k}_i \mathcal{H} \int \frac{d^3\mathbf{p}}{(2\pi)^3} \frac{\partial \mathcal{F}}{\partial p} p^2 \hat{p}^i = 4 i \hat{k}_i \mathcal{H} \int \frac{d^3\mathbf{p}}{(2\pi)^3} \mathcal{F} p \hat{p}^i = 4 \mathcal{H} \rho (1 + w) V \ . \quad (5.45)$$

The similar integration performed with $\bar{f} p \hat{p}^i$ in the Φ' term is vanishing.

4. Finally, the fourth and last integration to be performed is the following:

$$k \hat{k}_i \hat{k}_l \Psi \int \frac{d^3\mathbf{p}}{(2\pi)^3} \frac{\partial \bar{f}}{\partial p} E p \hat{p}^l \hat{p}^i \ . \quad (5.46)$$

Since $\bar{f}$ does not depend on $\hat{p}^i$, the angular integration yields a $\delta^{il}/3$. Hence, integrating by parts, we arrive at the following result:

$$k \hat{k}_i \hat{k}_l \Psi \int \frac{d^3\mathbf{p}}{(2\pi)^3} \frac{\partial \bar{f}}{\partial p} E p \hat{p}^l \hat{p}^i = -k\Psi \int \frac{d^3\mathbf{p}}{(2\pi)^3} \bar{f} \left(E + \frac{p^2}{3E} \right) = -k\Psi \rho (1 + w) \ . \quad (5.47)$$

We can finally put together the four contributions and the first moment of the perturbed Boltzmann equation for CDM is the following:

$$\boxed{V' + \mathcal{H}(1 - 3w)V + \frac{w'}{1 + w} V - \frac{\delta P / \delta \rho}{1 + w} k \delta - \frac{k \hat{k}_i \hat{k}_l \pi^{il}}{\rho (1 + w)} - k\Psi = 0} \quad (5.48)$$

This equation corresponds to the Euler equation.

Exercise 5.7 Find the above two equations starting now from:

$$\nabla_\mu \delta T^{\mu\nu} = 0 \ , \quad (5.49)$$

i.e. using the fluid approximation directly, without passing through kinetic theory.

As we can appreciate, at each moment we take new variables appear, such as V, δP and $\hat{k}_i \hat{k}_l \pi^{il}$, and this demands to take further moments unless we decide to truncate this procedure. This is possible if the particle velocity is very small, which amounts to say a very large mass if the particle is in a thermal bath, as we showed in Chap. 3.[1] This is the case for CDM, for which we shall use the truncated equations:

---

[1]In Piattella et al. (2013, 2016) the CDM velocity dispersion is taken into account and the second moment of the Boltzmann equation is computed.

$$\boxed{\delta_c' + kV_c + 3\Phi' = 0} \tag{5.50}$$

and

$$\boxed{V_c' + \mathcal{H}V_c - k\Psi = 0} \tag{5.51}$$

We shall employ the same technique of taking momenta of the Boltzmann equation also for baryons (free electrons). We shall find the same results there as the ones found here but with a source term in the equation for V, coming from the Thomson collisional term which couples photons with free electrons.

Exercise 5.8 Find Eqs. (5.50) and (5.51) in the synchronous gauge. Show that one can exploit the residual gauge freedom in order to choose $V_c^{\text{syn}} = 0$. Therefore, in the synchronous gauge CDM is described by a single equation only.

5.4 The Perturbed Boltzmann Equation for Massless Neutrinos

We do not discuss massive neutrinos here, though at least one of their families does have a mass. On the other hand, for sufficiently early times, e.g. before recombination, they certainly behave as relativistic particles. For a treatment of the Boltzmann equation for massive neutrinos see Ma and Bertschinger (1995).

Specializing Eq. (5.20) for massless neutrinos, i.e. setting $p = E$ and neglecting the collisional term, we have:

$$\mathcal{F}_\nu' + ik\mu\mathcal{F}_\nu - \mathcal{H}p\frac{\partial\mathcal{F}_\nu}{\partial p} + p\left(ik\mu\frac{h_{00}}{2} - \frac{1}{2}h_{ij}'\hat{p}^i\hat{p}^j\right)\frac{\partial\bar{f}_\nu}{\partial p} = 0. \tag{5.52}$$

In order to eliminate the partial derivative with respect to p, we multiply the above equation by p and integrate it in the proper momentum modulus, using the definition:

$$\int \frac{dp\, p^2}{2\pi^2} p\mathcal{F}_\nu \equiv 4\rho_\nu(\eta)\mathcal{N}(\eta, \mathbf{k}, \hat{p}). \tag{5.53}$$

The scalar part of $\mathcal{N}(\eta, \mathbf{k}, \hat{p})$ corresponds to the $F_\nu/4$ defined in Ma and Bertschinger (1995).

We could have started tackling Eq. (5.52) by taking its moments as we did in the previous section for CDM. However, since massless neutrinos are relativistic, we are not able to truncate the procedure at some point because $p/E = 1$, i.e. p/E never is a small parameter. For this reason it is more convenient to build a hierarchy from

$\mathcal{N}$, as we shall see in this section. For massive neutrinos we also need a hierarchy because their mass is very small, see Ma and Bertschinger (1995).

Performing the dp integration and using the definition of Eq. (5.53) in the Boltzmann equation (5.52), we get:

$$(4\rho_\nu\mathcal{N})' + 4ik\mu\rho_\nu\mathcal{N} + 16\mathcal{H}\rho_\nu\mathcal{N} - 4\rho_\nu\left(ik\mu\frac{h_{00}}{2} - \frac{1}{2}h'_{ij}\hat{p}^i\hat{p}^j\right) = 0 . \quad (5.54)$$

Using the background evolution of the density:

$$\rho'_\nu + 4\mathcal{H}\rho_\nu = 0 , \quad (5.55)$$

we can finally cast the neutrino Boltzmann equation as follows:

$$\boxed{\mathcal{N}' + ik\mu\mathcal{N} - \left(ik\mu\frac{h_{00}}{2} - \frac{1}{2}h'_{ij}\hat{p}^i\hat{p}^j\right) = 0} \quad (5.56)$$

Any calculations we would do here for neutrinos shall be repeated almost step-by-step later for photons. Indeed, the two Boltzmann equations are identical being the only (huge) difference that for neutrinos we do not have a collisional term or polarisation.

5.4.1 Scalar Perturbations

For scalar perturbations the neutrino Boltzmann equation becomes:

$$\boxed{\mathcal{N}^{(S)'} + ik\mu\mathcal{N}^{(S)} + \Phi' + ik\mu\Psi = 0} \quad (5.57)$$

Inspecting Eq. (5.57) we note that it contains a differential operator which depends just on μ. Hence, if the initial condition on $\mathcal{N}^{(S)}$ is axisymmetric, i.e. also depends only on μ, at any time we have $\mathcal{N}^{(S)}(\eta, \mathbf{k}, \mu)$. This shall be our hypothesis.

We then expand $\mathcal{N}^{(S)}(\eta, \mathbf{k}, \mu)$ in Legendre polynomials, or partial waves, following the convention of Ma and Bertschinger (1995):

$$\mathcal{N}^{(S)}(\eta, \mathbf{k}, \mu) = \sum_{\ell=0}^{\infty}(-i)^\ell(2\ell + 1)\mathcal{N}_\ell^{(S)}(\eta, \mathbf{k})\mathcal{P}_\ell(\mu) , \quad (5.58)$$

$$\mathcal{N}_\ell^{(S)}(\eta, \mathbf{k}) = \frac{1}{(-i)^\ell}\int_{-1}^{1}\frac{d\mu}{2}\mathcal{P}_\ell(\mu)\mathcal{N}^{(S)}(\eta, \mathbf{k}, \mu) . \quad (5.59)$$

So, applying Eq. (5.59) to Boltzmann equation (5.57) we obtain the following:

$$\mathcal{N}_\ell^{(S)\prime} + \frac{ik}{(-i)^\ell} \int_{-1}^{1} \frac{d\mu}{2} \mu \mathcal{P}_\ell \mathcal{N}^{(S)} + \frac{\Phi'}{(-i)^\ell} \int_{-1}^{1} \frac{d\mu}{2} \mathcal{P}_\ell + \frac{ik\Psi}{(-i)^\ell} \int_{-1}^{1} \frac{d\mu}{2} \mu \mathcal{P}_\ell = 0 \,.$$
(5.60)

Using the orthogonality relation of the Legendre polynomials:

$$\int_{-1}^{1} \frac{dx}{2} \mathcal{P}_\ell(x) \mathcal{P}_{\ell'}(x) = \frac{\delta_{\ell\ell'}}{2\ell+1} \,,$$
(5.61)

the integrals of the above equation are easily computed as follows:

$$\int_{-1}^{1} \frac{d\mu}{2} \mathcal{P}_\ell = \frac{\delta_{\ell 0}}{2\ell+1} \,, \quad \int_{-1}^{1} \frac{d\mu}{2} \mu \mathcal{P}_\ell = \frac{\delta_{\ell 1}}{2\ell+1} \,.$$
(5.62)

Therefore, we must distinguish 3 cases: when only one of the above integrals contributes or none of them does (for $\ell \geq 2$). Moreover, we employ the following recurrence relation:

$$(\ell+1)\mathcal{P}_{\ell+1}(\mu) = (2\ell+1)\mu\mathcal{P}_\ell(\mu) - \ell\mathcal{P}_{\ell-1}(\mu) \,,$$
(5.63)

which allows us to write:

$$\frac{ik}{(-i)^l} \int_{-1}^{1} \frac{d\mu}{2} \mu \mathcal{P}_l \mathcal{N}^{(S)} = \frac{k(l+1)}{2l+1} \mathcal{N}_{l+1}^{(S)} - \frac{kl}{2l+1} \mathcal{N}_{l-1}^{(S)} \,.$$
(5.64)

Exercise 5.9 Show that the hierarchy of Boltzmann equations for neutrinos is the following:

$$\boxed{(2\ell+1)\mathcal{N}_\ell^{(S)\prime} + k\left[(\ell+1)\mathcal{N}_{\ell+1}^{(S)} - \ell\mathcal{N}_{\ell-1}^{(S)}\right] = 0 \quad (\ell \geq 2) \,,}$$
(5.65)

$$\boxed{3\mathcal{N}_1^{(S)\prime} + 2k\mathcal{N}_2^{(S)} - k\mathcal{N}_0^{(S)} = k\Psi \quad (\ell = 1) \,,}$$
(5.66)

$$\boxed{\mathcal{N}_0^{(S)\prime} + k\mathcal{N}_1^{(S)} = -\Phi' \quad (\ell = 0) \,.}$$
(5.67)

This infinite set of equations is called hierarchy because the ℓ equation is sourced by the $\ell+1$ and $\ell-1$ multipoles. Of course, it is not possible to solve numerically an infinite number of equations so some truncation it is necessary at a certain ℓ_{max}.

Using the definitions (5.53) and (5.59), the **monopole** $\mathcal{N}_0^{(S)}$ can be written as:

$$4\mathcal{N}_0^{(S)} = \frac{1}{\rho_\nu} \int \frac{d\mu}{2} \int \frac{dp\,p^2}{2\pi^2} p\mathcal{F}_\nu = \frac{1}{\rho_\nu} \int \frac{d^3\mathbf{p}}{(2\pi)^3} p\mathcal{F}_\nu = \frac{\delta\rho_\nu}{\rho_\nu} = \delta_\nu \,.$$
(5.68)

Hence $4\mathcal{N}_0^{(S)} = \delta_\nu$, i.e. the monopole is proportional to the density contrast.[2] In Chap. 6 we shall see that indeed the primordial mode excited is just the monopole, making thus reliable our assumption of axial symmetry. For the **dipole** $\mathcal{N}_1^{(S)}$:

$$4\mathcal{N}_1^{(S)} = \frac{i}{\rho_\nu} \int \frac{d\mu}{2} \mu \int \frac{dp\, p^2}{2\pi^2} p\mathcal{F}_\nu = \frac{i\hat{k}^l}{\rho_\nu} \int \frac{d^3\mathbf{p}}{(2\pi)^3} p\hat{p}_l \mathcal{F}_\nu = \frac{(\rho_\nu + P_\nu)}{\rho_\nu} V_\nu = \frac{4}{3} V_\nu \ . \tag{5.69}$$

Hence, $3\mathcal{N}_1^{(S)} = V_\nu$. Finally, for the **quadrupole**:

$$4\mathcal{N}_2^{(S)} = -\frac{1}{\rho_\nu} \int \frac{d\mu}{2} \mathcal{P}_2(\mu) \int \frac{dp\, p^2}{2\pi^2} p\mathcal{F}_\nu = -\frac{3}{2\rho_\nu} \int \frac{d^3\mathbf{p}}{(2\pi)^3} p(\hat{k}_l \hat{p}^l \hat{k}_m \hat{p}^m - 1/3)\mathcal{F}_\nu$$

$$= -\frac{3}{2\rho_\nu} \hat{k}_l \hat{k}_m \left(\delta T_\nu^{lm} - \frac{1}{3} \delta^{lm} \delta T_{\nu i}^i \right) = -\frac{3}{2\rho_\nu} \hat{k}_l \hat{k}_m \pi_\nu^{lm} \ . \tag{5.70}$$

Therefore, $\hat{k}_l \hat{k}_m \pi_\nu^{lm} = -4\rho_\nu \mathcal{N}_2^{(S)}/3$. Similar expressions hold true for photons.

The above equations can be compared with those in Ma and Bertschinger (1995) by making the identification $4\mathcal{N}_0^{(S)} = \delta_\nu$, $3k\mathcal{N}_1^{(S)} = \theta_\nu$ and $2\mathcal{N}_2^{(S)} = \sigma_\nu$.

Exercise 5.10 Show that Eq. (5.57) can be formally integrated as follows:

$$\mathcal{N}^{(S)}(\eta_0, \mathbf{k}, \mu) = \mathcal{N}^{(S)}(\eta_i, \mathbf{k}, \mu)e^{ik\mu(\eta_i - \eta_0)} - \int_{\eta_i}^{\eta_0} d\eta(\Phi' + ik\mu\Psi)e^{ik\mu(\eta - \eta_0)} \ , \tag{5.71}$$

where η_i is some initial conformal time.

This result is called **line-of-sight integral** (Seljak and Zaldarriaga 1996). It is a simple formal integration but it is very effective for the numerical calculation of the evolution of CMB anisotropies. We shall see this in Chap. 10. Note that the gravitational potentials Φ and Ψ couple only with $\mathcal{N}_{0,1,2}^{(S)}$ via the Einstein equations. Hence, a way of avoiding the truncation of the hierarchy of Boltzmann equation is to solve Eq. (5.71) together with its integrals which give $\mathcal{N}_{0,1,2}^{(S)}$. See Weinberg (2006).

We now introduce the expansion of a plane wave into spherical harmonics:

$$e^{i\mathbf{k}\cdot\mathbf{r}} = 4\pi \sum_{\ell=0}^{\infty} \sum_{m=-\ell}^{\ell} i^\ell Y_\ell^{m*}(\hat{k}) Y_\ell^m(\hat{r}) j_\ell(kr) \ , \tag{5.72}$$

where j_ℓ is a spherical Bessel function and Y_ℓ^m is a spherical harmonic. The latter is very often used in cosmology and especially in CMB physics because it is the natural basis over which to expand quantities defined on the sphere (the celestial sphere, in astronomy). Since they are extremely important for our work to come, Sect. 12.5 is dedicated to them as a reminder.

---

[2] It is redundant to write $\mathcal{N}_0^{(S)}$ since the monopole has only the scalar contribution.

Recalling that $\mu = \hat{k} \cdot \hat{p}$, we introduce in Eq. (5.71) the following plane wave expansion:

$$e^{-ik\hat{k}\cdot\hat{p}(\eta_0-\eta)} = 4\pi \sum_{\ell'=0}^{\infty} \sum_{m'=-\ell'}^{\ell'} (-i)^{\ell'} Y_{\ell'}^{m'*}(\hat{k}) Y_{\ell'}^{m'}(\hat{p}) j_{\ell'}(kr) \qquad (5.73)$$

where we have defined:

$$r(\eta) \equiv \eta_0 - \eta \qquad (5.74)$$

Using the addition theorem for spherical harmonics, the above result can be written as:

$$e^{ik\mu(\eta-\eta_0)} = \sum_{\ell} (-i)^{\ell} (2\ell+1) \mathcal{P}_{\ell}(\mu) j_{\ell}(kr) , \qquad (5.75)$$

where $r \equiv \eta_0 - \eta$ and $j_{\ell}(kr)$ is a spherical Bessel function. Now, note that:

$$-ik\mu\Psi e^{ik\mu(\eta-\eta_0)} = \Psi \frac{d}{d\eta_0} e^{ik\mu(\eta-\eta_0)} . \qquad (5.76)$$

Therefore, Eq. (5.71) can be written as follows:

$$\mathcal{N}^{(S)}(\eta_0, \mathbf{k}, \mu) = \sum_{\ell} (-i)^{\ell} (2\ell+1) \mathcal{P}_{\ell}(\mu)$$

$$\left[\mathcal{N}^{(S)}(\eta_i, \mathbf{k}, \mu) j_{\ell}(kr_i) - \int_{\eta_i}^{\eta_0} d\eta \left(\Phi' - \Psi \frac{d}{d\eta_0} \right) j_{\ell}(kr) \right] . \qquad (5.77)$$

We shall see in Chap. 6 that at early times ($\eta_i \to 0$) the monopole contribution is dominant, hence we may approximate $\mathcal{N}^{(S)}(\eta_i, \mathbf{k}, \mu) \approx \mathcal{N}_0^{(S)}(\eta_i, \mathbf{k})$ and thus write:

$$\mathcal{N}_{\ell}^{(S)}(\eta_0, \mathbf{k}) = \mathcal{N}_0^{(S)}(\eta_i, \mathbf{k}) j_{\ell}(k\eta_0) - \int_0^{\eta_0} d\eta \left[\Phi' j_{\ell}(kr) - \Psi j_{\ell}'(kr) \right] . \qquad (5.78)$$

Of course we cannot really set $\eta_i = 0$ since this is the cosmological singularity. Such equation should be understood as $\eta_i \to 0$. Knowing the gravitational potentials allows us to determine $\mathcal{N}_{\ell}^{(S)}$ without solving the hierarchy of Boltzmann equations. The advantage is that Φ and Ψ are determined from the Einstein equations, which couple only with $\mathcal{N}_{0,1,2}^{(S)}$.

Exercise 5.11 Write the Boltzmann equation for neutrinos in the cases of tensor and vector perturbations. It might be useful to first study those for photons in the next section.

5.5   The Perturbed Boltzmann Equation for Photons

The Boltzmann equation for photons has a collisional term, coming from Thomson scattering among photons and free electrons. Thomson scattering is an excellent approximation as long as the energy of the photon is much smaller than the electron mass, i.e. $h\nu \ll 511 \, \text{keV}$ which corresponds to a redshift $\approx 10^{10}$ (the same order as the BBN one). So, as long as we deal with much smaller redshifts, our approximation is fine. If not, then the Klein–Nishina cross-section should be taken into account.

The scales of cosmological interest today were well outside the particle horizon for those large redshifts and we shall see that the evolution of perturbations on these super-horizon regimes is somewhat peculiar. In order to be convinced of this, consider a scale k today which, in order to be observationally interesting, must be much smaller than the particle horizon, which is proportional to H_0. Thus $k \gg H_0$. At some time in the past, this condition becomes $k \ll \mathcal{H}$, since $\mathcal{H}$ diverges for $\eta \to 0$. To be more quantitative, using Friedmann equation in the radiation-dominated epoch one has:

$$\frac{k}{\mathcal{H}} \approx \frac{ka}{H_0\sqrt{\Omega_{r0}}} \approx \frac{10^2}{(1+z)} \frac{k}{H_0} \, . \tag{5.79}$$

So, even a scale $k = 100 H_0$, which is today of order of $40 \, \text{Mpc}$ (already in the non-linear regime of evolution) for redshifts larger than 10^4 was outside the horizon.

An important feature of Thomson scattering is that the photon energy remains unchanged. This means that $C[\mathcal{F}_\gamma]$ does not depend on the photon energy p. Hence, as we did in the massless neutrino case, we can integrate the whole Boltzmann equation with respect to p, and define:

$$\boxed{\int \frac{dp\, p^2}{2\pi^2} p\mathcal{F}_\gamma \equiv 4\rho_\gamma(\eta)\Theta(\eta, \mathbf{k}, \hat{p})} \tag{5.80}$$

where Θ corresponds to $F_\gamma/4$ of Ma and Bertschinger (1995). The treatment of the Boltzmann equation for photons is more complicated with respect to the neutrino case because of the presence of a collisional term which depends on polarisation. We have therefore to develop three Boltzmann equations: One for each of Θ, Q and U. The latter are the two Stokes parameters related to linear polarisation. Θ is also a Stokes parameter, related to the intensity of light, i.e. its energy density, and for this reason it is the only parameter which couples to the metric. Thomson scattering does not produces circular polarisation and thus $V = 0$. If the reader is not familiar with polarisation of light and Stokes parameters, Sect. 12.7 offers a brief reminder.

In the next subsection we explicitly compute the collisional term for the Boltzmann equation for photons without considering polarisation.

5.5.1 Computing the Collisional Term Neglecting Polarisation

Consider scattering among photons and free electrons:

$$e^-(\mathbf{q}) + \gamma(\mathbf{p}) \leftrightarrow e^-(\mathbf{q}') + \gamma(\mathbf{p}') , \tag{5.81}$$

where the the photon energy is unchanged, i.e. $p = p'$, but its direction has been modified. The right hand side of Eq. (5.28) can be written as follows:

$$\mathcal{C}_{e\gamma} \equiv \frac{a}{p} C[\mathcal{F}_\gamma(\mathbf{p})] = \frac{a}{p} \int \frac{d^3 q}{(2\pi)^3 2 E_e(q)} \int \frac{d^3 q'}{(2\pi)^3 2 E_e(q')} \int \frac{d^3 p'}{(2\pi)^3 2p'} |\mathcal{M}|^2$$
$$(2\pi)^4 \delta^{(3)}(\mathbf{p} + \mathbf{q} - \mathbf{p}' - \mathbf{q}') \delta[p + E_e(q) - p' - E_e(q')]$$
$$\left[\bar{f}_e(q') \mathcal{F}_\gamma(\mathbf{p}') + \mathcal{F}_e(\mathbf{q}') \bar{f}_\gamma(p') - \bar{f}_e(q) \mathcal{F}_\gamma(\mathbf{p}) - \mathcal{F}_e(\mathbf{q}) \bar{f}_\gamma(p) \right] , \tag{5.82}$$

Evidently, the zeroth order term

$$\bar{f}_e(q') \bar{f}_\gamma(p') - \bar{f}_e(q) \bar{f}_\gamma(p) , \tag{5.83}$$

yields a vanishing integration because the energies do not change. Hence this collisional term is a perturbative quantity.

At recombination Θ is of order 10^{-5}, so the linear approximation is reasonable. However, there is another term which we must take into account: the electron velocity q/m_e, which causes a Doppler shift in the CMB photons. Therefore, we should keep track also of that when working on Eq. (5.82). In particular, being the electron non-relativistic and in the thermal bath with photons, then $q/m_e \sim \sqrt{T/m_e}$. For thermal energies of $\sim 1\,\mathrm{eV}$ then $q/m_e \sim 10^{-3}$, so it is a contribution that we definitely want to take into account.

Let us start our task. Expand the electron energy as usual:

$$E_e(q) = \sqrt{q^2 + m_e^2} = m_e + \frac{q^2}{2m_e} + \cdots , \tag{5.84}$$

and write Eq. (5.82) in the following way:

$$\mathcal{C}_{e\gamma} = \frac{a\pi}{4m_e^2 p} \int \frac{d^3 q}{(2\pi)^3} \int \frac{dp' p' d^2 \hat{p}'}{(2\pi)^3} |\mathcal{M}(\hat{p}, \hat{p}')|^2 \delta \left[p + \frac{q^2}{2m_e} - p' - \frac{(\mathbf{p} + \mathbf{q} - \mathbf{p}')^2}{2m_e} \right]$$
$$\left[\bar{f}_e(q) \mathcal{F}_\gamma(\mathbf{p}') + \mathcal{F}_e(\mathbf{q}) \bar{f}_\gamma(p') - \bar{f}_e(q) \mathcal{F}_\gamma(\mathbf{p}) - \mathcal{F}_e(\mathbf{q}) \bar{f}_\gamma(p) \right] . \tag{5.85}$$

Here, we have used the 3-momentum Dirac delta in order to eliminate the integration with respect to $\mathbf{q}'$ and at the denominator of the volume elements we have used just

$E_e = m_e$, because we need to keep track only of terms of the first order in $\mathbf{q}/m_e$. Indeed:

$$\frac{1}{E_e} = \frac{1}{m_e + \frac{q^2}{2m_e}} \approx \frac{1 - q^2/(2m_e^2)}{m_e} \,, \tag{5.86}$$

and the q^2/m_e^2 correction to the collisional term is totally negligible (it is something of order 10^{-12}). We have also put in evidence the $(\hat{p}, \hat{p}')$ dependence of the probability amplitude and used

$$\boxed{f_e(\mathbf{p} + \mathbf{q} - \mathbf{p}') \approx f_e(\mathbf{q})} \tag{5.87}$$

Physically, this happens because the electron mass-energy is so much larger than that of the average photon (something like 6 orders of magnitude) that the latter is unable to deviate the former from its path. It is like deviating a truck with a tennis ball.

More quantitatively, we can prove the goodness of the above approximation by using the 4-momentum conservation:

$$p_\gamma^\mu + q_e^\mu = p_\gamma'^\mu + q_e'^\mu \,. \tag{5.88}$$

The zero component gives us the energy conservation. Since the photon energy E does not change, so does not the electron energy E_e.

Exercise 5.12 In Eq. (5.88), put the photon 4-momenta on the same side and the electron ones on the other side and square. Show that:

$$E^2(1 - \cos\alpha_\gamma) = q^2(1 - \cos\alpha_e) \,, \tag{5.89}$$

where α_γ is the angle between the photon directions and α_e is the angle between the electron directions.

Being in a thermal bath, the photon energy is of order $E \sim T$, whereas the electron momentum is of order $q \sim \sqrt{m_e T}$. Therefore:

$$\frac{T}{m_e}(1 - \cos\alpha_\gamma) \sim 1 - \cos\alpha_e \,. \tag{5.90}$$

Since $T \ll m_e$ (because the electron is non-relativistic) one can see that $\alpha_e \sim \sqrt{T/m_e}$, i.e. the electron is only slightly deviated upon the scattering.

The electron final energy can be expanded as follows:

$$\frac{(\mathbf{p} + \mathbf{q} - \mathbf{p}')^2}{2m_e} = \frac{q^2}{2m_e} + \frac{\mathbf{q} \cdot (\mathbf{p} - \mathbf{p}')}{m_e} + \frac{(\mathbf{p} - \mathbf{p}')^2}{2m_e} \,, \tag{5.91}$$

and using this expansion in the energy Dirac delta in Eq. (5.85) we obtain:

$$\delta\left[p - p' + \frac{\mathbf{q} \cdot (\mathbf{p'} - \mathbf{p})}{m_e}\right], \tag{5.92}$$

where we have neglected the $(\mathbf{p} - \mathbf{p'})^2/m_e$ contribution, because $|\mathbf{p} - \mathbf{p'}|$ is at most $2p$ and the typical energy of the photon is that of the thermal bath, i.e. $T \ll m_e$.

The Dirac delta can be formally Taylor-expanded as follows:

$$\delta\left[p - p' + \frac{\mathbf{q} \cdot (\mathbf{p'} - \mathbf{p})}{m_e}\right] = \delta(p - p') + \frac{\partial\delta(p - p')}{\partial p'} \frac{\mathbf{q}}{m_e} \cdot (\mathbf{p} - \mathbf{p'}) . \tag{5.93}$$

and the collisional term (5.85) is cast as follows:

$$\mathcal{C}_{e\gamma} = \frac{a\pi}{4m_e^2 p} \int \frac{d^3\mathbf{q}}{(2\pi)^3} \int \frac{dp' p' d^2\hat{p}'}{(2\pi)^3} |\mathcal{M}(\hat{p}, \hat{p}')|^2 \times$$
$$\times \left[\delta(p - p') + \frac{\partial\delta(p - p')}{\partial p'} \frac{\mathbf{q}}{m_e} \cdot (\mathbf{p} - \mathbf{p'})\right] \times$$
$$\times \left[\bar{f}_e(q)\mathcal{F}_\gamma(\mathbf{p'}) + \mathcal{F}_e(\mathbf{q})\bar{f}_\gamma(p') - \bar{f}_e(q)\mathcal{F}_\gamma(\mathbf{p}) - \mathcal{F}_e(\mathbf{q})\bar{f}_\gamma(p)\right] . \tag{5.94}$$

Exercise 5.13 Show that, exploiting the two Dirac deltas, we obtain:

$$\mathcal{C}_{e\gamma} = \frac{a n_e}{32\pi^2 m_e^2} \int d^2\hat{p}' |\mathcal{M}(\hat{p}, \hat{p}')|^2 \left[\mathcal{F}_\gamma(\mathbf{p'}) - \mathcal{F}_\gamma(\mathbf{p})\right]$$
$$+ \frac{a n_e \mathbf{v_b}}{32\pi^2 m_e^2 p} \cdot \int p' dp' d^2\hat{p}' |\mathcal{M}(\hat{p}, \hat{p}')|^2 \frac{\partial\delta(p - p')}{\partial p'} (\mathbf{p} - \mathbf{p'}) \left[\bar{f}_\gamma(p') - \bar{f}_\gamma(p)\right] , \tag{5.95}$$

where

$$n_e \equiv \int \frac{d^3\mathbf{q}}{(2\pi)^3} \bar{f}_e(q) , \qquad n_e m_e \mathbf{v_b} \equiv \int \frac{d^3\mathbf{q}}{(2\pi)^3} \mathcal{F}_e(\mathbf{q})\mathbf{q} . \tag{5.96}$$

Note that we have assumed that $\mathbf{v}_e = \mathbf{v}_p = \mathbf{v_b}$, i.e. the velocity of the electron fluid to be equal to that of the proton one and then we have called it baryon fluid velocity. This is justified by the fact that Coulomb scattering tightly couples electrons and protons.

For Thomson scattering we have that (see Sect. 12.8):

$$\int d^2\hat{p}' |\mathcal{M}(\hat{p}, \hat{p}')|^2 = 32\pi^2 m_e^2 \sigma_T , \qquad \int d^2\hat{p}' |\mathcal{M}(\hat{p}, \hat{p}')|^2 \hat{p}' = 0 , \tag{5.97}$$

where the latter equation simply establishes the fact that Thomson scattering has not a a priori preferred direction. Therefore, integrating by parts the Dirac delta derivative, we get:

$$\mathcal{C}_{e\gamma} = \tau' \mathcal{F}_\gamma(p\hat{p}) + \frac{an_e}{32\pi^2 m_e^2} \int d^2\hat{p}' |\mathcal{M}(\hat{p}, \hat{p}')|^2 \mathcal{F}_\gamma(p\hat{p}') + p\frac{\partial \bar{f}_\gamma}{\partial p} \tau' \mathbf{v}_b \cdot \hat{p} \,,$$

$$(5.98)$$

where we have introduced **the optical depth**:

$$\boxed{\tau(\eta) \equiv \int_\eta^{\eta_0} d\eta' n_e \sigma_T a \,, \qquad \tau' = -n_e \sigma_T a}$$

$$(5.99)$$

As we did for neutrinos, since $|\mathcal{M}|^2$ does not depend on p, we can multiply the full Boltzmann equation by p and integrate in the modulus of the proper momentum.

The Boltzmann equation for photons thus becomes:

$$4\rho_\gamma \left(\Theta' + ik\mu\Theta - ik\mu\frac{h_{00}}{2} + \frac{1}{2}h'_{ij}\hat{p}^i\hat{p}^j \right) = \int \frac{dp\, p^2}{2\pi^2} p\, \mathcal{C}_{e\gamma} \,, \qquad (5.100)$$

and substituting the collisional term of Eq. (5.98) we find:

$$\Theta' + ik\mu\Theta - ik\mu\frac{h_{00}}{2} + \frac{1}{2}h'_{ij}\hat{p}^i\hat{p}^j = \tau'\Theta + \frac{an_e}{32\pi^2 m_e^2} \int d^2\hat{p}' |\mathcal{M}(\hat{p}, \hat{p}')|^2 \Theta(\hat{p}') - \tau'\hat{p}\cdot\mathbf{v}_b \,.$$

$$(5.101)$$

The term containing the Thomson scattering amplitude can be split as follows:

$$\frac{an_e}{32\pi^2 m_e^2} \int d^2\hat{p}' |\mathcal{M}|^2 \Theta(\hat{p}') = -\tau' \int \frac{d^2\hat{p}'}{4\pi} \Theta(\hat{p}') + \frac{an_e}{32\pi^2 m_e^2} \int d^2\hat{p}' |\mathcal{M}'|^2 \Theta(\hat{p}') \,.$$

$$(5.102)$$

The last contribution of the above equation has the following form:

$$\frac{an_e}{32\pi^2 m_e^2} \int d^2\hat{p}' |\mathcal{M}'|^2 \Theta(\hat{p}') = -\frac{\tau'}{10} \sum_{m=-2}^{2} Y_2^m(\hat{p}) \int d^2\hat{p}' Y_2^{m*}(\hat{p}') \Theta(\hat{p}') \,.$$

$$(5.103)$$

We now introduce the contribution from polarisation. We work out the collisional term in Sect. 12.8.

5.5.2 Full Boltzmann Equation Including Polarisation

The original derivation of the Thomson scattering matrix can be found in Chandrasekhar (1960). The theoretical framework for the CMB polarisation can be found e.g. in Kosowsky (1996), but also in Weinberg (2008). Here we write already the final equation that we are going to analyse, following Hu and White (1997) and Tram and Lesgourgues (2013), and derive the collisional term in Sect. 12.8.

Following Tram and Lesgourgues (2013), the three combined Boltzmann equations for Θ, Q and U can be written as:

$$\left(\frac{\partial}{\partial\eta}+ik\mu\right)\begin{pmatrix}\Theta\\Q\\iU\end{pmatrix}-\tau'\begin{pmatrix}\Theta-\int\frac{d^2\hat{p}'}{4\pi}\Theta(\hat{p}')-\hat{p}\cdot\mathbf{v_b}\\Q\\iU\end{pmatrix}$$

$$+\begin{pmatrix}-ik\mu\frac{h_{00}}{2}+\frac{1}{2}h'_{ij}\hat{p}^i\hat{p}^j\\0\\0\end{pmatrix}=$$

$$-\frac{\tau'}{10}\sum_{m=-2}^{2}\begin{bmatrix}Y_2^m(\hat{p})\\\frac{1}{2}\mathcal{E}^m(\hat{p})\\\frac{1}{2}\mathcal{B}^m(\hat{p})\end{bmatrix}\int d^2\hat{p}'\begin{bmatrix}Y_2^{m*}\Theta-\sqrt{\frac{3}{2}}\mathcal{E}^{m*}Q'-\sqrt{\frac{3}{2}}\mathcal{B}^{m*}iU\\-\sqrt{6}Y_2^{m*}\Theta+3\mathcal{E}^{m*}Q+3\mathcal{B}^{m*}iU\\-\sqrt{6}Y_2^{m*}\Theta+3\mathcal{E}^{m*}Q+3\mathcal{B}^{m*}iU\end{bmatrix}(\hat{p}'),$$

$$(5.104)$$

where we have used the definition:

$$\mathcal{E}^m\equiv{}_2Y_2^m+{}_{-2}Y_2^m, \qquad \mathcal{B}^m\equiv{}_2Y_2^m-{}_{-2}Y_2^m, \tag{5.105}$$

where $_{\pm2}Y_2^m$ are the spin-2 weighted spherical harmonics. Their explicit form is given in Tables 5.1 and 5.2.

Inspecting the above trio of Boltzmann equations in Eq. (5.104), one can see that iU and $\mathcal{B}^m Q/\mathcal{E}^m$ satisfy the same Boltzmann equation. Hence, since they have the same initial condition (which is zero, as we shall see in Chap. 6), we just need a single polarisation hierarchy. This is the principal result of Tram and Lesgourgues (2013), which contributes to make the CLASS code faster. Choosing the Q one, we have thus for the temperature fluctuation:

Table 5.1 Explicit functional form of the spin-0 and spin-2 spherical harmonics. Note that the spin-2 spherical harmonics can be obtained by the spin-2 by spatial inversion, i.e. $_{-2}Y_2^m(\hat{p})={}_2Y_2^m(-\hat{p})$. In these notes, we omit the Condon–Shortley phase

m	Y_2^m	$_2Y_2^m$
0	$\frac{1}{4}\sqrt{\frac{5}{\pi}}(3\cos^2\theta-1)$	$\frac{3}{4}\sqrt{\frac{5}{6\pi}}\sin^2\theta$
±1	$\frac{1}{2}\sqrt{\frac{15}{2\pi}}\sin\theta\cos\theta e^{\pm i\phi}$	$\frac{1}{4}\sqrt{\frac{5}{\pi}}\sin\theta(1\mp\cos\theta)e^{\pm i\phi}$
±2	$\frac{1}{4}\sqrt{\frac{15}{2\pi}}\sin^2\theta e^{\pm2i\phi}$	$\frac{1}{8}\sqrt{\frac{5}{\pi}}(1\mp\cos\theta)^2 e^{\pm2i\phi}$

Table 5.2 Explicit functional form of $\mathcal{E}^m$ and $\mathcal{B}^m$

m	$\mathcal{E}^m$	$\mathcal{B}^m$
0	$\sqrt{\frac{15}{8\pi}}\sin^2\theta$	0
±1	$-\frac{1}{2}\sqrt{\frac{5}{\pi}}\sin\theta\cos\theta e^{\pm i\phi}$	$\frac{1}{2}\sqrt{\frac{5}{\pi}}\sin\theta e^{\pm i\phi}$
±2	$\frac{1}{4}\sqrt{\frac{5}{\pi}}(1+\cos^2\theta)e^{\pm2i\phi}$	$-\frac{1}{2}\sqrt{\frac{5}{\pi}}\cos\theta e^{\pm2i\phi}$

$$\left(\frac{\partial}{\partial\eta}+ik\mu\right)\Theta-\tau'\left[\Theta-\int\frac{d^2\hat{p}'}{4\pi}\Theta(\hat{p}')-\hat{p}\cdot\mathbf{v_b}\right]-ik\mu\frac{h_{00}}{2}+\frac{1}{2}h'_{ij}\hat{p}^i\hat{p}^j=$$
$$-\frac{\tau'}{10}\sum_{m=-2}^{2}Y_2^m(\hat{p})\int d^2\hat{p}'\left[Y_2^{m*}\Theta-\sqrt{\frac{3}{2}}\mathcal{E}^{m*}Q-\sqrt{\frac{3}{2}}\frac{(\mathcal{B}^{m*})^2}{\mathcal{E}^{m*}}Q\right](\hat{p}')\,,$$

$$(5.106)$$

and for polarisation:

$$\left(\frac{\partial}{\partial\eta}+ik\mu\right)Q-\tau'Q=$$
$$\frac{\tau'}{10}\sum_{m=-2}^{2}\sqrt{\frac{3}{2}}\mathcal{E}^m(\hat{p})\int d^2\hat{p}'\left[Y_2^{m*}\Theta-\sqrt{\frac{3}{2}}\mathcal{E}^{m*}Q-\sqrt{\frac{3}{2}}\frac{(\mathcal{B}^{m*})^2}{\mathcal{E}^{m*}}Q\right](\hat{p}')\,.$$

$$(5.107)$$

Note that on the left hand side of Eq. (5.106) the dependence on the photon direction $\hat{p}$ enters through $\mu=\hat{k}\cdot\hat{p}$ and through $h'_{ij}\hat{p}^i\hat{p}^j$, which also depends on μ for scalar perturbations, cf. Eq. (5.28), and on μ and and the azimuthal angle ϕ for tensor and vector perturbations if $\hat{k}=\hat{z}$, cf. Eqs. (5.32) and (5.35). Instead, on the right hand side of Eq. (5.106) the dependence on the photon direction is $Y_2^m(\hat{p})$.

In order to match the angular dependences on the two sides, making the equation easier to manipulate, and also in order to easily express the metric contribution for tensor and vector perturbations, it is convenient to set $\hat{k}=\hat{z}$. In order to recall this, we define:

$$\Theta_P(k\hat{z})\equiv Q(k\hat{z})\,.\qquad(5.108)$$

As we shall see in Chap. 10, before performing the anti-Fourier transform in order to recover the physical quantities in the real space, we shall have to apply first a spatial rotation in order to recover a generic direction for $\hat{k}$. This rotation compensates if one computes straightaway the angular power spectra, since they are rotational-invariant quantities. In this case one can use at once the results of the equations of this section.

Choosing then a frame in which $\hat{k}=\hat{z}$, comparing $Y_2^m(\mu,\phi)$ with Eqs. (5.32) and (5.35) allows us to see that the $m=0$ contribution of the sum on the right hand side of Eq. (5.106) couples to scalar perturbations only, the $m=\pm2$ contributions couple to tensor perturbations and the $m=\pm1$ to vector perturbations.

Note that for scalar perturbations one has that $U=0$ in the reference frame $\hat{k}=\hat{z}$, since $\mathcal{B}^0=0$.

Unless differently stated, in the following the functional dependences of Θ and Θ_P are on η, $\mathbf{k}=k\hat{z}$ and μ and ϕ. The metric quantities do not depend on μ and ϕ.

5.5.3 Scalar Perturbations

For scalar perturbations we have that:

$$- ik\mu \frac{h_{00}}{2} + \frac{1}{2} h'_{ij} \hat{p}^i \hat{p}^j = \Phi' + ik\mu\Psi , \tag{5.109}$$

$$(\hat{p} \cdot \mathbf{v_b})^{(S)} = -i\mu V_b , \tag{5.110}$$

since recall that the scalar part of $\mathbf{v_b}$ is $\mathbf{v}_b^{(S)} = -i\hat{k}V_b$. Note that, as in the case of neutrinos, since no azimuthal dependence appears in the differential equation, it is natural to assume axial symmetry in $\Theta^{(S)}$, i.e. $\Theta^{(S)}(\eta, k\hat{z}, \mu)$.

As we shall see in Chap. 10, the scalar contribution $\Theta^{(S)}$ is the one which dominates the CMB temperature fluctuations because it is sourced by scalar perturbations in the metric, which are the strongest since they can grow (they are compressional modes).

The same goes for $\Theta_P^{(S)}(\eta, k\hat{z}, \hat{p})$, which is not sourced by metric fluctuations and therefore its angular dependence is less constrained. Nonetheless, we assume that the scalar part also depends only on μ. Note the dependence on $\mathbf{k} = k\hat{z}$ as a reminder that $\hat{k} = \hat{z}$.

Exercise 5.14 Perform the integration in $d\phi'$ for the $m = 0$ contribution of Eqs. (5.106) and (5.107) and using the results of Tables 5.1 and 5.2 show that:

$$\left(\frac{\partial}{\partial\eta} + ik\mu \right) \Theta^{(S)} - \tau' \left[\Theta^{(S)} - \int \frac{d\mu'}{2} \Theta^{(S)}(\mu') + i\mu V_b \right] + \Phi' + ik\mu\Psi =$$
$$-\frac{\tau'}{2} \mathcal{P}_2(\mu) \int \frac{d\mu'}{2} \left[\mathcal{P}_2 \Theta^{(S)} - (1 - \mathcal{P}_2)\Theta_P^{(S)} \right](\mu') , \tag{5.111}$$

and

$$\left(\frac{\partial}{\partial\eta} + ik\mu \right) \Theta_P^{(S)} - \tau'\Theta_P^{(S)} =$$
$$-\frac{\tau'}{2}[1 - \mathcal{P}_2(\mu)] \int \frac{d\mu'}{2} \left[-\mathcal{P}_2 \Theta^{(S)} + (1 - \mathcal{P}_2)\Theta_P^{(S)} \right](\mu') . \tag{5.112}$$

Recall now the partial wave expansions already used for the neutrino distribution:

$$\Theta_\ell^{(S)} = \frac{1}{(-i)^\ell} \int_{-1}^1 \frac{d\mu}{2} \mathcal{P}_\ell(\mu)\Theta^{(S)}(\mu) , \quad \Theta_{P\ell}^{(S)} = \frac{1}{(-i)^\ell} \int_{-1}^1 \frac{d\mu}{2} \mathcal{P}_\ell(\mu)\Theta_P^{(S)}(\mu) , \tag{5.113}$$

Exercise 5.15 Show that the Boltzmann equations for the scalar contribution to the photon temperature and polarisation are:

$$\Theta^{(S)'} + ik\mu\Theta^{(S)} + \Phi' + ik\mu\Psi = -\tau' \left[\Theta_0^{(S)} - \Theta^{(S)} - i\mu V_b - \frac{1}{2}\mathcal{P}_2(\mu)\Pi \right]$$

(5.114)

$$\Theta_P^{(S)'} + ik\mu\Theta_P^{(S)} = -\tau' \left[-\Theta_P^{(S)} + \frac{1}{2}[1 - \mathcal{P}_2(\mu)]\Pi \right]$$

(5.115)

where Π is defined as follows:

$$\Pi \equiv \Theta_2^{(S)} + \Theta_{P2}^{(S)} + \Theta_{P0}^{(S)}$$

(5.116)

The anisotropic nature of Thomson scattering and polarisation were neglected by Peebles and Yu (1970) whereas the former only was included by Wilson and Silk (1981). Polarisation was considered in Bond and Efstathiou (1984).

The above two equations are usually expanded in Legendre polynomials, as we did for the neutrino equation. Using then Eq. (5.113) applied to Eq. (5.114), we obtain the following equation:

$$\Theta_\ell^{(S)'} + \frac{ik}{(-i)^\ell} \int_{-1}^1 \frac{d\mu}{2} \mu \mathcal{P}_\ell \Theta^{(S)} = -\frac{\Phi'}{(-i)^\ell} \int_{-1}^1 \frac{d\mu}{2} \mathcal{P}_\ell + \frac{ik\Psi}{(-i)^\ell} \int_{-1}^1 \frac{d\mu}{2} \mu \mathcal{P}_\ell$$

$$-\tau' \left[\frac{\Theta_0^{(S)}}{(-i)^\ell} \int_{-1}^1 \frac{d\mu}{2} \mathcal{P}_\ell - \Theta_l^{(S)} - \frac{iV_b}{(-i)^\ell} \int_{-1}^1 \frac{d\mu}{2} \mu \mathcal{P}_\ell - \frac{\Pi}{2(-i)^\ell} \int_{-1}^1 \frac{d\mu}{2} \mathcal{P}_2 \mathcal{P}_\ell \right].$$

(5.117)

Using the orthogonality relation of the Legendre polynomials, cf. Eq. (5.61), the integrals of the above equation are easily computed as follows:

$$\int_{-1}^1 \frac{d\mu}{2} \mathcal{P}_\ell = \frac{\delta_{\ell 0}}{2\ell + 1}, \quad \int_{-1}^1 \frac{d\mu}{2} \mu \mathcal{P}_\ell = \frac{\delta_{\ell 1}}{2\ell + 1}, \quad \int_{-1}^1 \frac{d\mu}{2} \mathcal{P}_2 \mathcal{P}_\ell = \frac{\delta_{\ell 2}}{2\ell + 1}.$$

(5.118)

Therefore, we must distinguish among 4 cases, i.e. when one of the above integrals contributes or none of them does (for $\ell > 2$). We shall make use again of the recurrence relation of Eq. (5.63), which allows us to write:

$$(2\ell + 1)\Theta_\ell^{(S)'} + k\left[(\ell + 1)\Theta_{\ell+1}^{(S)} - \ell\Theta_{\ell-1}^{(S)}\right] = \tau'(2\ell + 1)\Theta_\ell^{(S)}, \qquad \ell > 2$$

(5.119)

The equation for the quadrupole $\ell = 2$:

$$10\Theta_2^{(S)'} + 2k\left(3\Theta_3^{(S)} - 2\Theta_1^{(S)}\right) = 10\tau'\Theta_2^{(S)} - \tau'\Pi$$

(5.120)

The equation for the dipole $\ell = 1$:

$$3\Theta_1^{(S)'} + k\left(2\Theta_2^{(S)} - \Theta_0^{(S)}\right) = k\Psi + \tau'\left(3\Theta_1^{(S)} - V_b\right)$$

(5.121)

Finally, the equation for the monopole $\ell = 0$:

$$\Theta_0^{(S)'} + k\Theta_1^{(S)} = -\Phi'$$

(5.122)

For the polarisation equation the steps to be performed are the same. Therefore:

$$\Theta_{P\ell}^{(S)'} + \frac{k(\ell + 1)}{2\ell + 1}\Theta_{P(\ell+1)}^{(S)} - \frac{k\ell}{2\ell + 1}\Theta_{P(\ell-1)}^{(S)} =$$
$$-\tau'\left[-\Theta_{P\ell}^{(S)} + \frac{\Pi}{2(-i)^\ell}\int_{-1}^1 \frac{d\mu}{2}\mathcal{P}_\ell - \frac{\Pi}{2(-i)^\ell}\int_{-1}^1 \frac{d\mu}{2}\mathcal{P}_\ell\mathcal{P}_2(\mu)\right].$$

(5.123)

So, the equation for $\ell > 2$ is the following:

$$(2\ell + 1)\Theta_{P\ell}^{(S)'} + k\left[(\ell + 1)\Theta_{P(\ell+1)}^{(S)} - \ell\Theta_{P(\ell-1)}^{(S)}\right] = \tau'(2\ell + 1)\Theta_{P\ell}^{(S)}, \qquad \ell > 2$$

(5.124)

The equation for the quadrupole $\ell = 2$:

$$10\Theta_{P2}^{(S)'} + 2k\left(3\Theta_{P3}^{(S)} - 2\Theta_{P1}^{(S)}\right) = 10\tau'\Theta_{P2}^{(S)} - \tau'\Pi$$

(5.125)

The equation for the dipole $\ell = 1$:

$$3\Theta_{P1}^{(S)'} + k\left(2\Theta_{P2}^{(S)} - \Theta_{P0}^{(S)}\right) = 3\tau'\Theta_{P1}^{(S)}$$

(5.126)

Finally, the equation for the monopole $\ell = 0$:

$$2\Theta_{P0}^{(S)'} + 2k\Theta_{P1}^{(S)} = 2\tau'\Theta_{P0}^{(S)} - \tau'\Pi$$

(5.127)

In order to compare with Ma and Bertschinger (1995), one has to make the identification $4\Theta_0^{(S)} = \delta_\gamma$, $3k\Theta_1^{(S)} = \theta_\gamma$, $2\Theta_2^{(S)} = \sigma_\gamma$ and $kV_b = \theta_b$.

Note that the monopole and the dipole are thus related to the density contrast and to the photon fluid velocity, hence they are gauge-dependent. In particular, the monopole can be reabsorbed into the determination of the background temperature whereas the dipole by a suitable boost. On the other hand, the quadrupole is related to the anisotropic stress, so it is gauge-invariant as well as the higher-order multipoles.

5.5.4 Tensor Perturbations

We derive in this section the hierarchy of equations describing the evolution of fluctuations in the photon distribution caused by tensor perturbations, i.e. by GW. Consider the tensor contribution of Eq. (5.106). Using Eq. (5.32) we have:

$$\left(\frac{\partial}{\partial\eta} + ik\mu\right)\Theta^{(T)} - \tau'\Theta^{(T)} + \frac{h'_+}{2}(1-\mu^2)\cos 2\phi + \frac{h'_\times}{2}(1-\mu^2)\sin 2\phi =$$

$$-\frac{\tau'}{10}\sum_{m=-2}^{2} Y_2^m(\hat{p})\int d^2\hat{p}'\left[Y_2^{m*}\Theta^{(T)} - \sqrt{\frac{3}{2}}\mathcal{E}^{m*}\Theta_P^{(T)} - \sqrt{\frac{3}{2}}\frac{(\mathcal{B}^{m*})^2}{\mathcal{E}^{m*}}\Theta_P^{(T)}\right](\hat{p}'),$$

$$(5.128)$$

whereas for the polarisation part:

$$\left(\frac{\partial}{\partial\eta} + ik\mu\right)\Theta_P^{(T)} - \tau'\Theta_P^{(T)} =$$

$$\frac{\tau'}{10}\sum_{m=-2}^{2}\sqrt{\frac{3}{2}}\mathcal{E}^m\int d^2\hat{p}'\left[Y_2^{m*}\Theta^{(T)} - \sqrt{\frac{3}{2}}\mathcal{E}^{m*}\Theta_P^{(T)} - \sqrt{\frac{3}{2}}\frac{(\mathcal{B}^{m*})^2}{\mathcal{E}^{m*}}\Theta_P^{(T)}\right](\hat{p}').$$

$$(5.129)$$

Note that in Eq. (5.106) the monopole term $\int d\hat{p}\, \Theta/4\pi$ is a pure scalar and therefore does not provide any tensor contribution. The scalar product $\mathbf{v}_b \cdot \hat{p}$ has a scalar contribution, used in the previous subsection, which is $i\mu V_b$, and it has also a vector contribution which can be written as:

$$(\mathbf{v}_b \cdot \hat{p})^{(V)} = U_b^1\sin\theta\cos\phi + U_b^2\sin\theta\sin\phi,$$ $$(5.130)$$

where we have used Eq. (5.31) and the fact that $k_i U_b^i = 0$ because of the vector nature of U_b^i and the choice of having $k^3 = k$, i.e. $\hat{k} = \hat{z}$. So, $\mathbf{v}_b \cdot \hat{p}$ does have an azimuthal dependence, but not in the form of $e^{i2\phi}$ and thus cannot contribute to the tensor Boltzmann equation (it contributes to the vector one).

Mimicking the azimuthal dependence produced by tensor perturbations in the metric, we can split the tensor contribution to the temperature anisotropy as follows, following Polnarev (1985) and Crittenden et al. (1993):

$$\Theta^{(T)}(\mu, \phi) = \Theta_+^{(T)}(\mu)(1 - \mu^2) \cos 2\phi + \Theta_\times^{(T)}(\mu)(1 - \mu^2) \sin 2\phi$$

$$= 4\sqrt{\frac{\pi}{15}} \sum_{\lambda = \pm 2} \Theta_\lambda^{(T)}(\mu) Y_2^\lambda(\mu, \phi) \quad (5.131)$$

where we have reproduced the sum over the helicities of Eq. (5.32), and similarly for the polarisation field:

$$\Theta_P^{(T)}(\mu, \phi) = \Theta_{P+}^{(T)}(\mu)(1 + \mu^2) \cos 2\phi + \Theta_{P\times}^{(T)}(\mu)(1 + \mu^2) \sin 2\phi$$

$$= 4\sqrt{\frac{\pi}{15}}\sqrt{\frac{3}{2}} \sum_{\lambda = \pm 2} \Theta_{P\lambda}^{(T)}(\mu) \mathcal{E}^\lambda(\mu, \phi) \quad (5.132)$$

So we have to select $m = \pm 2$ in sum on the right hand side of Eq. (5.106), and we get:

$$\left(\frac{\partial}{\partial \eta} + ik\mu - \tau'\right) \Theta_\lambda^{(T)} + \frac{h_\lambda'}{2} =$$

$$-\frac{\tau'}{10} \int d^2 \hat{p}' \left[Y_2^{\lambda *} \Theta_\lambda^{(T)} Y_2^\lambda - \frac{3}{2} \mathcal{E}^{\lambda *} \Theta_{P\lambda}^{(T)} \mathcal{E}^{\lambda *} - \frac{3}{2} \frac{(\mathcal{B}^{\lambda *})^2}{\mathcal{E}^{\lambda *}} \Theta_{P\lambda}^{(T)} \mathcal{E}^\lambda \right] (\hat{p}'), \quad (5.133)$$

whereas for the polarisation part:

$$\left(\frac{\partial}{\partial \eta} + ik\mu - \tau'\right) \Theta_{P\lambda}^{(T)} =$$

$$\frac{\tau'}{10} \int d^2 \hat{p}' \left[Y_2^{\lambda *} \Theta_\lambda^{(T)} Y_2^\lambda - \frac{3}{2} \mathcal{E}^{\lambda *} \Theta_{P\lambda}^{(T)} \mathcal{E}^\lambda - \frac{3}{2} \frac{(\mathcal{B}^{\lambda *})^2}{\mathcal{E}^{\lambda *}} \Theta_{P\lambda}^{(T)} \mathcal{E}^\lambda \right] (\hat{p}'), \quad (5.134)$$

where $\lambda = \pm 2$ and the equations are identical for the two choices. Let us work out the right hand sides.

Exercise 5.16 With the help of Tables 5.1 and 5.2 show that the integral on the right hand sides becomes:

$$-\frac{\tau'}{10} \int d\mu d\phi \frac{15}{32\pi} \left[(1 - \mu^2)^2 \Theta_\lambda^{(T)} - (1 + 6\mu^2 + \mu^4) \Theta_{P\lambda}^{(T)} \right]. \quad (5.135)$$

Let us introduce an expansion in Legendre polynomials for $\Theta_\lambda^{(T)}(\mu)$ and $\Theta_{P,\lambda}^{(T)}(\mu)$ similar to that in Eq. (5.113):

$$\Theta_{\lambda,\ell}^{(T)} = \frac{1}{(-i)^\ell} \int_{-1}^1 \frac{d\mu}{2} \mathcal{P}_\ell(\mu) \Theta_\lambda^{(T)}(\mu) , \quad \Theta_{P\lambda,\ell}^{(T)} = \frac{1}{(-i)^\ell} \int_{-1}^1 \frac{d\mu}{2} \mathcal{P}_\ell(\mu) \Theta_{P\lambda}^{(T)}(\mu) .$$

(5.136)

Exercise 5.17 Rewrite the contributions $(1 - \mu^2)^2$ and $(1 + 6\mu^2 + \mu^4)$ in terms of Legendre polynomials, and show that:

$$-\frac{3\tau'}{16} \int \frac{d\mu}{2} \left[\frac{8}{35} \mathcal{P}_4(\mu) - \frac{80}{105} \mathcal{P}_2(\mu) + \frac{8}{15} \mathcal{P}_0(\mu) \right] \Theta_\lambda^{(T)}$$

$$+\frac{3\tau'}{16} \int \frac{d\mu}{2} \left[\frac{8}{35} \mathcal{P}_4(\mu) + \frac{32}{7} \mathcal{P}_2(\mu) + \frac{16}{5} \mathcal{P}_0(\mu) \right] \Theta_{P\lambda}^{(T)} .$$

(5.137)

Hence, using Eq. (5.136), we can write the tensor Boltzmann equation for photons as follows (Crittenden et al. 1993):

$$\left(\frac{\partial}{\partial \eta} + ik\mu - \tau' \right) \Theta_\lambda^{(T)} + \frac{1}{2} h_\lambda' =$$

$$-\tau' \left[\frac{3}{70} \Theta_{\lambda,4}^{(T)} + \frac{1}{7} \Theta_{\lambda,2}^{(T)} + \frac{1}{10} \Theta_{\lambda,0}^{(T)} - \frac{3}{70} \Theta_{P\lambda,4}^{(T)} + \frac{6}{7} \Theta_{P\lambda,2}^{(T)} - \frac{3}{5} \Theta_{P\lambda,0}^{(T)} \right] \quad (5.138)$$

and for polarisation:

$$\left(\frac{\partial}{\partial \eta} + ik\mu - \tau' \right) \Theta_{P\lambda}^{(T)} =$$

$$\tau' \left[\frac{3}{70} \Theta_{\lambda,4}^{(T)} + \frac{1}{7} \Theta_{\lambda,2}^{(T)} + \frac{1}{10} \Theta_{\lambda,0}^{(T)} - \frac{3}{70} \Theta_{P\lambda,4}^{(T)} + \frac{6}{7} \Theta_{P\lambda,2}^{(T)} - \frac{3}{5} \Theta_{P\lambda,0}^{(T)} \right] . (5.139)$$

The same equations hold true also for $\Theta_{+,\times}^{(T)}$ and $\Theta_{P+,\times}^{(T)}$. The combination of terms between square brackets in the above equations is sometimes dubbed as Ψ in the literature. Since for us Ψ is already used as one of the Bardeen potentials, in Chap. 10 we shall use another notation.

5.5.5 Vector Perturbations

For completeness, we present here the Boltzmann equation for photons sourced by vector perturbations in the metric, though we shall not use it. Using Eqs. (5.35) and (5.130), we can write:

$$\left(\frac{\partial}{\partial\eta} + ik\mu - \tau'\right)\Theta^{(V)} - \frac{\tau'}{\cos\theta}\sqrt{\frac{2\pi}{15}}(Y_2^1 U_{b,-} + Y_2^{-1} U_{b,+})$$

$$-\frac{i}{2}\sqrt{\frac{2\pi}{15}}(Y_2^1 F_-' + Y_2^{-1} F_+') =$$

$$-\frac{\tau'}{10}\sum_{m=-2}^{2} Y_2^m(\hat{p}) \int d^2\hat{p}' \left[Y_2^{m*}\Theta^{(V)} - \sqrt{\frac{3}{2}}\mathcal{E}^{m*}\Theta_P^{(V)} - \sqrt{\frac{3}{2}}\frac{(\mathcal{B}^{m*})^2}{\mathcal{E}^{m*}}\Theta_P^{(V)} \right](\hat{p}'),$$

$$(5.140)$$

where

$$U_{b,\pm} \equiv U_{b1} \pm i U_{b2}, \qquad (5.141)$$

and for the polarisation:

$$\left(\frac{\partial}{\partial\eta} + ik\mu - \tau'\right)\Theta_P^{(V)} =$$

$$\frac{\tau'}{10}\sum_{m=-2}^{2}\sqrt{\frac{3}{2}}\mathcal{E}^m \int d^2\hat{p}' \left[Y_2^{m*}\Theta^{(V)} - \sqrt{\frac{3}{2}}\mathcal{E}^{m*}\Theta_P^{(V)} - \sqrt{\frac{3}{2}}\frac{(\mathcal{B}^{m*})^2}{\mathcal{E}^{m*}}\Theta_P^{(V)} \right](\hat{p}').$$

$$(5.142)$$

Introduce the vector contribution to temperature anisotropy as follows:

$$\boxed{\Theta^{(V)}(\mu,\phi) = \frac{i}{\cos\theta}\sqrt{\frac{2\pi}{15}}\sum_{\lambda=\pm1}\Theta_\lambda^{(V)}(\mu)Y_2^{-\lambda}(\mu,\phi)} \qquad (5.143)$$

where the factor $1/\cos\theta$ is due to the fact that in Eq. (5.130) only a $\sin\theta$ appears and thus it is not proportional to $Y_2^{\pm1}$. Similarly, for the polarisation field:

$$\boxed{\Theta_P^{(V)}(\mu,\phi) = -\sqrt{\frac{\pi}{5}}\sum_{\lambda=\pm1}\Theta_{P\lambda}^{(V)}(\mu)\mathcal{E}^{-\lambda}(\mu,\phi)} \qquad (5.144)$$

The vector Boltzmann equation for the temperature can then be written as:

$$\left(\frac{\partial}{\partial\eta} + ik\mu - \tau'\right)\Theta_\lambda^{(V)} + i\tau' U_{b^\cdot} - \frac{1}{2}F_\lambda'\mu =$$

$$\frac{i\tau'}{10}\mu \int d^2\hat{p}' \left[Y_2^{-\lambda*}\Theta_\lambda^{(V)}\frac{i}{\mu'}Y_2^{-\lambda} + \frac{3}{2}\mathcal{E}^{-\lambda*}\Theta_{P\lambda}^{(V)}\mathcal{E}^{-\lambda} + \frac{3}{2}\frac{(\mathcal{B}^{-\lambda*})^2}{\mathcal{E}^{-\lambda*}}\Theta_{P\lambda}^{(V)}\mathcal{E}^{-\lambda} \right],$$

$$(5.145)$$

and for polarisation:

$$\left(\frac{\partial}{\partial\eta}+ik\mu-\tau'\right)\Theta_{P\lambda}^{(V)}=$$

$$-\frac{\tau'}{10}\int d^2\hat{p}'\left[Y_2^{-\lambda*}\Theta_{\lambda}^{(V)}\frac{i}{\mu'}Y_2^{-\lambda}+\frac{3}{2}\mathcal{E}^{-\lambda*}\Theta_{P\lambda}^{(V)}\mathcal{E}^{-\lambda}+\frac{3}{2}\frac{(\mathcal{B}^{-\lambda*})^2}{\mathcal{E}^{-\lambda*}}\Theta_{P\lambda}^{(V)}\mathcal{E}^{-\lambda}\right],$$

$$(5.146)$$

With the help of Table 5.1 we have then to work out the integral:

$$-\frac{\tau'}{10}\int d\mu d\phi\frac{15}{8\pi}\left[\mu(1-\mu^2)i\Theta_{\lambda}^{(V)}+(1-\mu^4)\Theta_{P\lambda}^{(V)}\right],\qquad(5.147)$$

which, written in Legendre polynomial, becomes:

$$-\frac{3\tau'}{4}\int\frac{d\mu}{2}\left[\frac{2}{5}\mathcal{P}_1(\mu)-\frac{2}{5}\mathcal{P}_3(\mu)\right]i\Theta_{\lambda}^{(V)}$$

$$-\frac{3\tau'}{4}\int\frac{d\mu}{2}\left[\frac{4}{5}\mathcal{P}_0(\mu)-\frac{4}{7}\mathcal{P}_2(\mu)-\frac{8}{35}\mathcal{P}_4(\mu)\right]\Theta_{P\lambda}^{(V)}.\qquad(5.148)$$

Hence, using the usual Legendre expansion, we can write the vector Boltzmann equation for photons as follows:

$$\left(\frac{\partial}{\partial\eta}+ik\mu-\tau'\mu\right)\Theta_{\lambda}^{(V)}+i\tau'U_{b,\cdot}-\frac{1}{2}F_{\lambda}'=$$

$$i\tau'\mathcal{P}_1(\mu)\left(\frac{3}{10}\Theta_{\lambda,1}^{(V)}+\frac{3}{10}\Theta_{\lambda,3}^{(V)}-\frac{6}{35}\Theta_{P\lambda,4}^{(V)}+\frac{3}{7}\Theta_{P\lambda,2}^{(V)}+\frac{3}{5}\Theta_{P\lambda,0}^{(V)}\right).\quad(5.149)$$

and for polarisation:

$$\left(\frac{\partial}{\partial\eta}+ik\mu-\tau'\mu\right)\Theta_{P\lambda}^{(V)}=$$

$$-\tau'\left(\frac{3}{10}\Theta_{\lambda,1}^{(V)}+\frac{3}{10}\Theta_{\lambda,3}^{(V)}-\frac{6}{35}\Theta_{P\lambda,4}^{(V)}+\frac{3}{7}\Theta_{P\lambda,2}^{(V)}+\frac{3}{5}\Theta_{P\lambda,0}^{(V)}\right).\quad(5.150)$$

We end here our treatment of vector modes, dealing just with the scalar and tensor ones from now on.

5.6 Boltzmann Equation for Baryons

As baryons, we refer here generically to electrons and protons, neglecting helium nuclei. The latter can be straightforwardly included, as we are going to comment throughout the derivation. Electrons and protons interact via Coulomb scattering:

$$e + p \leftrightarrow e + p \,, \tag{5.151}$$

and in turn recall that electrons are also coupled to photons via Thomson scattering. Electrons also interact with Helium nuclei via Coulomb scattering. We assume the interactions to be so efficient that:

$$\delta_e = \delta_p = \delta_b \,, \qquad \mathbf{v}_e = \mathbf{v}_p = \mathbf{v}_b \,. \tag{5.152}$$

In other words, we do not expect to have more electrons here and more protons there, thereby generating cosmic dipoles. So, baryons together form the so-called **baryonic plasma**. Baryons are heavy and non-relativistic in the epochs of interest and thus there is no exchange of energy via Thomson scattering with photons. For this reason we expect that their density contrast is governed by an equation similar to Eq. (5.50).

We can check this by considering the Boltzmann equation for electrons:

$$\frac{d\mathcal{F}_e(\eta, \mathbf{x}, \mathbf{q})}{d\eta} = \langle c_{ep} \rangle_{QQ'q'} + \langle c_{e\gamma} \rangle_{pp'q'} \,, \tag{5.153}$$

where $\langle c_{ep} \rangle_{QQ'q'}$ is the collisional term relative to Coulomb scattering:

$$e(\mathbf{q}) + p(\mathbf{Q}) \leftrightarrow e(\mathbf{q}') + p(\mathbf{Q}') \,, \tag{5.154}$$

whereas $\langle c_{e\gamma} \rangle_{pp'q'}$ is the collisional term relative to Thomson scattering:

$$e(\mathbf{q}) + \gamma(\mathbf{p}) \leftrightarrow e(\mathbf{q}') + \gamma(\mathbf{p}') \,. \tag{5.155}$$

For protons, the Boltzmann equation is similar:

$$\frac{d\mathcal{F}_p(\eta, \mathbf{x}, \mathbf{Q})}{d\eta} = \langle c_{ep} \rangle_{qq'Q'} \,, \tag{5.156}$$

the only difference being that we neglect their interaction with photons, since the Thomson cross-section goes as $\propto 1/m^2$ and therefore it is 10^6 less important for protons rather than for electrons. A Boltzmann equation for Helium nuclei is similar to the one for protons. The brackets mean integration in the phase space:

$$\langle\langle (\cdots) \rangle\rangle_{qq'Q'} \equiv \int \frac{d^3\mathbf{q}}{(2\pi)^3} \int \frac{d^3\mathbf{q}'}{(2\pi)^3} \int \frac{d^3\mathbf{Q}'}{(2\pi)^3} (\cdots) \,, \tag{5.157}$$

whereas the integrands are similar to those in Eq. (5.82).

$$c_{e\gamma} \equiv \frac{a}{E(q)} \frac{(2\pi)^4 \delta^4 (q + p - q' - p') |\mathcal{M}|^2 [f_e(q') f_\gamma(p') - f_e(q) f_\gamma(p)]}{8 E(p') E(q)(p) E_e(q')} \,, \tag{5.158}$$

with a similar expression, but with different $|\mathcal{M}|^2$, for c_{ep}. Note the $a/E_e(q)$ factor in the above definition of $c_{e\gamma}$ that we have left in evidence in order to recall that it comes comes from changing the affine parameter derivative for the conformal time derivative in the Boltzmann equation, cf. Eq. (5.27).

An important approximation that we are making is the following: we are considering all the electron ionised (hence the name baryonic plasma). This is fine before recombination, of course, but it does not work after. After recombination, electrons and protons (and also helium) can be considered altogether as collisionless baryons. Therefore, their evolution is governed by equations identical to the ones we have developed earlier for CDM.

We are going to exploit again the fact that electrons and protons are non-relativistic, as we did for CDM, and take moments of the perturbed Boltzmann equations.

The calculations of the integrals of Eqs. (5.153) and (5.156) multiplied by the respective energies follow the same steps that we have used for Eqs. (5.40) and (5.48):

$$\int \frac{d^3\mathbf{q}}{(2\pi)^3} \frac{d\mathcal{F}_e}{d\eta} = \delta'_e + (1 + w_e)kV_e + 3\mathcal{H}\left(\frac{\delta P_e}{\delta \rho_e} - w_e\right)\delta_e + 3(1 + w_e)\Phi' ,$$

(5.159)

$$\int \frac{d^3\mathbf{q}}{(2\pi)^3} \frac{d\mathcal{F}_p}{d\eta} = \delta'_p + (1 + w_p)kV_p + 3\mathcal{H}\left(\frac{\delta P_p}{\delta \rho_p} - w_p\right)\delta_p + 3(1 + w_p)\Phi' .$$

(5.160)

These integrations are equal to the corresponding ones of the collisional terms, which are very simple and are the following:

$$\int \frac{d^3\mathbf{q}}{(2\pi)^3}[\langle c_{ep}\rangle_{QQ'q'} + \langle c_{e\gamma}\rangle_{pp'q'}] = 0 , \quad \int \frac{d^3\mathbf{Q}}{(2\pi)^3}\langle c_{ep}\rangle_{Q'q'q} = 0 , \quad (5.161)$$

i.e. vanishing because the scattering processes do not change the number of baryons but only reshuffle their momenta. Mathematically, this can be proved by noticing that $|\mathcal{M}|^2$ is symmetric with respect to the change of initial and final momenta.

Therefore, using the conditions stated earlier in Eq. (5.152), the zero-moment equations that we find for electrons and protons are identical if we set $w_e = w_p = 0$ and no effective speeds of sound. Neglecting these, we obtain an equation similar to Eq. (5.50), which we found for CDM:

$$\boxed{\delta' + kV_b + 3\Phi' = 0}$$

(5.162)

where, again, $kV_b = \theta_b$ in the notation of Ma and Bertschinger (1995).

Taking care of the first moments of the perturbed Boltzmann equations for electrons and protons, as we did for Eq. (5.48), leaves us with:

$$ i\hat{k}_i \int \frac{d^3\mathbf{q}}{(2\pi)^3} q\hat{q}^i \frac{d\mathcal{F}_e}{d\eta} = \rho_e (V'_e + \mathcal{H}V_e - k\Psi) , \tag{5.163} $$

$$ i\hat{k}_i \int \frac{d^3\mathbf{Q}}{(2\pi)^3} Q\hat{Q}^i \frac{d\mathcal{F}_p}{d\eta} = \rho_p (V'_p + \mathcal{H}V_p - k\Psi) , \tag{5.164} $$

where we have already neglected pressure and effective speed of sound. The first moments of the Boltzmann equations are thus:

$$ \int \frac{d^3\mathbf{q}}{(2\pi)^3} \frac{q\hat{q}^i}{E_e} \frac{df_e(\eta, \mathbf{x}, \mathbf{q})}{d\eta} = \int \frac{d^3\mathbf{q}}{(2\pi)^3} \frac{q\hat{q}^i}{E_e} [\langle c_{ep}\rangle_{QQ'q'} + \langle c_{e\gamma}\rangle_{pp'q'}] , \tag{5.165} $$

$$ \int \frac{d^3\mathbf{Q}}{(2\pi)^3} \frac{Q\hat{Q}^i}{E_p} \frac{df_p(\eta, \mathbf{x}, \mathbf{Q})}{d\eta} = \int \frac{d^3\mathbf{Q}}{(2\pi)^3} \frac{Q\hat{Q}^i}{E_p} \langle c_{ep}\rangle_{qq'Q'} . \tag{5.166} $$

Consider the sum of the two above Liouville operator, and using $V_e = V_p = V_b$, we can write:

$$ (\rho_e + \rho_p)(V'_b + \mathcal{H}V_b - k\Psi) = $$
$$ i\hat{k}_i \int \frac{d^3\mathbf{q}}{(2\pi)^3} q^i [\langle c_{ep}\rangle_{QQ'q'} + \langle c_{e\gamma}\rangle_{pp'q'}] + i\hat{k}_i \int \frac{d^3\mathbf{Q}}{(2\pi)^3} Q^i \langle c_{ep}\rangle_{qq'Q'} . \tag{5.167} $$

We write the sum of the background energies as

$$ \rho_e + \rho_p = n_e m_e + n_p m_p = n_b m_e + n_b m_p \approx n_b m_p \equiv \rho_b , \tag{5.168} $$

since the proton mass is much larger than the electron one, and hence

$$ V'_b + \mathcal{H}V_b - k\Psi = \frac{i\hat{k}_i}{\rho_b} [\langle c_{ep}(q^i + Q^i)\rangle_{QQ'q'q} + \langle c_{e\gamma}q^i\rangle_{pp'qq'}] . \tag{5.169} $$

Now we have to calculate the terms on the right hand side. The first one is:

$$ \langle c_{ep}(q^i + Q^i)\rangle_{QQ'q'q} = a \int \frac{d^3\mathbf{q}}{(2\pi)^3} \int \frac{d^3\mathbf{q}'}{(2\pi)^3} \int \frac{d^3\mathbf{Q}}{(2\pi)^3} \int \frac{d^3\mathbf{Q}'}{(2\pi)^3} (q^i + Q^i) $$
$$ \frac{(2\pi)^4 \delta^4(q + Q - q' - Q')|\mathcal{M}|^2 [f_e(q')f_p(Q') - f_e(q)f_p(Q)]}{8m_e^2 m_p^2} , $$

$$ \tag{5.170} $$

which is vanishing because $q^i + Q^i$ is the total initial 3-momentum, so it is equal to the final one $q'^i + Q'^i$ because it is conserved. Another way to see this is noticing that also $(q^i + Q^i)|\mathcal{M}|^2$ is symmetric with respect to the change of the initial momenta with the final ones and this implies that the integration is zero. Physically, being conserved the total 3-momentum is unaffected by the scattering.

The only survivor is the second term on the right hand side, which comes from Thomson scattering. By 3-momentum conservation we can write:

$$\langle c_{e\gamma}(q^i + p^i)\rangle_{pp'qq'} = 0 \quad \Rightarrow \quad \langle c_{e\gamma}q^i\rangle_{pp'qq'} = -\langle c_{e\gamma}p^i\rangle_{pp'qq'} . \tag{5.171}$$

That is, we have changed the average with respect to the photon momentum. We have then:

$$V_b' + \mathcal{H}V_b - k\Psi = -\frac{i}{\rho_b}\langle c_{e\gamma}\mu p\rangle_{pp'qq'} = -\frac{i}{\rho_b}\int_{-1}^{1}\frac{d\mu}{2}\mu\int\frac{dp\,p^2}{2\pi^2}p\,\mathcal{C}_{e\gamma} . \tag{5.172}$$

Part of the right hand side has already been calculated for photons, when we integrated over the modulus p and defined $\Theta(\eta, \mathbf{k}, \mu)$. This integration gives a $4\rho_\gamma$ contribution and leaves us with the same collisional term of Eq. (5.114) integrated over $d\mu/2$ and without the polarisation contribution:

$$V_b' + \mathcal{H}V_b - k\Psi = \frac{4i\tau'\rho_\gamma}{\rho_b}\int_{-1}^{1}\frac{d\mu}{2}\mu\left[\Theta_0^{(S)} - \Theta^{(S)} - i\mu V_b - \frac{1}{2}\mathcal{P}_2(\mu)\Theta_2^{(S)}\right] . \tag{5.173}$$

Performing the $d\mu$ integration and introducing the dipole $\Theta_1^{(S)}$, we obtain:

$$\boxed{V_b' + \mathcal{H}V_b - k\Psi = \frac{4\tau'\rho_\gamma}{3\rho_b}\left(V_b - 3\Theta_1^{(S)}\right)} \tag{5.174}$$

This equation can be also obtained exploiting momentum conservation for the two-fluid system baryonic plasma plus photons and using Eq. (5.48). In fact, momentum is not conserved individually for baryons and photon, but of course it is for the photon-baryon fluid. Let us write then Eq. (5.48) for the latter:

$$\rho_b\left(V_b' + \mathcal{H}V_b - k\Psi\right) + \frac{4}{3}\rho_\gamma V_\gamma' - \frac{k}{3}\rho_\gamma\delta_\gamma - k\hat{k}_l\hat{k}_m\pi_\gamma^{lm} - \frac{4k}{3}\rho_\gamma\Psi = 0 , \tag{5.175}$$

where we have neglected pressure and anisotropic stress for baryons, being them non-relativistic. Now, Eqs. (5.68), (5.69) and (5.70) hold true also for photons. So we have:

$$4\Theta_0^{(S)} = \delta_\gamma , \qquad 4\Theta_1^{(S)} = \frac{4}{3}V_\gamma , \qquad 4\Theta_2^{(S)} = -\frac{3}{2\rho_\gamma}\hat{k}_l\hat{k}_m\pi_\gamma^{lm} . \tag{5.176}$$

Exercise 5.18 Using the above definitions and Eq. (5.121) obtain Eq. (5.174).

We have thus obtained all the relevant equations which describe small fluctuations in the components of the universe. In Chap. 6 we shall tackle the issue of which initial conditions employ to our set of differential equations.

If we compare the above Eq. (5.174) with the corresponding one in Ma and Bertschinger (1995) we shall notice in this reference an extra term $k^2 c_s^2 \delta_b$, coming from the fact that the authors did not neglect the effective speed of sound contribution for baryons, cf. Eq. (5.48).

References

Bond, J.R., Efstathiou, G.: Cosmic background radiation anisotropies in universes dominated by nonbaryonic dark matter. Astrophys. J. **285**, L45–L48 (1984)

Chandrasekhar, S.: Radiative Transfer. Dover, New York (1960)

Crittenden, R., Bond, J.R., Davis, R.L., Efstathiou, G., Steinhardt, P.J.: The imprint of gravitational waves on the cosmic microwave background. Phys. Rev. Lett. **71**, 324–327 (1993)

Hu, W., White, M.J.: CMB anisotropies: total angular momentum method. Phys. Rev. D **56**, 596–615 (1997)

Kosowsky, A.: Cosmic microwave background polarization. Ann. Phys. **246**, 49–85 (1996)

Ma, C.-P., Bertschinger, E.: Cosmological perturbation theory in the synchronous and conformal Newtonian gauges. Astrophys. J. **455**, 7–25 (1995)

Peebles, P.J.E., Yu, J.T.: Primeval adiabatic perturbation in an expanding universe. Astrophys. J. **162**, 815–836 (1970)

Piattella, O.F., Rodrigues, D.C., Fabris, J.C., de Freitas Pacheco, J.A.: Evolution of the phase-space density and the Jeans scale for dark matter derivedfrom the Vlasov-Einstein equation. JCAP **1311**, 002 (2013)

Piattella, O.F., Casarini, L., Fabris, J.C., de Freitas Pacheco, J.A.: Dark matter velocity dispersion effects on CMB and matter power spectra. JCAP **1602**(02), 024 (2016)

Polnarev, A.G.: Polarization and anisotropy induced in the microwave background by cosmological gravitational waves. Sov. Astron. **29**, 607–613 (1985)

Seljak, U., Zaldarriaga, M.: A Line of sight integration approach to cosmic microwave background anisotropies. Astrophys. J. **469**, 437–444 (1996)

Tram, T., Lesgourgues, J.: Optimal polarisation equations in FLRW universes. JCAP **1310**, 002 (2013)

Weinberg, S.: A no-truncation approach to cosmic microwave background anisotropies. Phys. Rev. D **74**, 063517 (2006)

Weinberg, S.: Cosmology. Oxford University Press, Oxford (2008)

Wilson, M.L., Silk, J.: On the Anisotropy of the cosomological background matter and radiation distribution. 1. The Radiation anisotropy in a spatially flat universe. Astrophys. J. **243**, 14–25 (1981)

Chapter 6
Initial Conditions

Finirai per trovarla la Via... se prima hai il coraggio di perderti
(You will eventually find your Way... if you have first the courage
of lose yourself)

Tiziano Terzani, Un altro giro di giostra

In Chaps. 4 and 5 we have derived the evolution equations for small fluctuations about the homogeneous and isotropic FLRW background. These equations are differential and so, in order to have a well-posed Cauchy problem, we need to know which initial conditions to use. This is the topic of this chapter, where we shall be concerned with scalar perturbations only. Many papers in the literature are concerned about initial conditions in cosmology, but we shall mainly refer to Ma and Bertschinger (1995), Bucher et al. (2000).

We shall discuss primordial modes for scalar perturbations only, leaving the tensor one in Chap. 8, when we shall discuss of inflation. For this reason in this Chapter we drop the superscript (S).

6.1 Initial Conditions

We start by summarizing here the relevant evolution equations that we have found in Chaps. 4 and 5. For the photon temperature fluctuations, Eqs. (5.119)–(5.122):

$$(2\ell + 1)\Theta_\ell' + k\left[(\ell + 1)\Theta_{\ell+1} - \ell\Theta_{\ell-1}\right] = \tau'(2\ell + 1)\Theta_\ell , \qquad (\ell > 2) , \quad (6.1)$$

$$10\Theta_2' + 2k\left(3\Theta_3 - \frac{2}{3}V_\gamma\right) = 10\tau'\Theta_2 - \tau'\Pi , \quad (6.2)$$

$$V_\gamma' + k\left(2\Theta_2 - \frac{\delta_\gamma}{4}\right) = k\Psi + \tau'\left(V_\gamma - V_b\right) , \quad (6.3)$$

$$\delta_\gamma' + \frac{4}{3}kV_\gamma = -4\Phi' , \quad (6.4)$$

© Springer International Publishing AG, part of Springer Nature 2018
O. Piattella, *Lecture Notes in Cosmology*, UNITEXT for Physics,
https://doi.org/10.1007/978-3-319-95570-4_6

where we have used $4\Theta_0 = \delta_\gamma$ and $3\Theta_1 = V_\gamma$. For the polarisation field, Eqs. (5.124)–(5.127):

$$(2\ell + 1)\Theta'_{P\ell} + k\left[(\ell + 1)\Theta_{P(\ell+1)} - \ell\Theta_{P(\ell-1)}\right] = \tau'(2\ell + 1)\Theta_{P\ell}, \qquad (\ell > 2), \tag{6.5}$$

$$10\Theta'_{P2} + 2k\left(3\Theta_{P3} - 2\Theta_{P1}\right) = 10\tau'\Theta_{P2} - \tau'\Pi, \tag{6.6}$$

$$3\Theta'_{P1} + k\left(2\Theta_{P2} - \Theta_{P0}\right) = 3\tau'\Theta_{P1}, \tag{6.7}$$

$$2\Theta'_{P0} + 2k\Theta_{P1} = 2\tau'\Theta_{P0} - \tau'\Pi, \tag{6.8}$$

with

$$\Pi = \Theta_2 + \Theta_{P2} + \Theta_{P0}. \tag{6.9}$$

For neutrinos, Eqs. (5.65)–(5.67):

$$(2\ell + 1)\mathcal{N}'_\ell + k\left[(\ell + 1)\mathcal{N}_{\ell+1} - \ell\mathcal{N}_{\ell-1}\right] = 0, \qquad (\ell > 1), \tag{6.10}$$

$$V'_\nu + 2k\mathcal{N}_2 - k\frac{\delta_\nu}{4} = k\Psi, \tag{6.11}$$

$$\delta'_\nu + \frac{4}{3}kV_\nu = -4\Phi', \tag{6.12}$$

where we have used $4\mathcal{N}_0 = \delta_\nu$ and $3\mathcal{N}_1 = V_\nu$.

For CDM, Eqs. (5.50) and (5.51):

$$\delta'_c + kV_c + 3\Phi' = 0, \qquad V'_c + \mathcal{H}V_c - k\Psi = 0. \tag{6.13}$$

For baryons, Eqs. (5.162) and (5.174):

$$\delta'_b + kV_b + 3\Phi' = 0, \qquad V'_b + \mathcal{H}V_b - k\Psi = \frac{4\tau'\rho_\gamma}{3\rho_b}\left(V_b - V_\gamma\right), \tag{6.14}$$

Finally, we consider the Einstein equations (4.179) and (4.184):

$$3\mathcal{H}\left(\Phi' - \mathcal{H}\Psi\right) + k^2\Phi = 4\pi Ga^2\left(\rho\delta + \rho_b\delta_b + \rho_\gamma\delta_\gamma + \rho_\nu\delta_\nu\right), \tag{6.15}$$

$$k^2(\Phi + \Psi) = -32\pi Ga^2\left(\rho_\gamma\Theta_2 + \rho_\nu\mathcal{N}_2\right), \tag{6.16}$$

where we have expressed the photons and neutrinos anisotropic stresses in terms of their quadrupole moments.

Now we have to understand when to set our initial conditions. These should be values for the above quantities Θ_ℓ, $\mathcal{N}_\ell$, δ_c, δ_b, V_c, V_b, Φ and Ψ at a certain initial instant. This instant cannot be $\eta = 0$, i.e. the Big Bang, evidently, because it is a singularity, and so it should be some small $\eta_i > 0$.

Now, any scale k, being η growing, shall pass from a $\eta \ll 1/k$ regime called **super-horizon evolution**, to a $\eta \sim 1/k$ regime called **horizon crossing**, finally to a $\eta \gg 1/k$ called **sub-horizon evolution**. The mentioned horizon is the particle one, since recall from Eq. (2.137) that:

$$\chi_p(\eta) = \int_0^t \frac{dt'}{a(t')} = \int_0^\eta d\eta' = \eta , \qquad (6.17)$$

i.e. the conformal time represents the comoving distance travelled by a photon since the Big Bang.

Therefore, the initial values for the perturbative quantities are set when $k\eta \ll 1$ and are also called **primordial modes** for the scales of observational interest today because are deep into the radiation-dominated epoch. For this reason, we shall frequently use the result $a \propto \eta$.

6.2 Evolution Equations in the $k\eta \ll 1$ Limit

All the equations that we shall use here are valid only at early times, in the radiation-dominated epoch, and on very large scales, i.e. $k\eta \ll 1$.

What one does is basically an expansion with respect to $k\eta$ of the perturbative variables. Let us see what happens for the case of the photon monopole Eq. (6.4), just to fix the ideas. Suppose that we have the following expansions:

$$\delta_\gamma = \sum_{n=0}^\infty \delta_\gamma^{(n)}(k\eta)^n , \quad V_\gamma = \sum_{n=0}^\infty V_\gamma^{(n)}(k\eta)^n , \quad \Phi = \sum_{n=0}^\infty \Phi^{(n)}(k\eta)^n . \qquad (6.18)$$

Insert them into Eq. (6.4), and find:

$$\sum_{n=0}^\infty \delta_\gamma^{(n)} n (k\eta)^n \frac{1}{\eta} + \frac{4}{3} k \sum_{n=0}^\infty V_\gamma^{(n)}(k\eta)^n = -4 \sum_{n=0}^\infty \Phi^{(n)} n (k\eta)^n \frac{1}{\eta} . \qquad (6.19)$$

This equation can be cast as follows:

$$\sum_{n=0}^\infty \delta_\gamma^{(n)} n (k\eta)^n + \frac{4}{3} \sum_{n=0}^\infty V_\gamma^{(n)}(k\eta)^{n+1} = -4 \sum_{n=0}^\infty \Phi^{(n)} n (k\eta)^n . \qquad (6.20)$$

So, we see that $V_\gamma^{(0)}$ couples with $\delta_\gamma^{(1)}$ and $\Phi^{(1)}$ in the expansion. In other words, V_γ is of order $(k\eta)\delta_\gamma$ and $(k\eta)\Phi$, hence is subdominant with respect to δ_γ and Φ. This means that, at the lowest order, Eq. (6.4) becomes:

$$\boxed{\delta'_\gamma = -4\Phi'} \tag{6.21}$$

This procedure is equivalent to neglecting a contribution proportional to k with respect to a time derivative.

Exercise 6.1 Apply the same reasoning which brought us to Eq. (6.21) to the neutrinos monopole and to CDM and baryons density contrasts, showing that at the lowest order of approximation we have:

$$\boxed{\delta'_\nu = -4\Phi'} \tag{6.22}$$

$$\boxed{\delta'_c = -3\Phi'} \tag{6.23}$$

$$\boxed{\delta'_b = -3\Phi'} \tag{6.24}$$

Integrating the above equations for the monopoles, we get:

$$\delta_\gamma = -4\Phi + 4C_\gamma \,, \tag{6.25}$$

$$\delta_\nu = -4\Phi + C_\nu \,, \tag{6.26}$$

$$\delta_c = -3\Phi + C_c \,, \tag{6.27}$$

$$\delta_b = -3\Phi + C_b \,, \tag{6.28}$$

where we have introduced 4 integration constants. Cast the above equations as follows:

$$\delta_\gamma = -4\Phi + 4C_\gamma \,, \tag{6.29}$$

$$\delta_\nu = \delta_\gamma + S_\nu \,, \tag{6.30}$$

$$\delta_c = \frac{3}{4}\delta_\gamma + S_c \,, \tag{6.31}$$

$$\delta_b = \frac{3}{4}\delta_\gamma + S_b \,, \tag{6.32}$$

for a reason that will be clearer later. We have defined:

$$S_\nu \equiv C_\nu - 4C_\gamma \,, \qquad S_c \equiv C_c - 3C_\gamma \,, \qquad S_b \equiv C_b - 3C_\gamma \,, \tag{6.33}$$

and these are called **density isocurvature modes** or **entropy modes**. Be careful that the C's and thus the S's are not actually constants, but functions of **k**. Hereafter we will call inappropriately as "constants" those quantities which are time-independent.

6.2.1   Multipoles in the $k\eta \ll 1$ Limit

From the hierarchy of the equations for photons we can see that each multipole Θ_ℓ enters the differential equation for $\Theta_{\ell+1}$ multiplied by k. This means that $\Theta_{\ell+1} \sim (k\eta)\Theta_\ell$, i.e. each multipole is subdominant with respect the previous one. This is also true for $\Theta_{P\ell}$ and for $\mathcal{N}_\ell$ and it is an important fact that we shall employ afterwards.

Applying the limit $k\eta \ll 1$ to the photon equations we get, at the dominant order:

$$(2\ell + 1)\Theta'_\ell - k\ell\Theta_{\ell-1} = \tau'(2\ell + 1)\Theta_\ell , \qquad \ell > 2 , \tag{6.34}$$

$$10\Theta'_2 - \frac{4}{3}kV_\gamma = \tau'\left(9\Theta_2 - \Theta_{P0} - \Theta_{P2}\right) , \tag{6.35}$$

$$V'_\gamma - k\frac{\delta_\gamma}{4} = k\Psi + \tau'\left(V_\gamma - V_b\right) . \tag{6.36}$$

Similarly, for the polarization field:

$$(2\ell + 1)\Theta'_{P\ell} - k\ell\Theta_{P(\ell-1)} = \tau'(2\ell + 1)\Theta_{P\ell} , \qquad \ell > 2 , \tag{6.37}$$

$$10\Theta'_{P2} - 4k\Theta_{P1} = \tau'\left(9\Theta_{P2} - \Theta_{P0} - \Theta_2\right) , \tag{6.38}$$

$$3\Theta'_{P1} - k\Theta_{P0} = 3\tau'\Theta_{P1} , \tag{6.39}$$

$$2\Theta'_{P0} = \tau'\left(\Theta_{P0} - \Theta_{P2} - \Theta_2\right) . \tag{6.40}$$

Here comes a very useful approximation, based on the fact that:

$$-\tau' = n_e\sigma_T a \propto a^{-2} \propto \eta^{-2} . \tag{6.41}$$

That is, the optical depth time derivative, which is the photon-electron interaction rate, diverges for $\eta \to 0$ as $1/\eta^2$. This is logical, since the universe becomes denser the more we go backwards in time and so more interactions take place. This is also called **tight-coupling** and we will use it as an approximation closer to recombination in Chap. 10.

But then, in order to prevent the perturbations from diverging, the quantities multiplying τ' must vanish for $\eta \to 0$, i.e.:

$$\Theta_\ell = \Theta_{P\ell} = 0 , \qquad \ell > 2 , \tag{6.42}$$

$$9\Theta_2 - \Theta_{P0} - \Theta_{P2} = 0 , \tag{6.43}$$

$$9\Theta_{P2} - \Theta_{P0} - \Theta_2 = 0 , \tag{6.44}$$

$$V_\gamma = V_b , \qquad \Theta_{P1} = 0 , \tag{6.45}$$

$$\Theta_{P0} = \Theta_{P2} + \Theta_2 . \tag{6.46}$$

Exercise 6.2 Combine the above equations and show that:

$$\boxed{\Theta_2 = \Theta_{P2} = \Theta_{P0} = 0} \tag{6.47}$$

Hence, the coupling between electrons and photons is so efficient that it washes out the quadrupole and higher moments of the temperature fluctuations and all the polarisation moments. Moreover, it tightly couples photons and baryons in a single fluid with velocity $V_\gamma = V_b$.

The $k\eta \ll 1$ neutrino equations are the same as the photon ones, but with no interaction terms:

$$(2\ell + 1)\mathcal{N}_\ell' - k\ell\mathcal{N}_{\ell-1} = 0 , \qquad \ell \geq 2 , \tag{6.48}$$

$$V_\nu' - k\frac{\delta_\nu}{4} = k\Psi . \tag{6.49}$$

Again, we can see that $V_\nu \sim (k\eta)\delta_\nu, \mathcal{N}_2 \sim (k\eta)^2\delta_\nu$, and so on. Nothing can make the multipoles of the neutrino temperature to vanish, as Thomson scattering for photons, therefore we need initial conditions for all of them.

Exercise 6.3 Suppose that:

$$\mathcal{N}_\ell = c_\ell(k\eta)^\ell , \tag{6.50}$$

at the lowest order, where c_ℓ is some constant (in this instance a true constant, a number). From Eq. (6.48) show that:

$$c_\ell = \frac{1}{2\ell + 1}c_{\ell-1} , \tag{6.51}$$

and thus:

$$\mathcal{N}_\ell = \frac{k\eta}{2\ell + 1}\mathcal{N}_{\ell-1} , \qquad \ell \geq 2 . \tag{6.52}$$

Therefore, once we know the initial condition on $\mathcal{N}_1$, we can determine the initial conditions for all the subsequent multipoles from the above equation. But, what about the initial condition on V_ν?

Exercise 6.4 Combining Eqs. (6.21), (6.22), (6.36) and (6.49) show that:

$$V_\gamma'' = V_\nu'' . \tag{6.53}$$

This does not necessarily means that $V_\gamma = V_\nu$, but in general:

$$V_\gamma = V_\nu + q_\nu , \tag{6.54}$$

where q_ν is at most a linear function of η and is the **neutrino velocity isocurvature mode** or **relative neutrino heat flux**.

6.2.2 CDM and Baryons Velocity Equations

Consider now the velocity equations for CDM and baryons. At the lowest order in $k\eta$ they become:

$$\eta V_c' + V_c = (k\eta)\Psi , \tag{6.55}$$

$$\eta V_b' + V_b = (k\eta)\Psi + \frac{4\eta\tau' \rho_\gamma}{3\rho_b}(V_b - V_\gamma) . \tag{6.56}$$

First of all, $\eta\tau'\rho_\gamma/\rho_b \propto 1/a^2 \propto 1/\eta^2$. Thus, in order for V_b not to diverge one has to have $V_\gamma = V_b$, consistently with what we have found earlier for photons. With this condition holding true, the equation for V_b is similar to the one for V_c:

$$\eta V_c' + V_c = (k\eta)\Psi , \tag{6.57}$$

$$\eta V_b' + V_b = (k\eta)\Psi . \tag{6.58}$$

From these equations we can conclude that V_c and V_b are subdominant with respect to Ψ.

6.2.3 The $k\eta \ll 1$ Limit of the Einstein Equations

Noting that:

$$4\pi Ga^2 = \frac{3\mathcal{H}^2}{2\rho_{\text{tot}}} = \frac{3}{2\rho_{\text{tot}}\eta^2} , \tag{6.59}$$

where ρ_{tot} is the total energy density, the generalised Poisson equation can be written as:

$$2\eta\Phi' - 2\Psi + 2(k\eta)^2\Phi = \frac{\rho_\gamma}{\rho_{\text{tot}}}\delta_\gamma + \frac{\rho_\nu}{\rho_{\text{tot}}}\delta_\nu + \frac{\rho_c}{\rho_{\text{tot}}}\delta_c + \frac{\rho_b}{\rho_{\text{tot}}}\delta_b . \tag{6.60}$$

Neglect now $(k\eta)^2$ and define the density fraction $R_i \equiv \rho_i/\rho_{\text{tot}}$ for each component, such that:

$$R_\gamma + R_\nu + R_c + R_b = 1 . \tag{6.61}$$

Using the solutions found for the monopoles, we arrive at:

$$2\eta\Phi' - 2\Psi = -4\Phi\left(R_\gamma + R_\nu + \frac{3}{4}R_c + \frac{3}{4}R_b\right) + 4R_\gamma C_\gamma + R_\nu C_\nu + R_c C_c + R_b C_b \ . \tag{6.62}$$

Note that

$$4\left(R_\gamma + R_\nu + \frac{3}{4}R_c + \frac{3}{4}R_b\right) = 3 + R_\gamma + R_\nu \ , \tag{6.63}$$

and since we are deep into the radiation dominated epoch, photons and neutrinos dominate and thus:

$$R_\gamma + R_\nu \approx 1 \ . \tag{6.64}$$

Exercise 6.5 Show that R_ν is constant at early times and its value is:

$$R_\nu = \frac{\rho_\nu}{\rho_\nu + \rho_\gamma} = \frac{N_{\text{eff}}(7/8)(4/11)^{4/3}}{1 + N_{\text{eff}}(7/8)(4/11)^{4/3}} = 0.4089 \ , \tag{6.65}$$

the last number coming from choosing $N_{\text{eff}} = 3.046$.

Equation (6.62) can be thus written as follows:

$$\boxed{2\eta\Phi' + (3 + R_\gamma + R_\nu)\Phi - 2\Psi = (3 + R_\gamma + R_\nu)C_\gamma + R_\nu S_\nu + R_c S_c + R_b S_b} \tag{6.66}$$

where we have also introduced the density isocurvature modes.

Since $\Theta_2 = 0$ for $k\eta \ll 1$, the Einstein equation for the anisotropic stress becomes:

$$\boxed{k^2\eta^2(\Phi + \Psi) = -12R_\nu \mathcal{N}_2} \tag{6.67}$$

This equation tells us that, consistently, $\mathcal{N}_2 \sim (k\eta)^2\delta_\nu$, since $\Phi \sim \delta_\nu$. However, because of that factor $(k\eta)^2$, we have to keep track of the time-evolution of $\mathcal{N}_2$, which is of the same order. From the neutrino equations that we have derived earlier it is not difficult to obtain that, upon differentiating Eq. (6.67):

$$2k^2\eta(\Phi + \Psi) + k^2\eta^2(\Phi' + \Psi') = -12R_\nu \mathcal{N}_2' = -\frac{8}{5}R_\nu k V_\nu \ . \tag{6.68}$$

Differentiate again, and obtain:

$$2k^2(\Phi + \Psi) + 4k^2\eta(\Phi' + \Psi') + k^2\eta^2(\Phi'' + \Psi'') = -\frac{8}{5}R_\nu k V_\nu' =$$
$$-\frac{8}{5}R_\nu k^2\left(\Psi + \frac{\delta_\nu}{4}\right) \ . \tag{6.69}$$

Now the k^2 can be simplified and differentiating again, one obtains:

$$6(\Phi' + \Psi') + 6\eta(\Phi'' + \Psi'') + \eta^2(\Phi''' + \Psi''') = -\frac{8}{5}R_\nu\left(\Psi' - \Phi'\right) \quad (6.70)$$

where we used $\delta_\nu' = -4\Phi'$ and the fact that R_ν is constant when radiation dominates.

We have obtained a closed equation for Φ and Ψ, despite of the fact that it is of third order. Combining it with the generalised Poisson equation, one can obtain an equation of fourth order for Φ or Ψ only.

Now we solve the equations found in this section in 5 cases, i.e. when only one of the 5 constants is different from zero.

6.3 The Adiabatic Primordial Mode

The adiabatic mode is defined by $S_\nu = S_c = S_b = q_\nu = 0$. Only $C_\gamma \neq 0$. With this condition, we have for the density contrasts:

$$\frac{1}{3}\delta_c = \frac{1}{3}\delta_b = \frac{1}{4}\delta_\gamma = \frac{1}{4}\delta_\nu = -\Phi + C_\gamma \quad (6.71)$$

As for the gravitational potentials, Eq. (6.66) becomes:

$$\eta\Phi' + 2\Phi - \Psi = 2C_\gamma , \quad (6.72)$$

and we have to combine this together with Eq. (6.70) in order to obtain a fourth order equation for Φ or Ψ. The result is, for Φ:

$$\eta^3\Phi'''' + 12\eta^2\Phi''' + 4\left(9 + \frac{2}{5}R_\nu\right)\eta\Phi'' + 8\left(3 + \frac{2}{5}R_\nu\right)\Phi' = 0 . \quad (6.73)$$

This equation can be solved exactly, recalling that R_ν is constant. The fact that only derivatives of Φ appear already tells us that $\Phi = $ constant is a solution and indeed it is the growing mode that we are going to consider. The other 3 independent solutions of the above equation are:

$$\Phi \propto \eta^{-\frac{5}{2} \pm \frac{1}{2}\sqrt{1 - \frac{32R_\nu}{5}}} , \qquad \Phi \propto \frac{1}{\eta} . \quad (6.74)$$

Hence, these are decaying modes, the first two even complex if $R_\nu > 5/32$, and we neglect them.

Since C_γ and Φ are constant, then from Eq. (6.72) we deduce that Ψ is also constant. In order to determine their values in terms of C_γ we have to solve the system of Eqs. (6.72) and (6.69):

$$2\Phi - \Psi = 2C_\gamma \,, \tag{6.75}$$

$$\Phi + \Psi = -\frac{4}{5} R_\nu \left(\Psi - \Phi + C_\gamma \right) \,. \tag{6.76}$$

which allows us to establish:

$$\boxed{\Phi = \frac{2C_\gamma(5 + 2R_\nu)}{15 + 4R_\nu} \,, \quad \Psi = -\frac{10C_\gamma}{15 + 4R_\nu} \,, \quad \Phi = -\Psi\left(1 + \frac{2}{5}R_\nu\right)} \tag{6.77}$$

We shall later show in Chap. 8 how the inflationary mechanism is able to establish a value for C_γ. Note how the presence of neutrinos via R_ν prevents $\Phi = -\Psi$.

The C_γ that we are using here corresponds to the $-2C$ of Ma and Bertschinger (1995) and in order to obtain the result of Bucher et al. (2000) one has to set $C_\gamma = -1$.[1]

From Eq. (6.71) we can write:

$$\delta_\gamma = \frac{20C_\gamma}{15 + 4R_\nu} = -2\Psi \,. \tag{6.78}$$

Using Eq. (6.67), the primordial mode of the neutrino quadrupole moment is:

$$\boxed{\mathcal{N}_2 = -\frac{k^2\eta^2(\Phi + \Psi)}{12R_\nu} = \frac{k^2\eta^2}{30}\Psi} \tag{6.79}$$

What about the velocities? We have just seen that $\Psi \sim (k\eta)^0$ so if also $V_c \sim V_b \sim (k\eta)^0$, then Eqs. (6.57) and (6.58) become, at the lowest order:

$$\eta V_c' + V_c = 0 \,, \tag{6.80}$$

$$\eta V_b' + V_b = 0 \,. \tag{6.81}$$

The solutions are $V_c \propto V_b \propto 1/\eta$. These solutions are not admissible because they diverge for $\eta \to 0$. This means that $V_c \propto V_b \sim (k\eta)$, at the lowest order. Substituting this ansatz again in Eqs. (6.57) and (6.58) it is easy to find:

$$V_c = V_b = \frac{k\eta\Psi}{2} \,. \tag{6.82}$$

---

[1]There must be a typo in Eq. 28 of Bucher et al. (2000), since the gravitational potentials appear to be equal but they cannot be because of the presence of neutrinos and the quadrupole moment of their distribution.

Since $q_\nu = 0$, then $V_\nu = V_\gamma$ and recalling that $V_\gamma = V_b$, we have for the velocities primordial modes:

$$V_\nu = V_\gamma = V_b = V_c = \frac{k\eta\Psi}{2} \tag{6.83}$$

All the above modes depend on the same unique constant C_γ, which is *the* primordial perturbation, in the sense that its non-vanishing generates the fluctuations for all the matter components.

Exercise 6.6 Compare the solutions that we have found for the adiabatic mode with those of Refs. Ma and Bertschinger (1995), Bucher et al. (2000).

6.3.1 Why "Adiabatic"?

Now we are going to understand the reason for the adjective "adiabatic". From the thermodynamical relation:

$$T dS = dU + p dV , \tag{6.84}$$

we can write:

$$\frac{T dS}{V} = d\rho_{tot} - \frac{\rho_{tot} + P_{tot}}{n_{tot}} dn_{tot} . \tag{6.85}$$

Recast this as follows:

$$\frac{T dS}{V} = \rho_c \delta_c + \rho_\gamma \delta_\gamma + \rho_b \delta_b + \rho_\nu \delta_\nu - \frac{\rho_{tot} + P_{tot}}{n_{tot}} (\delta n_c + \delta n_\gamma + \delta n_b + \delta n_\nu) . \tag{6.86}$$

Using:

$$\delta_c = \frac{\delta n_c}{n_c} , \qquad \delta_b = \frac{\delta n_b}{n_b} , \qquad \delta_\gamma = \frac{4}{3}\frac{\delta n_\gamma}{n_\gamma} , \qquad \delta_\nu = \frac{4}{3}\frac{\delta n_\nu}{n_\nu} , \tag{6.87}$$

we finally have:

$$\frac{T dS}{V} = \left(\rho_c - n_c \frac{\rho_{tot} + P_{tot}}{n_{tot}}\right)\left(\delta_c - \frac{3}{4}\delta_\gamma\right) + \left(\rho_b - n_b \frac{\rho_{tot} + P_{tot}}{n_{tot}}\right)\left(\delta_b - \frac{3}{4}\delta_\gamma\right)$$
$$\left(\rho_\nu - \frac{3}{4}n_\nu \frac{\rho_{tot} + P_{tot}}{n_{tot}}\right)\left(\delta_\nu - \delta_\gamma\right) . \tag{6.88}$$

Therefore, $\delta_c/3 = \delta_\gamma/4 = \delta_\nu/4 = \delta_b/3$, i.e. $S_c = S_\nu = S_b = 0$, implies $dS = 0$ and hence adiabaticity. When one of S_ν, S, S_b and q_ν is different from zero we have isocurvature modes, respectively divided into **isocurvature neutrino density**, **isocurvature CDM**, **isocurvature baryons** and **isocurvature neutrino velocity**. However, it might be more appropriate to call them **entropy perturbations**, on the basis of Eq. (6.88).

The name "isocurvature" comes from the fact that when these modes are present it is possible to have $\mathcal{R} = 0$ or $\zeta = 0$, from Eqs. (4.131) and (4.132). In the Newtonian gauge, we have:

$$\mathcal{R} = \Phi + \mathcal{H}v_{\text{tot}}, \qquad \zeta = \Phi + \frac{R_\gamma \delta_\gamma + R_\nu \delta_\nu + R_c \delta_c + R_b \delta_b}{3 + R_\gamma + R_\nu}. \tag{6.89}$$

Some care must be used when computing v_{tot} in the above formula for $\mathcal{R}$ since it is the total one but it is not the sum of the single components' v's. This happens because the SVT decomposition is made not directly on v_i but rather on the perturbed energy-momentum tensor $\delta T^0{}_i$ and the latter is, from Eq. (4.73):

$$\delta T^0{}_i(\text{tot}) = (\rho_{\text{tot}} + P_{\text{tot}})v_i(\text{tot}). \tag{6.90}$$

Hence, v_{tot} is defined through a weighted average with weight $\rho + P$, i.e.

$$v_{\text{tot}} = \frac{\sum_s (\rho_s + P_s)v_s}{\rho_{\text{tot}} + P_{\text{tot}}}, \tag{6.91}$$

which, in our 4-components model, gives:

$$v_{\text{tot}} = \frac{4R_\gamma v_\gamma + 4R_\nu v_\nu + 3R_c v_c + 3R_b v_b}{3 + R_\gamma + R_\nu}. \tag{6.92}$$

Now, using the relation $v = -V/k$ and the condition found in Eq. (6.83), one can finally write:

$$\mathcal{R} = \Phi - \frac{4R_\gamma V_\gamma + 4R_\nu V_\nu + 3R_c V_c + 3R_b V_b}{k\eta(3 + R_\gamma + R_\nu)} = \Phi - \frac{\Psi}{2}. \tag{6.93}$$

One finds the same result for ζ, when substituting Eq. (6.66) in (6.89) (and considering the constant mode). Hence, we have that:

$$\boxed{\mathcal{R} = \zeta = \Phi - \frac{\Psi}{2} = C_\gamma} \tag{6.94}$$

Hence, the C_γ is related to the curvature perturbations and we have also proved that these are equal and constant on large scales for adiabatic perturbations. We shall prove this important result again in Sect. 12.4.

6.4 The Neutrino Density Isocurvature Primordial Mode

Neglect CDM and baryons in Eq. (6.89).

Exercise 6.7 Since $R_{b,c} \to 0$ at early times, use Eq. (6.30) in order to find:

$$\zeta = C_\gamma + \frac{R_\nu S_\nu}{4} . \tag{6.95}$$

If $S_\nu = 0$ we recover the adiabatic case. In order to make $\zeta = 0$, we need $4C_\gamma = -R_\nu S_\nu$. Therefore, the density contrasts are related as follows:

$$\boxed{\frac{1}{3}\delta = \frac{1}{3}\delta_b = \frac{1}{4}\delta_\gamma = \frac{1}{4}\delta_\nu - \frac{S_\nu}{4} = -\Phi - \frac{R_\nu S_\nu}{4}} \tag{6.96}$$

In order to determine the gravitational potentials, from Eq. (6.66) we have:

$$\eta\Phi' + 2\Phi - \Psi = 0 , \tag{6.97}$$

and combining it with Eq. (6.70) one finds the very same Eq. (6.73) for Φ that we obtained in the adiabatic case. Hence, we choose the constant Φ solution, which allows us to establish:

$$\boxed{\Psi = 2\Phi , \qquad \Phi = \frac{R_\nu^2 S_\nu}{15 + 4R_\nu}} \tag{6.98}$$

In order to obtain the same results of Bucher et al. (2000) one has to set $R_\nu S_\nu = -1$. Using Eq. (6.67), the initial condition on the neutrino quadrupole moment is:

$$\boxed{\mathcal{N}_2 = -\frac{k^2\eta^2(\Phi + \Psi)}{12R_\nu} = -\frac{k^2\eta^2\Phi}{4R_\nu}} \tag{6.99}$$

Since $q_\nu = 0$, then $V_\nu = V_\gamma$ and the initial conditions on the velocities are again:

$$\boxed{V_\nu = V_\gamma = V_b = V = \frac{k\eta\Psi}{2} = k\eta\Phi} \tag{6.100}$$

6.5   The CDM and Baryons Isocurvature Primordial Modes

The treatment for CDM or baryons isocurvature modes is the same, therefore we present explicitly the CDM case only. Suppose that only $S_c \neq 0$, then we have for ζ:

$$\zeta = \frac{R_c S_c}{4} . \tag{6.101}$$

Since we are deep into the radiation-dominated epoch:

$$R_c \equiv \frac{\rho_c}{\rho_{\text{tot}}} \sim \frac{\rho_{c0}}{\rho_{\text{tot},0}} a = \Omega_{c0} a . \tag{6.102}$$

Then R_c goes to zero and so thus ζ, denoting thus correctly an isocurvature mode. The density contrasts are related as follows:

$$\boxed{\frac{\delta_\gamma}{4} = \frac{\delta_\nu}{4} = \frac{\delta_b}{3} = \frac{\delta_c - S_c}{3} = -\Phi} \tag{6.103}$$

and all of them vanishes for $\eta \to 0$, except $\delta_c = S_c$, which sources all the perturbations. Equation (6.66) tells us that:

$$2\eta\Phi' - 2\Psi + 4\Phi = R_c S_c , \tag{6.104}$$

and using Eq. (6.102), we write:

$$\eta\Phi' - \Psi + 2\Phi = S_c \Omega_{c0} \eta , \tag{6.105}$$

where we have incorporated the factor $1/2$ and the proportionality constant of $a \propto \eta$ into S. This can be done, of course, without losing generality since S is itself a constant. Now, since the right hand side of the above equation depends linearly on η, and $\Phi \to 0$ for $\eta \to 0$, the only possibility is that the gravitational potentials are linearly dependent on the time. Using the ansatz:

$$\Phi = \bar{\Phi}\eta , \qquad \Psi = \bar{\Psi}\eta , \tag{6.106}$$

Equations (6.105) and (6.69) become:

$$3\bar{\Phi} - \bar{\Psi} = S_c \Omega_{c0} \tag{6.107}$$

$$3(\bar{\Phi} + \bar{\Psi}) = -\frac{4}{5} R_\nu \left(\bar{\Psi} - \bar{\Phi}\right) . \tag{6.108}$$

Solving this system of two equations, we find the CDM isocurvature modes:

$$\Phi = \frac{(15 + 4R_\nu)S_c\Omega_{c0}}{4(15 + 2R_\nu)}\eta \,, \quad \Psi = \frac{(4R_\nu - 15)S_c\Omega_{c0}}{4(15 + 2R_\nu)}\eta \,, \quad \Phi = -\Psi\frac{15 + 4R_\nu}{15 - 4R_\nu}$$

$$(6.109)$$

Again as we expected, the absence of neutrinos causes $\Phi = -\Psi$ since there is no other source of anisotropic stress. Using Eq. (6.67), the primordial mode of the neutrino quadrupole moment is:

$$\mathcal{N}_2 = -\frac{k^2\eta^2(\Phi + \Psi)}{12R_\nu} = -\frac{k^2\eta^3}{6(15 + 2R_\nu)}S_c\Omega_{c0} \qquad (6.110)$$

Again for the velocities:

$$V_\nu = V_\gamma = V_b = V_c = \frac{k\Psi}{2} \qquad (6.111)$$

The isocurvature modes for baryons have exactly the same form, just change Ω_{c0} with Ω_{b0}.

6.6   The Neutrino Velocity Isocurvature Primordial Mode

These modes are problematic in the conformal Newtonian gauge because they diverge for $\eta \to 0$. On the other hand, they are well-defined in the synchronous gauge Bucher et al. (2000). Let us see what is the problem. When we set $C_\gamma = S_\nu = S_c = S_b = 0$, we can straightforwardly find:

$$\zeta = 0 \,, \quad \frac{\delta_\gamma}{4} = \frac{\delta_\nu}{4} = \frac{\delta_c}{3} = \frac{\delta_b}{3} = -\Phi \qquad (6.112)$$

and Eq. (6.66) becomes:

$$\eta\Phi' + 2\Phi - \Psi = 0 \,. \qquad (6.113)$$

Here a constant mode seems to be viable, just as in the case of the neutrino density isocurvature mode. However, we need here also Eq. (6.69), which has the following form, taking into account $\delta_\nu = -4\Phi$:

$$2(\Phi + \Psi) + 4\eta(\Phi' + \Psi') + \eta^2(\Phi'' + \Psi'') = -\frac{8}{5}R_\nu(\Psi - \Phi) \,. \qquad (6.114)$$

If we choose the constant mode, we get:

$$2\Phi = \Psi \ , \tag{6.115}$$

$$\Phi + \Psi = -\frac{4}{5} R_\nu \left(\Psi - \Phi \right) \ . \tag{6.116}$$

From this system we clearly see that the only solution is the trivial one. Therefore, from Eq. (6.113) one sees that we are left with the only following possibility:

$$\boxed{\Phi = \frac{\bar{\Phi}}{\eta}} \qquad \boxed{\Psi = \frac{\bar{\Psi}}{\eta}} \tag{6.117}$$

which is problematic in the limit $\eta \to 0$.

Substituting this ansatz into the generalised Poisson equation we can easily determine that:

$$\bar{\Phi} = \bar{\Psi} \ , \tag{6.118}$$

i.e. the two gravitational potentials are equal. Substituting this result into Eq. (6.68), one obtains:

$$k\bar{\Phi} = -\frac{4}{5} R_\nu V_\nu \ , \tag{6.119}$$

From the dipole neutrino and the dipole photon equations we can readily establish that:

$$V_\gamma' = V_\nu' = 0 \ , \tag{6.120}$$

i.e. V_γ and V_ν are constant and their difference is indeed our neutrino velocity isocurvature mode q_ν. Thus:

$$k\bar{\Phi} = \frac{4}{5} R_\nu \left(V_\gamma - q_\nu \right) \ . \tag{6.121}$$

We can determine V_γ in terms of Φ considering the tight coupling with baryons, which implies $V_\gamma = V_b$ and the baryon velocity Eq. (6.58), which now is:

$$\eta V_b' + V_b = k\eta \Psi = k\bar{\Phi} \ . \tag{6.122}$$

The right hand side is a constant and so V_b must be. In particular:

$$V_b = k\bar{\Phi} = V_\gamma \ . \tag{6.123}$$

This relation, inserted into Eq. (6.121), gives:

$$k\bar{\Phi} = \frac{4}{5} R_\nu \left(k\bar{\Phi} - q_\nu \right) \ . \tag{6.124}$$

Finally, we have for the potentials:

$$\boxed{\Phi = \Psi = -\frac{4R_\nu q_\nu}{5 - 4R_\nu}\frac{1}{k\eta}} \tag{6.125}$$

and for the velocities:

$$\boxed{V = V_b = V_\gamma = -\frac{R_\nu q_\nu}{5 - 4R_\nu}} \tag{6.126}$$

The neutrino velocity and quadrupole moment are:

$$\boxed{V_\nu = q_\nu - V_\gamma} \qquad \boxed{\mathcal{N}_2 = -\frac{k^2\eta^2(\Phi + \Psi)}{12R_\nu} = \frac{2k\eta q_\nu}{3(5 - 4R_\nu)}} \tag{6.127}$$

6.7   Planck Constraints on Isocurvature Modes

In Ade et al. (2016a, c) the isocurvature mode are constrained through the Planck data. These tests are related also to inflation, since this is a mechanism of production of primordial fluctuations, as we shall see later in Chap. 8. In particular, the most successful and simple inflationary models are those based on a single scalar field and they predict adiabatic initial perturbations. If other fields are present, then isocurvature modes can be produced. Depending on the mechanism of production, adiabatic and isocurvature modes can be correlated. In Ade et al. (2016a) for example the totally correlated case:

$$S_m = \text{sgn}(\alpha)\sqrt{\frac{|\alpha|}{1 - |\alpha|}}\,\zeta\,, \tag{6.128}$$

is considered, where

$$S_m = S_c + \frac{R_b}{R_c}S_b\,, \tag{6.129}$$

is the total matter isocurvature mode. The fact that S_m is proportional to ζ means that the totally-correlated case is being considered (because of this the modes are described by the same parameters and hence put in the above form of proportionality). The constraint found on α is:

$$\alpha = 0.0003^{+0.0016}_{-0.0012}\,, \tag{6.130}$$

at 95% CL, using also polarisation data. A much more detailed analysis, including possible correlations, is performed in Ade et al. (2016c) with the result

$$|\alpha_{\text{non-adi}}| < 1.9\%, 4.0\%, 2.9\%\,, \tag{6.131}$$

for CDM, neutrino density and neutrino velocity isocurvature modes respectively. In the above equation, $\alpha_{\mathrm{non-adi}}$ is the non-adiabatic contribution to the CMB temperature power spectrum. How the different primordial modes affect the latter will be shown in Chap. 10.

References

Ade, P.A.R., et al.: Planck 2015 results XIII. Cosmological parameters. Astron. Astrophys. **594**, A13 (2016a)

Ade, P.A.R., et al.: Planck 2015 results XX. Constraints on inflation. Astron. Astrophys. **594**, A20 (2016c)

Bucher, M., Moodley, K., Turok, N.: The general primordial cosmic perturbation. Phys. Rev. D **62**, 083508 (2000)

Ma, C.-P., Bertschinger, E.: Cosmological perturbation theory in the synchronous and conformal Newtonian gauges. Astrophys. J. **455**, 7–25 (1995)

Chapter 7
Stochastic Properties of Cosmological Perturbations

> *The more the universe seems comprehensible, the more it also seems pointless*
>
> Steven Weinberg, The first three minutes

The attentive reader must have noticed that, in fact, we have given no actual initial conditions on cosmological perturbations in Chap. 6. What we have done is to give the form of the primordial modes and show how, in the 5 different cases that we investigated, all scalar perturbations are sourced by a single scalar potential. For example, in the adiabatic case we showed in Eq. (6.77) how $\Phi(\mathbf{k})$ is related to $C_\gamma(\mathbf{k})$, which we proved to be equal to $\zeta(\mathbf{k})$ and $\mathcal{R}(\mathbf{k})$.

Note the following important point: all the evolution equations that we derived for the perturbations, i.e. all the evolution equations for photons, neutrinos, CDM and baryons, plus Einstein equations, depend only on the modulus k and not on $\mathbf{k}$. This means that only the initial conditions depend on the latter. We could solve the evolution equation for some initial condition in which $\mathcal{R}(\eta_i, \mathbf{k}) = 1$, i.e. normalised to unity, and then the result will depend only on k: $\mathcal{R}(\eta, k)$. This is usually called **transfer function**. Multiplying it by the true initial condition $\mathcal{R}(\eta_i, \mathbf{k})$ we obtain the correct result. Of course, this reasoning applies to any perturbation, not only $\mathcal{R}$.

In this chapter we discuss a very important point of cosmology, both from the viewpoint of theory and especially from the observational one: the stochastic nature of cosmological perturbations, embedded in the stochastic character of the initial conditions. This means that $\mathcal{R}(\eta_i, \mathbf{k})$ shall be regarded as a random variable.

This chapter mainly follows Weinberg (2008) and Lyth and Liddle (2009).

© Springer International Publishing AG, part of Springer Nature 2018
O. Piattella, *Lecture Notes in Cosmology*, UNITEXT for Physics,
https://doi.org/10.1007/978-3-319-95570-4_7

7.1 Stochastic Cosmological Perturbations and Power Spectrum

First of all, what do cosmological perturbations have to do with statistics? After all, we have derived differential equations describing the evolutions of the perturbative variables and, provided that we are able to solve them, we should determine such variables exactly with no room for randomness.

Consider the following point: the set of differential equations that we have found are ODE which only describe the time evolution of the perturbative variables, with their spatial (or scale) dependences to be provided as initial conditions. For example, focus on the CDM density contrast $\delta_c(\eta, \mathbf{x})$, for which the evolution equations Eqs. (5.50) and (5.51) are the simplest:

$$\delta_c' + kV_c + 3\Phi' = 0 , \qquad V_c + \mathcal{H}V_c - k\Psi = 0 . \tag{7.1}$$

These equations depend only on k, therefore the $\mathbf{k}$-dependence comes from the initial condition, which indeed we know in the adiabatic case from Eq. (6.78) to be:

$$\delta_c(\mathbf{k}) = \frac{15}{15 + 4R_\nu}\zeta(\mathbf{k}) . \tag{7.2}$$

Do we have such initial conditions, i.e. do we know the scale dependence of the primordial curvature perturbation? Not exactly. According to our current understanding, the universe had a quantum origin which we are not yet able to describe in detail because of the lack of a quantum theory of gravity and also because we have no experiments penetrating the trans-Planckian scales, allowing us at least to see what is happening there.

The evolution equations which we had found start to be valid well after the Planck scale, when gravity is safely classical. However, the initial conditions that we use are coming from a quantum phase and as such they carry a probabilistic feature. For example, we shall investigate the inflationary paradigm and we shall see that it provides a prediction not on $\zeta(\mathbf{k})$ itself but rather on its probability distribution. In particular, if it is a Gaussian one hence all the information is contained in its variance, which is called **power spectrum**.

There is also another important point which stresses how useful is to treat cosmological perturbations as random variables. Observationally, it is impossible to determine $\delta(\eta, \mathbf{x})$ for any given time because we receive signals from our past light-cone only. It is impossible thus to recover the initial $\zeta(\mathbf{x})$ from what we observe. Even if we would be able to concoct a theory predicting the initial $\zeta(\mathbf{x})$, we would not be able to fully test it since we cannot determine $\delta(\eta, \mathbf{x})$ completely from observation.

Moreover, determining exactly $\delta(\eta, \mathbf{x})$ is an awful lot of information. For example, it would mean that we are able to predict the position of a certain galaxy at a certain time. Though this might be interesting to some extent, we are rather more concerned with averaged quantities, such as the average distance among galaxies, because these contain information on gravity and the expanding universe.

Hence, promoting (or rather demoting) $\delta(\eta, \mathbf{x})$ to a random variable, we are then able to infer informations about its expectation value, variance or other high-order correlators from the observation limited to our past light-cone. This allows us to make contact with the **descriptive statistics** that we are able to make on the distribution of structures in the sky.

7.2 Random Fields

A function $G(\mathbf{x})$ is a random field if its values are random variables for any $\mathbf{x}$ and distribution functions

$$F_{1,2,\ldots,n}(g_1, g_2, \ldots, g_n) = F[G(\mathbf{x}_1) < g_1, G(\mathbf{x}_2) < g_2, \ldots, G(\mathbf{x}_n) < g_n] , \quad (7.3)$$

exist for any n. We denote with capital G the random field, so it has $\mathbf{x}$ dependence, and with g a certain value that it can assume at a given $\mathbf{x}$ among all the possible ones which form the **ensemble**. The subscripts in $g_1, g_2, \ldots, g_n$ refer to the different points $\mathbf{x}_1, \mathbf{x}_2, \ldots, \mathbf{x}_n$ in which the random field is evaluated.

In particular, at a given point $\mathbf{x}_1$ some probability density functional exists:

$$p_1(g_1)dg_1 , \quad (7.4)$$

describing how probable is for G to assume a certain value g_1 in $\mathbf{x}_1$. In terms of the distribution function, $p(g_1)$ is defined as:

$$p_1(g_1) = \frac{dF_1(g_1)}{dg_1} . \quad (7.5)$$

Since F is a cumulative probability then $F_1(-\infty) = 0$ and $F_1(\infty) = 1$. In cosmology, $G(\mathbf{x})$ is a perturbative quantity, such as the density contrast $\delta(\mathbf{x})$.

As usual, an expectation value of the random field is defined via an **ensemble average**:

$$\boxed{\langle G(\mathbf{x}_1) \rangle \equiv \int_\Omega g_1 p_1(g_1)dg_1} \quad (7.6)$$

where Ω denotes the ensemble.

In general, one has:

$$p_1(g_1) \neq p_2(g_2) , \quad (7.7)$$

i.e. the probability distribution of the values which G may assume in $\mathbf{x}_1$ is in general different from the one of the values which G may assume in $\mathbf{x}_2$. When this does not happen, i.e. the probability of the realisation is translationally invariant, the random field is said to be **statistically homogeneous**. For a statistically homogeneous random field then Eq. (7.6) is independent of $\mathbf{x}$

$$\langle G \rangle \equiv \int_\Omega g p(g) dg \tag{7.8}$$

Now we can think of more complicated and richer configurations of the random field G asking for example what is the probability of $G(\mathbf{x}_1)$ and $G(\mathbf{x}_2)$ being g_1 and g_2, respectively. This is given by

$$p_{12}(g_1, g_2) dg_1 dg_2 , \tag{7.9}$$

which can be written again as a derivative of the distribution function F_{12}. In general:

$$p_{12}(g_1, g_2) \neq p_1(g_1) p_2(g_2) , \tag{7.10}$$

unless the realisations are independent, in which case the random process is said to be Poissonian. The 2-dimensional probability density allows us to define the **2-point correlation function** as follows:

$$\xi(\mathbf{x}_1, \mathbf{x}_2) \equiv \langle G(\mathbf{x}_1) G(\mathbf{x}_2) \rangle \equiv \int_\Omega g_1 g_2 p_{12}(g_1, g_2) dg_1 dg_2 \tag{7.11}$$

In a similar fashion, one can define N-point correlation functions as

$$\xi^{(N)}(\mathbf{x}_1, \mathbf{x}_2, \ldots, \mathbf{x}_N) \equiv \langle G(\mathbf{x}_1) G(\mathbf{x}_2) \ldots G(\mathbf{x}_N) \rangle$$
$$\equiv \int_\Omega g_1 g_2 \ldots g_N \, p_{12\ldots N}(g_1, g_2, \ldots, g_N) dg_1 dg_2 \ldots g_N . \tag{7.12}$$

The order of the points matters, i.e. in general $p_{12\ldots N} \neq p_{21\ldots N}$ for example, and we are using the same ensemble Ω for each point.

Exercise 7.1 If the random field is statistically homogeneous, show that the 2-point correlation function has the following property:

$$\xi(\mathbf{x}_1, \mathbf{x}_2) = \xi(\mathbf{x}_1 - \mathbf{x}_2) . \tag{7.13}$$

Given a rotation matrix R, define $\mathbf{x}_{R1} = R\mathbf{x}_1$. The random field is **statistically isotropic** if

$$p_1(g_1) = p_{R1}(g_{R1}) \qquad \forall R , \tag{7.14}$$

i.e. the probability of the realisation is rotationally invariant.

Exercise 7.2 If the random field is statistically homogeneous and statistically isotropic, show that the 2-point correlation function has the following property:

$$\xi(\mathbf{x}_1, \mathbf{x}_2) = \xi(\mathbf{x}_1 - \mathbf{x}_2) = \xi(r_{12}) , \qquad (7.15)$$

where $r_{12} = |\mathbf{x}_1 - \mathbf{x}_2|$. That is, the 2-point correlation function depends only on the distance between the two points.

Following the standard definition that we learn in the first course of statistics, the **ensemble variance** of the random field is defined as:

$$\boxed{\sigma^2(\mathbf{x}_1, \mathbf{x}_2) \equiv \langle G(\mathbf{x}_1)G(\mathbf{x}_2)\rangle - \langle G(\mathbf{x}_1)\rangle\langle G(\mathbf{x}_2)\rangle} \qquad (7.16)$$

Therefore, if the random field is statistically homogeneous and isotropic, we get:

$$\sigma^2(r_{12}) = \xi(r_{12}) - \langle G\rangle^2 , \qquad (7.17)$$

where recall that $\langle G\rangle^2$ does not depend on the position. Note that if random process is Poissonian, i.e. $p_{12}(g_1, g_2) = p_1(g_1)p_2(g_2)$, then the variance is zero, since:

$$\xi(r_{12}) = \langle G\rangle^2 . \qquad (7.18)$$

If G is a variable representing the distribution of galaxies, we do not expect it to be a Poissonian random variable because gravity turns the odds for the galaxies to be closer to each other rather than farther. Hence, we attribute deviations from the Poissonian behaviour to gravity and for this reason it is very important to study the 2-point correlation function.

In observational cosmology we are able to compute just a spatial average of the field $G(\mathbf{x})$, since we have just a single realisation, i.e. our universe. So, consider the spatial average on a given volume V:

$$\bar{G} \equiv \frac{1}{V} \int_V d^3\mathbf{x}\, G(\mathbf{x}) . \qquad (7.19)$$

What can we say about $X \equiv \bar{G} - \langle G\rangle$? We use this as an estimator of the error which we are making by exchanging the ensemble average with the spatial one.

Exercise 7.3 Show that:

$$\langle X\rangle = \frac{1}{V} \int_V d^3\mathbf{x}\, \langle G(\mathbf{x})\rangle - \langle G\rangle = 0 , \qquad (7.20)$$

and the variance is:

$$\langle X^2 \rangle = \frac{1}{V^2} \int_V d^3\mathbf{x}_1 \int_V d^3\mathbf{x}_2 \langle G(\mathbf{x}_1)G(\mathbf{x}_2)\rangle - \langle G \rangle^2 . \tag{7.21}$$

Assuming statistical homogeneity we can then obtain

$$\langle X^2 \rangle = \frac{1}{V^2} \int_V d^3\mathbf{x}_1 \int_V d^3\mathbf{x}_2 \, \xi(\mathbf{x}_1 - \mathbf{x}_2) - \langle G \rangle^2 . \tag{7.22}$$

We see that for a Poissonian process this variance is vanishing. In fact, if the galaxies are randomly distributed, any volume is a good sample of the realisation.

The above integral can be written as

$$\langle X^2 \rangle = \frac{1}{V} \int_V d^3\mathbf{r} \, \xi(\mathbf{r}) - \langle G \rangle^2 . \tag{7.23}$$

In Sect. 7.8 we show that this variance goes to zero if $V \to \infty$, a fact known as **ergodic theorem** and based on a couple of reasonable assumptions. On the other hand, in practice the volume is finite and thus $\langle X^2 \rangle$ is in general different from zero. In cosmology, it is called **cosmic variance**. Assuming statistical isotropy in the above integral and using a spherical volume for convenience, we can write:

$$\langle X^2 \rangle = \frac{3}{R^3} \int_0^R dr \, r^2 \xi(r) - \langle G \rangle^2 . \tag{7.24}$$

This is almost all we need to know about random fields in order to apply them to cosmology, at least from the theoretical point of view. How to compute correlators from the observational data is another issue which is not addressed in these notes.

Note that in the above description of a random field G we have used the configuration space whereas we have mostly used Fourier-transformed quantities representing our cosmological perturbations. We assume that the FT of a random field is also a random field, and all the above described properties apply.

Observationally, we are able to probe a realisation in a certain finite volume, say a box of volume L^3. Hence, the FT is defined here as a Fourier series:

$$G(\mathbf{x}) = \frac{1}{L^3} \sum_n G_n e^{i\mathbf{k}_n \cdot \mathbf{x}} \tag{7.25}$$

where the coefficients are given by:

$$G_n = \int d^3\mathbf{x} \, G(\mathbf{x}) e^{-i\mathbf{k}_n \cdot \mathbf{x}} , \tag{7.26}$$

and the wavenumbers are quantised, because of the periodic boundary conditions that we must impose, which are equivalent to ask that G vanishes on the sides of the box:

$$k_n = \frac{2\pi}{L}n \, , \tag{7.27}$$

where $\mathbf{n}$ is a generic vector whose components are integers.

If the side of the box goes to infinity, we recover the usual FT, i.e.

$$G(\mathbf{x}) = \int \frac{d^3k}{(2\pi)^3} \, \tilde{G}(\mathbf{k})e^{i\mathbf{k}\cdot\mathbf{x}} \, , \qquad \tilde{G}(\mathbf{k}) = \int d^3x \, G(\mathbf{x})e^{-i\mathbf{k}\cdot\mathbf{x}} \, . \tag{7.28}$$

Note that if $G(\mathbf{x})$ is a real field, then $\tilde{G}(-\mathbf{k}) = \tilde{G}^*(\mathbf{k})$, because of the following:

Exercise 7.4 Compare:

$$G(\mathbf{x}) = \int \frac{d^3k}{(2\pi)^3} \, \tilde{G}(\mathbf{k})e^{i\mathbf{k}\cdot\mathbf{x}} \, , \qquad G^*(\mathbf{x}) = \int \frac{d^3k}{(2\pi)^3} \, \tilde{G}^*(\mathbf{k})e^{-i\mathbf{k}\cdot\mathbf{x}} \, . \tag{7.29}$$

The two expressions are equal if $G(\mathbf{x})$ is real. Therefore, show that $\tilde{G}(-\mathbf{k}) = \tilde{G}^*(\mathbf{k})$.

The relation $\tilde{G}(-\mathbf{k}) = \tilde{G}^*(\mathbf{k})$, is called **reality condition** and we shall exploit it in the following Chapters for the predictions on the observed spectra.

7.3 Power Spectrum and Gaussian Random Fields

Consider the 2-point correlation function for the Fourier transform of $G(\mathbf{x})$:

$$\langle \tilde{G}(\mathbf{k})\tilde{G}^*(\mathbf{k}')\rangle = \int d^3x \int d^3x' \langle G(\mathbf{x})G(\mathbf{x}')\rangle e^{-i\mathbf{k}\cdot\mathbf{x}}e^{i\mathbf{k}'\cdot\mathbf{x}'} \, . \tag{7.30}$$

Assume statistical homogeneity. Then:

$$\langle \tilde{G}(\mathbf{k})\tilde{G}^*(\mathbf{k}')\rangle = \int d^3x \int d^3x' \xi_G(\mathbf{x}' - \mathbf{x})e^{-i\mathbf{k}\cdot\mathbf{x}}e^{i\mathbf{k}'\cdot\mathbf{x}'} \, , \tag{7.31}$$

and changing variable from $\mathbf{x}'$ to $\mathbf{z} = \mathbf{x}' - \mathbf{x}$ we get:

$$\langle \tilde{G}(\mathbf{k})\tilde{G}^*(\mathbf{k}')\rangle = \int d^3x \, e^{-i(\mathbf{k}-\mathbf{k}')\cdot\mathbf{x}} \int d^3z \, \xi_G(\mathbf{z})e^{i\mathbf{k}'\cdot\mathbf{z}} \, . \tag{7.32}$$

Now, using the representation of the Dirac delta we get

$$\boxed{\langle \tilde{G}(\mathbf{k})\tilde{G}^*(\mathbf{k}')\rangle = (2\pi)^3\delta^{(3)}(\mathbf{k} - \mathbf{k}')P_G(\mathbf{k})} \tag{7.33}$$

with the **power spectrum** defined as:

$$P_G(\mathbf{k}) \equiv \int d^3\mathbf{x}\, \xi_G(\mathbf{x}) e^{-i\mathbf{k}\cdot\mathbf{x}} \qquad (7.34)$$

i.e. as the FT of the 2-point correlation function. If also statistical isotropy is assumed, one has:

$$P_G(k) = 2\pi \int_0^\infty dr\, r^2\, \xi_G(r) \int_{-1}^1 du\, e^{-ikru} , \qquad (7.35)$$

where u is the cosine of the angle between $\mathbf{k}$ and $\mathbf{x}$, which we have used as variable in the integration. Because of statistical isotropy then the power spectrum depends only on the modulus of k. Performing the u integration, one gets:

$$P_G(k) = 4\pi \int_0^\infty dr\, r^2\, \xi_G(r) \frac{\sin(kr)}{kr} . \qquad (7.36)$$

7.3.1   Definition of a Gaussian Random Field

Now we define a Gaussian random field, from the point of view of its Fourier components first. FT Gaussian random fields are defined by uncorrelated modes, i.e. precisely by Eq. (7.33). So, we have proved that Gaussianity implies statistical homogeneity. Moreover, Gaussian random field are characterised by the fact that the expectation value and all the correlators of odd order are vanishing, i.e.

$$\langle \tilde{G}(\mathbf{k}) \rangle = \langle \tilde{G}(\mathbf{k}_1)\tilde{G}(\mathbf{k}_2)\tilde{G}(\mathbf{k}_3) \rangle = \cdots = 0 . \qquad (7.37)$$

All the correlators of even order can be written in terms of the second-order correlators, i.e. the power spectrum, which thus contains all the information about the random field. For example, the 4-point correlator is:

$$\langle \tilde{G}(\mathbf{k}_1)\tilde{G}(\mathbf{k}_2)\tilde{G}(\mathbf{k}_3)\tilde{G}(\mathbf{k}_4) \rangle = \langle \tilde{G}(\mathbf{k}_1)\tilde{G}(\mathbf{k}_2) \rangle \langle \tilde{G}(\mathbf{k}_3)\tilde{G}(\mathbf{k}_4) \rangle$$
$$+ \langle \tilde{G}(\mathbf{k}_1)\tilde{G}(\mathbf{k}_3) \rangle \langle \tilde{G}(\mathbf{k}_2)\tilde{G}(\mathbf{k}_4) \rangle + \langle \tilde{G}(\mathbf{k}_1)\tilde{G}(\mathbf{k}_4) \rangle \langle \tilde{G}(\mathbf{k}_2)\tilde{G}(\mathbf{k}_3) \rangle . \qquad (7.38)$$

Since all the Fourier modes are uncorrelated, their superposition is Gaussian-distributed by virtue of the central limit theorem and hence the reason why they are called Gaussian perturbations. So the g's are distributed as a Gaussian:

$$p(g) = \frac{1}{\sqrt{2\pi}\sigma_g} e^{-g^2/2\sigma_g^2} , \qquad (7.39)$$

where we are assuming a vanishing expectation value and the variance has been defined in Eq. (7.16). We have used g here instead of $G(\mathbf{x})$ because of statistical homogeneity.

7.3.2 Estimator of the Power Spectrum and Cosmic Variance

Theoretically, we shall provide predictions on $P_G(k)$ whereas observationally we determine $\xi_G(r)$, so it is more appropriate to invert the above FT. So, starting from Eq. (7.11), we have:

$$
\xi_G(r) = \langle G(\mathbf{x})G(\mathbf{x}+\mathbf{r})\rangle = \int \frac{d^3k}{(2\pi)^3} \int \frac{d^3k'}{(2\pi)^3} \langle \tilde{G}(\mathbf{k})\tilde{G}^*(\mathbf{k}')e^{i\mathbf{k}\cdot\mathbf{x}-i\mathbf{k}'\cdot(\mathbf{x}+\mathbf{r})}\rangle .
$$
(7.40)

We have included the Fourier modes in the average in order to investigate the difference between doing the spatial average or the ensemble one.

Doing the ensemble average. Using Eq. (7.33) in Eq. (7.40), the average must be carried on the G's:

$$
\xi_G(r) = \int \frac{d^3k}{(2\pi)^3} \int d^3k' P_G(k)\delta^{(3)}(\mathbf{k}-\mathbf{k}')e^{i\mathbf{k}\cdot\mathbf{x}-i\mathbf{k}'\cdot(\mathbf{x}+\mathbf{r})} ,
$$
(7.41)

and gives us:

$$
\xi_G(r) = \int \frac{d^3k}{(2\pi)^3} P_G(k)e^{-i\mathbf{k}\cdot\mathbf{r}} .
$$
(7.42)

Note that, since ξ is dimensionless, then $P_G(k)$ has dimension of a volume. Working out the angular integration, we get:

$$
\xi_G(r) = \int \frac{dk\, k^2}{(2\pi)^2} P_G(k) \int_{-1}^{1} du\, e^{-ikru} ,
$$
(7.43)

where again u is the cosine of the angle between $\mathbf{k}$ and $\mathbf{r}$. Integrating for the last time, we get:

$$
\xi_G(r) = \int_0^\infty dk \frac{k^2 P_G(k)}{2\pi^2} \frac{\sin(kr)}{kr}
$$
(7.44)

It is customary then to define a dimensionless power spectrum as

$$
\Delta_G^2(k) \equiv \frac{k^3 P_G(k)}{2\pi^2}
$$
(7.45)

so that Eq. (7.44) becomes:

$$\xi_G(r) = \int_0^\infty \frac{dk}{k} \, \Delta_G^2(k) \frac{\sin(kr)}{kr} \tag{7.46}$$

Doing the spatial average. Suppose that none of the quantities in the integrand of Eq. (7.40) is a stochastic variable and that $\langle \dots \rangle$ is a spatial average, which is the one relevant from the point of view of observation. We must rewrite $\xi_G(r)$ as follows:

$$\hat{\xi}_G(r) = \frac{1}{V} \int_V d^3x \sum_{n,m} \frac{1}{V^2} G_n G_m^* e^{i\mathbf{k}_n \cdot \mathbf{x} - i\mathbf{k}_m \cdot (\mathbf{x}+\mathbf{r})} \,, \quad G_n = \tilde{G}(\mathbf{k}_n) \,, \quad \mathbf{k}_n = \frac{2\pi}{L}\mathbf{n} \,, \tag{7.47}$$

where we have used the Fourier series expansion because we are in a finite volume, that we have chosen to be a cube of side L and hence $V = L^3$. This volume can be thought as the survey volume. Note that the summation over n and m is intended as a sum over the three components of $\mathbf{n}$ and $\mathbf{m}$ since we cannot use statistical isotropy now. The spatial integration gives:

$$\frac{1}{L^3} \int_V d^3\mathbf{x} \, e^{i(\mathbf{k}_n - \mathbf{k}_m)\cdot\mathbf{x}} = \delta_{nm} \,. \tag{7.48}$$

Exercise 7.5 Show in one dimension, for simplicity, that:

$$\frac{1}{L} \int_{-L/2}^{L/2} dx \, e^{i(k_n - k_m)x} = \frac{2\sin[(k_n - k_m)L/2]}{(k_n - k_m)L} \,. \tag{7.49}$$

Since k is quantised, the difference $k_n - k_m$ is always a multiple of $2\pi/L$, thereby making the sine to vanish. The only exception is when the difference is zero, i.e. for $n = m$.

So, we have for the correlation function:

$$\hat{\xi}_G(r) = \sum_n \frac{1}{V^2} |G_n|^2 e^{-i\mathbf{k}_n \cdot \mathbf{r}} \,, \tag{7.50}$$

from which we infer the power spectrum as:

$$P_n \equiv P(\mathbf{k}_n) = \frac{|G_n|^2}{V} \,. \tag{7.51}$$

Since the survey volume is finite we saw that the smallest wavenumber that we can build is $2\pi/L$ and the smallest cell in the wavenumber space has thus volume $(2\pi/L)^3$.

Exercise 7.6 Prove that the number of independent modes between k and $k + dk$ is[1]:

$$N_k = 4\pi k^2 \left(\frac{L}{2\pi}\right)^3 dk = \frac{1}{2\pi^2}(kL)^3 \frac{dk}{k} . \tag{7.52}$$

From this number of independent modes, we then know that the cosmic variance of the power spectrum is:

$$\frac{\sigma_P(k)}{P(k)} \simeq \frac{1}{\sqrt{N_k}} \simeq \frac{1}{r_k^{1/2}(kL)^{3/2}} , \tag{7.53}$$

where we have defined $r_k \equiv dk/k$ as the resolution of the survey. If we assume infinite resolution then we have to derive again the above result since the dimension is reduced by one.

Exercise 7.7 Show that the number of independent modes on the sphere of radius k in the wavenumber space are:

$$N_k = 4\pi k^2 \left(\frac{L}{2\pi}\right)^2 = \frac{(kL)^2}{\pi} . \tag{7.54}$$

In this case, one gets:

$$\frac{\sigma_P(k)}{P(k)} \simeq \frac{1}{\sqrt{N_k}} \simeq \frac{1}{kL} , \tag{7.55}$$

for the cosmic variance, which is the result obtained in Weinberg (2008).

These formulae for the cosmic variance tell us that the latter is negligible if the scale probed is much smaller that the dimension L of the survey, i.e. $kL \gg 1$.

7.4 Non-Gaussian Perturbations

For non-Gaussian perturbations the odd-order correlators are non-vanishing. For example:

$$\langle \tilde{G}(\mathbf{k}_1)\tilde{G}(\mathbf{k}_2)\tilde{G}(\mathbf{k}_3) \rangle = (2\pi)^3 \delta^{(3)}(\mathbf{k}_1 + \mathbf{k}_2 + \mathbf{k}_3) B_G(k_1, k_2, k_3) , \tag{7.56}$$

where $B_G(k_1, k_2, k_3)$ is called **bispectrum** and can be rewritten as:

---

[1] This calculation is essentially identical to the one which leads to the definition of the Fermi momentum of a gas of fermions, which might be more familiar to the reader. See e.g. Huang (1987)

$$B_G(k_1, k_2, k_3) = \mathcal{B}_G(k_1, k_2, k_3) \left[P_G(k_1, k_2) + P_G(k_2, k_3) + P_G(k_1, k_3) \right] \ , \tag{7.57}$$

i.e. in terms of a purely FT of a 3-point correlation function, called **reduced bispectrum**, multiplied by all the possible combinations of the FT transforms of the 2-point correlation functions (i.e. the PS). This formula can be obtained using the following expansion:

$$G(\mathbf{x}) = G_G(\mathbf{x}) + f_{NL} \left(G_G^2(\mathbf{x}) - \langle G_G^2(\mathbf{x}) \rangle \right) + \dots \ , \tag{7.58}$$

where $\langle G_G^2(\mathbf{x}) \rangle \equiv \sigma_G^2$. This expansion is called **local type non-Gaussianity** and is based on the fact that the square of a Gaussian random field $G_G(\mathbf{x})$ is not Gaussian. The amount of non-Gaussianity is indicated by the free parameter f_{NL}, which Planck has constrained to be (Ade et al. 2016):

$$\boxed{f_{NL} = 2.5 \pm 5.7} \quad \text{(68\% CL, statistical)} \tag{7.59}$$

The huge relative error shows how difficult is to extract this kind of information and at the same time how Gaussianity ($f_{NL} = 0$) is fully consistent with data. Note that we are talking here of primordial non-Gaussianity. In the structure formation process, non-Gaussianity naturally arises in the non-linear regime of evolution.

7.5 Matter Power Spectrum, Transfer Function and Stochastic Initial Conditions

All the formula derived in the previous section can be in principle directly adapted to the matter density contrast field $\delta(\eta, \mathbf{k})$,[2] but how do we deal with the time dependence of the latter? In principle, one just needs to substitute $G(\mathbf{x})$ for $\delta(\eta, \mathbf{x})$, where we are leaving on purpose the time-dependence. We then have:

$$\xi_\delta(\eta, r) = \int_0^\infty \frac{dk}{k} \Delta_\delta^2(\eta, r) \frac{\sin kr}{kr} \ , \qquad \Delta_\delta^2(\eta, r) = \frac{k^3 P_\delta(\eta, k)}{2\pi^2} \ , \tag{7.60}$$

and, from Eq. (7.33) we have

$$\langle \delta(\eta, \mathbf{k}) \delta^*(\eta, \mathbf{k}') \rangle = (2\pi)^3 \delta^{(3)}(\mathbf{k} - \mathbf{k}') P_\delta(\eta, k) \ . \tag{7.61}$$

In the above equation we have assumed Gaussianity in the initial conditions. In the linear case the modes evolve independently and hence Gaussianity is preserved, for this reason the above equation is valid even if it is considered at any time η.

---

[2]We drop here the subscript of δ in order to keep a light notation. The treatment can be in principle applied to any δ, but the most interested case is that for matter, i.e. CDM plus baryons.

As we have anticipated, the stochastic character of $\delta(\eta, \mathbf{k})$ is due just to its initial value $\delta(\mathbf{k})$. Hence, it is customary to write:

$$\boxed{P_\delta(\eta, k) = T^2(\eta, k) P_\delta(k)} \tag{7.62}$$

where $T(\eta, k)$ is the **transfer function** and $P_\delta(k)$ is the **primordial power spectrum**, i.e.

$$\langle \delta(\mathbf{k})\delta^*(\mathbf{k}')\rangle = (2\pi)^3 \delta^{(3)}(\mathbf{k} - \mathbf{k}') P_\delta(k) , \tag{7.63}$$

where $\delta(\mathbf{k})$ is the initial condition of δ. The primordial power spectrum is a prediction of inflation that we shall compute in Chap. 8 whereas the transfer function is the solution of the evolution equations that we have found in Chaps. 4 and 5, with the primordial modes found in Chap. 6 as initial conditions. These are characterised by constants multiplied by a function of $\mathbf{k}$, which becomes our stochastic initial condition. We call this $\alpha(\mathbf{k})$ for scalar modes and $\beta(\mathbf{k}, \lambda)$ for tensor modes, borrowing the notation used in Weinberg (2008). Also, in the rest of these notes we shall drop the explicit use of the transfer function $T(\eta, k)$, by adopting the renormalisation:

$$\delta(\eta, \mathbf{k}) \to \alpha(\mathbf{k})\delta(\eta, k) , \qquad \Theta(\eta, \mathbf{k}, \hat{p}) \to \alpha(\mathbf{k})\Theta(\eta, k, \hat{p}) , \tag{7.64}$$

e.g. for the density contrast and for Θ, when it shall be necessary.

Let us see a practical example of how to calculate $T^2(\eta, k)$ and its normalised initial conditions. Consider adiabaticity. Then, all perturbations are sourced by $\alpha(\mathbf{k}) = \zeta(\mathbf{k}) = \mathcal{R}(\mathbf{k}) = C_\gamma(\mathbf{k})$. If we normalise the initial condition $C_\gamma(\mathbf{k}) = 1$, then we have from Eq. (6.71) and the subsequent ones:

$$\frac{1}{3}\delta_c = \frac{1}{3}\delta_b = \frac{1}{4}\delta_\gamma = \frac{1}{4}\delta_\nu = -\Phi + 1 , \quad \Psi = -\frac{10}{15 + 4R_\nu} , \quad \Phi = -\Psi\left(1 + \frac{2}{5}R_\nu\right) . \tag{7.65}$$

With this collection of initial conditions we can compute the transfer functions for each variable, square them and propagate the primordial power spectrum forward to any time. Hence, for matter, we can write:

$$P_\delta(\eta, k) = T^2(\eta, k) P_\delta(k) = T^2(\eta, k)\left(\frac{15}{15 + 4R_\nu}\right)^2 P_\zeta(k) . \tag{7.66}$$

Finally, a comment on the following. Note that in Eq. (7.63) we have the product of two δ's in $\langle \delta(\mathbf{k})\delta^*(\mathbf{k}')\rangle$. Is not this of second order and thus negligible? Locally it is, but the ensemble average can be seen as a spatial average, cf. Eq. (7.47), and thus we have a vanishing quantity integrated over an infinite volume giving a finite result.

7.6   CMB Power Spectra

CMB photons do not come from a localised source but from the whole celestial sphere and from a finite distance, the last scattering surface, which corresponds to a redshift of $z \sim 1100$. Hence, one usually approaches the CMB power spectrum differently from the matter one (this coming from the observation of galaxies).

The relative fluctuation in the temperature $\delta T / T$ is the same Θ that we defined in Eq. (5.80) if we suppose that the perturbed distribution function $\mathcal{F}_\gamma$ does not depend on the photon energy p. Since this assumption is justified by the fact that Thomson scattering with electrons is the relevant physical process generating Θ and in Thomson scattering the energy of the photon is unchanged, we shall use $\delta T / T = \Theta$.

Note that $\Theta = \Theta(\eta, \mathbf{x}, \hat{p})$, but from the point of view of observation, we are just interested in temperature fluctuations here and today, i.e. for $\eta = \eta_0$ and $\mathbf{x} = \mathbf{x}_0$, where the latter is the position of our laboratory (i.e. Earth). Moreover, photons travel on the light-cone (they come from our past light-cone) and from the fixed distance to the last-scattering surface. Hence:

$$\mathbf{x}_* = -(\eta_0 - \eta_*)\hat{p} = -r_*\hat{p} , \qquad (7.67)$$

where η_0 is the present conformal time, η_* the recombination one and $r_* \equiv \eta_0 - \eta_*$ is the comoving distance to recombination. Only photons satisfying this relation can be detected because their momentum is in the "correct" direction (i.e. towards us) and they free stream only after recombination. Therefore, of the 6 variables upon which $\Theta(\eta, \mathbf{x}, \hat{p})$ in principle depends only 2 are observationally relevant, since we observe

$$\Theta(\eta_0, \mathbf{x}_0, \hat{n}) , \qquad (7.68)$$

where $\hat{n} = -\hat{p}$ are indeed coordinates on the celestial sphere. The direction to the photon towards us is $\hat{p}$, so our line of sight unit vector has the opposite sign.

Now we omit the $(\eta_0, \mathbf{x}_0)$ dependence of Θ. It is customary to expand a function on the sphere in spherical harmonics $Y_\ell^m(\theta, \phi)$. Hence, for our temperature fluctuation we have[3]:

$$\boxed{\Theta(\hat{n}) = \sum_{\ell=0}^{\infty} \sum_{m=-\ell}^{\ell} a_{T,\ell m} Y_\ell^m(\theta, \phi)} \qquad (7.69)$$

where we have written $\hat{n} = (\sin\theta \cos\phi, \sin\theta \sin\phi, \cos\theta)$ (i.e. employed spherical coordinates) and $Y_\ell^m(\theta, \phi)$ are the spherical harmonics. Recall that these are revised in some detail in Sect. 12.5.

In these notes we adopt a normalisation which guarantees orthormality, i.e.

---

[3]The expansion is usually considered for ΔT, the temperature fluctuation. In this case, the $a_{T,\ell m}$'s carry dimensions of temperature. The only difference between the coefficients of the two expansions is a factor T_0, i.e. the temperature of the CMB.

$$\int d^2\hat{n}\ Y_\ell^m(\hat{n})Y_{\ell'}^{m'*}(\hat{n}) = \delta_{\ell\ell'}\delta_{mm'}\ . \tag{7.70}$$

Note that the spherical harmonics Y_ℓ^m are in general complex[4] and so the $a_{\ell m}$ must also be such, in order for $\Theta(\theta,\phi)$ to be real.

Exercise 7.8 According to our conventions established in Sect. 12.5 show that

$$Y_\ell^{m*}(\hat{n}) = Y_\ell^{-m}(\hat{n})\ , \tag{7.71}$$

and hence, in order for $\Theta(\hat{n})$ to be real, show that:

$$a_{T,\ell m}^* = a_{T,\ell,-m}\ , \tag{7.72}$$

which is a **reality condition** for the coefficients $a_{T,\ell m}$.

Note that the expansion (7.69) can be done at each point of spacetime, but then the angular dependence changes because the (θ,ϕ) measured from one point in space correspond to different celestial coordinates as seen from another spot in space.

The expansion of Eq. (7.69) can be inverted as follows:

$$a_{T,\ell m} = \int d^2\hat{n}\ Y_\ell^{*m}(\hat{n})\Theta(\hat{n})\ , \tag{7.73}$$

and the $a_{\ell m}$'s are promoted to stochastic variable, just as Θ is. Again, it is the initial conditions for cosmological perturbations which are actual stochastic variables for which inflation predicts a power spectrum, so we shall introduce a transfer function. For Gaussian perturbations, the expectation value and variance of the $a_{T,\ell m}$'s are:

$$\boxed{\langle a_{T,\ell m}\rangle = 0} \qquad \boxed{\langle a_{T,\ell m}a_{T,\ell'm'}^*\rangle = \delta_{\ell\ell'}\delta_{mm'}C_{TT,\ell}} \tag{7.74}$$

and the $C_{TT,\ell} = \langle |a_{T,\ell m}|^2\rangle$ form the CMB power spectrum. Introducing the Fourier transform, we have:

$$C_{TT,\ell} = \langle\int \frac{d^3k}{(2\pi)^3}\int \frac{d^3k'}{(2\pi)^3}e^{i(k-k')\cdot x_0}\int d^2\hat{n}\ Y_\ell^{*m}(\hat{n})\Theta(\mathbf{k},\hat{n})\int d^2\hat{n}'\ Y_\ell^m(\hat{n}')\Theta^*(\mathbf{k}',\hat{n}'))\ . \tag{7.75}$$

The ensemble average acts on the temperature fluctuations, which we renormalise to some primordial mode $\alpha(\mathbf{k})$:

$$\langle\Theta(\mathbf{k},\hat{n})\Theta^*(\mathbf{k}',\hat{n}')\rangle = \langle\alpha(\mathbf{k})\alpha^*(\mathbf{k}')\rangle\Theta(k,\hat{n})\Theta^*(k',\hat{n}')\ . \tag{7.76}$$

---

[4]The notation $Y_{\ell m}$ usually denotes spherical harmonics in the real form, obtained by combining Y_ℓ^m in a suitable way in order to trade the imaginary exponential $\exp(im\phi)$ for a sine or cosine function of $m\phi$.

We should have perhaps written $T_\Theta(k, \hat{n})$, stressing the introduction of a transfer function, but we have decided to keep a simple notation.

For scalar perturbations we saw that, assuming axially symmetric initial conditions, the dependence is only on $\hat{k} \cdot \hat{p} = \mu = -\hat{k} \cdot \hat{n}$ and we have used the multipole expansion in Eq. (5.113), which can be easily inverted as follows:

$$\Theta^{(S)}(k, \mu) = \sum_\ell (-i)^\ell (2\ell + 1) \mathcal{P}_\ell(\mu) \Theta_\ell^{(S)}(k) . \tag{7.77}$$

The evolution of the multipoles $\Theta_\ell^{(S)}(\eta, k)$ until η_0 (today) is given by the hierarchy of Eqs. (5.119)–(5.122) which in fact depend only on the modulus k.

Assuming adiabatic Gaussian perturbations, i.e. $\alpha(\mathbf{k}) = \mathcal{R}(\mathbf{k})$ and

$$\langle \mathcal{R}(\mathbf{k}) \mathcal{R}^*(\mathbf{k}') \rangle = (2\pi)^3 \delta^{(3)}(\mathbf{k} - \mathbf{k}') P_\mathcal{R}(k) , \tag{7.78}$$

we have from Eq. (7.75):

$$C_{TT,\ell}^S = \int \frac{d^3\mathbf{k}}{(2\pi)^3} |\Theta_\ell^{(S)}(\eta_0, k)|^2 P_\mathcal{R}(k) \int d^2\hat{n} \, Y_\ell^{m*}(\hat{n}) \int d^2\hat{n}' \, Y_\ell^m(\hat{n}')$$
$$\sum_{\ell'} (-i)^{\ell'} (2\ell' + 1) \mathcal{P}_{\ell'}(\mu) \sum_{\ell''} i^{\ell''} (2\ell'' + 1) \mathcal{P}_{\ell''}(\mu) . \tag{7.79}$$

Note that a similar result can be obtained from Eq. (7.75) if we perform as spatial average, i.e. an average over $\mathbf{x}_0$, as we saw in the case of the three-dimensional power spectrum. In this case the square modulus $|\Theta(\mathbf{k}, \hat{n})|^2$ appears, as our estimator of the angular power spectrum.

Recall that the Legendre polynomial is proportional to Y_ℓ^0 and thus from the orthogonality of the spherical harmonics and the addition theorem we get:

$$\int d^2\hat{n} \, \mathcal{P}_{\ell'}(\mu) Y_\ell^{m*}(\hat{n}) = \int d^2\hat{n} \, \mathcal{P}_{\ell'}(-\hat{k} \cdot \hat{n}) Y_\ell^{m*}(\hat{n}) = \frac{4\pi}{2\ell + 1} Y_\ell^{m*}(\hat{k}) \delta_{\ell\ell'} . \tag{7.80}$$

Hence we have:

$$\boxed{C_{TT,\ell}^S = \frac{2}{\pi} \int dk \, k^2 |\Theta_\ell^{(S)}(\eta_0, k)|^2 P_\mathcal{R}(k) = 4\pi \int \frac{dk}{k} |\Theta_\ell^{(S)}(\eta_0, k)|^2 \Delta_\mathcal{R}^2(k)} \tag{7.81}$$

where we have used:

$$\int d^2\hat{k} \, |Y_\ell^m(\hat{k})|^2 = 1 . \tag{7.82}$$

We shall discuss the tensor contribution to the CMB TT spectrum in Chap. 10.

Not only temperature anisotropies are measurable in the CMB sky, but also polarisation ones. Hence, we have more correlation functions and spectra than the temperature-temperature (TT) one. As we discussed in Chap. 5, Thomson scattering provides only linear polarisation, which is described by the two Stokes parameters Q and U. It is customary to use the combinations $Q \pm iU$ since these have helicity 2, i.e. under a rotation of an angle θ in the polarisation plane they transform as:

$$(Q \pm iU) \rightarrow e^{\pm 2i\theta}(Q \pm iU) . \tag{7.83}$$

Similar quantities were defined for gravitational waves. Now, $Q \pm iU$ represent fields of helicity 2 on the sphere and as such they can be expanded in spin 2-weighted spherical harmonics:

$$(Q + iU)(\hat{n}) = \sum_{\ell=2}^{\infty} \sum_{m=-\ell}^{\ell} a_{P,\ell m} \, {}_2Y_\ell^m(\hat{n}) , \tag{7.84}$$

$$(Q - iU)(\hat{n}) = \sum_{\ell=2}^{\infty} \sum_{m=-\ell}^{\ell} a_{P,\ell m}^* \, {}_2Y_\ell^{m*}(\hat{n}) . \tag{7.85}$$

The spin-weighted spherical harmonics do not satisfy simple reality conditions as the spin-0 ones (the usual spherical harmonics) do, i.e. $Y_\ell^{m*} = Y_\ell^{-m}$, and therefore it is convenient to use the following combinations of $a_{P,\ell m}$:

$$a_{E,\ell m} = -(a_{P,\ell m} + a_{P,\ell-m}^*)/2 , \qquad a_{B,\ell m} = i(a_{P,\ell m} - a_{P,\ell-m}^*)/2 . \tag{7.86}$$

These satisfy reality conditions and have the following properties under spatial inversion: $a_{E,\ell m}$ gains a factor $(-1)^\ell$, as the $a_{\ell m}$ of the temperature-temperature correlation, whereas $a_{B,\ell m}$ gain an extra -1 factor.

If we assume thus stochastic initial conditions which are invariant under spatial inversion, the only four spectra that we can build from CMB temperature and polarisation measurement are thus:

$$\boxed{\langle a_{T,\ell m} a_{T,\ell' m'}^* \rangle = \delta_{\ell\ell'} \delta_{mm'} C_{TT,\ell}} \quad \boxed{\langle a_{T,\ell m}^* a_{E,\ell' m'} \rangle = \delta_{\ell\ell'} \delta_{mm'} C_{TE,\ell}} \tag{7.87}$$

$$\boxed{\langle a_{E,\ell m} a_{E,\ell' m'}^* \rangle = \delta_{\ell\ell'} \delta_{mm'} C_{EE,\ell}} \quad \boxed{\langle a_{B,\ell m} a_{B,\ell' m'}^* \rangle = \delta_{\ell\ell'} \delta_{mm'} C_{BB,\ell}} \tag{7.88}$$

Note that there is no dependence on m in the power spectra. This is again due to statistical isotropy. Violation of the latter is related to the so-called **CMB anomalies**. These are unexpected features in the CMB sky, such as the *axis of evil* and the large cold spot in the southern hemisphere (Schwarz et al. 2016). The explanation of these anomalies is an open issue of modern cosmology.

Statistical isotropy is motivated by the Copernican principle and predicted by many inflationary theories.[5]

---

[5] According to the Copernican principle we do not occupy a special position in the universe. Therefore, its average properties that we are able to determine should be the same as those determined by any other observer.

7.6.1   Cosmic Variance of Angular Power Spectra

In this section we compute the cosmic variance of the angular power spectra. The result can be applied to any field defined on the sphere and for Gaussian perturbations, which are characterised by a correlation function which depends only on the angular separation and thus the power spectrum is C_ℓ, depending only on ℓ.

In order to compute the cosmic variance, let us do first an estimate. Our objective is to determine C_ℓ observationally, in order to compare it with our theoretical prediction. In order to do that, we probe $\langle a_{\ell m} a_{\ell m}^* \rangle$ for different values of m, which are $2\ell + 1$. Hence, we have $2\ell + 1$ possible sampling of C_ℓ for any given ℓ and a sampling error

$$\Delta C_\ell \propto \sqrt{2\ell + 1} . \tag{7.89}$$

The relative error associated to the sampling, i.e. the cosmic variance, is thus:

$$\sigma_{C_\ell} = \frac{\Delta C_\ell}{2\ell + 1} \propto \frac{1}{\sqrt{2\ell + 1}} . \tag{7.90}$$

Now, let us make a more precise calculation following (Weinberg 2008). Consider the temperature-temperature correlation function, as an example:

$$\langle \Theta(\hat{n}) \Theta(\hat{n}') \rangle = \sum_{\ell m \ell' m'} \langle a_{\ell m} a_{\ell' m'} \rangle Y_\ell^m(\hat{n}) Y_{\ell'}^{m'}(\hat{n}') = \sum_{\ell m} C_\ell Y_\ell^m(\hat{n}) Y_\ell^{-m}(\hat{n}') , \tag{7.91}$$

where $\cos \theta \equiv \hat{n} \cdot \hat{n}'$ and we have performed the ensemble average assuming Gaussian perturbations. The $-m$ of the second spherical harmonics comes from the reality condition by which $a_{\ell m}^* = a_{\ell, -m}$.

Using the addition theorem of the spherical harmonics, we can do the sum over m in the above formula and obtain:

$$C(\theta) \equiv \langle \Theta(\hat{n}) \Theta(\hat{n}') \rangle = \sum_\ell C_\ell \frac{2\ell + 1}{4\pi} P_\ell(\hat{n} \cdot \hat{n}') = \sum_\ell C_\ell \frac{2\ell + 1}{4\pi} P_\ell(\cos \theta) . \tag{7.92}$$

So we have explicitly proved that for Gaussian perturbations the correlation function depends only on the angle between the two directions. Inverting this relation using the orthonormality of the Legendre polynomials, we get:

$$C_\ell = \frac{1}{4\pi} \int d^2\hat{n} \, d^2\hat{n}' \, P_\ell(\hat{n} \cdot \hat{n}') \langle \Theta(\hat{n}) \Theta(\hat{n}') \rangle , \tag{7.93}$$

which is Eq. (7.75), without introducing the FT, which we do not need here.

The integral is of course on the whole sky. These are the theoretical C_ℓ's, and the average is the ensemble one. Observationally, the only average that we can do is the angular one, i.e.

$$C_\ell^{\text{obs}} = \frac{1}{4\pi} \int d^2\hat{n} \, d^2\hat{n}' \, \mathcal{P}_\ell(\hat{n} \cdot \hat{n}') \Theta(\hat{n}) \Theta(\hat{n}') . \tag{7.94}$$

Exercise 7.9 Show that, substituting the spherical harmonics expansions, we have:

$$C_\ell^{\text{obs}} = \frac{1}{4\pi} \sum_{LML'M'} \int d^2\hat{n} \, d^2\hat{n}' \, \mathcal{P}_\ell(\hat{n} \cdot \hat{n}') a_{LM} Y_L^M(\hat{n}) a_{L'M'} Y_{L'}^{M'}(\hat{n}')$$

$$= \frac{1}{2\ell + 1} \sum_m a_{\ell m} a_{\ell, -m} . \tag{7.95}$$

Here it appears more clearly that for each value of ℓ we have $2\ell + 1$ possible realisations and thus we expect that the counting error is $\sqrt{2\ell + 1}$. We can calculate this exactly, and the cosmic variance is the following ensemble average:

$$\sigma_{C_\ell}^2 = \left\langle \left(\frac{C_\ell - C_\ell^{\text{obs}}}{C_\ell} \right)^2 \right\rangle = 1 - 2\frac{\langle C_\ell^{\text{obs}} \rangle}{C_\ell} + \frac{1}{C_\ell^2} \langle C_\ell^{\text{obs}2} \rangle . \tag{7.96}$$

Of course, the ensemble average of $\langle C_\ell^{\text{obs}} \rangle$ is C_ℓ, just look at the integral which defines it, and the ensemble average of C_ℓ is C_ℓ, since it is already averaged. Therefore, let us focus on

$$\langle C_\ell^{\text{obs}2} \rangle = \frac{1}{(2\ell + 1)^2} \sum_{mm'} \langle a_{\ell m} a_{\ell, -m} a_{\ell m'} a_{\ell, -m'} \rangle . \tag{7.97}$$

Since the perturbations are Gaussian, this 4-point correlator can be split into the following sum, by Eq. (7.38):

$$\langle a_{\ell m} a_{\ell, -m} a_{\ell m'} a_{\ell, -m'} \rangle = \langle a_{\ell m} a_{\ell, -m} \rangle \langle a_{\ell m'} a_{\ell, -m'} \rangle + \langle a_{\ell m} a_{\ell m'} \rangle \langle a_{\ell, -m} a_{\ell, -m'} \rangle$$

$$+ \langle a_{\ell m} a_{\ell, -m'} \rangle \langle a_{\ell m'} a_{\ell, -m} \rangle . \tag{7.98}$$

Using Eq. (7.74), we finally get:

$$\boxed{\sigma_{C_\ell}^2 = \frac{2}{2\ell + 1}} \tag{7.99}$$

as expected.

7.7   Power Spectrum for Tensor Perturbations

When we compute the power spectrum for GW, using the decomposition of helicities of Eq. (4.211), we get:

$$\langle h_{ij}^T(\eta, \mathbf{k}) h_{lm}^{T*}(\eta, \mathbf{k}')\rangle = \sum_{\lambda, \lambda'=\pm 2} e_{ij}(\hat{k}, \lambda) e_{lm}^*(\hat{k}', \lambda')\langle h(\eta, \mathbf{k}, \lambda) h^*(\eta, \mathbf{k}', \lambda')\rangle,$$

(7.100)

the stochastic behaviour being carried by the initial condition on $h(\eta, \mathbf{k}, \lambda)$, which we then renormalise as:

$$\boxed{h(\eta, \mathbf{k}, \lambda) = \beta(\mathbf{k}, \lambda) h(\eta, k)}$$

(7.101)

Since the evolution equation for tensor perturbations depends only on k and does not depend on the helicity λ, we incorporate the latter in the stochastic initial condition.

Asking translational invariance (Gaussian perturbations) and rotational invariance, we have:

$$\langle \beta(\mathbf{k}, \lambda) \beta^*(\mathbf{k}', \lambda')\rangle = (2\pi)^3 P_h(k) \delta_{\lambda\lambda'} \delta^{(3)}(\mathbf{k} - \mathbf{k}').$$

(7.102)

Therefore, in the ensemble average of $\langle h_{ij}^T(\eta, \mathbf{k}) h_{lm}^{T*}(\eta, \mathbf{k}')\rangle$ it appears the sum over the helicities:

$$\Pi_{ij,lm}(\hat{k}) \equiv \sum_{\lambda=\pm 2} e_{ij}(\hat{k}, \lambda) e_{lm}^*(\hat{k}, \lambda).$$

(7.103)

We can make this sum explicit as follows. Consider the fact that $\Pi_{ij,lm}(\hat{k})$ is a tensor which depends on $\hat{k}$. In order to have the correct combination ij, lm of indices it must be a combination of the only independent tensors that are around, i.e. δ_{ij} and $\hat{k}_m$. In order to have the correct combination ij, lm of indices, we can combine two δ_{ij}, one δ_{ij} and two $\hat{k}_m$ or four $\hat{k}_m$. So we can put forward the following ansatz:

$$\Pi_{ij,lm}(\hat{k}) = A(\delta_{il}\delta_{jm} + \delta_{im}\delta_{jl}) + B\delta_{ij}\delta_{lm}$$
$$+C\delta_{ij}\hat{k}_l\hat{k}_m + D\delta_{lm}\hat{k}_i\hat{k}_j + E(\delta_{il}\hat{k}_j\hat{k}_m + \delta_{jm}\hat{k}_i\hat{k}_l + \delta_{im}\hat{k}_j\hat{k}_l + \delta_{jl}\hat{k}_i\hat{k}_m)$$
$$+F\hat{k}_i\hat{k}_j\hat{k}_l\hat{k}_m,$$ (7.104)

where we have grouped together all the terms which must be symmetrised in order to respect the symmetry of $\Pi_{ij,lm}(\hat{k})$, which is symmetric in ij and in lm separately. The coefficient introduced above are, in principle, complex.

Exercise 7.10 Contract the above ansatz for $\Pi_{ij,lm}(\hat{k})$, with $\hat{k}^i$, with $\hat{k}^l$ and with δ^{ij}. Since the result must be zero, show that have the following conditions on the coefficients:

$$A = -B = C = D = -E = F .$$

(7.105)

Thus, we can write:

$$\Pi_{ij,lm}(\hat{k}) = A(\delta_{il}\delta_{jm} + \delta_{im}\delta_{jl} - \delta_{ij}\delta_{lm}$$
$$+\delta_{ij}\hat{k}_l\hat{k}_m + \delta_{lm}\hat{k}_i\hat{k}_j - \delta_{il}\hat{k}_j\hat{k}_m - \delta_{jm}\hat{k}_i\hat{k}_l - \delta_{im}\hat{k}_j\hat{k}_l - \delta_{jl}\hat{k}_i\hat{k}_m$$
$$+\hat{k}_i\hat{k}_j\hat{k}_l\hat{k}_m) .$$

(7.106)

Now, since $\Pi^*_{ij,lm}(\hat{k}) = \Pi_{lm,ij}(\hat{k})$, i.e. complex conjugation amounts to exchange the couple ij with lm, it is not difficult to conclude that A is a real number. In order to determine which number, consider the normalisation used in Eq. (4.206). If we set $\hat{k} = \hat{z}$ in our improved ansatz above and choose $i = j = l = m = 1$, we get:

$$\Pi_{11,11}(\hat{k}) = \sum_{\lambda=\pm 2} e_{11}(\hat{k}, \lambda)e^*_{11}(\hat{k}, \lambda) = 1 = A .$$

(7.107)

Therefore, we can conclude that:

$$\Pi_{ij,lm}(\hat{k}) = \delta_{il}\delta_{jm} + \delta_{im}\delta_{jl} - \delta_{ij}\delta_{lm}$$
$$+\delta_{ij}\hat{k}_l\hat{k}_m + \delta_{lm}\hat{k}_i\hat{k}_j - \delta_{il}\hat{k}_j\hat{k}_m - \delta_{jm}\hat{k}_i\hat{k}_l - \delta_{im}\hat{k}_j\hat{k}_l - \delta_{jl}\hat{k}_i\hat{k}_m$$
$$+\hat{k}_i\hat{k}_j\hat{k}_l\hat{k}_m .$$

(7.108)

7.8 Ergodic Theorem

Following Weinberg (2008), here we briefly presents the ergodic theorem, which allows us to exchange ensemble and spatial average under certain conditions.

Consider a random variable $G(\mathbf{x})$ in a D-dimensional Euclidean space and assume statistical homogeneity, i.e.

$$\langle G(\mathbf{x}_1)G(\mathbf{x}_2)\dots G(\mathbf{x}_n)\rangle = \langle G(\mathbf{x}_1 + \mathbf{z})G(\mathbf{x}_2 + \mathbf{z})\dots G(\mathbf{x}_n + \mathbf{z})\rangle ,$$

(7.109)

i.e. when we calculate the correlator, the result does not change upon translation of the fields.

Also, let us make the following reasonable assumption. When we calculate a correlator where a certain amount of points are very far from another set, then the correlator breaks. That is:

$$\langle G(\mathbf{x}_1 + \mathbf{u})G(\mathbf{x}_2 + \mathbf{u}) \dots G(\mathbf{y}_1 - \mathbf{u})G(\mathbf{y}_2 - \mathbf{u}) \dots \rangle$$
$$\to_{|\mathbf{u}| \to \infty} \langle G(\mathbf{x}_1 + \mathbf{u})G(\mathbf{x}_2 + \mathbf{u}) \dots \rangle \langle G(\mathbf{y}_1 - \mathbf{u})G(\mathbf{y}_2 - \mathbf{u}) \dots \rangle$$
$$= \langle G(\mathbf{x}_1)G(\mathbf{x}_2) \dots \rangle \langle G(\mathbf{y}_1)G(\mathbf{y}_2) \dots \rangle \,,$$

i.e., if we think of a simple 2-point correlation function, the points become uncorrelated on a large distance and so it is like having independent ensembles, which is the key in order to exchange the spatial average with the ensemble one. In the last line of the above equation we have used statistical homogeneity.

Now define what in some sense is closely related to the cosmic variance:

$$\sigma_R^2(\mathbf{x}_1, \mathbf{x}_2, \dots) \equiv$$
$$\left\langle \left[\left(\int d^D \mathbf{z} N_R(\mathbf{z}) G(\mathbf{x}_1 + \mathbf{z}) G(\mathbf{x}_2 + \mathbf{z}) \dots \right) - \langle G(\mathbf{x}_1)G(\mathbf{x}_2) \dots \rangle \right]^2 \right\rangle ,$$
$$(7.110)$$

where the integral is a spatial average about a point $\mathbf{z}_0$, recall that the dimension of the space is D, and the window function or filter is for example:

$$N_R(\mathbf{z}) \equiv \frac{1}{(\sqrt{\pi}R)^D} e^{-|\mathbf{z} - \mathbf{z}_0|^2 / R^2} \,, \qquad (7.111)$$

which, as it can be checked, is normalised to unity, i.e. $\int d^D \mathbf{z} \, N_R(\mathbf{z}) = 1$, the integration being over all the space. This function is basically a filter and its form is not important as long it is almost constant for $|\mathbf{z} - \mathbf{z}_0|^2 < R^2$ and decays rapidly for $|\mathbf{z} - \mathbf{z}_0|^2 > R^2$. It introduces a scale R which is the limiting one until which we are able to do a spatial average and so we want to check how this variance depends on R. We can think of R as the size of a survey, for example.

The ergodic theorem states that

$$\sigma_R^2(\mathbf{x}_1, \mathbf{x}_2, \dots) \sim R^{-D} \qquad \text{for } R \to \infty \,. \qquad (7.112)$$

In order to prove it, expand the square in the variance:

$$\sigma_R^2 = \left\langle \int d^D \mathbf{z} N_R(\mathbf{z}) \int d^D \mathbf{w} N_R(\mathbf{w}) (G(\mathbf{x}_1 + \mathbf{z}) G(\mathbf{x}_2 + \mathbf{z}) \cdots - \langle G(\mathbf{x}_1)G(\mathbf{x}_2) \dots \rangle) \right.$$
$$\left. \times \, (G(\mathbf{x}_1 + \mathbf{w}) G(\mathbf{x}_2 + \mathbf{w}) \cdots - \langle G(\mathbf{x}_1)G(\mathbf{x}_2) \dots \rangle) \right\rangle \,. \qquad (7.113)$$

We can incorporate the ensemble average into the integral because of $\int d^D z \, N_R(z) = 1$. Now apply the average and use statistic homogeneity:

$$\sigma_R^2 = \int d^D \mathbf{z} N_R(\mathbf{z}) \int d^D \mathbf{w} N_R(\mathbf{w}) (\langle G(\mathbf{x}_1 + \mathbf{z}) G(\mathbf{x}_2 + \mathbf{z}) \dots G(\mathbf{x}_1 + \mathbf{w}) G(\mathbf{x}_2 + \mathbf{w}) \dots \rangle$$
$$- \langle G(\mathbf{x}_1)G(\mathbf{x}_2) \dots \rangle^2) \,. \qquad (7.114)$$

Now change integration variables:

$$\mathbf{u} = (\mathbf{z} - \mathbf{w})/2 , \qquad \mathbf{v} = (\mathbf{z} + \mathbf{w})/2 , \tag{7.115}$$

and, remembering to take into account the Jacobian of the transformation, which is 2^D, and the expression we have adopted for the window function:

$$\sigma_R^2 = \left(\frac{2}{\pi R^2}\right)^D \int d^D \mathbf{v} \int d^D \mathbf{u} \; e^{-|\mathbf{u}+\mathbf{v}-\mathbf{z}_0|^2/R^2 - |\mathbf{v}-\mathbf{u}-\mathbf{z}_0|^2/R^2}$$
$$(\langle G(\mathbf{x}_1 + \mathbf{u} + \mathbf{v})G(\mathbf{x}_2 + \mathbf{u} + \mathbf{v}) \ldots G(\mathbf{x}_1 + \mathbf{v} - \mathbf{u})G(\mathbf{x}_2 + \mathbf{v} - \mathbf{u}) \ldots \rangle -$$
$$\langle G(\mathbf{x}_1)G(\mathbf{x}_2) \ldots \rangle^2) . \tag{7.116}$$

Now, the $\mathbf{v}$ in the fields is always summed to the coordinate so we can use again statistic homogeneity and eliminate $\mathbf{v}$ from the average, getting:

$$\sigma_R^2 = \left(\frac{2}{\pi R^2}\right)^D \int d^D \mathbf{v} \int d^D \mathbf{u} \; e^{-2|\mathbf{u}|^2/R^2} e^{-|\mathbf{v}-\mathbf{z}_0|^2/R^2}$$
$$(\langle G(\mathbf{x}_1 + \mathbf{u})G(\mathbf{x}_2 + \mathbf{u}) \ldots G(\mathbf{x}_1 - \mathbf{u})G(\mathbf{x}_2 - \mathbf{u}) \ldots \rangle - \langle G(\mathbf{x}_1)G(\mathbf{x}_2) \ldots \rangle^2) . \tag{7.117}$$

The integration in $\mathbf{v}$ can be immediately performed, being a Gaussian integral equal to $(\pi/2)^{D/2} R^D$:

$$\sigma_R^2 = \left(\frac{2}{\pi R^2}\right)^{D/2} \int d^D \mathbf{u} \; e^{-2|\mathbf{u}|^2/R^2}$$
$$(\langle G(\mathbf{x}_1 + \mathbf{u})G(\mathbf{x}_2 + \mathbf{u}) \ldots G(\mathbf{x}_1 - \mathbf{u})G(\mathbf{x}_2 - \mathbf{u}) \ldots \rangle - \langle G(\mathbf{x}_1)G(\mathbf{x}_2) \ldots \rangle^2) . \tag{7.118}$$

Note that $\mathbf{z}_0$ is no more present, as it should be because of statistical homogeneity. The term between parenthesis goes to zero for large $|\mathbf{u}|$, on the basis of the hypothesis made at the beginning of our discussion, therefore the $|\mathbf{u}|$ integral is finite and above all, its value does not depend on R because the difference among the ensemble averages does not depend on our choice of R. Then, we can conclude that:

$$\boxed{\sigma_R = \mathcal{O}(R^{-D/2})} \tag{7.119}$$

References

Ade, P.A.R., et al.: Planck 2015 results. XVII. Constraints on primordial non-Gaussianity. Astron. Astrophys. **594**, A17 (2016)

Huang, K.: Statistical Mechanics, 2nd edn. Wiley-VCH, New York (1987)

Lyth, D.H., Liddle, A.R.: The Primordial Density Perturbation: Cosmology, Inflation and the Origin of Structure (2009)

Schwarz, D.J., Copi, C.J., Huterer, D., Starkman, G.D.: CMB anomalies after Planck. Class. Quant. Grav. **33**(18), 184001 (2016)

Weinberg, S.: Cosmology. Oxford University Press, Oxford (2008)

Chapter 8
Inflation

> *And if inflation is wrong, then God missed a good trick. But, of course, we've come across a lot of other good tricks that nature has decided not to use*
>
> Jim Peebles, interview at Princeton (1994)

We dedicate this chapter to inflation, a model of the primordial universe in which an almost constant H provides a scale factor a growing exponentially with the cosmic time. Inflation is able to solve some puzzles related to background cosmology and also to provide a testable prediction of the power spectrum of primordial fluctuations. Among the first pioneering works on inflation there are Starobinsky (1979), Guth (1981), Linde (1982) and Albrecht and Steinhardt (1982).

An alternative to inflation which has risen some interest is the so-called **bouncing cosmology** in which the universe is eternal and whose evolution is characterised by a contracting phase followed by an expanding one, through a bouncing phase whose physical details are governed by quantum cosmology. See Novello and Bergliaffa (2008) and Brandenberger and Peter (2017) for recent reviews on the subject which, though interesting, we shall not tackle here.

8.1 The Flatness Problem

Recall the definition of the curvature density parameter that we gave in Eq. (2.69):

$$\Omega_K \equiv -\frac{K}{H^2 a^2} \, , \tag{8.1}$$

and of its observed value coming from the latest Planck data (Ade et al. 2016b):

$$\Omega_{K0} = 0.0008^{+0.0040}_{-0.0039} \, . \tag{8.2}$$

© Springer International Publishing AG, part of Springer Nature 2018
O. Piattella, *Lecture Notes in Cosmology*, UNITEXT for Physics,
https://doi.org/10.1007/978-3-319-95570-4_8

at the 95% confidence level. It is with a lot of confidence, much larger that 95%, that we can conclude that:

$$|\Omega_{K0}| < 1 . \tag{8.3}$$

But then notice the following:

$$|\Omega_K| = \left| -\frac{K}{H^2 a^2} \right| = \left| -\frac{K}{H^2 a^2} \frac{H_0^2 a_0^2}{H^2 a^2} \right| = |\Omega_{K0}| \frac{H_0^2}{H^2 a^2} < \frac{H_0^2}{H^2 a^2} , \tag{8.4}$$

where we have used $a_0 = 1$. Now, how does the function on the right hand side scale? When $a \to 0$, since radiation dominates, we have that:

$$\frac{H_0^2}{H^2 a^2} \sim a^2 \to 0 . \tag{8.5}$$

That is, the more in the past we go the closer to zero Ω_K gets. And closer means *really* closer. Indeed:

Exercise 8.1 Show that at the radiation-matter equivalence, i.e. at $z = 10^4$, $\Omega_K < 10^{-4}$. Show that at BBN, i.e. at $z = 10^{10}$, $\Omega_K < 10^{-16}$. Then show that the Planck time corresponds roughly to $z = 10^{32}$, and then there we have $\Omega_K < 10^{-60}$.

The Planck time is the farthest we can extrapolate our classical theory and 10^{-60} is something *really* close to zero. What is the problem here? The problem is that if by some reason $\Omega_K \sim 10^{-59}$ at the Planck era, then today Ω_{K0} would be ten times larger and in complete disagreement with observation. In order to match what we observe today, Ω_K has to be determined at the Planck scale with a precision of 60 significative digits! This is an example of **fine-tuning**.

Fine tunings might not be actual problems for theories, in the same sense that falsified predictions rule out wrong theories, but denote something *ad hoc*, unnatural and therefore something that we would like to explain in another way if possible.

As for the flatness problem the very simple idea put forward in Guth (1981) is the following. Regard Eq. (8.1). If H is constant, then $\Omega_K \propto 1/a^2$. Thus, if before the radiation-dominated era the very early universe experienced a phase in which H is constant, this could explain why Ω_K is so small to begin with, without recurring to any fine-tuning. This primordial phase with H constant, or almost constant, is **inflation**.

Inflation must take place in a primordial phase of evolution, prior to radiation domination, in order not to spoil the successful predictions of the Hot Big Bang model.

Exercise 8.2 Show that a constant H implies that:

$$H = \text{constant} \Rightarrow a \propto e^{Ht} , \tag{8.6}$$

i.e. an exponential growth with time of the scale factor.

Let us see quantitatively how inflation can solve the flatness problem. Suppose that

$$\frac{|K|}{a_i^2 H_i^2} = \mathcal{O}(1) , \tag{8.7}$$

at the beginning of the inflationary phase, to which we refer with a subscript i. That is, at the beginning of inflation the spatial curvature might be a relevant fraction of the total energy density content of the universe.

Now, at the end of inflation, which we denote with a subscript I, the scale factor a has grown of a factor e^N. The number N is called **e-folds number**. Then we have:

$$\frac{|K|}{a_I^2 H_I^2} = \frac{|K|}{a_i^2 H_i^2} e^{-2N} \approx e^{-2N} . \tag{8.8}$$

Therefore, today we have that:

$$|\Omega_{K0}| = \frac{|K|}{H_0^2} = \frac{|K|}{a_I^2 H_I^2} \left(\frac{a_I H_I}{H_0} \right)^2 \approx e^{-2N} \left(\frac{a_I H_I}{H_0} \right)^2 . \tag{8.9}$$

In order to have $|\Omega_{K0}| < 1$, we need

$$\boxed{\frac{a_I H_I}{H_0} < e^N} \tag{8.10}$$

In order to estimate the ratio on the left hand side, let us suppose that inflation ends into the radiation-dominated epoch, so that:

$$H_I^2 \approx H_0^2 \Omega_{r0}/a_I^4 , \tag{8.11}$$

and thus:

$$a_I \approx \Omega_{r0}^{1/4} \sqrt{\frac{H_0}{H_I}} . \tag{8.12}$$

We have then:

$$e^N > \Omega_{r0}^{1/4} \sqrt{\frac{H_I}{H_0}} = \Omega_{r0}^{1/4} \left(\frac{\rho_I}{\rho_0}\right)^{1/4} = \frac{\rho_I^{1/4}}{0.037 \, h \, \text{eV}} \, , \quad (8.13)$$

where ρ_I is the energy scale at which inflation ends, but since H is constant during the inflationary phase, then ρ_I is energy density scale of inflation *tout court*.

Certainly we do not want to spoil BBN, therefore ρ_I must be larger than $1\,\text{MeV}^4$. With this constraint we get $N > 17$. Choosing the Planck scale, one gets $N > 68$ and choosing the GUT energy scale one gets $N > 62$.

8.2   The Horizon Problem

The **horizon problem** is an issue that appears when we calculate the angular size of the particle horizon at recombination and notice that it is only a small portion of the CMB sky. Then, how is it possible that the latter is so isotropic if no causal process could have provided the conditions to be so?

Let us again consider this issue more quantitatively. The proper particle-horizon distance is the following:

$$d_{\text{H}} = a(t) \int_0^t \frac{dt'}{a(t')} = a \int_0^a \frac{da'}{H(a')a'^2} \, . \quad (8.14)$$

In an universe dominated by matter and radiation the above expression becomes:

$$d_{\text{H}} = a \int_0^a \frac{da'}{H_0\sqrt{\Omega_{m0}a' + \Omega_{r0}}} = \frac{2a}{H_0\Omega_{m0}} \left(\sqrt{\Omega_{m0}a + \Omega_{r0}} - \sqrt{\Omega_{r0}}\right) \, . \quad (8.15)$$

Now, recall that the angular diameter distance has the following form:

$$d_{\text{A}} = a(t) \int_t^{t_0} \frac{dt'}{a(t')} = a \int_a^1 \frac{da'}{H(a')a'^2} \, . \quad (8.16)$$

Again, in an universe dominated by matter and radiation the above expression becomes:

$$d_{\text{A}} = a \int_a^1 \frac{da'}{H_0\sqrt{\Omega_{m0}a' + \Omega_{r0}}} = \frac{2a}{H_0\Omega_{m0}} \left(\sqrt{\Omega_{m0} + \Omega_{r0}} - \sqrt{\Omega_{m0}a + \Omega_{r0}}\right) \, . \quad (8.17)$$

The ratio $d_{\text{H}}/d_{\text{A}}$ represents the angular radius of the particle horizon at a given scale factor. From the above calculations we obtain:

$$\frac{d_{\text{H}}}{d_{\text{A}}} = \frac{\sqrt{\Omega_{m0}a + \Omega_{r0}} - \sqrt{\Omega_{r0}}}{\sqrt{\Omega_{m0} + \Omega_{r0}} - \sqrt{\Omega_{m0}a + \Omega_{r0}}} \, . \quad (8.18)$$

This ratio tends to zero for $a \to 0$ and at recombination it is equal to:

$$\frac{d_H}{d_A}(a_{rec} = 10^{-3}) = 0.018 , \tag{8.19}$$

which corresponds to about $1°$ in the CMB sky. Therefore, we have roughly $4\pi/(0.018)^2 \approx 10^4$ causally disconnected regions in the sky.

Another way to see this problem is to compare d_H with the radius of the universe, say R, which is proportional to the scale factor a, and calculating thus the number of causally disconnected regions, i.e.:

$$N_{cdr}^{1/3} \equiv \frac{R}{d_H} = \frac{\Omega_{m0} H_0 R_i}{2a_i \left(\sqrt{\Omega_{m0}a + \Omega_{r0}} - \sqrt{\Omega_{r0}}\right)} , \tag{8.20}$$

where R_i is some initial radius of the universe corresponding to an initial scale factor a_i.

Exercise 8.3 Typically one assumes that $N_{cdr} \sim 10^{90}$ at the Planck scale, i.e. of the order or larger than the observed number of particles in the universe. See e.g. Linde (2017) for a recent discussion on the issue of the initial conditions of the universe.

Given $N_{cdr} \sim 10^{90}$ at $a = 10^{-32}$ determine then $H_0 R_i/a_i$ in Eq. (8.20) and deduce the value of N_{cdr} at recombination. Show that, again, $N_{cdr} \approx 10^4$.

So, the particle horizon at recombination is a very small fraction of the CMB sky. If no causal process could have taken place beyond 1 degree then what did cause the high isotropy of the CMB temperature?

This issue can be solved in pretty much the same way as we did for the flatness problem. Assume an initial inflationary phase such that:

$$a(t) = a_i e^{H_I(t-t_i)} = a_I e^{-H_I(t_I-t)} . \tag{8.21}$$

Then, the proper particle-horizon distance that we have calculated in Eq. (8.15) acquires the following contribution for very small times:

$$d_H \approx a \int_{t_i}^{t_I} dt \frac{e^{H_I(t_I-t)}}{a_I} , \tag{8.22}$$

which we actually put equal to the same d_H because we know that it has to dominate it in order to solve the horizon problem. Therefore:

$$d_H \approx a \int_{t_i}^{t_I} dt \frac{e^{H_I(t_I-t)}}{a_I} = \frac{a}{a_I H_I}(e^N - 1) . \tag{8.23}$$

Since $d_A \approx a/H_0$ for small scale factors, we can conclude that:

$$\frac{d_{\mathrm{H}}}{d_{\mathrm{A}}} \approx \frac{H_0}{a_I H_I} e^N \, , \tag{8.24}$$

and so, in order to have $d_{\mathrm{H}} > d_{\mathrm{A}}$, we obtain the condition:

$$\boxed{\frac{a_I H_I}{H_0} < e^N} \tag{8.25}$$

which is the same as in Eq. (8.10). The same solution for two different problems seems to be a good sign and a good point in favour of inflation.

Usually a third problem concerning standard cosmology goes along with the above two and it is the one related to the abundance of unwanted relics, such as magnetic monopoles, which are produced via some symmetry breaking mechanism in the very early universe and are not observed today. We do not treat this issue here, see Weinberg (2013) for more details, but it is clear that inflation provides a mechanism for diluting these unwanted relics beyond the possibility of observation.

We now discuss how inflation can be realised.

8.3 Single Scalar Field Slow-Roll Inflation

We present here the possibility of implementing inflation via a single canonical scalar field φ, called **inflaton**, which is subject to some potential $V(\varphi)$ with the property that the scalar field initially rolls slowly down it attaining its minimum after the end of inflation. This kind of behaviour is called **slow-roll inflation** and it proves to be very successful. We now see in some detail how the inflationary phase occurs.

The Lagrangian of a canonical scalar field is:

$$\mathcal{L} = \frac{1}{2} g^{\mu\nu} \partial_\mu \varphi \partial_\nu \varphi + V(\varphi) \, , \tag{8.26}$$

where the plus sign might be deceiving (we are used in Lagrangian mechanics to a difference between the kinetic term and the potential one) if we do not recall that the signature in use is $(-, +, +, +)$.

The energy-momentum tensor of a canonical scalar field has the following form:

$$T^\alpha{}_\beta = g^{\alpha\nu} \partial_\nu \varphi \partial_\beta \varphi - \delta^\alpha{}_\beta \left[\frac{1}{2} g^{\mu\nu} \partial_\mu \varphi \partial_\nu \varphi + V(\varphi) \right] \, , \tag{8.27}$$

where $V(\varphi)$ is some potential. In the background FLRW metric $ds^2 = -dt^2 + a(t)^2 \delta_{ij} dx^i dx^j$, the scalar field has to depend only on t, and thus the energy density and pressure are:

$$\rho_\varphi = -T^0_{\ 0} = \frac{1}{2}\dot{\varphi}^2 + V(\varphi) , \qquad P_\varphi = \frac{1}{3}\delta^i_{\ j} T^j_{\ i} = \frac{1}{2}\dot{\varphi}^2 - V(\varphi) . \qquad (8.28)$$

Exercise 8.4 Derive the Klein–Gordon equation for φ using the continuity equation. Show that:

$$\ddot{\varphi} + 3H\dot{\varphi} + V_{,\varphi} = 0 , \qquad (8.29)$$

where $V_{,\varphi} \equiv dV/d\varphi$. Show that in the conformal time

$$\varphi'' + 2\mathcal{H}\varphi' + a^2 V_{,\varphi} = 0 . \qquad (8.30)$$

Now, Friedmann and the acceleration equations for this scalar field as the unique content of the universe are written as:

$$H^2 = \frac{8\pi G}{3}\left[\frac{1}{2}\dot{\varphi}^2 + V(\varphi)\right] , \qquad \frac{\ddot{a}}{a} = -\frac{8\pi G}{3}\left[\dot{\varphi}^2 - V(\varphi)\right] . \qquad (8.31)$$

Exercise 8.5 Show that the acceleration equation can be written in the following form:

$$\dot{H} = -4\pi G\dot{\varphi}^2 . \qquad (8.32)$$

In order to have an accelerated phase of expansion we need that H varies slowly, i.e. $\dot{H} \sim 0$, but how much slowly? Since the only time scale present in the problem is $1/H$ itself, we need that

$$\frac{|\dot{H}|}{H^2} \ll 1 , \qquad (8.33)$$

because this is the only dimensionless combination possible with $\dot{H}$. It means that, during an expansion time $1/H$, the relative variation of H is much less than unity. Using Friedmann equation and the expression for $\dot{H}$ we can write the above condition as:

$$\boxed{\dot{\varphi}^2 \ll V(\varphi)} \qquad (8.34)$$

which is our first **slow-roll condition**. When the kinetic term of the scalar field is negligible with respect to the potential one, one has from Eq. (8.28):

$$P_\varphi \approx -\rho_\varphi \approx -V(\varphi) \approx \text{constant} . \qquad (8.35)$$

That is, the scalar field potential, when it dominates over the kinetic term, behaves as a cosmological constant.

Note that condition (8.34) does not really demand the potential $V(\varphi)$ to be flat, as for example the Coleman–Weinberg (Erick, not Steven) potential (Coleman and Weinberg 1973) used in the "new" inflationary scenario (Albrecht and Steinhardt 1982). What counts is the kinetic term to be very small compared with the potential one and this can be achieved even for $V(\varphi) \propto \varphi^2$ or φ^4 potentials. We shall see this in more detail later.

The condition given in Eq. (8.33) is usually reformulated in term of a parameter ϵ, called **slow-roll parameter** and defined as the ratio in Eq. (8.33), i.e.

$$\epsilon \equiv -\frac{\dot{H}}{H^2} = \frac{d}{dt}\left(\frac{1}{H}\right) \qquad (8.36)$$

and it is the derivative of the Hubble radius. For an exactly exponential expansion H is constant (de Sitter space), and thus $\epsilon = 0$. The slow-roll condition is thus realised by demanding that $|\epsilon| \ll 1$. During the radiation-dominated epoch, $a \propto \sqrt{t}$ and thus $1/H = 2t$ and $\epsilon = 2$.

Exercise 8.6 The minimal requirement for inflation is that it must produce an acceleration, i.e. $\ddot{a} > 0$. For the limiting case in which we have $\ddot{a} = 0$, show that $\epsilon = 1$.

In Fig. 8.1 the dashed lines represents generic physical scales $\lambda^{\text{phys}} \propto a/k$ growing proportionally to the scale factor and exiting the Hubble radius during inflation, since $1/H$ stays almost constant (it is really constant in the figure), and re-entering after inflation ends, during the radiation-dominated epoch. Of course, the last scale to exit is also the first one to re-enter, whereas the observable scale which first exits the horizon, is the one which re-enters the horizon today. According to the normalisation used in Fig. 8.1, this scale is a and exited the horizon when:

$$a_{\text{exit}} = \frac{H_0}{H_I} . \qquad (8.37)$$

Using Eq. (8.13), we can thus write down the maximum number of e-folds to which we could have access observationally as:

$$N_{\text{max}} = \ln\left(\frac{a_I}{a_{\text{exit}}}\right) = \ln\left(\frac{a_I H_I}{H_0}\right) = \ln\left(\frac{\rho_I^{1/4}}{0.037\, h\, \text{eV}}\right) . \qquad (8.38)$$

For the energy scale of GUT, i.e. 10^{16} GeV, one gets $N_{\text{max}} \approx 61$. Note that this is not the duration of inflation. It can last much longer, e.g. $N \approx 145$ as computed in Bolliet et al. (2017), but we can observationally probe only the last 61 e-folds (depending on the energy scale of inflation).

We shall see later that when a perturbative scale crosses the Hubble horizon during inflation, it gets a specific value which remains constant during it super-horizon

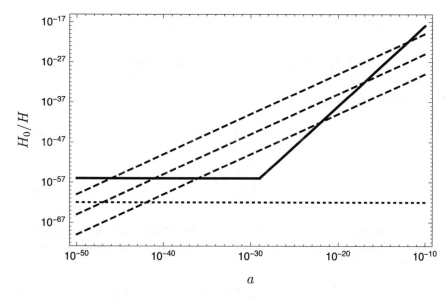

Fig. 8.1 The solid line represents the evolution of the Hubble radius $1/H$, normalised to the present one $1/H_0$ for the ΛCDM model. The dashed lines represents different scales, which grow as a, exiting the Hubble radius and then re-entering again during the radiation-dominated epoch. The dotted line represents the Planck scale. We have chosen inflation to end at $a = 10^{-29}$, corresponding to the GUT scale 10^{16} GeV

evolution and then serves as initial condition when re-entering the horizon during the radiation-dominated epoch.

In Fig. 8.1 there is also a horizontal dotted line. This represents the Planck scale. Note that scales which re-enter the horizon after the end of inflation, and are thus observable by us for example in the CMB, were smaller than the Planck scale. Since we do not know how gravitation behaves on such smaller scales, the above is known as **Trans-Planckian problem**. See e.g. Brandenberger and Martin (2013) for a recent review on the subject.

8.3.1 More Slow-Roll Parameters

It is not only important that the inflaton slow-rolls, but also that it does so for a sufficiently long time in order to provide at least $N = 60$ e-folds. In order to make this claim more quantitative, let us investigate how ϵ varies by computing its time-derivative. Using the definition (8.36) we get:

$$\dot{\epsilon} = \frac{2\dot{H}^2}{H^3} - \frac{\ddot{H}}{H^2} , \tag{8.39}$$

and from Eq. (8.32) we get:

$$\dot{\epsilon} = 2H\epsilon^2 + \frac{8\pi G\dot{\varphi}}{H^2}\ddot{\varphi} = 2H\epsilon^2 - 2\frac{\dot{H}\ddot{\varphi}}{H^2\dot{\varphi}} = 2H\epsilon^2 + 2\epsilon\frac{\ddot{\varphi}}{\dot{\varphi}} \,. \tag{8.40}$$

Let us write this expression as:

$$\dot{\epsilon} = 2H\epsilon(\epsilon - \eta) \,, \tag{8.41}$$

where

$$\boxed{\eta \equiv -\frac{1}{H}\frac{\ddot{\varphi}}{\dot{\varphi}}} \tag{8.42}$$

is our second slow-roll parameter, not to be confused with the conformal time. The smallness of η then guarantees that ϵ varies slowly. Indeed, $\eta \ll 1$ gives us, from the KG equation (8.29) the condition

$$\boxed{3H\dot{\varphi} \approx -V_{,\varphi}} \tag{8.43}$$

Considering $\epsilon, \eta \ll 1$ as first-order quantities, the time-derivative $\dot{\epsilon}$ is thus a second-order quantity, cf. Eq. (8.41).

Nothing prevents us from considering now the time-derivative of η and then to define a third slow-roll parameter α, of which we could consider again the time-derivative, defining a fourth slow-roll parameter β and so on, constructing a hierarchy of slow-roll parameters. This was done in Liddle et al. (1994), but we limit ourselves here to ϵ and η, upon which the predictions on the scalar and tensor spectral indices depend, as we shall see. It must be stressed, on the other hand, that future observation might be able to constrain with great precision the runnings of the spectral indices, which depend on the higher-order slow-roll parameters, see Muñoz et al. (2017).

Since it is the scalar field that triggers the inflationary phase, it is useful to express ϵ and η in terms of quantities related to the scalar field itself, i.e. the potential $V(\varphi)$ and its derivatives. This can be done as follows. Combining the definition of ϵ in Eq. (8.36) with Friedmann equation and Eq. (8.32), one gets:

$$\epsilon = \frac{3\dot{\varphi}^2}{2V + \dot{\varphi}^2} = \frac{3\dot{\varphi}^2}{2V} + \mathcal{O}\left[\left(\frac{\dot{\varphi}^2}{2V}\right)^2\right] \,, \tag{8.44}$$

where we are implementing the slow-roll condition (8.34). Using now Eq. (8.43), we find:

$$\boxed{\epsilon \approx \frac{1}{16\pi G}\left(\frac{V_{,\varphi}}{V}\right)^2 \equiv \epsilon_V} \tag{8.45}$$

at the lowest order in $\dot{\varphi}^2/2V$. In this equation we have defined ϵ_V as a quantity describing the steepness of the inflaton potential.

Since η depends on the second time derivative of the scalar field, we expect it to depend on the second derivative of the potential, i.e. $V_{,\varphi\varphi}$, in the slow-roll limit. Using Eq. (8.43) and the above definition (8.42), we can write:

$$\eta \approx \frac{V_{,\varphi\varphi} + 3\dot{H}}{3H^2} . \tag{8.46}$$

Now, recalling the definition of ϵ and that, at the lowest order in the slow-roll approximation, $3H^2 \approx 8\pi GV$, we can conclude that:

$$\boxed{\eta + \epsilon \approx \frac{1}{8\pi G} \frac{V_{,\varphi\varphi}}{V} \equiv \eta_V} \tag{8.47}$$

It exists thus a hierarchy of slow-roll parameters based on the derivative of the Hubble factor and another one based on those of the potential. They of course can be put in relation, as we did above for the first two slow-roll parameters.

So, in order to have $\epsilon_V, \eta_V \ll 1$ and thus trigger inflation not necessarily V has to be constant but rather its first and second derivatives must be much smaller than the value of V itself.

We can relate the number of e-folds to the slow-roll parameter ϵ as follows. Recalling that the number of e-folds is $N = \ln a$, we can immediately write that:

$$\dot{N} = H \quad \Rightarrow \quad \Delta N_{12} = \int_{t_1}^{t_2} H\, dt , \tag{8.48}$$

where $t_1 < t_2$ are two generic instants during the inflationary phase. Changing variable in favor of the inflaton field, we can write:

$$\Delta N_{12} = \int_{\varphi_1}^{\varphi_2} \frac{H}{\dot{\varphi}} d\varphi , \tag{8.49}$$

where $\varphi_1 \equiv \varphi(t_1)$ and $\varphi_2 \equiv \varphi(t_2)$. Using the slow-roll conditions presented earlier one has:

$$\Delta N_{12} = 8\pi G \int_{\varphi_2}^{\varphi_1} \frac{V}{V_\varphi} d\varphi . \tag{8.50}$$

Since V/V_φ is almost constant and very large during inflation (its square is proportional to $1/\epsilon$), we can pull it out of the integral and approximate the above equation as:

$$\Delta N_{12} \approx \frac{8\pi GV}{V_\varphi}(\varphi_1 - \varphi_2) = \frac{V}{V_\varphi} \frac{\varphi_1 - \varphi_2}{M_{\mathrm{Pl}}^2} , \tag{8.51}$$

where we have introduced the Planck mass M_{Pl} instead of $1/\sqrt{8\pi G}$. In order to produce a large N it might be possible that $|\varphi_1 - \varphi_2| > M_{\text{Pl}}$, but this not necessarily leads to a trans-Planckian problem because the energy scale of inflation is $V(\varphi)$, not the field itself, and so it is the potential which has to be smaller than the Planck scale in order for a classical treatment to be valid. This can be achieved, for example, if there is a sufficiently small coupling constant.

Using Eq. (8.45), we can write in general that:

$$\Delta N = \frac{1}{\sqrt{2\epsilon}} \frac{\Delta\varphi}{M_{\text{Pl}}} . \tag{8.52}$$

Suppose that φ_1 and φ_2 are the value of the inflaton field for which the wavenumbers corresponding to the CMB multipole $\ell = 1$ and $\ell = 100$ exit the horizon. Since, as we are going to see in Chap. 10, $\ell \propto k$, then:

$$\Delta N = \ln 100 = 4.6 . \tag{8.53}$$

The above formula then gives a bound between the variation of the inflaton field and ϵ, based on the observable scales of the CMB. Such bound is known as **Lyth bound** (Lyth 1997). See also Di Marco (2017) for a recent perspective on the Lyth bound.

8.3.2 Reheating

When ϵ_V and η_V cease to be very small and attain values of order unity, inflation ends. Close to the minimum say φ_0 the inflaton potential can be expanded in the following way:

$$V(\varphi) = V_0 + \frac{1}{2} \left. V_{,\varphi\varphi} \right|_{\varphi=\varphi_0} (\varphi - \varphi_0)^2 + \cdots \approx \frac{1}{2} m_\varphi^2 (\varphi - \varphi_0)^2 , \tag{8.54}$$

where $V_0 \equiv V(\varphi_0)$ is the minimum of the potential which we assume to be very small, in fact negligible, in order not to generate an important vacuum energy contribution (which dominates only at late times). We have also introduced the inflaton mass as it is customary, i.e. as the second derivative of the potential evaluated at φ_0.

The above approximated potential is the one of a harmonic oscillator with proper frequency m_φ and so we expect the inflaton to perform oscillations about the minimum, damped by the Hubble flow H but not only. Since inflation needs to end and there the radiation-dominated epoch must start, we need to couple the inflaton to other matter fields in order for the inflaton to lose energy in favour of the latter. Phenomenologically, one can write:

$$\dot\rho_\varphi + 3H\rho_\varphi(1 + w_\varphi) = -\Gamma\rho_\varphi , \tag{8.55}$$

$$\dot\rho_{\text{M}} + 3H\rho_{\text{M}}(1 + w_{\text{M}}) = \Gamma\rho_\varphi , \tag{8.56}$$

where Γ is some scattering rate governing the decay of the inflaton and with M we refer to matter in general which has to be relativistic in order to give rise to a radiation-dominated epoch and thus $w_M \approx 1/3$.

Therefore, the inflaton oscillations are also damped by the presence of Γ. This final phase of inflation, transiting to the radiation-dominated epoch is called **reheating**.

Exercise 8.7 Assuming w_φ constant, show that:

$$\rho_\varphi(t) = \rho_\varphi(t_I) \left[\frac{a(t_I)}{a(t)}\right]^{3(1+w_\varphi)} \exp\left(-\int_{t_I}^t dt' \Gamma(t')\right) , \qquad (8.57)$$

where t_I is the time at which inflation ends.

With this formal solution, we can find another formal one for the matter part.

Exercise 8.8 Assuming $w_M = 1/3$, show that:

$$\rho_M(t) = \frac{\rho_\varphi(t_I)a(t_I)^{3(1+w_\varphi)}}{a(t)^4} \int_{t_I}^t dt' \Gamma(t')a(t')^{1-3w_\varphi} \exp\left(-\int_{t_I}^{t'} dt'' \Gamma(t'')\right) . \qquad (8.58)$$

From the above equation one sees that the energy density of the matter fields is zero at the end of inflation, then it rises more or less abruptly (depending on Γ) and then finally decreases again as $1/a^4$ after the inflation has given up all its energy. Now, assume Γ to be constant and $w_\varphi = 0$, for simplicity. We get for the matter density:

$$\rho_M(t) = \frac{\rho_\varphi(t_I)\Gamma a(t_I)^3}{a(t)^4} \int_{t_I}^t dt' a(t') \exp\left[-\Gamma(t' - t_I)\right] . \qquad (8.59)$$

Integrating once by parts and considering the limit $\Gamma(t - t_I) \gg 1$, i.e. a very large decay rate, we get:

$$\rho_M(t) = \frac{\rho_\varphi(t_I)a(t_I)^4}{a(t)^4} , \qquad (8.60)$$

i.e. the decay takes place so rapidly that all the energy of the inflaton is passed to the matter.

If, on the other hand, $\Gamma(t - t_I) \ll 1$, i.e. a very small decay rate, we can approximate the matter density as:

$$\rho_M(t) \approx \frac{\rho_\varphi(t_I)\Gamma a(t_I)^3}{a(t)^4} \int_{t_I}^t dt' a(t') , \qquad (8.61)$$

and since Γ is so small, the inflaton is still dominating the dynamics giving thus for the scale factor

$$a(t) = a(t_I)(t/t_I)^{2/3} , \tag{8.62}$$

because we have chosen $w_\varphi = 0$. Integrating, we have thus:

$$\rho_M(t) \approx \frac{3}{5} \frac{\rho_\varphi(t_I)\Gamma t_I}{(t/t_I)^{8/3}} \left[(t/t_I)^{5/3} - 1 \right] , \tag{8.63}$$

and the maximum is attained at $t_{\max}/t_I = (8/3)^{3/5}$ and its value is:

$$\rho_{M,\max}(t) \approx 0.139 \frac{\Gamma}{H(t_I)} \rho_\varphi(t_I) . \tag{8.64}$$

Hence, in this case the energy density of matter is much smaller than that of the inflaton. Most of the latter is still spent driving the expansion of the universe instead of generating matter, because of the small Γ.

We have seen how to produce an inflationary phase and how to quantify it, through the slow-roll parameters. Now we are going to see what inflation has to say about quantum fluctuations. We hypothesise that before inflation the universe was quantum and quantum fluctuations were turned into classical ones by inflation itself, though we do not address the details of the quantum-to-classical transition of the primordial fluctuations. About the latter topic, see e.g. Kiefer and Polarski (2009).

8.4   Production of Gravitational Waves During Inflation

We are going to quantize the equation that we found for the evolution of gravitational waves, i.e. from Eq. (4.191):

$$h_{ij}^{T''} + 2\mathcal{H}h_{ij}^{T'} - \nabla^2 h_{ij}^T = 0 , \tag{8.65}$$

where recall that h_{ij}^T is traceless and transverse, i.e.

$$h_{ii}^T = 0 , \qquad \partial^j h_{ij}^T = 0 , \tag{8.66}$$

and no source term π_{ij}^T appears in the GW equation since this is vanishing for a scalar field. This can be understood intuitively, since the inflaton φ is a scalar and thus unable of producing a tensor perturbation. Mathematically, just look Eq. (8.117) and realise that no anisotropic stresses appear in the $\delta T^i{}_j$ of a scalar field, since this is diagonal.

The above Eq. (8.65) has no source term, is gauge-invariant and paves the way for a similar treatment that we shall do for scalar perturbations. For this reason we tackle the tensor ones first.

The Fourier transform of the above equation is Eq. (4.197), which we write here in the following compact form:

$$h'' + 2\mathcal{H}h' + k^2 h = 0 , \tag{8.67}$$

where $h(\eta, k)$ represents either $h_{+,\times}$ or $h(\lambda = \pm 2)$. Since the above equation only contains the modulus k, only the initial condition on h shall have a dependence on $\mathbf{k}$, cf. Eq. (7.101).

It shall be useful to cast Eq. (8.67) in the form of a harmonic oscillator one with no damping term by defining:

$$\boxed{g \equiv \frac{ah}{\sqrt{32\pi G}} = \frac{M_{\mathrm{Pl}}}{2} ah} \tag{8.68}$$

The normalisation comes in order to give g dimensions of a mass. Indeed, h is dimensionless and $\sqrt{8\pi G} = 1/M_{\mathrm{Pl}}$. This is necessary in order to give the correct dimensionality, a cube length or inverse cube mass, to the power spectrum.

A very direct way to see that $\sqrt{32\pi G}$ is the correct normalisation is e.g. to look to Mukhanov's calculations in Mukhanov et al. (1992). He finds the action which gives Eq. (8.65) by perturbing up to the second order a $f(R)$ action about a generic FLRW background. In the Einstein-Hilbert case $f(R) = R$ and for a spatially flat FLRW background one has:

$$\delta_2 S = \frac{1}{64\pi G} \int a^2 \left(h^{Ti'}{}_k h^{Tk'}{}_i - \partial_l h^{Ti}{}_k \partial^l h^{Tk}{}_i \right) d^4 x . \tag{8.69}$$

In Mukhanov et al. (1992) the corresponding second order action for the scalar field is also found and the factor outside the integral is $1/2$. Hence, in order to properly compare tensor and scalar fluctuations we must rescale the former as:

$$h^{Ti}{}_j \rightarrow \frac{h^{Ti}{}_j}{\sqrt{32\pi G}} . \tag{8.70}$$

Exercise 8.9 Use Eq. (8.68) into Eq. (8.67) in order to find:

$$\boxed{g'' + \left(k^2 - \frac{a''}{a} \right) g = 0} \tag{8.71}$$

The GW wave can be written as:

$$h_{ij}^T(\eta, \mathbf{x}) = \sum_{\lambda=\pm 2} \int \frac{d^3\mathbf{k}}{(2\pi)^3} \left[h(\eta, k)e^{i\mathbf{k}\cdot\mathbf{x}} a(\mathbf{k}, \lambda)e_{ij}(\hat{k}, \lambda) \right.$$

$$\left. h^*(\eta, k)e^{-i\mathbf{k}\cdot\mathbf{x}} a^*(\mathbf{k}, \lambda)e_{ij}^*(\hat{k}, \lambda) \right] , \qquad (8.72)$$

where the sum is over the helicity of the GW and $e_{ij}(\hat{k}, \lambda)$ is the polarisation tensor defined in Sect. 4.6. We have put the initial dependence on $\mathbf{k}$ of the GW in $a(\mathbf{k}, \lambda)$ and its complex conjugate, which is introduced in order to guarantee that $h_{ij}^T(\eta, \mathbf{x})$ is real.

Now, we assume the initial state of the GW field to be a quantum one on very small scales $k \gg aH$. Thus, we promote h_{ij}^T and a to operators and impose the canonical commutation relations:

$$\left[a(\mathbf{k}, \lambda), a(\mathbf{k}', \lambda') \right] = 0 , \qquad \left[a(\mathbf{k}, \lambda), a^\dagger(\mathbf{k}', \lambda') \right] = (2\pi)^3 \delta^{(3)}(\mathbf{k} - \mathbf{k}')\delta_{\lambda\lambda'} , \qquad (8.73)$$

which tell us that $a^\dagger(\mathbf{k}, \lambda)$ creates a graviton of momentum-energy k and helicity λ, whereas $a(\mathbf{k}, \lambda)$ destroys it.

The quantum state of the universe during inflation is the vacuum $|0\rangle$, which by definition is annihilated by $a(\mathbf{k}, \lambda)$. The expectation value on the vacuum:

$$\langle 0|h_{ij}^T(\eta, \mathbf{x})h_{lm}^T(\eta, \mathbf{x}')|0\rangle , \qquad (8.74)$$

is what shall define our primordial spectrum. Two comments are in order here. First of all, we could choose another quantum state for the universe which is not necessarily the vacuum one. Second, the above is a vacuum expectation value which we expect to become an ensemble average over classical random fields well outside the horizon.

Exercise 8.10 Compute the vacuum expectation value in Eq. (8.74) using the plane-wave expansion (8.72) and the commutation relations (8.73). Show that:

$$\langle 0|h_{ij}^T(\eta, \mathbf{x})h_{lm}^T(\eta, \mathbf{x}')|0\rangle = \int \frac{d^3\mathbf{k}}{(2\pi)^3} |h(\eta, k)|^2 e^{i\mathbf{k}\cdot(\mathbf{x}-\mathbf{x}')} \Pi_{ij,lm}(\hat{k}) , \qquad (8.75)$$

where

$$\Pi_{ij,lm}(\hat{k}) \equiv \sum_{\lambda=\pm 2} e_{ij}(\hat{k}, \lambda)e_{lm}^*(\hat{k}, \lambda) , \qquad (8.76)$$

is the sum over the helicities, defined in Eq. (7.10.3).

Comparing the result found in the exercise with Eq. (7.42) it is quite straightforward to see that the tensor quantum perturbations are Gaussian with power spectrum:

$$P_h(\eta, k) \propto |h(\eta, k)|^2 , \tag{8.77}$$

hence we need to solve Eq. (8.67), or rather Eq. (8.71), in order to determine $|h(\eta, k)|^2$.

Inspect Eq. (8.71). There are two regimes of interest. The first is for very short wavelength, i.e. $k^2 \gg a''/a$, for which the equation becomes:

$$g'' + k^2 g = 0 , \qquad (k^2 \gg a''/a) , \tag{8.78}$$

which is the usual harmonic oscillator equation. This means that the details of the inflationary evolution, encoded in $a(\eta)$, are not important on very small scales. This was not unexpected since on very small scales the expansion of the universe is irrelevant. Moreover, from the QFT point of view, we can then look at the quantisation as that of a free field in Minkowski space and thus determine:

$$g(\eta, k) = \frac{1}{\sqrt{2k}} e^{-ik\eta} , \quad g^*(\eta, k) = \frac{1}{\sqrt{2k}} e^{ik\eta} , \qquad (k^2 \gg a''/a) , \tag{8.79}$$

where the normalisation $1/\sqrt{2k}$ comes from the quantisation procedure in Minkowski space. We can take this solution as the "initial condition" in solving Eq. (8.71) therefore addressing only those modes which satisfy the condition $k^2 \gg a''/a$ during the inflationary epoch.

The second regime of evolution is given by the condition $k^2 \ll |a''/a|$, i.e. for very long wavelength, for which Eq. (8.71) becomes:

$$g'' - \frac{a''}{a} g = 0 , \qquad (k^2 \ll |a''/a|) . \tag{8.80}$$

This equation has the following formal solution:

$$g(\eta, k) = C_1(k)a + C_2(k)a \int^\eta \frac{d\bar{\eta}}{a(\bar{\eta})^2} , \qquad (k^2 \ll |a''/a|) , \tag{8.81}$$

which contains the solution

$$\frac{g}{a} = \frac{h}{\sqrt{32\pi G}} = \text{constant} , \tag{8.82}$$

representing thus the constant value that h reaches outside the horizon and that will eventually become the initial condition at the re-entering, during the radiation-dominated epoch.

Let us try to be more quantitative.

Exercise 8.11 Assume that during inflation $H = a'/a^2$ is a constant say H_Λ, i.e. consider a de Sitter space. Show that:

$$a(\eta) = -\frac{1}{H_\Lambda \eta} \ .$$

(8.83)

Since the scale factor and H_Λ are positive, then η is negative during inflation.

With the solution of Eq. (8.83), which we could take valid also in the case of slowly varying H, show that the condition used in Eq. (8.80) can be written as:

$$k^2 \ll |a''/a| = 2H_\Lambda^2 a^2 \ .$$

(8.84)

This is a condition for the physical wavenumber k/a to be much smaller than the Hubble factor, i.e. it is a condition for super-Hubble scales. In general $|a''/a| \propto 1/\eta^2$, for dimensional reasons, hence the condition in Eq. (8.80) can also be understood as:

$$k\eta \ll 1 \ ,$$

(8.85)

which means super-horizon scales, since η is the comoving particle horizon.

In this approximation then Eq. (8.71) becomes:

$$g'' + \left(k^2 - \frac{2}{\eta^2}\right) g = 0 \ .$$

(8.86)

Exercise 8.12 Show that the solution of Eq. (8.86) can be written as:

$$g(\eta, k) = \frac{C_1(k)}{\eta} [k\eta \cos(k\eta) - \sin(k\eta)] + \frac{C_2}{\eta} [k\eta \sin(k\eta) + \cos(k\eta)] \ .$$

(8.87)

Now, let us select the $\exp(-ik\eta)$ mode. Show that in this case the solution becomes:

$$g(\eta, k) = C_1(k)ke^{-ik\eta} \left(1 - \frac{i}{k\eta}\right) \ .$$

(8.88)

Now suppose an initial condition in $\eta = \eta_i$, such that:

$$k^2 \gg \frac{a''}{a} = \frac{2}{\eta_i^2} \ , \quad \Rightarrow \quad k|\eta_i| \gg 1 \ ,$$

(8.89)

so that we can do the matching with Eq. (8.79) and obtain:

$$C_1(k)ke^{-ik\eta_i} = \frac{1}{\sqrt{2k}}e^{-ik\eta_i} \ , \quad \Rightarrow \quad C_1(k)k = \frac{1}{\sqrt{2k}} \ , \quad (k|\eta_i| \gg 1) \ .$$

(8.90)

and so we can write:

$$g(\eta, k) = \frac{e^{-ik\eta}}{\sqrt{2k}} \left(1 - \frac{i}{k\eta}\right), \quad (k|\eta_i| \gg 1). \tag{8.91}$$

Using now Eq. (8.68), we get for the power spectrum:

$$P_h(\eta, k) = \frac{32\pi G}{a(\eta)^2} P_g(\eta, k) = \frac{16\pi G}{a(\eta)^2 k} \left(1 + \frac{1}{k^2\eta^2}\right) = \frac{16\pi G}{k} H_\Lambda^2 \eta^2 \left(1 + \frac{1}{k^2\eta^2}\right), \tag{8.92}$$

and the dimensionless one, according to the definition of Eq. (7.45), is:

$$\Delta_h^2(\eta, k) = \frac{k^3 P_h(\eta, k)}{2\pi^2} = \frac{8\pi G}{\pi^2} H_\Lambda^2 \left[1 + (k^2\eta^2)\right] = \frac{H_\Lambda^2}{\pi^2 M_{\text{Pl}}^2} \left[1 + (k^2\eta^2)\right], \tag{8.93}$$

and recall that this result is valid only for those scales for which $k|\eta_i| \gg 1$.

This spectrum depends on time, so which η should we choose? We have seen, through the discussion culminating in Eq. (8.84), that when a scale k exits the horizon, its corresponding $h(\eta, k)$ becomes constant. Its value will be the initial condition during the radiation-dominated epoch, when the scale re-enters the horizon. Therefore, for $k\eta \to 0$ we have:

$$\Delta_h^2(\eta, k) = \frac{H_\Lambda^2}{\pi^2 M_{\text{Pl}}^2}, \tag{8.94}$$

which does not depend on the scale k anymore, i.e. it is a **scale-invariant** power spectrum. The only caveat is that this spectrum is valid only for those scales for which $k|\eta_i| \gg 1$, i.e. scales which exit the horizon:

$$a_{\text{exit}}^2 H_\Lambda^2 \eta_i^2 \gg 1, \quad \Rightarrow \quad \eta_i^2 \gg \eta_{\text{exit}}^2, \tag{8.95}$$

well after η_i. Recall that η is negative during inflation, so the above condition makes sense.

Note that for a constant H_Λ the slow-roll parameter ϵ is vanishing. In order to describe the inflationary phase more realistically, we should consider a small but non-vanishing ϵ. In this case H varies and a k-dependence is gained by the power spectrum.

Let us write the definition of ϵ in Eq. (8.36), but using the conformal time:

$$\epsilon = \left(\frac{a}{\mathcal{H}}\right)' \frac{1}{a} = 1 - \frac{\mathcal{H}'}{\mathcal{H}^2}. \tag{8.96}$$

Exercise 8.13 Solve the above differential equation for $\mathcal{H}$ assuming a constant ϵ. Show that:

$$\mathcal{H} = -\frac{1}{(1 - \epsilon)\eta}. \tag{8.97}$$

From this solution, it is not difficult to show that:

$$\frac{a''}{a} = \mathcal{H}' + \mathcal{H}^2 = \frac{1}{(1-\epsilon)\eta^2} + \frac{1}{(1-\epsilon)^2\eta^2} \approx \frac{2+3\epsilon}{\eta^2} \,, \qquad (8.98)$$

where in the last approximation we have considered $\epsilon \ll 1$, as it should be during inflation, and kept ϵ to first order.

Substituting into Eq. (8.71), we get then:

$$g'' + \left(k^2 - \frac{2+3\epsilon}{\eta^2}\right)g = 0 \,, \qquad (8.99)$$

which can be solved exactly, giving as general solution a combination of Hankel functions:

$$g(\eta, k) = C_1(k)\sqrt{-\eta}H_\nu^{(1)}(-k\eta) + C_2(k)\sqrt{-\eta}H_\nu^{(2)}(-k\eta) \,, \quad \nu = \frac{\sqrt{3}}{2}\sqrt{3+4\epsilon} \,, \qquad (8.100)$$

where the minus sign must be introduced in order to account for the fact that $\eta < 0$. We have chosen to express the solution in terms of the Hankel functions since it is easier to see which one of them matches the initial condition of Eq. (8.79). Indeed, consider the asymptotic expansion, cf. Abramowitz and Stegun (1972), of $H_\nu^{(1)}(-k\eta)$:

$$H_\nu^{(1)}(-k\eta) \sim \sqrt{\frac{2}{-\pi k\eta}}e^{-ik\eta-i\nu\pi/2-i\pi/4} \,, \qquad (k|\eta| \gg 1) \,. \qquad (8.101)$$

Hence, this is the right behavior at very small scales for g and the integration constant must be:

$$C_1(k) = \frac{\sqrt{\pi}}{2}e^{i\nu\pi/2+i\pi/4} \,, \qquad (8.102)$$

i.e. indeed a (complex) constant, since it bears no dependence from k. The solution we look for is then:

$$g(\eta, k) = \frac{\sqrt{\pi}}{2}e^{i\nu\pi/2+i\pi/4}\sqrt{-\eta}H_\nu^{(1)}(-k\eta) \,, \quad \nu = \frac{\sqrt{3}}{2}\sqrt{3+4\epsilon} \,, \qquad (8.103)$$

and the power spectrum is thus:

$$P_h(\eta, k) = \frac{4}{M_{\text{Pl}}^2 a(\eta)^2}|g(\eta, k)|^2 = \frac{\pi|\eta|}{M_{\text{Pl}}^2 a(\eta)^2}|H_\nu^{(1)}(-k\eta)|^2 \,. \qquad (8.104)$$

For $k\eta \to 0$, i.e. on very large scales, one has that

$$H_\nu^{(1)}(-k\eta) \sim -i\frac{\Gamma(\nu)}{\pi}\left(\frac{k|\eta|}{2}\right)^{-\nu} \,, \qquad (k|\eta| \to 0) \,, \qquad (8.105)$$

and hence the dimensionless power spectrum becomes:

$$\Delta_h^2(\eta, k) = \frac{k^3 |\eta| \Gamma(\nu)^2}{2\pi^3 M_{\text{Pl}}^2 a^2} \left(\frac{k|\eta|}{2} \right)^{-2\nu} , \qquad (k|\eta| \to 0) . \tag{8.106}$$

Considering ϵ at first order in the exponent of $k|\eta|$ and negligible elsewhere, we can write:

$$\Delta_h^2(\eta, k) = \frac{k^3 |\eta| \Gamma(\nu)^2}{2\pi^3 M_{\text{Pl}}^2 a^2} \left(\frac{k|\eta|}{2} \right)^{-3-2\epsilon} = \frac{1}{\pi^2 M_{\text{Pl}}^2 a^2 |\eta|^2} (k|\eta|)^{-2\epsilon} . \tag{8.107}$$

Now, the power spectrum is no more scale-invariant but gained a small k-dependence: a power law with exponent -2ϵ. The latter is usually named as **tensor spectral index** and denoted as n_T. We shall see a little more detail about it later.

On large scales, h is time-independent and so its power spectrum is. Hence, we can choose any convenient value of η when to evaluate Δ_h^2. It is customary to use the time η_k at which a given scale k crosses the horizon, i.e.

$$k = \mathcal{H}(\eta_k) , \tag{8.108}$$

because the details of the k-dependence of the power spectrum depend on the inflationary model but, as we saw in Eqs. (8.78) and (8.80), they manifest themselves only at horizon-crossing.

Using Eq. (8.97), the horizon crossing condition gives the following relation between $|\eta|$ and k:

$$k = \mathcal{H}(\eta_k) = \frac{1}{(1 - \epsilon)|\eta_k|} , \tag{8.109}$$

which at fist order in ϵ gives:

$$k|\eta_k| = 1 + \epsilon . \tag{8.110}$$

Using again Eq. (8.97), we can write the power spectrum evaluated at horizon crossing as:

$$\Delta_h^2(k) = \frac{1}{\pi^2 M_{\text{Pl}}^2} \frac{\mathcal{H}(\eta_k)^2}{a^2(\eta_k)} , \tag{8.111}$$

which is usually put as:

$$\Delta_h^2(k) = \frac{H^2}{\pi^2 M_{\text{Pl}}^2} \bigg|_{k=aH} , \tag{8.112}$$

i.e. it has the same form as the one in the de Sitter case, computed in Eq. (8.94), but allows for a time-dependent Hubble factor.

It is very interesting to notice that if we could directly measure the gravitational waves background, we would be able to determine the energy scale of inflation, i.e. H.

8.5 Production of Scalar Perturbations During Inflation

Consider our FLRW metric perturbed with scalar perturbations only and written in the conformal Newtonian gauge, i.e.

$$ds^2 = a(\eta)^2 \left[-(1 + 2\Psi)\, d\eta^2 + (1 + 2\Phi)\, \delta_{ij} dx^i dx^j \right] . \tag{8.113}$$

Accordingly, consider also perturbations of our inflaton field:

$$\boxed{\varphi = \bar{\varphi}(\eta) + \delta\varphi(\eta, \mathbf{x})} \tag{8.114}$$

Exercise 8.14 Using Eqs. (8.27) and (8.114), write down the perturbed energy-momentum tensor. Show that:

$$T^0{}_0 = -\frac{1}{2a^2}\bar{\varphi}'^2 - V(\varphi) - \frac{1}{a^2}\bar{\varphi}'\delta\varphi' + \frac{1}{a^2}\Psi\bar{\varphi}'^2 - V_{,\varphi}\delta\varphi , \tag{8.115}$$

$$T^0{}_i = -\frac{1}{a^2}\bar{\varphi}'\partial_i\delta\varphi , \tag{8.116}$$

$$T^i{}_j = \delta^i{}_j \left[\frac{1}{2a^2}\bar{\varphi}'^2 - V(\varphi) \right] + \delta^i{}_j \left(\frac{1}{a^2}\bar{\varphi}'\delta\varphi' - \frac{1}{a^2}\Psi\bar{\varphi}'^2 - V_{,\varphi}\delta\varphi \right) . \tag{8.117}$$

The perturbed contribution in the above expressions is grouped to the right. Note the following important fact: $T^i{}_j$ is diagonal and thus no anisotropic stresses can be sourced by a single scalar field. We already know from Eq. (4.184) that this fact implies $\Psi = -\Phi$.

Exercise 8.15 Obtain the perturbed Klein–Gordon equation. Start from:

$$\nabla_\mu T^\mu{}_\nu = 0 , \tag{8.118}$$

and then put $\nu = 0$. Show that:

$$\boxed{\delta\varphi'' + 2\mathcal{H}\delta\varphi' + \left(k^2 + V_{,\varphi\varphi}a^2\right)\delta\varphi = 2\Phi V_{,\varphi}a^2 - 4\Phi'\bar{\varphi}'} \tag{8.119}$$

where we have already used the conformal time and $\Psi = -\Phi$. Compare this equation with the corresponding one found by Mukhanov and co-authors in Mukhanov et al. (1992). It is also useful to consider that found in Weinberg (2008) and recover the above one by transforming from the cosmic time to the conformal one.

Unfortunately, the contributions $V_{,\varphi\varphi}a^2$, $2\Phi V_{,\varphi}a^2$ and $4\Phi'\bar{\varphi}'$ make life more difficult. Indeed, if they were absent, we would get the following equation:

$$\delta\varphi'' + 2\mathcal{H}\delta\varphi' + k^2\delta\varphi = 0 \,, \tag{8.120}$$

which is formally identical to Eq. (8.67), and therefore all the calculations performed for the GW case would follow in the same fashion.

Even if we could neglect $V_{,\varphi\varphi}$ in Eq. (8.119) during the slow-roll phase, there would be no a priori reason to neglect the whole right hand side of Eq. (8.119). Moreover, we do not really want to follow the evolution of $\delta\varphi$, for at least two reasons. First, the inflaton supposedly decays into ordinary matter during reheating, and we do not want to address this problem. Second, as we saw in Eq. (6.94), what we really need is to know $\mathcal{R}$ or ζ on very large scales when radiation starts to dominate.

Let us work with $\mathcal{R}$. It turns out that $\mathcal{R}$ is conserved (i.e. is a constant) on large scales and for adiabatic perturbations. We have already seen this in Chap. 6 and we show this in detail again in Chap. 12.

It turns out that, as it happened for h, when inflation brings a scale k outside the horizon, its corresponding $\mathcal{R}(\eta, k)$ attains a constant value which shall serve as initial condition during the radiation-dominated epoch. There shall be a crucial difference with respect to the case of GW, which we shall meet shortly.

Exercise 8.16 Express v in Eq. (4.131) in terms of the inflaton field velocity potential. From Eq. (4.73) we know that:

$$T^{0(\varphi)}{}_i = \left(\bar{\rho}_\varphi + \bar{P}_\varphi\right) v_i^{(\varphi)} = \frac{\bar{\varphi}'^2}{a^2} v_i^{(\varphi)} \,. \tag{8.121}$$

Using Eq. (8.116), show that:

$$v_i^{(\varphi)} = -\frac{\partial_i \delta\varphi}{\bar{\varphi}'} \,. \tag{8.122}$$

Thus, writing

$$v_i^{(\varphi)} = \partial_i v^{(\varphi)} \,, \tag{8.123}$$

we can identify the velocity potential of the inflation field (we drop the superscript φ now) as:

$$v = -\frac{\delta\varphi}{\bar{\varphi}'} \,. \tag{8.124}$$

Hence, using the result of the exercise, we can write:

$$\mathcal{R} = \Phi - \frac{\mathcal{H}}{\bar{\varphi}'}\delta\varphi \,. \tag{8.125}$$

We need to find an evolution equation for $\mathcal{R}$. In principle, combining the above expression with the Klein–Gordon equation (8.119) we can eliminate $\delta\varphi$. Then, Φ can be dealt with via the use of the Einstein equations, which are:

$$3\mathcal{H}(\Phi' + \mathcal{H}\Phi) + k^2\Phi = 4\pi G\left(\bar{\varphi}'\delta\varphi' + \Phi\bar{\varphi}'^2 + V_{,\varphi}a^2\delta\varphi\right) , \quad (8.126)$$

$$\Phi' + \mathcal{H}\Phi = -4\pi G\bar{\varphi}'\delta\varphi , \quad (8.127)$$

$$\Phi'' + 3\mathcal{H}\Phi' + (2\mathcal{H}' + \mathcal{H}^2)\Phi = -4\pi G\left(\bar{\varphi}'\delta\varphi' + \Phi\bar{\varphi}'^2 - V_{,\varphi}a^2\delta\varphi\right) . \quad (8.128)$$

obtained combining Eq. (4.179) with Eq. (8.115), Eq. (4.185) with Eqs. (8.123) and (8.124), Eq. (4.189) with Eq. (8.117). All with $\Phi = -\Psi$, of course.

Though in principle possible, it is a mountain of calculations which we can at least limit by choosing a different gauge. In particular, following Weinberg (2008), let us consider the following gauge:

$$ds^2 = -a^2(1 + E)d\eta^2 + 2a^2 F_{,i}d\eta dx^i + a^2(1 + A)\delta_{ij}dx^i dx^j , \quad \delta\hat{\varphi} = 0 , \quad (8.129)$$

i.e. a gauge in which the perturbed scalar field (which we denote with a hat, in the new gauge) is zero. Thus in this gauge

$$\mathcal{R} = \frac{A}{2} , \quad (8.130)$$

and so our objective is to find a closed equation for A. The price to pay in order to put the perturbed scalar field equal to zero is to introduce one more scalar perturbation in the metric. Now, let us calculate the energy-momentum tensor in the new gauge.

Exercise 8.17 Using Eqs. (8.27) and (8.129), write down the perturbed energy-momentum tensor. Show that:

$$\hat{T}^0{}_0 = -\frac{1}{2a^2}(1 - E)\bar{\varphi}'^2 - V(\varphi) , \quad (8.131)$$

$$\hat{T}^0{}_i = 0 , \quad \hat{T}^i{}_0 = \frac{1}{a^2}F_{,i}\bar{\varphi}'^2 , \quad (8.132)$$

$$\hat{T}^i{}_j = \delta^i{}_j\left[\frac{1}{2a^2}(1 - E)\bar{\varphi}'^2 - V(\varphi)\right] . \quad (8.133)$$

The above is a pretty simple energy-momentum tensor and indeed we now are going to exploit its simplicity. First of all, since $\delta\hat{T}^0{}_i = 0$, we have from Eq. (4.43) that:

$$-\mathcal{H}E + A' = 0 , \quad (8.134)$$

i.e. a simple algebraic relation between two of the three scalar potentials.

The second relation that we are going to us is the continuity equation.

Exercise 8.18 Compute the continuity equation $\nabla_\nu T^\nu{}_0 = 0$ using the energy-momentum tensor given in Eqs. (8.131)–(8.133). Compute from the metric in Eq. (8.129) the only necessary Christoffel symbol:

$$\Gamma^i_{0j} = \mathcal{H}\delta^i{}_j + \frac{A'}{2}\delta^i{}_j \ . \tag{8.135}$$

Then show that:

$$\frac{a^2}{2}\left[\frac{E}{a^2}(\mathcal{H}^2 - \mathcal{H}')\right]' + (\mathcal{H}^2 - \mathcal{H}')\left(\nabla^2 F + 3\mathcal{H}E - \frac{3}{2}A'\right) = 0 \ , \tag{8.136}$$

where we have used the relation $\mathcal{H}^2 - \mathcal{H}' = 4\pi G\bar{\varphi}'^2$.

As last relation, we shall use δR_{ij} computed from the metric in Eq. (8.129). So, using Eq. (4.34), we have:

$$\delta R_{ij} = -\frac{1}{2}E_{,ij} - \frac{\mathcal{H}}{2}E'\delta_{ij} - (2\mathcal{H}^2 + \mathcal{H}')E\delta_{ij} - \frac{1}{2}(\nabla^2 A\delta_{ij} + A_{,ij})$$
$$+\frac{1}{2}A''\delta_{ij} + \frac{5}{2}\mathcal{H}A'\delta_{ij} + (2\mathcal{H}^2 + \mathcal{H}')A\delta_{ij}$$
$$-\mathcal{H}\nabla^2 F\delta_{ij} - F'_{ij} - 2\mathcal{H}F_{,ij} \ . \tag{8.137}$$

Through the Einstein equations, we know that:

$$\delta R_{ij} = 8\pi G\left(\delta T_{ij} - a^2 A\delta_{ij}\frac{\bar{T}}{2} - a^2\delta_{ij}\frac{\delta T}{2}\right) \ , \tag{8.138}$$

where $\bar{T}$ and δT are the background and perturbed trace, respectively, of the energy-momentum tensor.

Exercise 8.19 Calculate the right hand side of the above equation. Show that:

$$\delta R_{ij} = 8\pi Ga^2 A\delta_{ij}V = (2\mathcal{H}^2 + \mathcal{H}')A\delta_{ij} \ . \tag{8.139}$$

Extracting thus only the part of δR_{ij} which is proportional to δ_{ij}, we finally get:

$$-\frac{\mathcal{H}}{2}E' - (2\mathcal{H}^2 + \mathcal{H}')E - \frac{1}{2}\nabla^2 A + \frac{1}{2}A'' + \frac{5}{2}\mathcal{H}A' - \mathcal{H}\nabla^2 F = 0 \ . \tag{8.140}$$

Exercise 8.20 Combine Eqs. (8.134), (8.136) and (8.140), eliminating E and F and thus finding the following equation for $\mathcal{R}$:

$$\mathcal{R}'' + \frac{2}{\mathcal{H}}\mathcal{R}'\left[\mathcal{H}^2 - \mathcal{H}' + \frac{\mathcal{H}}{2}\frac{(\mathcal{H}^2 - \mathcal{H}')'}{\mathcal{H}^2 - \mathcal{H}'}\right] + k^2\mathcal{R} = 0 . \tag{8.141}$$

Show that this equation can be written in a more compact form as:

$$\boxed{\mathcal{R}'' + 2\frac{z'}{z}\mathcal{R}' + k^2\mathcal{R} = 0} \tag{8.142}$$

with

$$z \equiv \frac{a\bar{\varphi}'}{\mathcal{H}} . \tag{8.143}$$

Equation (8.142) is called **Mukhanov-Sasaki equation**, cf. Mukhanov (1985) and Sasaki (1986). Notice that this equation has the very same structure as Eq. (8.67) if one makes the change $a \to z$. Therefore, a similar analysis applies and a constant solution for $\mathcal{R}$ is allowed when:

$$k^2 \ll |z''/z| , \tag{8.144}$$

which is a similar, but not identical, condition as $k^2 \ll |a''/a|$. In fact, note that:

$$z^2 = a^2\left(\frac{\bar{\varphi}'}{\mathcal{H}}\right)^2 = a^2\frac{3\bar{\varphi}'^2}{8\pi G(\bar{\varphi}'^2/2 + Va^2)} = a^2\frac{\epsilon}{4\pi G} , \tag{8.145}$$

where we have used Eq. (8.44), written in the conformal time. The above relation between z and a is exact and it calls into play the slow roll parameter ϵ.

Now, the procedure that we need in order to obtain the power spectrum for $\mathcal{R}$ is the same as the one used for h. We promote $\mathcal{R}$ to a quantum field and compute its vacuum expectation value, which will become the power spectrum itself. The vacuum expectation value is computed first by solving the Mukhanov-Sasaki equation, choosing the correct Minkowski behaviour at large k.

Finally, we can use the same result of Eq. (8.112), remembering to divide by the $32\pi G$ factor that we have introduced in order to give dimensionality to h, and multiplying by $a^2/z^2 = 4\pi G/\epsilon$, in order to recover the $2z'/z$ factor in front of $\mathcal{R}'$, instead of $2\mathcal{H}$.

Thus, we can finally write down the scalar power spectrum as:

$$\boxed{P_{\mathcal{R}} = \frac{H^2}{4M_{\text{Pl}}^2\epsilon k^3}\Big|_{k=aH}} \qquad \boxed{\Delta_{\mathcal{R}}^2 = \frac{H^2}{8\pi^2 M_{\text{Pl}}^2\epsilon}\Big|_{k=aH}} \tag{8.146}$$

This is the most important result of this chapter, since it provides a prediction which can be (and indeed is) tested observationally. Using Eq. (6.94), one can relate this power spectrum to those of the other quantities which become relevant at horizon re-entering during the radiation-dominated epoch. For example, using Eq. (6.77), we can write that:

$$\Delta_\Phi^2 = \frac{4(5 + 2R_\nu)^2}{(15 + 4R_\nu)^2} \Delta_\mathcal{R}^2 \ . \tag{8.147}$$

In general, Φ is not constant on large scales during inflation, but it is indeed constant on large scales during radiation domination, as we saw in Chap. 6. The above thus constitutes the initial condition, on the power spectrum, at horizon re-entering.

Now, let us perform a calculation similar to that in the previous section, culminated in Eq. (8.112).

Exercise 8.21 Employing Eq. (8.41) written in the conformal time, show that:

$$\frac{z'}{z} = \mathcal{H}(1 + \epsilon - \eta) \ . \tag{8.148}$$

Afterwards, assume ϵ and η to be constant. Then, use Eq. (8.97) and recast the Mukhanov-Sasaki equation as follows:

$$\mathcal{R}'' - \frac{2(1 + 2\epsilon - \eta)}{\tau}\mathcal{R}' + k^2\mathcal{R} = 0 \ , \tag{8.149}$$

where we are now employing τ as conformal time, in order to avoid confusion with the second slow-roll parameter η. Finally, show that the above equation can be written as:

$$(z^2\mathcal{R})'' + \left(k^2 - \frac{2 + 6\epsilon - 3\eta}{\tau^2}\right)(z^2\mathcal{R}) = 0 \ . \tag{8.150}$$

This equation has the same form as Eq. (8.99), and hence a similar analysis applies. In particular, its solution is of the form:

$$z^2(\tau)\mathcal{R}(\tau, k) = C_1(k)\sqrt{-\tau}H_\nu^{(1)}(-k\tau) + C_2(k)\sqrt{-\tau}H_\nu^{(2)}(-k\tau) \ , \tag{8.151}$$

with order

$$\nu = \frac{\sqrt{3}}{2}\sqrt{3 + 8\epsilon - 4\eta} \ . \tag{8.152}$$

Exercise 8.22 Following the same steps that we saw in the tensor case, show that:

$$\Delta_\mathcal{R}^2(\tau, k) = \frac{1}{8\pi^2 M_{\text{Pl}}^2 \epsilon} \frac{1}{a^2\tau^2}(k|\tau|)^{-4\epsilon+2\eta} \ . \tag{8.153}$$

For the scalar case a different k-dependence of the spectrum appears, involving the second slow-roll parameter. The combination $-4\epsilon + 2\eta$ is the first order expression of the **scalar spectral index**. Evaluating the above spectrum at horizon crossing, we get the result already shown in Eq. (8.146).

8.6 Spectral Indices

It is customary to cast the dimensionless scalar and tensor power spectra as follows:

$$\Delta_S^2 \equiv \Delta_{\mathcal{R}}^2 \equiv \frac{k^3 P_{\mathcal{R}}(k)}{2\pi^2} = \left.\frac{H^2}{8\pi^2 M_{\text{Pl}}^2 \epsilon}\right|_{k=aH} \equiv A_S \left(\frac{k}{k_*}\right)^{n_S(k)-1} , \quad (8.154)$$

$$\Delta_T^2 \equiv 2\Delta_h^2 \equiv \frac{k^3 P_h(k)}{\pi^2} = \left.\frac{2H^2}{\pi^2 M_{\text{Pl}}^2}\right|_{k=aH} \equiv A_T \left(\frac{k}{k_*}\right)^{n_T(k)} , \quad (8.155)$$

where the general k-dependence (given by the specific model of inflation) is embedded in $n_S(k)$ and $n_T(k)$, which are known as **scalar spectral index** and **tensor spectral index**.

We have introduced here the pivot scale k_*, which is usually taken to be 0.002 Mpc^{-1} or 0.05 Mpc^{-1} (Ade et al. 2016a) and the factor 2 in $\Delta_T^2 \equiv 2\Delta_h^2$ takes into account the two polarisations of the GW. The **spectral amplitudes** A_S and A_T, of which only the first is constrained by observation since we do not detect yet the primordial GW background, are related to the energy scale of inflation.

Since the spectral indices are not necessarily constant, the spectra can be written in the following more general form:

$$\ln \frac{\Delta_S^2}{A_S} = \left[n_S - 1 + \frac{1}{2}\frac{dn_S}{d\ln k}\ln\frac{k}{k_*} + \frac{1}{6}\frac{d^2 n_S}{d(\ln k)^2}\left(\ln\frac{k}{k_*}\right)^2 + \cdots \right]\ln\frac{k}{k_*} , \quad (8.156)$$

$$\ln \frac{\Delta_T^2}{A_T} = \left[n_T + \frac{1}{2}\frac{dn_T}{d\ln k}\ln\frac{k}{k_*} + \cdots \right]\ln\frac{k}{k_*} . \quad (8.157)$$

The derivative of the spectral index with respect to $\ln k$ is called **running of the spectral index**. We do not consider runnings in detail here, for simplicity, and Planck has not constrained them very tightly, as we shall see. However, they (at least the first one) will probably become of great interest in future CMB experiments (Muñoz et al. 2017).

From the above Eq. (8.156) at first order, it is straightforward to write:

$$n_S - 1 = \frac{d\ln \Delta_S^2}{d\ln k} , \qquad n_T = \frac{d\ln \Delta_T^2}{d\ln k} . \quad (8.158)$$

For the tensor case:

$$n_T = 2\frac{k}{H}\frac{dH}{dk}\bigg|_{aH=k}. \tag{8.159}$$

Recall that H does not depend on the scale in general, being it a background quantity, but it is the evaluation at $aH = k$ that makes H depending on k. This is because different scales cross the horizon at different times during inflation. In particular, recall from Eq. (8.83) that:

$$\eta = \int_0^a \frac{da}{Ha^2} \approx \frac{1}{H}\int_0^a \frac{da}{a^2} = -\frac{1}{Ha}, \tag{8.160}$$

since H is almost constant during inflation. Now, when we evaluate $aH = k$, we have that $\eta = -1/k$. Therefore, we obtain for the tensor spectral index:

$$n_T = 2\frac{k}{H}\frac{dH}{d\eta}\frac{d\eta}{dk}\bigg|_{aH=k} = 2\frac{1}{kH}\frac{dH}{d\eta}\bigg|_{aH=k}. \tag{8.161}$$

By definition of ϵ, cf. Eq. (8.36), we know that:

$$\frac{dH}{d\eta} = -aH^2\epsilon, \tag{8.162}$$

so that we can finally determine:

$$\boxed{n_T = -2\epsilon} \tag{8.163}$$

which is the result already derived in Eq. (8.107).

For the scalar spectral index is pretty much the same calculation:

$$n_S - 1 = \frac{d\ln(H^2/\epsilon)}{d\ln k}\bigg|_{aH=k} = -2\epsilon - \frac{d\ln\epsilon}{d\ln k}\bigg|_{aH=k}. \tag{8.164}$$

We can deal with the last term as follows:

$$\frac{d\ln\epsilon}{d\ln k}\bigg|_{aH=k} = \frac{k}{\epsilon}\epsilon'\frac{d\eta}{dk}\bigg|_{aH=k} = \frac{1}{k\epsilon}\epsilon'\bigg|_{aH=k}. \tag{8.165}$$

Using Eq. (8.41) one has:

$$\epsilon' = 2aH\epsilon(\epsilon - \eta), \tag{8.166}$$

and the scalar spectral index is thus written as:

$$\boxed{n_S - 1 = -4\epsilon + 2\eta = -6\epsilon_V + 2\eta_V} \tag{8.167}$$

again in agreement with the result found in Eq. (8.153).

Finally, define the **tensor-to-scalar ratio**:

$$r_* \equiv \frac{\Delta_T^2(k_*)}{\Delta_S^2(k_*)} = \frac{A_T}{A_S} = 16\epsilon = -8n_T \tag{8.168}$$

It is customary (Ade et al. 2016a) to use r in order to express the energy scale of Inflation at the time when the pivot scale exits the Hubble radius. From Eq. (8.155) we have:

$$H_*^2 = \frac{\pi^2 M_{\text{Pl}}^2}{2} \Delta_T^2(k_*) . \tag{8.169}$$

Using the definition of r_* and Eq. (8.154), one gets:

$$H_*^2 = \frac{\pi^2 M_{\text{Pl}}^2}{2} r_* \Delta_S^2(k_*) = \frac{\pi^2 M_{\text{Pl}}^2}{2} r_* A_S , \tag{8.170}$$

from which:

$$V_* = \frac{3\pi^2 M_{\text{Pl}}^4}{2} r_* A_S \tag{8.171}$$

8.7 Observational Results

In Ade et al. (2016a) one can find recent constraints on the inflationary parameters. They slightly change with respect to the different datasets considered. We report here the spectral index and its running and running of the running at 68% CL:

$$n_S = 0.9586 \pm 0.0056 , \tag{8.172}$$

$$\frac{dn_S}{d \ln k} = 0.009 \pm 0.010 , \tag{8.173}$$

$$\frac{d^2 n_S}{d(\ln k)^2} = 0.025 \pm 0.013 , \tag{8.174}$$

using the pivot scale $k_* = 0.05$ Mpc^{-1}. For the scalar amplitude at 68% CL:

$$\ln(10^{10} A_S) = 3.094 \pm 0.034 . \tag{8.175}$$

For the tensor-to-scalar ratio:

$$r_{0.002} < 0.10 , \tag{8.176}$$

at 95% confidence level. From these numbers, we can write the energy scale of Inflation at the time when the pivot scale exits the Hubble radius as follows:

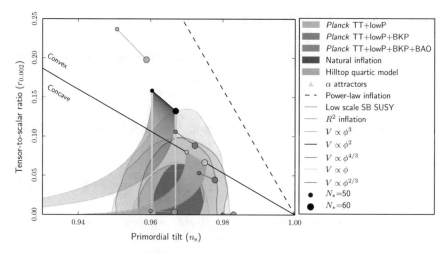

Fig. 8.2 This is Fig. 12 of Ade et al. (2016a) and shows the marginalised joint 68% and 95% CL regions for n_s and r at $k = 0.002$ Mpc^{-1} from Planck compared to the theoretical predictions of selected inflationary models. Note that the marginalised joint 68 and 95% CL regions have been obtained by assuming $dn_s/d \ln k = 0$

$$V_* = \left(1.88 \times 10^{16} \text{ GeV}\right)^4 \frac{r}{0.10}, \qquad (8.177)$$

and therefore obtain an upper bound, which corresponds to GUT energy scale.

In Fig. 8.2 we borrow (with permission, of course) Fig. 12 of Ade et al. (2016a) showing the constraints on some inflationary models in the parameter space n_s vs. r. Note here that N_* is the number of e-folds to the end of inflation, which is limited in the range 50–60 because the observed scales, in particular the pivot one $k = 0.002$ Mpc^{-1} must have had time to leave the horizon and then to re-enter.

8.8 Examples of Models of Inflation

We conclude this Chapter presenting some important models of inflations, some of those in Fig. 8.2, limiting ourselves to the class of single field models. There is a huge number of inflationary models, in Martin et al. (2014) 193 of them are analysed, and it seems that data favour the simplest category of single field slow-roll inflation that we have presented in this chapter. See also Tsujikawa (2014) for a very nice review on many inflationary models.

8.8.1   General Power Law Potential

A very simple model of inflation consists in an inflaton field described by the potential:

$$V(\varphi) = \lambda_n \frac{\varphi^n}{n} \, , \tag{8.178}$$

from which one can easily determine the slow-roll parameters from Eqs. (8.45) and (8.47):

$$\epsilon_V = \frac{n^2}{16\pi G \varphi^2} = \frac{n^2 M_{\mathrm{Pl}}^2}{16\pi \varphi^2} \, , \qquad \eta_V = \frac{n(n-1)}{8\pi G \varphi^2} = \frac{n(n-1) M_{\mathrm{Pl}}^2}{8\pi \varphi^2} \, . \tag{8.179}$$

In order for these parameters to be very small and thus trigger inflation, one has to have $\varphi \gg M_{\mathrm{Pl}}$ and note that this condition does not depend on the coupling. We can put the predictions on the spectral indices as functions of the number of e-folds to the end of inflation by defining the value of the field at which inflation ends as $\epsilon_V(\varphi_f) = 1$.

For the general power law potential, this condition becomes:

$$\frac{n^2}{16\pi G \varphi_f^2} = 1 \quad \Rightarrow \quad \varphi_f = \frac{n}{\sqrt{16\pi G}} \, . \tag{8.180}$$

The number of e-folds to the end of inflation can be obtained exactly from Eq. (8.50):

$$N = \frac{8\pi G}{n} \int_{\varphi_f}^{\varphi_i} \varphi \, d\varphi = \frac{4\pi G}{n} \left(\varphi_i^2 - \frac{n^2}{16\pi G} \right) \, , \tag{8.181}$$

from which the initial value of the field is:

$$\varphi_i^2 = \frac{n}{4\pi G} \left(N + \frac{n}{4} \right) \, . \tag{8.182}$$

From this, the slow-roll parameters can be written as:

$$\epsilon_V = \frac{n}{4N + n} \, , \qquad \eta_V = \frac{2(n-1)}{4N + n} \, , \tag{8.183}$$

and the scalar spectral index and the tensor-to-scalar ratio are the following:

$$n_S - 1 = -6\epsilon_V + 2\eta_V = -\frac{2(n+2)}{4N + n} \, , \qquad r = 16\epsilon_V = \frac{16n}{4N + n} = \frac{8n}{n+2}(1 - n_S) \, . \tag{8.184}$$

If we substitute Eq. (8.182) into Eq. (8.178), we get:

$$V(\varphi) \approx \lambda_n n^{n/2-1} N^{n/2} M_{\mathrm{Pl}}^n \, . \tag{8.185}$$

In order for the classical treatment to be valid, we need $V(\varphi) \ll M_{\mathrm{Pl}}^4$ and hence:

$$\lambda_n \ll \frac{M_{\mathrm{Pl}}^{4-n}}{n^{n/2-1} N^{n/2}} . \tag{8.186}$$

For $n = 4$ and $N = 60$ we have $\lambda_4 \ll 10^{-5}$.

8.8.2 The Starobinsky Model

The Starobinsky model is actually a $f(R)$ model, hence a modification of GR, whose action is

$$S = \frac{M_{\mathrm{Pl}}^2}{2} \int d^4x \sqrt{-g} f(R) , \tag{8.187}$$

where $f(R)$ is a generic function of the Ricci scalar. When $f(R) = R$ we recover GR. The above action can be recast as follows:

$$S = \int d^4x \sqrt{-g} \left[\frac{M_{\mathrm{Pl}}^2}{2} \varphi R - V(\varphi) \right] , \tag{8.188}$$

i.e. as a **non-minimally coupled scalar-tensor theory**, where the scalar field φ and its potential are defined as:

$$\varphi \equiv M_{\mathrm{Pl}} \frac{df}{dR} , \qquad V(\varphi) \equiv \frac{M_{\mathrm{Pl}}^2}{2} \left(R \frac{df}{dR} - f \right) . \tag{8.189}$$

This kind of theory has a non-minimal coupling because of the term φR (the minimal coupling to geometry occurs just with $\sqrt{-g}$). The action written as in Eq. (8.188) is also said to be in the **Jordan frame** and corresponds to the Brans–Dicke theory (Brans and Dicke 1961) with a special choice for its free parameter (i.e. $\omega = 0$).

Exercise 8.23 Performing the **conformal transformation**

$$\hat{g}_{\mu\nu} = \frac{df}{dR} g_{\mu\nu} = \frac{\varphi}{M_{\mathrm{Pl}}} g_{\mu\nu} , \tag{8.190}$$

show that:

$$\hat{S} = \int d^4x \sqrt{-\hat{g}} \left[\frac{M_{\mathrm{Pl}}^2}{2} \hat{R} - \frac{1}{2} \hat{g}^{\mu\nu} \partial_\mu \chi \partial_\nu \chi - U(\chi) \right] , \tag{8.191}$$

where $\hat{R}$ is the Ricci scalar computed from $\hat{g}_{\mu\nu}$ and:

$$U \equiv \frac{V M_{\text{Pl}}^2}{\varphi^2} = \frac{M_{\text{Pl}}^2}{2(df/dR)^2} \left(R \frac{df}{dR} - f \right) , \qquad \chi \equiv \sqrt{\frac{3}{2}} M_{\text{Pl}} \ln \frac{\varphi}{M_{\text{Pl}}} . \qquad (8.192)$$

Note that, in order for the above definition of χ to make sense, $df/dR > 0$.

The action $\hat{S}$ is said to be in the **Einstein frame**. Thus, any $f(R)$ theory can be reformulated as a scalar-tensor theory, the scalar field being $\varphi \equiv M_{\text{Pl}} df/dR$. The $f(R)$ theories have been intensively studied in the past decades, see the comprehensive reviews Sotiriou and Faraoni (2010) and De Felice and Tsujikawa (2010).

Since any $f(R)$ can be conformally mapped in GR plus a canonical scalar field, it seems natural to investigate how this kind of modified gravity accounts for inflation. Indeed, probably the most successful application of a $f(R)$ theory is in the inflationary paradigm, as we are going to show in a moment.

Now it is χ in action (8.191) which plays the role of the inflaton. The Starobinsky model (Starobinsky 1979) is given by a quadratic R^2 correction to the usual Einstein-Hilbert term in the action for gravity:

$$f(R) = R + \frac{R^2}{6M^2} . \qquad (8.193)$$

Then, one can easily calculate from Eq. (8.192):

$$U = \frac{3 M_{\text{Pl}}^2 M^2 R^2}{4(R + 3M^2)^2} , \qquad \chi = \sqrt{\frac{3}{2}} M_{\text{Pl}} \ln \frac{R + 3M^2}{3M^2} . \qquad (8.194)$$

Exercise 8.24 Invert the above relation for $\chi(R)$, obtaining thus a $R(\chi)$, and substitute it into the expression for $U(\chi)$, obtaining thus the expression for the inflaton potential in the Starobinsky model:

$$\boxed{U(\chi) = \frac{3}{4} M_{\text{Pl}}^2 M^2 \left(1 - e^{-\sqrt{2/3}\chi/M_{\text{Pl}}} \right)^2} \qquad (8.195)$$

In Fig. 8.3 we plot the Starobinsky potential $U(\chi)$ as function of the scalar field χ. The most important feature is the plateau at large values of the field which is the ideal feature for a slow-roll evolution of the inflaton field.

The slow-roll parameters are easily calculated from Eqs. (8.45), (8.47) and (8.195):

$$\epsilon_U = \frac{4}{3} \frac{e^{-2\sqrt{2/3}\chi/M_{\text{Pl}}}}{\left(1 - e^{-\sqrt{2/3}\chi/M_{\text{Pl}}}\right)^2} , \qquad \eta_U = \frac{4}{3} \frac{2e^{-2\sqrt{2/3}\chi/M_{\text{Pl}}} - e^{-\sqrt{2/3}\chi/M_{\text{Pl}}}}{\left(1 - e^{-\sqrt{2/3}\chi/M_{\text{Pl}}}\right)^2} . \qquad (8.196)$$

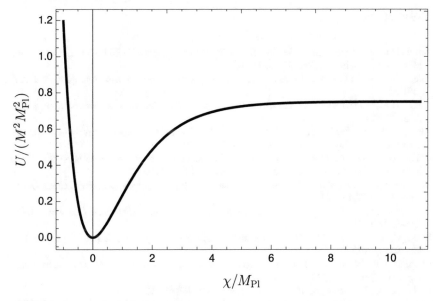

Fig. 8.3 Plot of the Starobinsky potential, in Eq. (8.195)

The number of e-folds to the end of inflation is obtained by solving the following integral:

$$N = \sqrt{\frac{3}{8}} \int_{\chi_f}^{\chi} \frac{d\chi}{M_{\text{Pl}}} \left(e^{\sqrt{2/3}\chi/M_{\text{Pl}}} - 1 \right) \approx \frac{3}{4} e^{\sqrt{2/3}\chi/M_{\text{Pl}}} , \tag{8.197}$$

so that we can write the slow-roll parameters as:

$$\epsilon_U = \frac{12}{(4N-3)^2} \approx \frac{3}{4N^2} , \qquad \eta_U = 4\frac{9-4N}{(4N-3)^2} \approx -\frac{1}{N} , \tag{8.198}$$

where we have kept the dominant contributions for large N. The scalar spectral index and the tensor-to-scalar ratio are thus written, using Eqs. (8.167) and (8.168), as follows:

$$n_S = 1 - \frac{2}{N} , \qquad r = \frac{12}{N^2} . \tag{8.199}$$

Substituting $N = 50$ and 60, the predictions obtained are in excellent agreement with the Planck constraints shown in Fig. 8.2, making thus the Starobinsky model quite successful.

References

Abramowitz, M., Stegun, I.A.: Handbook of Mathematical Functions with Formulas, Graphs, and Mathematical Tables. Dover, Illinois (1972)

Ade, P.A.R., et al.: Planck 2015 results. XX. constraints on inflation. Astron. Astrophys. **594**, A20 (2016a)

Ade, P.A.R., et al.: Planck 2015 results. XIII. cosmological parameters. Astron. Astrophys. **594**, A13 (2016b)

Albrecht, A., Steinhardt, P.J.: Cosmology for grand unified theories with radiatively induced symmetry breaking. Phys. Rev. Lett. **48**, 1220–1223 (1982)

Bolliet, B., Barrau, A., Martineau, K., Moulin, F.: Some clarifications on the duration of inflation in loop quantum cosmology. Class. Quant. Grav. **34**(14), 145003 (2017)

Brandenberger, R.H., Martin, J.: Trans-Planckian issues for inflationary cosmology. Class. Quant. Grav. **30**, 113001 (2013)

Brandenberger, R., Peter, P.: Bouncing cosmologies: progress and problems. Found. Phys. **47**(6), 797–850 (2017)

Brans, C., Dicke, R.H.: Mach's principle and a relativistic theory of gravitation. Phys. Rev. **124**, 925–935 (1961)

Coleman, S.R., Weinberg, E.J.: Radiative corrections as the origin of spontaneous symmetry breaking. Phys. Rev. D **7**, 1888–1910 (1973)

De Felice, A., Tsujikawa, S.: f(R) theories. Living Rev. Relativ. **13**, 3 (2010)

Di Marco, A.: Lyth bound, eternal inflation and future cosmological missions. Phys. Rev. D **96**(2), 023511 (2017)

Guth, A.H.: The inflationary universe: a possible solution to the horizon and flatness problems. Phys. Rev. D **23**, 347–356 (1981)

Kiefer, C., Polarski, D.: Why do cosmological perturbations look classical to us? Adv. Sci. Lett. **2**, 164–173 (2009)

Liddle, A.R., Parsons, P., Barrow, J.D.: Formalizing the slow roll approximation in inflation. Phys. Rev. D **50**, 7222–7232 (1994)

Linde, A.: On the problem of initial conditions for inflation. In: Black Holes, Gravitational Waves and Spacetime Singularities Rome, Italy, 9–12 May 2017

Linde, A.D.: A new inflationary universe scenario: a possible solution of the horizon, flatness, homogeneity, isotropy and primordial monopole problems. Phys. Lett. **108B**, 389–393 (1982)

Lyth, D.H.: What would we learn by detecting a gravitational wave signal in the cosmic microwave background anisotropy? Phys. Rev. Lett. **78**, 1861–1863 (1997)

Martin, J., Ringeval, C., Trotta, R., Vennin, V.: The best inflationary models after planck. JCAP **1403**, 039 (2014)

Mukhanov, V.F.: Gravitational instability of the Universe filled with a scalar field. JETP Lett. **41**, 493–496 (1985). Pisma Zh. Eksp. Teor. Fiz.41,402(1985)

Mukhanov, V.F., Feldman, H.A., Brandenberger, R.H.: Theory of cosmological perturbations. Part 1. Classical perturbations. Part 2. Quantum theory of perturbations. Part 3. Extensions. Phys. Rept. **215**, 203–333 (1992)

Muñoz, J.B., Kovetz, E.D., Raccanelli, A., Kamionkowski, M., Silk, J.: Towards a measurement of the spectral runnings. JCAP **1705**, 032 (2017)

Novello, M., Bergliaffa, S.E.P.: Bouncing cosmologies. Phys. Rept. **463**, 127–213 (2008)

Sasaki, M.: Large scale quantum fluctuations in the inflationary Universe. Prog. Theor. Phys. **76**, 1036 (1986)

Sotiriou, T.P., Faraoni, V.: f(R) theories of gravity. Rev. Mod. Phys. **82**, 451–497 (2010)

Starobinsky, A.A.: Spectrum of relict gravitational radiation and the early state of the universe. JETP Lett. **30**, 682–685 (1979). Pisma Zh. Eksp. Teor. Fiz.30,719(1979)

Tsujikawa, S.: Distinguishing between inflationary models from cosmic microwave background. PTEP **2014**(6), 06B104 (2014)

Weinberg, S.: The Quantum Theory of fields: vol. 2. Modern Applications. Cambridge University Press, Cambridge (2013)

Weinberg, S.: Cosmology. Oxford University Press, Oxford (2008)

Chapter 9
Evolution of Perturbations

*On peut braver les lois humaines, mais non résister aux lois
naturelles*
(One can challenge human laws, but not the natural ones)
Jules Verne, Vingt mille lieues sous le mers

In this Chapter we solve exactly some of the equations that we found in Chap. 4,
using some approximations. In particular, we can distinguish 4 cases of evolution:

1. On super-horizon scales,
2. In the matter-dominated epoch,
3. In the radiation-dominated epoch,
4. Deep inside the horizon,

for which it is possible to perform analytic calculations and thus gain a clearer
physical insight.

Our scope is to understand the shape of the matter power spectrum, plotted in
Fig. 9.1 from the analysis of the SDSS DR5 (Data Release 5) performed in Percival
et al. (2007). Since we already know the form of the primordial power spectrum,
cf. Eq. (8.154), the above task amounts to determine the matter, CDM plus baryons,
transfer function.

It must be noted that the data points in Fig. 9.1 are derived from the observation of
the distribution of galaxies in the sky and hence provide information on the galaxy
density contrast δ_g, which is in general a **biased tracer** of the underlying distribution
of matter in the sense that the galaxy correlation function is not equal to the total matter
correlation function (Kaiser 1984). At low redshift this bias is usually considered as
a constant

$$\delta_g = b\delta_m , \tag{9.1}$$

with δ_m being the matter density contrast (baryonic plus CDM), but for larger redshift
it might be a function of redshift and of the wavenumber, i.e. $b = b(k, z)$.

© Springer International Publishing AG, part of Springer Nature 2018
O. Piattella, *Lecture Notes in Cosmology*, UNITEXT for Physics,
https://doi.org/10.1007/978-3-319-95570-4_9

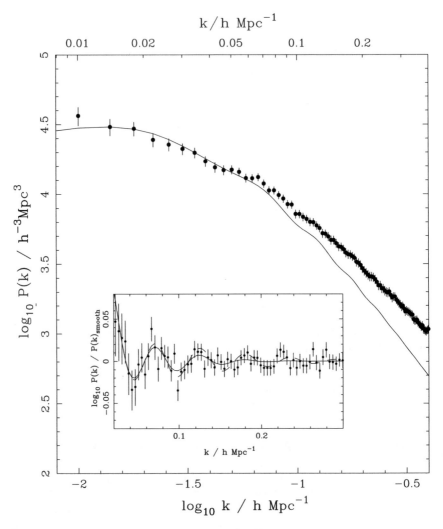

Fig. 9.1 Matter power spectrum, from Percival et al. (2007)

Beyond the cosmic variance affecting in a relevant way any large-scale observation, the determination of the power spectrum is also afflicted by another noise, which is called **shot noise** and is due to the fact that δ_g is given by a discrete distribution (that of galaxies) tracing a continuous one (that of the underlying matter).

Finally, spectra such as that in Fig. 9.1 are 3-d, in the sense that they are computed from the spatial distribution of galaxies. While determining the angular positions on the celestial sphere is not complicated, the only direct measure of distance that we have is the redshift. It is possible, of course, to transform the redshift into an actual distance (a proper distance, for example) through the cosmological model that we want to test, but determining redshift is time-consuming, especially if it is done via spectroscopy, and introduces extra errors due to peculiar motions and to photometry (if z is determined photometrically). Hence, it is perhaps more convenient to work with a 2-d power spectrum, the angular one $w(\theta)$, since angular positions on the celestial sphere are easily, precisely and rapidly determined.

9.1 Evolution on Super-Horizon Scales

A given comoving wavenumber k is super-horizon at a certain conformal time η, if

$$\boxed{k\eta \ll 1} \tag{9.2}$$

Since usually $\mathcal{H} \propto 1/\eta$, the above condition amounts to:

$$\boxed{k \ll \mathcal{H}} \tag{9.3}$$

which can be rewritten for the physical scale as follows:

$$\boxed{\frac{k}{a} \ll H} \tag{9.4}$$

The super-horizon regime is the same one that we used in Chap. 6 when we investigated the primordial modes. The main difference with what we are going to see here is that we do not limit ourselves to the radiation-dominated epoch, but investigate what happens through radiation-matter equality and also through matter-DE equality.

Thus, the above conditions can be written as follows in the epochs of interest:

$$k \ll \mathcal{H} = \frac{1}{\eta} \quad \Rightarrow \quad k\eta \ll 1 , \tag{9.5}$$

during the radiation-dominated epoch (for which $a \propto \eta$),

$$k \ll \mathcal{H} = \frac{2}{\eta} \quad \Rightarrow \quad k\eta \ll 2 , \tag{9.6}$$

during the matter-dominated epoch (for which $a \propto \eta^2$) and

$$k \ll \mathcal{H} = \frac{H_\Lambda}{1 + H_\Lambda(\eta_0 - \eta)} , \tag{9.7}$$

for the Λ-dominated era. This is similar to the inflationary phase, but with $\eta > 0$, hence the above expression for the conformal Hubble factor.

We use Eqs. (6.21)–(6.24):

$$\delta_\gamma' = -4\Phi' \,, \quad \delta_\nu' = -4\Phi' \,, \quad \delta_c' = -3\Phi' \,, \quad \delta_b' = -3\Phi' \,, \tag{9.8}$$

the Poisson equation, written using Eq. (6.59) as follows:

$$\frac{3}{\mathcal{H}}\left(\Phi' - \mathcal{H}\Psi\right) + \frac{k^2}{\mathcal{H}^2}\Phi = \frac{3}{2\rho_{\text{tot}}}\left(\rho_c\delta_c + \rho_b\delta_b + \rho_\gamma\delta_\gamma + \rho_\nu\delta_\nu\right) \,, \tag{9.9}$$

where we put in evidence the $k^2/\mathcal{H}^2$ factor that we are going to neglect, and the anisotropic stress equation (4.184):

$$k^2(\Phi + \Psi) = -32\pi G a^2 \rho_\nu \mathcal{N}_2 \,. \tag{9.10}$$

We could use again Eq. (6.59) in order to put in evidence the $k^2/\mathcal{H}^2$ term on the left, but it is not necessary because the neutrino quadrupole is of the same order and thus we need to perform differentiations as we did in Chap. 6 in order to get a closed equation for the potentials.

Note that we have neglected the photon quadrupole contribution. It is an approximation motivated by the fact that before recombination the tight coupling with electrons washes out Θ_2, whereas after radiation-matter equality R_γ becomes rapidly ($\propto 1/a \propto 1/\eta^2$) negligible.

When matter dominates, also R_ν is negligible, so we expect the potentials to become equal. Since our objective here is to perform an analytic calculation, we assume already $\Phi = -\Psi$. This is incorrect, strictly speaking, when considering the radiation-matter domination transition but it is fine when considering the matter-DE one. Therefore, let us rewrite Eq. (9.9) as follows:

$$3\mathcal{H}\left(\Phi' + \mathcal{H}\Phi\right) = \frac{3\mathcal{H}^2}{2\rho_{\text{tot}}}\left(\rho_c\delta_c + \rho_b\delta_b + \rho_\gamma\delta_\gamma + \rho_\nu\delta_\nu\right) \,. \tag{9.11}$$

This equation holds true also in presence of DE, provided that the latter does not cluster (e.g. it is Λ).

9.1.1   Evolution Through Radiation-Matter Equality

In presence of radiation and matter only, Friedmann equation can be written as follows:

$$\mathcal{H}^2 = \frac{8\pi G a^2}{3}(\rho_m + \rho_r) = \frac{8\pi G}{3}\rho_m\left(1 + \frac{1}{y}\right) \,, \tag{9.12}$$

where we have grouped together the species which evolve in the same way and indicated them with a subscript r, i.e. radiation (photons and neutrinos) and with a subscript m, i.e. matter (CDM and baryons). We have also employed the definition:

$$y \equiv \frac{\rho_m}{\rho_r} = \frac{a}{a_{eq}} , \qquad (9.13)$$

where a_{eq} is the equivalence scale factor.

Exercise 9.1 Assume adiabaticity, i.e.

$$\delta_c = \delta_b = \frac{3}{4}\delta_\gamma = \frac{3}{4}\delta_\nu \equiv \delta_m . \qquad (9.14)$$

Rewrite Eq. (9.11) using y as independent variable. Show that:

$$y\frac{d\Phi}{dy} + \Phi = \frac{4+3y}{6(y+1)}\delta_m . \qquad (9.15)$$

Exercise 9.2 We are looking now for a closed equation for Φ. Therefore, differentiate Eq. (9.15) with respect to y and use $\delta'_m = -3\Phi'$ in order to find:

$$\boxed{\frac{d^2\Phi}{dy^2} + \frac{(7y+8)(3y+4)+2y}{2y(y+1)(3y+4)}\frac{d\Phi}{dy} + \frac{1}{y(y+1)(3y+4)}\Phi = 0} \qquad (9.16)$$

Exercise 9.3 Quite unexpectedly, the above Eq. (9.16) can be solve exactly. Indeed, use the following transformation (Kodama and Sasaki 1984):

$$u \equiv \frac{y^3\Phi}{\sqrt{1+y}} , \qquad (9.17)$$

and show that

$$\frac{d^2u}{dy^2} + \left[-\frac{2}{y} + \frac{3}{2(y+1)} - \frac{3}{3y+4} \right]\frac{du}{dy} = 0 . \qquad (9.18)$$

Exercise 9.4 Integrate once Eq. (9.18) and show that:

$$\ln\frac{du}{dy} = C_1 + 2\ln y - \frac{3}{2}\ln(y+1) + \ln(3y+4) , \qquad (9.19)$$

where C_1 is an integration constant. By exponentiating and integrating again show that:

$$\frac{y^3 \Phi}{\sqrt{1+y}} = A \int_0^y dy \frac{y^2(3y+4)}{(1+y)^{3/2}} , \tag{9.20}$$

where A is an integration constant related to C_1 and assume that $y^3 \Phi \to 0$ for $y \to 0$.

Exercise 9.5 Solve the above integration and show that:

$$\Phi(y) = \frac{\Phi_P}{10y^3} \left(16\sqrt{1+y} + 9y^3 + 2y^2 - 8y - 16 \right) , \tag{9.21}$$

where Φ_P is the primordial gravitational potential, which we introduced in Eq. (6.77) in place of $C_\gamma = \mathcal{R} = \zeta$. Show that $\Phi(y \to 0) = \Phi_P$.

From the above solution we see that for $y \to \infty$ (which here means deep into the matter-dominated epoch) the gravitational potential drops of 10%, i.e.

$$\Phi \to \frac{9}{10} \Phi_P , \qquad \text{for } y \to \infty . \tag{9.22}$$

In Fig. 9.2 we display the evolution of Φ/Φ_P as function of y, as given by Eq. (9.22).

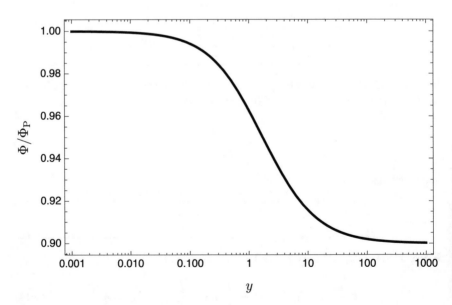

Fig. 9.2 Evolution of the gravitational potential Φ on super-horizon scales ($k = 0$) through radiation-matter equality. From Eq. (9.22)

It can be shown that Φ is constant on super-horizon scales for any background evolution with $w \neq -1$ constant and for adiabatic perturbations (Mukhanov et al. 1992).

Exercise 9.6 Setting $\Phi = -\Psi$, combine Eq. (4.179) with Eq. (4.189) and use Eq. 4.129. Show that:

$$\Phi'' + 3\mathcal{H}\left(1 + c_{ad}^2\right)\Phi' + \left[2\mathcal{H}' + \mathcal{H}^2\left(1 + 3c_{ad}^2\right)\right]\Phi + k^2 c_{ad}^2 \Phi = -4\pi G a^2 \Gamma .$$
(9.23)

where we have defined $c_{ad}^2 \equiv P'/\rho'$ as the **adiabatic speed of sound**. Show that this equation can be cast as follows:

$$\boxed{u'' + \left(k^2 c_a^2 - \frac{\theta''}{\theta}\right)u = -a^2(\rho + P)^{-1/2}\Gamma}$$
(9.24)

where

$$u \equiv \frac{\Phi}{4\pi G(\rho + P)^{1/2}}, \qquad \theta \equiv \frac{1}{a}\left(\frac{\rho}{\rho + P}\right)^{1/2}.$$
(9.25)

It is clear that the above transformation cannot treat the cosmological constant, for which $P = -\rho$. Nevertheless, it is interesting to see that Eq. (9.24) has a form similar to Eq. (8.71) and hence, for $k = 0$ and $\Gamma = 0$ (adiabatic perturbations) its general solution is:

$$u = C_1 \theta + C_2 \theta \int \frac{d\eta}{\theta^2} .$$
(9.26)

Exercise 9.7 Consider a single fluid model $P = w\rho$, with w constant and different from -1. Show, from solving the Friedmann equation, that:

$$a = (\eta/\eta_0)^{2/(1+3w)} ,$$
(9.27)

where $w \neq -1/3$ (in this case the solution grows exponentially with the conformal time). Show then that:

$$\theta \int \frac{d\eta}{\theta^2} \propto \frac{a}{\mathcal{H}} ,$$
(9.28)

and thus show, using Eq. (9.25), that Φ is constant.

Hence the 9/10 drop of Φ between the radiation-dominated era and the matter-dominated one can be extended to any kind of adiabatic fluid with $w \neq -1$ constant, using the constancy of $\mathcal{R}$ on large scales. See e.g. Mukhanov (2005). Indeed, from (12.33) we have

$$\mathcal{R} = \Phi + \mathcal{H}\frac{\Phi' - \mathcal{H}\Psi}{4\pi G a^2(\rho + P)} = \Phi + \frac{2}{3}\frac{\mathcal{H}^{-1}\Phi' + \Phi}{1 + w}. \tag{9.29}$$

Now, assume that w changes from a constant value w_i to another constant value w_f. For each of the two cases Φ is a constant, Φ_i and Φ_f, respectively. Then, taking advantage of the constancy of $\mathcal{R}$, we can say that:

$$\Phi_i + \frac{2}{3}\frac{\Phi_i}{1 + w_i} = \Phi_f + \frac{2}{3}\frac{\Phi_f}{1 + w_f}, \tag{9.30}$$

i.e

$$\Phi_f = \Phi_i\frac{5 + 3w_i}{5 + 3w_f}\frac{1 + w_f}{1 + w_i}, \tag{9.31}$$

with which we can easily check the result of Eq. (9.22).

9.1.2   Evolution in the Λ-Dominated Epoch

As already mentioned, we cannot use the above formulae for the case of greatest interest, which is for $w = -1$, i.e. the cosmological constant. In this case, we have to start directly from Eq. (9.23). Since $P = -\rho$ and constant, then $P' = 0$ and thus $c_{ad} = 0$.

Exercise 9.8 Using Eq. (9.7) into Eq. (9.23) with $c_{ad} = 0$ and $\Gamma = 0$, show that:

$$\Phi'' + \frac{3H_\Lambda}{1 + H_\Lambda(\eta_0 - \eta)}\Phi' + \frac{3H_\Lambda^2}{[1 + H_\Lambda(\eta_0 - \eta)]^2}\Phi = 0. \tag{9.32}$$

Note that η does not go to infinity, but to a maximum value $\eta_\infty = \eta_0 - 1/H_\Lambda$ for which the scale factor diverges. Moreover, note that this equation, and its solution, are valid also for small scales because for $c_{ad} = 0$ the k-dependence is suppressed.

Exercise 9.9 Find the relation between the cosmic time t and the conformal time using Eq. (9.7) and show that $\eta_\infty = \eta_0 - 1/H_\Lambda$ corresponds to an infinite t.

Exercise 9.10 Change variable to the scale factor in Eq. (9.32), and show that:

$$\frac{d^2\Phi}{da^2} + \frac{5}{a}\frac{d\Phi}{da} + \frac{3}{a^2}\Phi = 0. \tag{9.33}$$

Solve this equation, using a power-law ansatz, and show that:

$$\Phi = C_1 a^{-1} + C_2 a^{-3} . \tag{9.34}$$

Hence, when Λ dominates, Φ is not constant on super-horizon scales but vanishes rapidly, as $\Phi \propto 1/a$ or $\Phi \propto 1 + H_\Lambda(\eta_0 - \eta)$.

9.1.3 Evolution Through Matter-DE Equality

In order to study the transition between the matter-dominated epoch and the DE-dominated one, we consider Eq. (9.11) neglecting radiation:

$$3\mathcal{H}\left(\Phi' + \mathcal{H}\Phi\right) = \frac{3\mathcal{H}^2 \rho_m}{2\rho_{tot}} \delta_m , \tag{9.35}$$

where we have already considered adiabatic perturbations. Now, let us introduce the following variable:

$$x \equiv \frac{\rho_x}{\rho_m} = \frac{\tilde{\rho}(a/a_x)^{-3(1+w_x)}}{\tilde{\rho}(a/a_x)^{-3}} = \left(\frac{a}{a_x}\right)^{-3w_x} , \tag{9.36}$$

where ρ_x is a DE component with equation of state w_x, which we assume constant, and a_x is the scale factor at matter-DE equivalence, for which both the densities are equal to $\tilde{\rho}$. Note that $w_x < -1/3$, in order to have a useful DE (it has to produce an accelerated expansion) and recall that, in order for Eq. (9.35) to be valid, DE must not cluster.

Exercise 9.11 Following the same steps which brought us to Eq. (9.16), find a closed equation for Φ, using x as independent variable. Show that:

$$\frac{d^2\Phi}{dx^2} + \frac{6w_x(x+1) - 2x - 5}{6w_x x(x+1)} \frac{d\Phi}{dx} - \frac{1}{3w_x x(x+1)} \Phi = 0 . \tag{9.37}$$

This equation can be cast as a hypergeometric equation and thus solved exactly (Piattella et al. 2014). Only one of the two independent solutions is well-behaved for $x \to 0$, i.e. is a constant:

$$\Phi \propto -2w_x + \frac{4w_x(1+x)}{5} {}_2F_1\left(1, 1 - \frac{1}{3w_x}, 1 - \frac{5}{6w_x}, -x\right) \to_{x\to 0} -\frac{6w_x}{5} . \tag{9.38}$$

The integration constant has to be picked in order to match with $9\Phi_P/10$.

In Fig. 9.3 we display the evolution of Φ, computed for $w_x = -1.5, -1, -0.5$ from Eq. (9.38).

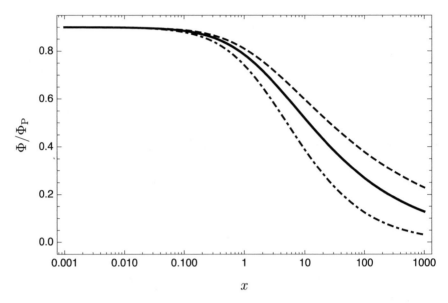

Fig. 9.3 Evolution of the gravitational potential Φ on super-horizon scales ($k = 0$) through matter-DE equality, from Eq. (9.38). The solid line represents the cosmological constant case $w_x = -1$, the dashed line $w_x = -1.5$ and the dash-dotted one $w_x = -0.5$

In Fig. 9.4 we display the evolution of the potentials Φ (solid line) and $-\Psi$ (dashed-line) for the ΛCDM model with adiabatic initial condition using CLASS. The wavenumber chosen here is $k = 10^{-4}\,\mathrm{Mpc}^{-1}$, which corresponds to a scale larger than the horizon today and hence which spent the whole evolution outside the horizon. Note how the two potentials display a difference at early times, due to the presence of neutrinos, which are of course taken into account in CLASS.

The investigation of the evolution of scales larger than the horizon today is not very useful because they are not observable. However, the scales that we do observe today were outside the horizon in the past. We shall see in the next section that in the matter-dominated regime the gravitational potential Φ is constant at all scales, even through horizon-crossing, and thus it is interesting to know the behaviour of super-horizon modes (the same behaviour is not shared by the density contrast). The same does not happen when radiation dominates, but instead the gravitational potential decays and oscillates rapidly for those scales which enter the horizon.

Since the behaviour in the radiation-dominated and in the matter-dominated epoch is so dramatically different, it is useful to introduce the so-called **equivalence wave number**, i.e. the wavenumber corresponding to a scale which enters the horizon at the equivalence epoch, and thus defined as:

$$k_{\mathrm{eq}} \equiv \mathcal{H}_{\mathrm{eq}} \,. \tag{9.39}$$

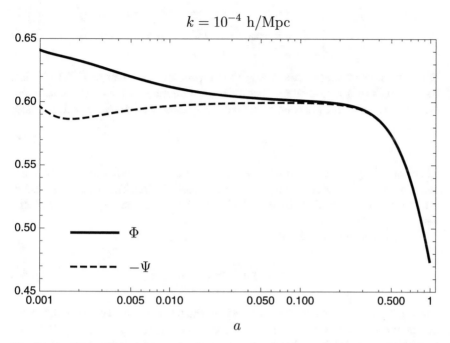

Fig. 9.4 Evolution of the potentials Φ (solid line) and $-\Psi$ (dashed-line) for the ΛCDM model with adiabatic initial condition using CLASS. The wavenumber chosen here is $k = 10^{-4}\,\mathrm{Mpc}^{-1}$

Neglecting DE, Friedmann equation in presence of radiation and matter is written as follows:

$$\mathcal{H}^2 = \frac{8\pi G}{3}(\rho_m + \rho_r)a^2 = H_0^2(\Omega_{m0}a^{-1} + \Omega_{r0}a^{-2}), \qquad (9.40)$$

and from this we can establish that, at equivalence, the conformal Hubble parameter has the following expression:

$$\mathcal{H}_{eq}^2 = \frac{16\pi G}{3}\frac{\rho_{R0}}{a_{eq}^2} = 2H_0^2\Omega_{r0}(1 + z_{eq})^2 = k_{eq}^2, \qquad (9.41)$$

or, using the matter density parameter:

$$\mathcal{H}_{eq}^2 = \frac{16\pi G}{3}\frac{\rho_{M0}}{a_{eq}} = 2H_0^2\Omega_{m0}(1 + z_{eq}) = k_{eq}^2, \qquad (9.42)$$

from which it is clear that:

$$1 + z_{eq} = \frac{\Omega_{m0}}{\Omega_{r0}}. \qquad (9.43)$$

Using the observed values for the density parameters, we have that:

$$\boxed{k_{\text{eq}} = \frac{\sqrt{2}H_0\Omega_{\text{m0}}}{\sqrt{\Omega_{\text{r0}}}} \approx 0.014 \, h \, \text{Mpc}^{-1}} \tag{9.44}$$

The behaviour of super-horizon modes is especially important in CMB physics. Indeed, in Chap. 10 we shall see that for a given multipole ℓ the CMB temperature correlation spectrum C_ℓ is determined mostly by those wavenumbers which satisfy:

$$\ell \approx k(\eta_0 - \eta_*) = kr_* \approx k\eta_0 \,, \tag{9.45}$$

where η_0 is the present conformal time, η_* is the one corresponding to recombination and $r_* \equiv \eta_0 - \eta_*$ is the comoving distance to recombination. The last approximation is motivated by the fact that, using CLASS and the ΛCDM model, $\eta_* \approx 3 \times 10^2 \, \text{Mpc}$ whereas $\eta_0 \approx 10^4 \, \text{Mpc}$.

The above expression can be manipulated as follows:

$$\ell \approx k\eta\frac{\eta_0}{\eta} \approx k\eta\frac{1}{\sqrt{a}} \,, \tag{9.46}$$

where η is a generic past conformal time in the matter-dominated epoch and, for this reason, we have used $a \propto \eta^2$. So, e.g. at radiation-matter equality, i.e. $a \approx 10^{-4}$, the super-horizon scales $k\eta < 1$ contribute to the monopoles $\ell \lesssim 100$ and at recombination $a_* \approx 10^{-3}$, $\ell \lesssim 30$.

9.2   The Matter-Dominated Epoch

Let us now neglect completely radiation. Being no photon and neutrino quadrupoles, then $\Phi = -\Psi$ and since matter dominates, $\delta P_{\text{tot}} = 0$.

Equation (4.189) then provides us straightaway with a closed equation for Φ:

$$\Phi'' + 3\mathcal{H}\Phi' + 2\mathcal{H}'\Phi + \mathcal{H}^2\Phi = -4\pi Ga^2\delta P_{\text{tot}} = 0 \,, \tag{9.47}$$

which can be solved exactly since, being $a \propto \eta^2$, we have:

$$\Phi'' + \frac{6}{\eta}\Phi' = 0 \,. \tag{9.48}$$

The general solution is:

$$\boxed{\Phi(k, \eta) = A(k) + B(k)(k\eta)^{-5} = A(k) + \hat{B}(k)a^{-5/2}} \tag{9.49}$$

where $A(k)$, $B(k)$ and $\hat{B}(k)$ are functions of k, being the latter equal to B times the proportionality factor between the conformal time and the scale factor, which are related by $a \propto \eta^2$. We have kept a $k\eta$ dependence above because it is dimensionless, as Φ, A and B are.

Now, let us neglect the decaying mode $(k\eta)^{-5}$ since it disappears very fast. The important result here is that the gravitational potential is constant at all scales, through horizon crossing, during matter domination. We shall see a very different behaviour when radiation dominates.

Hence, provided that we are deep into the matter-dominated epoch, at large scales, when $k\eta < 1$, we can match the results for Φ of this section and the previous one and see that:

$$A(k < 1/\eta) = \frac{9}{10}\Phi_P(k) , \qquad (\eta > \eta_{eq}) . \tag{9.50}$$

So, the gravitational potential Φ, for those scales which enter the horizon during the matter dominated epoch, is a constant with value:

$$\boxed{\Phi(k) = \frac{9}{10}\Phi_P(k) , \qquad (k\eta_{eq} < 1)} \tag{9.51}$$

The condition $k\eta_{eq} < 1$, which corresponds to $k < k_{eq}$ guarantees that the mode was outside the horizon at equivalence and hence that it entered during matter domination.

Considering the generalised Poisson equation with Φ constant, we have:

$$3\mathcal{H}^2\Phi + k^2\Phi = \frac{3\mathcal{H}^2}{2}\delta_m , \tag{9.52}$$

where δ_m is the density contrast of matter, which we define here as:

$$\delta_m \equiv (1 - \Omega_{b0})\delta_c + \Omega_{b0}\delta_b , \tag{9.53}$$

which comes from the fact that, being in the matter-dominated epoch, $\rho_{tot} \propto a^{-3}$ and $\Omega_{c0} + \Omega_{b0} = 1$ (of course we keep on considering a spatially flat universe).

We must be careful that even in the matter-dominated epoch, but before recombination, for those modes inside the horizon we can have that $\delta_c \gg \delta_b$ because baryons are tightly coupled to photons and thus δ_b cannot grow, whereas CDM fluctuations can. For these modes, soon after recombination, δ_b becomes equal to δ_c or, in other words, baryons fall into the potential wells of CDM. We shall see this in more detail later. For large scales, those which enter the horizon after recombination, we have $\delta_c = \delta_b = \delta_m$ (assuming, as usual, adiabaticity).

For the moment, let us focus on the evolution for δ_m. It follows immediately that:

$$\delta_m(k, \eta) = 2A(k)\left(1 + \frac{k^2}{3\mathcal{H}^2}\right) = \frac{9\Phi_P(k)}{5}\left(1 + \frac{k^2\eta^2}{12}\right) , \qquad (k < k_{eq}) . \tag{9.54}$$

Therefore, δ_m is constant on super-horizon scales, but when a scale crosses the horizon it starts to grow as $\delta \propto \eta^2 \propto a$, as the plot in Fig. 9.5 shows. Note that the

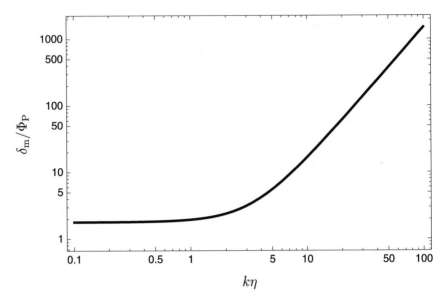

Fig. 9.5 Evolution of the matter density contrast δ_m normalised to the primordial potential, in the matter-dominated era. From Eq. (9.54)

first equality holds true at any scales, but the second one only for $k < k_{eq}$, because only in this regime we are allowed to use Eq. (9.50).

The power spectrum at any time during the matter dominated epoch can thus be put immediately in relation with the primordial one:

$$P_\delta(k, \eta) = \frac{81}{25} \left(1 + \frac{k^2\eta^2}{12}\right)^2 P_\Phi(k) , \qquad (k < k_{eq}) . \qquad (9.55)$$

For $n_S = 1$ the primordial power spectrum goes as $P_\Phi \propto 1/k^3$, cf. Eq. (8.154), and thus the above matter one grows linearly with k (when $k\eta > 1$). From the above equation we can read off the matter transfer function, i.e.:

$$T_\delta(k, \eta) = \frac{9}{5} \frac{k^2\eta^2}{12} , \qquad (1/\eta < k < k_{eq}) . \qquad (9.56)$$

From this solution we can infer that the larger k is, i.e. the smaller the scale under consideration is, the more it grows. This scenario is called **bottom-up** because smaller scales becomes non-linear before the larger ones. In other words, first small structures form and then these can merge in order to form larger structures. The bottom-up scenario is the contrary of the **top-down** scenario (Zeldovich 1984) by which first the largest structures form and then fragmentise in order to form the smaller ones.

In Eq. (9.56) we can appreciate that the k-dependence and the η one are separate. The latter is proportional to $\eta^2 \propto a$ and is usually called **growth factor**. The separation of the k and η dependences takes place because matter has vanishing adiabatic speed of sound and the k-dependence of the equation governing δ_m comes in multiplied by the gravitational potential, which is a constant (during matter domination).

We can see this in detail, by combining Eqs. (5.40) and (5.48).

Exercise 9.12 Combine Eqs. (5.40) and 5.48 after putting to zero w, δP and the anisotropic stresses, which are indeed vanishing for CDM and also for baryons, after recombination. Show that:

$$\boxed{\delta_m'' + \mathcal{H}\delta_m' = -k^2\Psi - 3\Phi'' - 3\mathcal{H}\Phi'} \qquad (9.57)$$

In the above equation, the k-dependence enters only through $k^2\Psi$ and then, in the matter-dominated epoch we have that:

$$\delta_m'' + \mathcal{H}\delta_m' = k^2\Phi \, , \qquad (9.58)$$

with Φ constant (if we neglect already the decaying mode).

Exercise 9.13 Check that the solution of the above equation is the same as Eq. (9.54).

The transfer function that we have determined in this section is valid for small values of k, i.e. $k < k_{eq} \approx 0.014\, h\, \mathrm{Mpc}^{-1}$, which correspond to very large scales which we do not actually observe or for which the errors and the cosmic variance are too large. Indeed, in Fig. 9.1 there are data points only for scales larger than k_{eq}. It is necessary therefore to understand how matter fluctuations behave during radiation-domination.

9.2.1 Baryons Falling into the CDM Potential Wells

We offer here a simple calculation which should convey the idea of how important CDM is for structure formation. This is often stated otherwise as the fact that after recombination, baryons fall into the gravitational potential wells of CDM.

As we anticipated earlier, before recombination baryons were tight-coupled to photons in the early-times plasma and, when they decouple and their over-densities are free to grow, in general we have $\delta_b \ll \delta_c$ for those modes which were well inside the horizon during recombination.

Let us see this more quantitatively. In the same fashion by which we obtained Eq. (9.58), we can write the following coupled equations for CDM and baryons:

$$\delta_c'' + \mathcal{H}\delta_c' = k^2\Phi \,, \tag{9.59}$$

$$\delta_b'' + \mathcal{H}\delta_b' = k^2\Phi \,, \tag{9.60}$$

of which the baryonic one is valid only after recombination. These equations are coupled since Φ is determined by both the components. Indeed, from the Poisson equation we have that:

$$(3\mathcal{H}^2 + k^2)\Phi = \frac{3\mathcal{H}^2}{2\rho_{tot}}(\rho_c\delta_c + \rho_b\delta_b) \equiv \frac{3\mathcal{H}^2}{2}\delta_m \,, \tag{9.61}$$

Now, the two Eqs. (9.59) and (9.60) have solutions (neglecting the decaying mode):

$$\delta_c(k, \eta) = C_1(k) + \frac{k^2\eta^2}{6}A(k) \,, \qquad \delta_b(k, \eta) = C_2(k) + \frac{k^2\eta^2}{6}A(k) \,, \tag{9.62}$$

where we have used the constant potential solution for the potential, in Eq. (9.49) (also here neglecting the decaying mode). Using this solution in Eq. (9.53) and comparing with Eq. (9.54), we can conclude that:

$$C_1(k) + \Omega_{b0}[C_2(k) - C_1(k)] = 2A(k) \,. \tag{9.63}$$

For large scales $k\eta \ll 1$ we already know that $\delta_c = \delta_b = \delta_m$, because of adiabaticity, and hence $C_1 = C_2 = 2A$.

For small scales, at recombination $\delta_c \gg \delta_b$ because baryons were tightly coupled to photons and thus δ_b could not grow. This, combined with the crucial fact that $\Omega_{b0} = 0.04$ is small, makes δ_m (and the gravitational potential) dominated by δ_c. Hence, again $\delta_c = \delta_b$ soon after recombination, the detailed transient coming from the decaying modes that we have neglected.

In other words, baryons fall in the potential wells already created by CDM. Without CDM, δ_b would grow proportionally to a, by a factor 10^3 by today, being of order 10^{-2}. This is way too small in order for account of the structures that we observe.

In Fig. 9.6 we plot the evolution of δ_c (solid line) and δ_b (dashed line) for $k = 1\,\text{Mpc}^{-1}$ computed with CLASS. Note the oscillations in δ_b, which are related to the **baryon acoustic oscillations** (BAO) of the matter power spectrum, cf. the small box inside Fig. 9.1, but are not the same thing. The oscillations in the plot of Fig. 9.6 are function of the time (scale factor) evolution, whereas those in the matter power spectrum of Fig. 9.1 are function of the wavenumber k.

The oscillations of Fig. 9.6 are caused by the tight coupling of baryons with photons before recombination, when structure formation is impossible. On the other hand, CDM can grow even when radiation dominates and thus, for the chosen scale $k = 1\,\text{Mpc}^{-1}$, the ratio δ_c/δ_b is of about 3 orders of magnitude.

In Fig. 9.7 we plot again the evolution of δ_c (solid line) and δ_b (dashed line) for $k = 1\,\text{Mpc}^{-1}$ computed with CLASS, but this time with a negligible amount of CDM ($\Omega_{c0}h^2 = 10^{-6}$). Note how δ_b grows six orders of magnitude less today than in the standard case.

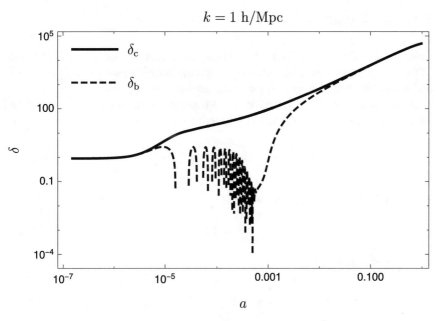

Fig. 9.6 Evolution of δ_c (solid line) and δ_b (dashed line) for $k = 1\,\mathrm{Mpc}^{-1}$ computed with CLASS

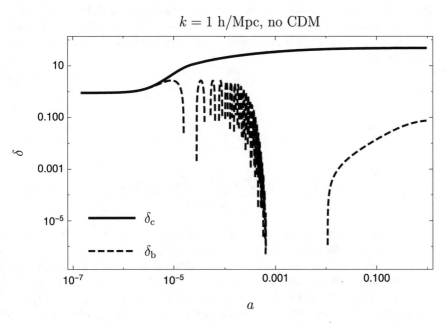

Fig. 9.7 Evolution of δ_c (solid line) and δ_b (dashed line) for $k = 1\,\mathrm{Mpc}^{-1}$ computed with CLASS with a negligible amount of CDM ($\Omega_{c0}h^2 = 10^{-6}$)

9.3 The Radiation-Dominated Epoch

Consider now the case of full radiation dominance and neglect δ_c and δ_b as source of the gravitational potentials. Moreover, assume adiabaticity, so that $\delta_\gamma = \delta_\nu = \delta_r$ and neglect the neutrino anisotropic stress, so that $\Phi = -\Psi$.

Since we are deep into the radiation-dominated epoch, then $w = c_{ad}^2 = 1/3$ and thus from Eq. (9.23) we can immediately write:

$$\Phi'' + \frac{4}{\eta}\Phi' + \frac{k^2}{3}\Phi = 0 \,, \tag{9.64}$$

where we have used $a \propto \eta$. Defining

$$u \equiv \Phi\eta \,, \tag{9.65}$$

the above equation becomes

$$u'' + \frac{2}{\eta}u' + \left(\frac{k^2}{3} - \frac{2}{\eta^2}\right)u = 0 \,. \tag{9.66}$$

This is a Bessel equation with solutions $j_1(k\eta/\sqrt{3})$ and $n_1(k\eta/\sqrt{3})$, i.e. the spherical Bessel functions of order 1. Since n_1 diverges for $k\eta \to 0$, we discard it as unphysical.

Recovering the gravitational potential Φ and using the fact that (Abramowitz and Stegun 1972)

$$j_1(x) = \frac{\sin x}{x^2} - \frac{\cos x}{x} \,, \tag{9.67}$$

and

$$\lim_{x\to 0}\frac{\sin x - x\cos x}{x^3} = \frac{1}{3} \,, \tag{9.68}$$

we can write

$$\Phi = 3\Phi_P \frac{\sin(k\eta/\sqrt{3}) - (k\eta/\sqrt{3})\cos(k\eta/\sqrt{3})}{(k\eta/\sqrt{3})^3} \,. \tag{9.69}$$

This solution shows that as soon as a mode k of the gravitational potential enters the horizon, it rapidly decays as $1/\eta^3$ or $1/a^3$ while oscillating.

In Eq. (9.69) we plot the evolution of the gravitational potential, according to Eq. (9.69). Note how Φ starts to decay right after $k\eta > 1$, i.e. after horizon crossing.

The goodness of the solution (9.69) plotted in Fig. 9.8 can be appreciated in Fig. 9.9 where both Φ (solid line) and $-\Psi$ (dashed line) are plotted. Outside the horizon the two potentials are constant with a difference due to the neutrino fraction R_ν. As soon as they enter the horizon they rapidly decay to zero.

Now, recall Eq. (9.57) that we derived for the matter density contrast. It can be used in the radiation-dominated epoch, but only for CDM:

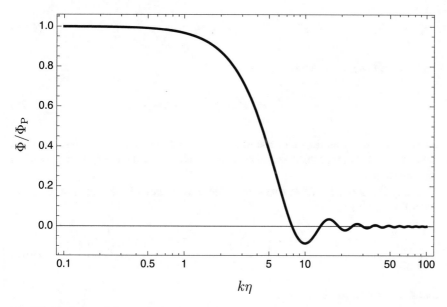

Fig. 9.8 Evolution of the gravitational potential Φ deep into the radiation-dominated era. From Eq. (9.69)

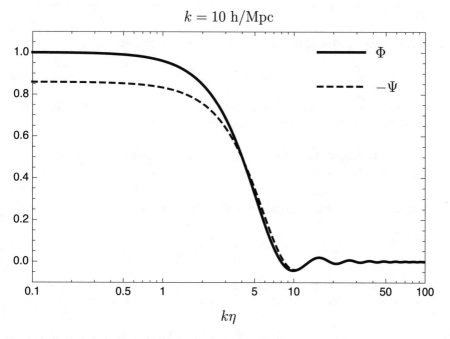

Fig. 9.9 Evolution of the gravitational potentials Φ (solid line) and $-\Psi$ (dashed line) deep into the radiation-dominated era computed with CLASS for the ΛCDM model and $k = 10 \, \mathrm{Mpc}^{-1}$. Adiabatic perturbations have been used and the initial values have been normalised to that of Φ

$$\delta_c'' + \frac{1}{\eta}\delta_c' = k^2\Phi - 3\Phi'' - \frac{3}{\eta}\Phi' \; . \tag{9.70}$$

Using Eq. (9.64), we can write

$$\delta_c'' + \frac{1}{\eta}\delta_c' = 2k^2\Phi + \frac{9}{\eta}\Phi' \equiv S(k,\eta) \; . \tag{9.71}$$

As stated at the beginning of the section, we are so deep into the radiation-dominated epoch that the matter density contrast does not contribute to the gravitational potential but only feels it.

Using Eq. (9.69) with $x \equiv k\eta$ as new independent variable, the function $S(x)$ has the following form:

$$\frac{S(x)}{k^2\Phi_P} = \frac{9\left[(27x - 2x^3)\cos\left(x/\sqrt{3}\right) + \sqrt{3}\left(5x^2 - 27\right)\sin\left(x/\sqrt{3}\right)\right]}{x^5} \; , \tag{9.72}$$

and the equation for δ_c becomes:

$$\frac{d}{dx}\left(x\frac{d\delta}{dx}\right) = 9\Phi_P \frac{\left[(27x - 2x^3)\cos\left(x/\sqrt{3}\right) + \sqrt{3}\left(5x^2 - 27\right)\sin\left(x/\sqrt{3}\right)\right]}{x^4} \; . \tag{9.73}$$

The homogeneous part of this equation has a simple solution:

$$\delta_{\text{hom}} = C_1 + C_2\ln x \; , \tag{9.74}$$

i.e. a constant C_1 times a logarithmic contribution. A particular solution is obtained by integrating twice the right-hand side, thus obtaining:

$$\delta_{\text{part}} = \frac{9\Phi_P\left[-x^3\text{Ci}\left(x/\sqrt{3}\right) + \sqrt{3}\left(x^2 - 3\right)\sin\left(x/\sqrt{3}\right) + 3x\cos\left(x/\sqrt{3}\right)\right]}{x^3} \; , \tag{9.75}$$

where $\text{Ci}(z)$ is the cosine integral function, defined as

$$\text{Ci}(z) \equiv -\int_z^\infty dt\,\frac{\cos t}{t} \; . \tag{9.76}$$

The above is a particular solution, therefore the integration constants which stem from the indefinite integration can be incorporated in C_1. The general solution for δ_c is then:

$$\delta_c = C_1 + C_2\ln x + \frac{9\Phi_P\left[-x^3\text{Ci}\left(x/\sqrt{3}\right) + \sqrt{3}\left(x^2 - 3\right)\sin\left(x/\sqrt{3}\right) + 3x\cos\left(x/\sqrt{3}\right)\right]}{x^3} \; . \tag{9.77}$$

For $x \to 0$, we can expand the above solution as follows:

$$\delta(x \to 0) = C_1 + C_2 \ln(x) + \Phi_P \left[-9\ln(x) - 9\gamma + 6 + \frac{9\ln(3)}{2} \right] + \mathcal{O}\left(x^2\right) ,$$
(9.78)

where γ is the Euler constant.

Since $\ln x$ is divergent for $x \to 0$ and we do not want δ_c to diverge, we have to ask:

$$C_2 = 9\Phi_P .$$
(9.79)

Moreover, we know that $\delta_c(x \to 0) = 3\Phi_P/2$ when we choose adiabatic initial conditions (and neglect neutrinos), cf. Eqs. (6.71) and (6.77), thus:

$$C_1 = -\frac{9}{2}\Phi_P \left[-2\gamma + 1 + \ln(3) \right] .$$
(9.80)

We plot the evolution of δ_c through horizon-crossing in Fig. 9.10.

For $x \gg 1$, deep inside the horizon, we can neglect the contribution δ_{part} to the solution for δ_c since it decays rapidly. Thus, we can write the density contrast as follows:

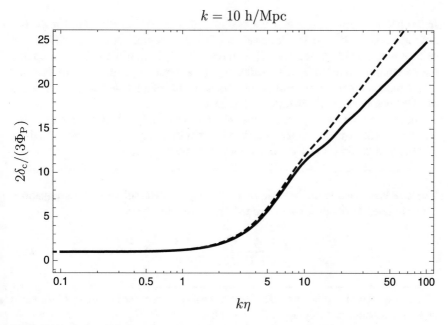

$$k = 10 \text{ h/Mpc}$$

Fig. 9.10 Evolution of δ_c deep into the radiation-dominated era computed from Eqs. (9.77), (9.79) and (9.80) (solid line) compared with the numerical calculation performed with CLASS for $k = 10\,\text{Mpc}^{-1}$ (dashed line). Note the semi-logarithmic scale employed

$$\delta_c = -\frac{9}{2}\Phi_P\left[-2\gamma + 1 + \ln(3)\right] + 9\Phi_P \ln x = A\Phi_P \ln(Bk\eta) , \qquad (9.81)$$

where

$$A = 9 , \qquad B = \exp\left[\gamma - \frac{1}{2} - \frac{\ln(3)}{2}\right] \approx 0.62 . \qquad (9.82)$$

9.4 Deep Inside The Horizon

The last domain in which it is possible to analytically solve the equations for the perturbations is when $k \gg \mathcal{H}$, i.e. deep inside the horizon. We shall neglect baryons in this calculation (this is imprecise and we shall see why numerically) and assume $\Phi = -\Psi$ (which is a good approximation, on the basis of the results of the previous section).

The relevant equations are thus the following ones:

$$\delta_c' + kV_c = -3\Phi' , \qquad (9.83)$$

$$V_c' + \mathcal{H}V_c = -k\Phi , \qquad (9.84)$$

$$k^2\Phi = 4\pi G a^2 \rho_c \delta_c . \qquad (9.85)$$

In the latter equation we have neglected all the potential terms except for the one accompanied by k^2 and we have also neglected radiation perturbations. It is not evident why should we neglect $\rho_r\delta_r$ with respect $\rho_c\delta_c$ even when $\rho_r \gg \rho_c$ deep into the radiation-dominated epoch. Neglecting δ_b, at least, is justified by the fact that before recombination it behaves as the fluctuation in radiation and afterwards as that in CDM, while being Ω_b always subdominant.

An explanation of why we can neglect $\rho_r\delta_r$ is offered by Weinberg who shows that new modes appear (dubbed *fast*) which rapidly decay and oscillate (Weinberg 2002). He also takes into account baryons at first order in Ω_{b0}.

Exercise 9.14 Use again the variable $y \equiv a/a_{eq}$ and manipulate the three equations above in order to obtain a single second-order equation for δ_c:

$$\frac{d^2\delta_c}{dy^2} + \frac{2 + 3y}{2y(y+1)}\frac{d\delta_c}{dy} - \frac{3}{2y(y+1)}\delta_c = 0 . \qquad (9.86)$$

This equation is known as **Mészáros equation** (Meszaros 1974). A solution can be found at once, multiplying by $2y(y+1)$:

$$2y(y+1)\frac{d^2\delta_c}{dy^2} + 2\frac{d\delta_c}{dy} + 3y\frac{d\delta_c}{dy} - 3\delta_c = 0 . \qquad (9.87)$$

A linear ansatz $\delta_c \propto y$ kills the second derivative and the last two terms on the left hand side. Therefore, the simple solution we looked for is:

$$\boxed{D_1(y) \equiv y + \frac{2}{3}} \tag{9.88}$$

This is also the growing mode. For $y \ll 1$, i.e. before matter-radiation equality, δ is practically constant, whereas for $y \gg 1$ we have the known growth linear with respect to the scale factor.

In order to find the other independent solution say D_2, we can use the Wronskian:

$$W(y) = D_1 \frac{dD_2}{dy} - \frac{dD_1}{dy} D_2 . \tag{9.89}$$

Exercise 9.15 Show that the Wronskian satisfies the simple first-order differential equation:

$$\frac{dW}{dy} = -\frac{2 + 3y}{2y(y + 1)} W , \tag{9.90}$$

from which one gets

$$W = \frac{1}{y\sqrt{1 + y}} . \tag{9.91}$$

Exercise 9.16 From the very definition of the Wronskian in Eq. (9.89), write a first-order equation also for D_2, which is the following:

$$(y + 2/3)^2 \frac{d}{dy}\left(\frac{D_2}{y + 2/3}\right) = \frac{1}{y\sqrt{1 + y}} . \tag{9.92}$$

Integrate it and show that the result is:

$$\boxed{D_2 = \frac{9}{2}\sqrt{1 + y} - \frac{9}{4}(y + 2/3) \ln\left(\frac{\sqrt{y + 1} + 1}{\sqrt{y + 1} - 1}\right)} \tag{9.93}$$

This mode grows logarithmically when $y \ll 1$, recovering the logarithmic solution of the previous section. It decays as $1/y^{3/2}$ for $y \gg 1$.

The complete solution for δ_c on small scales $k \gg \mathcal{H}$ and through radiation-matter equality is then:

$$\delta_c(k, a) = C_1(k)D_1(a) + C_2(k)D_2(a) . \tag{9.94}$$

The dependence of C_1 and C_2 from k can be established by matching this solution with the one of the previous section in Eq. (9.81).

9.5   Matching and CDM Transfer Function

As we discussed earlier and as it is clear from Fig. 9.1 the today observed scales in the matter power spectrum are those for which $k > k_{eq}$, i.e. those which entered the horizon before matter equality.

In the last two section we have obtained exact solution for δ_c deep into the radiation-dominated epoch for all scales and through radiation-matter equality at very small scales. There is then the possibility of matching the two solutions on very small scales and thus obtain in this regime the CDM transfer function until today.

Consider the following two solutions found in the previous sections, i.e.

$$\delta_c(k, \eta) = A\Phi_P(k)\ln(Bk\eta) \, , \tag{9.95}$$

$$\delta_c(k, a) = C_1(k)D_1(a) + C_2(k)D_2(a) \, , \tag{9.96}$$

which are valid on very small scales, i.e. $k\eta \gg 1$. The purpose is to find the functional forms of $C_1(k)$ and $C_2(k)$.

Using Eqs. (9.40) and (9.41), one can approximate the Hubble parameter deep into the radiation era as follows:

$$\mathcal{H} \approx \frac{\mathcal{H}_{eq}a_{eq}}{\sqrt{2}a} \, , \tag{9.97}$$

and solving using the conformal time, one has:

$$a = \frac{\mathcal{H}_{eq}a_{eq}}{\sqrt{2}}\eta \, . \tag{9.98}$$

The proportionality constant is the correct one which gives $\mathcal{H} = 1/\eta$ when substituted in the approximated formula for $\mathcal{H}$.

We can thus write the logarithmic solution for δ_c, deep into the radiation-dominated era, as follows:

$$\delta_c(k, a) = A\Phi_P(k)\ln\left(Bk\frac{\sqrt{2}a}{\mathcal{H}_{eq}a_{eq}}\right) \, . \tag{9.99}$$

Introducing $y \equiv a/a_{eq}$, the equivalence wavenumber $k_{eq} = \mathcal{H}_{eq}$ and the rescaled wavenumber:

$$\kappa \equiv \frac{\sqrt{2}k}{k_{eq}} = \frac{k\sqrt{\Omega_{r0}}}{H_0\Omega_{m0}} = \frac{k}{0.052\,\Omega_{m0}\,h^2\,\text{Mpc}^{-1}} \tag{9.100}$$

one has:

$$\delta_c(k, a) = A\Phi_P(k)\ln(B\kappa y) \, . \tag{9.101}$$

Recall that this solution is valid deep into the radiation-dominated epoch, thus $y \ll 1$. At the same time, it holds true only on very small scales, i.e. $k\eta \gg 1$. Using the above formulae, this means:

$$k\eta = k\frac{\sqrt{2}a}{\mathcal{H}_{eq}a_{eq}} = k\frac{\sqrt{2}a}{k_{eq}a_{eq}} = \kappa y \gg 1 \,. \tag{9.102}$$

Therefore, if we want to match radiation-domination solution with the solution of the Mészáros equation, we need to choose a suitable y_m at which performing the junction of the two solutions such that $1/\kappa \ll y_m \ll 1$.

Asking the equality of the two solutions and their derivatives at y_m implies to solve the following system:

$$A\Phi_P \ln(B\kappa y_m) = C_1 (y_m + 2/3) + C_2 D_2(y_m) \,, \tag{9.103}$$

$$\frac{A\Phi_P}{y_m} = C_1 + C_2 \left.\frac{dD_2}{dy}\right|_{y=y_m} \,. \tag{9.104}$$

Exercise 9.17 Solve the above system and take the dominant contribution for $y_m \to 0$ (because the junction condition has to be imposed for $y_m \ll 1$). Show that:

$$C_1 = \frac{3}{2}A\Phi_P \ln\left(4B\kappa e^{-3}\right) \,, \qquad C_2 = \frac{2}{3}A\Phi_P \,. \tag{9.105}$$

Therefore, the solution for δ_c valid at all time and at small scales is the following:

$$\delta_c(k, y) = \frac{3}{2}A\Phi_P \ln\left(4\sqrt{2}B\kappa e^{-3}\right)(y + 2/3) + \frac{2}{3}A\Phi_P D_2(y) \,, \quad (\kappa \gg 1) \,. \tag{9.106}$$

We can project this solution at late times, when matter dominates and thus neglect the decaying mode D_2 and write:

$$\delta_c(k, a) = \frac{3}{2}A\Phi_P \ln\left(\frac{4\sqrt{2}Be^{-3}k}{k_{eq}}\right)\frac{a}{a_{eq}} \,, \quad (a \gg a_{eq}, k \gg k_{eq}) \tag{9.107}$$

Note again the separated dependence from k and from the scale factor, being the growth function $D(a) = a$. This allows us to write the transfer function for CDM as follows, using Eq. (9.42) in order to eliminate a_{eq}:

$$T_\delta(k) = \frac{Ak_{eq}^2}{2H_0^2\Omega_{m0}} \ln\left(\frac{4\sqrt{2}Be^{-3}k}{k_{eq}}\right) D(a) \,, \quad (a \gg a_{eq}, k \gg k_{eq}) \tag{9.108}$$

Recall that the factor $3\Phi_P/2$ is the adiabatic initial condition on δ_c and thus enters the primordial power spectrum. The above transfer function can be generalised including DE, if the latter does not cluster. If it is the case, as for the cosmological constant, its effect enters only the growth factor and in k_{eq} since, being another component and having the fixed total $\Omega_{r0} + \Omega_{m0} + \Omega = 1$, the relative amount of radiation and matter has to change (usually the amount of radiation is fixed) and thus k_{eq} changes.

We can also determine the transfer function for the gravitational potential Φ at late times. From Eq. (9.85) we have that:

$$k^2\Phi = 4\pi Ga^2\rho_m\delta_m = \frac{3H_0^2\Omega_{m0}}{2a}\delta_m \,, \tag{9.109}$$

where we have included baryons, since we are considering late times and we have seen that after recombination $\delta_b = \delta_c$. We are neglecting any DE contribution in the expansion of the universe and considering a radiation plus matter model. Hence, $\Omega_0 \approx 1$.

We thus have for the gravitational potential, combining Eqs. (9.107) and (9.109):

$$\Phi(k) = \frac{9Ak_{eq}^2}{8k^2}\Phi_P(k) \ln\left(\frac{4\sqrt{2}Be^{-3}k}{k_{eq}}\right) \,, \qquad k \gg k_{eq} \tag{9.110}$$

The transfer function for the gravitational potential is usually normalised to $9\Phi_P/10$ and therefore:

$$T_\Phi(k) = \frac{5Ak_{eq}^2}{4k^2} \ln\left(\frac{4\sqrt{2}Be^{-3}k}{k_{eq}}\right) \,, \qquad k \gg k_{eq} \tag{9.111}$$

Exercise 9.18 Using $\Omega_{r0}h^2 = 4.15 \times 10^{-5}$ in Eq. (9.44) and defining

$$q \equiv k \times \frac{\text{Mpc}}{\Omega_{m0}h^2} \,, \tag{9.112}$$

show that we can cast the above transfer function in the following form:

$$T_\Phi(k) = \frac{\ln(2.40q)}{(4.07q)^2} \tag{9.113}$$

See also Weinberg (2008), p. 310.

We have written the transfer function as in Eq. (9.113) because it is simpler to compare it with the numerical fit of Bardeen, Bond, Kaiser and Szalay (BBKS) of the exact transfer function (Bardeen et al. 1986), which is the following:

$$T_{BBKS}(k) = \frac{\ln(1 + 2.34q)}{2.34q} \left[1 + 3.89q + (16.2q)^2 + (5.47q)^3 + (6.71q)^4\right]^{-1/4},$$

$$(9.114)$$

and see that for large q it goes as $\ln(2.34q)/(3.96q)^2$, which is in good agreement with our analytic estimate (9.113).

Given the transfer function T_Φ, δ_m can be written, from Eq. (9.109), as:

$$\delta_m(k, a) = \frac{2a}{3H_0^2\Omega_{m0}}k^2\Phi = \frac{3k^2}{5H_0^2\Omega_{m0}}\Phi_P(k)T_\Phi(k)a = \frac{3k^2}{5H_0^2\Omega_{m0}}\Phi_P(k)T_\Phi(k)D(a).$$

$$(9.115)$$

In the last equality we have recovered the growth factor $D(a)$ in order to provide a more general formula. From this solution we can obtain the power spectrum for δ_m starting from the primordial one for Φ, i.e.

$$P_\delta(k, a) = \frac{9k^4}{25H_0^4\Omega_{m0}^2}P_\Phi(k)T_\Phi^2(k)D^2(a). \tag{9.116}$$

Using Eq. 6.94, with $\Phi = -\Psi$ since we are neglecting neutrinos, $\Phi_P = 2\mathcal{R}/3$ and thus the primordial power spectrum of Φ can be traded with the $\mathcal{R}$ one, and we get:

$$P_\delta(k, a) = \frac{4k^4}{25H_0^4\Omega_{m0}^2}P_\mathcal{R}(k)T_\Phi^2(k)D^2(a). \tag{9.117}$$

From Eq. (8.154), we can write the explicit dependence of the primordial power spectrum of k as follows:

$$\boxed{P_\delta(k, a) = \frac{8\pi^2 k}{25H_0^4\Omega_{m0}^2}A_S\left(\frac{k}{k_*}\right)^{n_S-1}T_\Phi^2(k)D^2(a)} \tag{9.118}$$

The power spectrum $P_\delta(k, a)$ can be determined through the observation of the distribution of galaxies in the sky. Therefore, through the above formula we can probe many quantities of great interest such as the primordial tilt of the power spectrum.

Now, let us make a plot of the power spectrum today ($z = 0$) using CLASS and fixing all the parameters to the ΛCDM best fit values except for Ω_{m0}, which we let free. We show in Fig. 9.11 the shape of the matter power spectrum for $\Omega_{m0} = 0.1$ (blue) and $\Omega_{m0} = 0.3$ (yellow), $\Omega_{m0} = 0.7$ (green) and $\Omega_{m0} = 0.99$ (red).

The first interesting feature of the power spectrum is that it has a maximum. This maximum takes place roughly at k_{eq} for the following reason: entering the horizon at equivalence is the best time to do that in order for a matter fluctuation to grow more. In fact, as we have seen, scales that entered the horizon earlier (i.e. $k > k_{eq}$) are suppressed because radiation is dominating and that δ grows logarithmically during this epoch, cf. Eq. (9.77). These are the scales of great observational interest, as can be appreciated from Fig. 9.1.

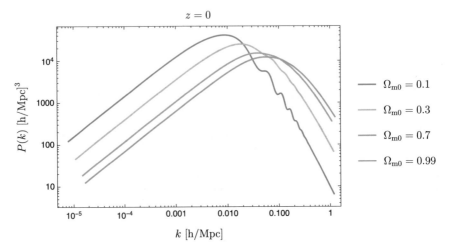

Fig. 9.11 Matter power spectra for (left to right, in case no colours are available) $\Omega_{m0} = 0.1$ (blue), $\Omega_{m0} = 0.3$ (yellow), $\Omega_{m0} = 0.7$ (green) and $\Omega_{m0} = 0.99$ (red)

On the other hand, the scales that entered after the equivalence grow proportionally to a, cf. Eq. (9.54). Evidently, the scale which entered at equivalence has had more time to grow than all the others, hence the the maximum or **turnover** in the power spectrum.

The second interesting feature of the power spectrum is that its maximum is shifted to the left when we reduce Ω_{m0}. This means that the equivalence wavenumber k_{eq} is smaller when Ω_{m0} is smaller and this can be immediately seen from Eq. (9.44) if we fix Ω_{r0}.

Remarkably, observation favours the line for $\Omega_{m0} = 0.3$ of Fig. 9.11, as can also be seen from Fig. 9.1. Since the radiation content is very well established from CMB observation (and our knowledge of neutrinos) as well as the spatial flatness of the universe (from the CMB, we shall see this in Chap. 10), and its age (at least a lower bound) and H_0 are also well-determined, the only way to make the total is to add a further component, which clearly is DE. The important point here is that the necessity for DE can already be seen by analysing the large scale structure of the universe (together with CMB) and this was realised well before type Ia supernovae started to be used as standard candles (Maddox et al. 1990; Efstathiou et al. 1990).

For completeness, in Fig. 9.12 we plot with CLASS the matter power spectrum at $z = 0$ for the ΛCDM model, with different choices of the initial conditions.

The transfer function thus tells us how the shape of the primordial power spectrum is changed through the cosmological evolution and through the analytic estimates that we have done in this Chapter we understood that the k-dependence is set up during radiation-domination.

However, we have made two important assumptions in our calculations: we have neglected baryons and neutrino anisotropic stress (we have taken into account neutrinos from the point of view of the background expansion). If we aim to precise

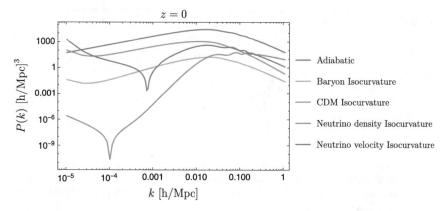

Fig. 9.12 Matter power spectra for different initial conditions: Adiabatic (blue), Baryon Isocurvature (yellow), CDM Isocurvature (green), Neutrino density Isocurvature (red), and Neutrino velocity Isocurvature (purple)

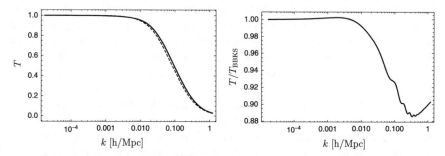

Fig. 9.13 *Left Panel.* Evolution of BBKS transfer function of Eq. (9.114) (solid line) compared with the numerical computation of CLASS, using $\Omega_{m0} = 0.95$ (which means negligible DE). *Right Panel.* Ratio between the numerical result and the BBKS transfer function

predictions, and we have to in order to keep the pace with the increasing sophistication of the observational techniques, we must take into account them. A more precise fitting formula (to numerical calculations performed with CMBFAST) taking into account baryons and neutrino anisotropic stress is given by Eisenstein and Hu (1998).

In Fig. 9.13 we compare the BBKS transfer function with the numerical calculation of CLASS, adopting $\Omega_{m0} = 0.95$ while leaving all the remaining cosmological parameters as in the ΛCDM model, except for Ω_Λ which is adjusted to the value $\Omega_\Lambda = 1.632908 \times 10^{-3}$ in order to match the correct budget of energy density. So, in practice, we are neglecting DE.

As can be appreciated from the plots, the BBKS transfer function overestimates of about 5% the correct transfer function for scales $k \gtrsim 0.1\, h\, \mathrm{Mpc}^{-1}$, see also Dodelson (2003), p. 208. The reason is that baryons behave like radiation before decoupling,

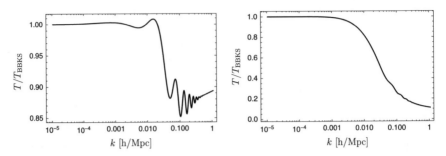

Fig. 9.14 *Left Panel.* Ratio between the numerical result computed with CLASS assuming the ΛCDM model, and the BBKS transfer function with $\Omega_{m0}h^2 = 0.12038$. *Right Panel.* Ratio between the numerical result computed with CLASS assuming the ΛCDM model, and the BBKS transfer function with $\Omega_{m0} = 1$

because of their tight coupling due to Thomson scattering. Therefore, they contribute further to thwart scales entering the horizon before the equivalence.

In Fig. 9.14 we display the ratio between the numerical transfer function computed with CLASS for the ΛCDM model and the BBKS transfer function in two cases. In the left panel, the same matter density parameter of the ΛCDM model is used for both the transfer functions. In the right panel, we used the BBKS transfer function with $\Omega_{m0} = 1$.

The latter choice was made in order to reproduce the plot of Dodelson (2003), p. 208 and thus to show the very large correction due to the cosmological constant which, if not taken into account, leads to a 80% error at the scale $k = 0.1\,h\,\mathrm{Mpc}^{-1}$.

When Λ is taken into account properly, the BBKS transfer function still overestimates the correct transfer function of at least a 10% on small scales, which implies an imprecision of 1% in the power spectrum (since this depends on the squared transfer function). Moreover, as it is obvious since it does not include baryons, the BBKS cannot describe the BAO, which appear as oscillations in the transfer function at about $k = 0.1\,h\,\mathrm{Mpc}^{-1}$ in Fig. 9.14.

9.6 The Transfer Function for Tensor Perturbations

Treating the evolution of tensor perturbations, which we have done in the context of inflation in Chap. 8 is much easier than for scalar modes because of two reasons. The first is that CDM and baryons are very non-relativistic and so do not possess anisotropic stress, which would source GW. The second reason is that, though photons and neutrinos do have anisotropic stresses this are always very small so that it is a good approximation to neglect them.

Therefore, what we have to do is to recover the calculations of Sect. 8.4 and solve Eq. (8.67) with a $\mathcal{H}$ given in the radiation- and matter-dominated epoch. As we already know, on very large scales, meaning $k^2 \ll a''/a$, h is a constant (we shall call it h_P) and this holds true for whatever background expansion we choose. We used this fact in order to determine the primordial power spectrum and in the present section we use it again to determine the initial value for solving Eq. (8.67).

Using Eqs. (9.40) and (9.41), we can write the conformal Hubble parameter as follows:

$$\mathcal{H} = \frac{\mathcal{H}_{eq}\sqrt{1+y}}{\sqrt{2y}}, \tag{9.119}$$

where we have introduced again $y \equiv \rho_m/\rho_r$ as new independent variable.

Exercise 9.19 Cast Eq. (8.67) using y as independent variable and the above form of the conformal Hubble factor. Show that:

$$(1+y)\frac{d^2h}{dy^2} + \left[\frac{1}{2} + \frac{2(1+y)}{y}\right]\frac{dh}{dy} + \kappa^2 h = 0 \tag{9.120}$$

using also Eq. (9.100).

The above equation cannot be solved analytically, but we can find an exact solution in the same four instances that we worked out for scalar perturbations. On super-horizon scales we already know that $h = h_P$ irrespective of the background evolution (this is relevant only for the decaying mode), so we skip this case.

9.6.1 Radiation-Dominated Epoch

In this case, we consider $y \ll 1$ in Eq. (9.120):

$$\frac{d^2h}{dy^2} + \frac{2}{y}\frac{dh}{dy} + \kappa^2 h = 0, \tag{9.121}$$

which can be rewritten as:

$$\frac{d^2(\kappa h y)}{d(\kappa y)^2} + \kappa h y = 0, \tag{9.122}$$

and hence is a harmonic oscillator equation for the quantity $\kappa h y$ with respect to the variable κy, with unitary frequency. Hence, the solution for h is:

$$h(k, y) = C_1(k)\frac{\sin(\kappa y)}{\kappa y} + C_2(k)\frac{\cos(\kappa y)}{\kappa y}, \tag{9.123}$$

where as usual C_1 and C_2 are generic k-dependent functions. For $y \to 0$, or equivalently on very large scales $\kappa y \to 0$ only the sine tends to a finite result and hence we must put C_2 to zero:

$$h(k, y) = h_{\mathrm{P}}(k)\frac{\sin(\kappa y)}{\kappa y} = h_{\mathrm{P}}(k) j_0(\kappa y) . \tag{9.124}$$

Equivalently, putting $\mathcal{H} = 1/\eta$ into the original equation (8.67) one finds:

$$h'' + \frac{2}{\eta}h' + k^2 h = 0 , \tag{9.125}$$

and so:

$$h(k, \eta) = h_{\mathrm{P}}(k)\frac{\sin(k\eta)}{k\eta} = h_{\mathrm{P}}(k) j_0(k\eta) , \tag{9.126}$$

since, indeed, $\kappa y = k\eta$ when radiation dominates, cf. Eq. (9.102).

9.6.2   Matter-Dominated Epoch

In this case, we consider $y \gg 1$ in Eq. (9.120):

$$y\frac{d^2 h}{dy^2} + \frac{5}{2}\frac{dh}{dy} + \kappa^2 h = 0 , \tag{9.127}$$

or, using Eq. (8.67) with $\mathcal{H} = 2/\eta$:

$$h'' + \frac{4}{\eta}h' + k^2 h = 0 . \tag{9.128}$$

This equation can be cast in the form of a Bessel equation.

Exercise 9.20 Introduce the new function:

$$h \equiv g(k\eta)^\alpha, \tag{9.129}$$

where α is a number to be determined. Then, show that:

$$\eta^2 g'' + (2\alpha + 4)\eta g' + g[k^2\eta^2 + \alpha^2 + 3\alpha] = 0 . \tag{9.130}$$

Then, we recover the form of the Bessel function for $\alpha = -3/2$, for which the order is $3/2$ and hence:

$$g \propto J_{3/2}(k\eta) = \sqrt{\frac{2k\eta}{\pi}} j_1(k\eta) \,, \tag{9.131}$$

where we have introduced the spherical Bessel function and neglected the one of second kind, since it diverges for $k\eta \to 0$. The GW amplitude thus evolves as:

$$h(k, \eta) = C_1(k) \frac{j_1(k\eta)}{k\eta} = C_1(k) \left[\frac{\sin(k\eta)}{(k\eta)^3} - \frac{\cos(k\eta)}{(k\eta)^2} \right] . \tag{9.132}$$

In the limit $k\eta \to 0$:

$$\frac{\sin(k\eta)}{(k\eta)^3} - \frac{\cos(k\eta)}{(k\eta)^2} \to \frac{1}{3} \,, \tag{9.133}$$

hence we can write:

$$h(k, \eta) = 3h_P(k) \left[\frac{\sin(k\eta)}{(k\eta)^3} - \frac{\cos(k\eta)}{(k\eta)^2} \right] \,, \tag{9.134}$$

but this is valid only for those modes which enters the horizon well deep into the matter-dominated epoch. Using the definition of κ in Eq. (9.100) and the fact that, deep in the matter-dominated epoch we have that:

$$\mathcal{H} = \frac{2}{\eta} = \frac{\mathcal{H}_{eq}}{\sqrt{2y}} \,, \tag{9.135}$$

we can conclude that:

$$k\eta = 2\kappa\sqrt{y} \,, \tag{9.136}$$

and thus the solution for $h(k, y)$ is:

$$h(k, y) = \frac{3h_P(k)}{4} \left[\frac{\sin(2\kappa\sqrt{y})}{2(\kappa\sqrt{y})^3} - \frac{\cos(2\kappa\sqrt{y})}{(\kappa\sqrt{y})^2} \right] \,, \tag{9.137}$$

which again is valid only for those modes $\kappa\sqrt{y} \ll 1$ and hence, since $y \gg 1$, then $\kappa \ll 1$, i.e. modes which entered the horizon well deep into the matter-dominated epoch.

9.6.3 Deep Inside The Horizon

In the case $k \gg \mathcal{H}$ we work directly on Eq. (8.67) and, following Weinberg (2008), we introduce a new variable

$$x \equiv \int \frac{d\eta}{a^2} \,, \tag{9.138}$$

with the purpose of eliminating the first-order derivative, and obtain thus the following equation:

$$\frac{d^2 h}{dx^2} + k^2 a^4 h = 0 .$$ (9.139)

The latter is not analytically solvable for a generic $a(x)$ but it is possible to find an approximated, WKB solution. Indeed, if $a(x)$ is approximately constant, the above equation becomes that of a harmonic oscillator and the solution is simply:

$$h(k, x) \propto e^{\pm i k a^2} .$$ (9.140)

Taking into account the x dependence of a, we can use the following ansatz:

$$h(k, x) = A(x) \exp\left(\pm i k \int a^2 dx\right) ,$$ (9.141)

and substitute it back into Eq. (9.139), obtaining:

$$\frac{d^2 A}{dx^2} + 2\frac{dA}{dx}(\pm i k a^2) + A\left(\pm 2 i k a \frac{da}{dx}\right) + k^2 a^4 A = 0 .$$ (9.142)

Equating the imaginary part to zero, we have then:

$$\frac{dA}{dx} a^2 + A a \frac{da}{dx} = 0 ,$$ (9.143)

for which the solution is:

$$A(x) \propto 1/a(x) .$$ (9.144)

Hence, the approximated WKB solution for the GW is:

$$h(k, \eta) \propto \frac{1}{a} \exp\left(\pm i k \int a^2 dx\right) = \frac{1}{a} \exp\left(\pm i k \eta\right) .$$ (9.145)

Note that this solution is valid for any background expansion, since we have made no assumptions on the latter. Moreover, note the oscillatory behaviour $\exp(\pm i k \eta)$ with amplitude damped with a factor $1/a$. Recall from Weinberg (1972) that the energy-momentum tensor of a GW is:

$$\langle t_{\mu\nu} \rangle = \frac{p_\mu p_\nu}{16\pi G} \left(|h_+|^2 + |h_\times|^2\right) ,$$ (9.146)

where $t_{\mu\nu}$ is the gravitational pseudo-tensor, the average on it is on a sufficiently large region of spacetime such that the oscillatory terms gives a constant result and p_μ is the physical momentum $p_\mu = dx_\mu/d\lambda$ of the GW (of the graviton). We now that in a FLRW metric for a massless particle $p_0 = p \propto 1/a$ and hence $\langle t_{00} \rangle \propto 1/a^4$ as we expected for the energy density of a relativistic species.

Now, we can match the above WKB solution (for $y \ll 1$) with the one we found in Eq. (9.126) (for large κy), where we use y instead of η because it is straightforward then to project the matching at late times. It is the same procedure we employed in order to find the transfer function for CDM. So, we have:

$$\frac{C_1(k)}{y a_{eq}} \sin(\kappa y) = h_P(k) \frac{\sin(\kappa y)}{\kappa y} , \qquad (9.147)$$

and thus

$$C_1(k) = h_P(k) \frac{a_{eq}}{\kappa} = h_P(k) \frac{a_{eq} H_0 \Omega_{m0}}{k \sqrt{\Omega_{r0}}} = h_P(k) \frac{H_0 \sqrt{\Omega_{r0}}}{k} . \qquad (9.148)$$

So, the solution for small scales is:

$$\boxed{h(k, \eta) = h_P(k) \frac{H_0 \sqrt{\Omega_{r0}}}{k a(\eta)} \sin(k\eta)} \qquad (9.149)$$

Note that this solution is valid for any content of the universe, since there is no more a y dependence appearing (any content after radiation-domination, since we have matched there the solutions). Close to the present time η_0 we can set as usual $a(\eta_0) = 1$ and then the gravitational wave solution for small scales is then:

$$\boxed{h(k, t) = h_P(k) \frac{H_0 \sqrt{\Omega_{r0}}}{k} \sin\left[k\eta_0 + k(t - t_0)\right]} \qquad (9.150)$$

for t close to t_0 and where we have used $d\eta = dt/a(t)$ close to present time. This GW profile is very different from those detected by LIGO for the merging of black holes, the first in Abbott et al. (2016), and from the spectacular event GW170817 of the merging of two neutron stars detected both by LIGO and VIRGO (Abbott et al. 2017) and whose electromagnetic counterpart was also seen as a short gamma-ray burst (GRB) (Abbott et al. 2017). The main characteristics of these profiles is their growth in frequency and amplitude for times close to the merging one. The cosmological GW profile is a simple sine function (for very small scales), with an amplitude containing cosmological information of great relevance such as the primordial amplitude, which we have seen to be related to the energy scale of inflation.

References

Abbott, B., et al.: GW170817: observation of gravitational waves from a binary neutron star inspiral. Phys. Rev. Lett. **119**(16), 161101 (2017)

Abbott, B.P., et al.: Observation of gravitational waves from a binary black hole merger. Phys. Rev. Lett. **116**(6), 061102 (2016)

Abbott, B.P., et al.: Multi-messenger observations of a binary neutron star merger. Astrophys. J. **848**(2), L12 (2017)

Abramowitz, M., Stegun, I.A.: Handbook of Mathematical Functions with Formulas, Graphs, and Mathematical Tables. Dover, New York (1972)

Bardeen, J.M., Bond, J.R., Kaiser, N., Szalay, A.S.: The statistics of peaks of Gaussian random fields. Astrophys. J. **304**, 15–61 (1986)

Dodelson, S.: Modern Cosmology. Academic Press, Amsterdam (2003)

Efstathiou, G., Sutherland, W.J., Maddox, S.J.: The cosmological constant and cold dark matter. Nature **348**, 705–707 (1990)

Eisenstein, D.J., Hu, W.: Baryonic features in the matter transfer function. Astrophys. J. **496**, 605 (1998)

Kaiser, N.: On the spatial correlations of Abell clusters. Astrophys. J. **284**, L9–L12 (1984)

Kodama, H., Sasaki, M.: Cosmological perturbation theory. Prog. Theor. Phys. Suppl. **78**, 1–166 (1984)

Maddox, S.J., Efstathiou, G., Sutherland, W.J., Loveday, J.: Galaxy correlations on large scales. Mon. Not. R. Astron. Soc. **242**, 43–49 (1990)

Meszaros, P.: The behaviour of point masses in an expanding cosmological substratum. Astron. Astrophys. **37**, 225–228 (1974)

Mukhanov, V.: Physical Foundations of Cosmology. Cambridge University Press, Cambridge (2005)

Mukhanov, V.F., Feldman, H.A., Brandenberger, R.H.: Theory of cosmological perturbations. Part 1. Classical perturbations. Part 2. Quantum theory of perturbations. Part 3. Extensions. Phys. Rept. **215**, 203–333 (1992)

Percival, W.J., et al.: The shape of the SDSS DR5 galaxy power spectrum. Astrophys. J. **657**, 645–663 (2007)

Piattella, O.F., Martins, D.L.A., Casarini, L.: Sub-horizon evolution of cold dark matter perturbations through dark matter-dark energy equivalence epoch. JCAP **1410**(10), 031 (2014)

Weinberg, S.: Gravitation and Cosmology: Principles and Applications of the General Theory of Relativity. Wiley, New York (1972)

Weinberg, S.: Cosmological fluctuations of short wavelength. Astrophys. J. **581**, 810–816 (2002)

Weinberg, S.: Cosmology. Oxford University Press, Oxford (2008)

Zeldovich, Y.B.: Structure of the Universe. Astrophys. Space Phys. Rev. **3**, 1 (1984)

Chapter 10
Anisotropies in the Cosmic Microwave Background

> Long the realm of armchair philosophers, the study of the
> origins and evolution of the universe became a physical science
> with falsifiable theories
>
> Wayne Hu, PhD Thesis

In this chapter we attack the hierarchy of Boltzmann equations that we have found for photons and present an approximate, semi-analytic solution which will allow us to understand the temperature correlation in the CMB sky and its relation with the cosmological parameters. Our scope is to understand the features of the angular, temperature-temperature power spectrum in Fig. 10.1.

Note that in this plot the definition

$$\mathcal{D}_\ell^{TT} \equiv \frac{\ell(\ell+1)C_{TT,\ell}}{2\pi} , \tag{10.1}$$

is used. We shall see the reason for the $\ell(\ell+1)$ normalisation, whereas the $C_{TT,\ell}$'s are given in Eq. (7.81) as functions of the multipole moments of the temperature distribution and the primordial power spectrum for scalar perturbations.

In Fig. 10.1 we can also see data points up to $\ell \approx 2500$. What can we say from this number about the angular sensitivity of *Planck*? It can be roughly computed as follows. For a given $\ell_{\max}$ how many realisations of $a_{\ell m}$ do we have?

Exercise 10.1 For each ℓ we have $2\ell + 1$ possible values of m, thus show that:

$$N_{\ell_{\max}} = \sum_{\ell=0}^{\ell_{\max}} (2\ell + 1) = (\ell_{\max} + 1)^2 . \tag{10.2}$$

© Springer International Publishing AG, part of Springer Nature 2018
O. Piattella, *Lecture Notes in Cosmology*, UNITEXT for Physics,
https://doi.org/10.1007/978-3-319-95570-4_10

Fig. 10.1 CMB TT
spectrum. Figure taken from
Ade et al. (2016). The red
solid line is the best fit
ΛCDM model

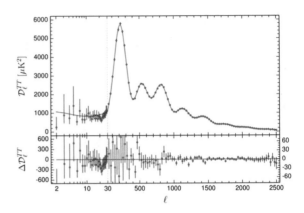

The full sky has:

$$4\pi \ \text{rad}^2 = \frac{4}{\pi}(180 \ \text{deg})^2 \approx 41000 \ \text{deg}^2 \ . \qquad (10.3)$$

If an experiment has sensitivity of 7 deg, then we can have at most

$$\frac{4}{\pi}(180/7)^2 \approx 842 \ , \qquad (10.4)$$

pieces of independent information and therefore we can determine as many $a_{\ell m}$.
This gives $\ell_{\max} \approx 28$ and it was the sensitivity of *CoBE*. For *Planck*, the angular
sensitivity was of 5 arcmin, which corresponds to

$$\frac{4}{\pi}(180 \times 60/5)^2 \approx 10^6 \ , \qquad (10.5)$$

pieces of independent information and then to $\ell_{\max} = 2436$.

In this chapter we omit the superscript S referring to the scalar perturbations
contribution to Θ, since most of the time we shall discuss of it. We shall only use T
in order to distinguish the tensor contribution.

10.1 Free-Streaming

It is convenient to start neglecting the collisional term in the Boltzmann equation and
considering thus the phase of photon **free-streaming**. The following discussion is
similar to the one in Sect. 5.4. Consider Eq. (5.27) for photons and with no collisional
term. Using the definition of Θ in Eq. (5.80), we can write for scalar perturbations:

$$\left(\frac{\partial}{\partial\eta} + \frac{dx^i}{d\eta}\frac{\partial}{\partial x^i}\right)(\Theta + \Psi) = \Psi' - \Phi' . \tag{10.6}$$

As we know from Boltzmann equation, the differential operator on the left hand side is a convective derivative, i.e. a derivative along the photon path:

$$\frac{d}{d\eta}(\Theta + \Psi) = \Psi' - \Phi' , \tag{10.7}$$

whose inversion is the basis of the **line-of-sight integration** approach to CMB anisotropies (Seljak and Zaldarriaga 1996), which is an alternative to attacking the hierarchy of coupled Boltzmann equations (which still must be attacked but can be truncated at much lower ℓ's) as it was done e.g. in Ma and Bertschinger (1995). We shall see this technique in some detail in Sect. 10.5.

For time-independent potentials, as they are in the matter-dominated epoch, the collisionless Boltzmann equation for photons tells us that $\Theta + \Psi$ is constant along the photons paths, i.e. along our past light-cone, since recombination.

Recall that the scalar-perturbed metric that we are using is given in Eq. (4.171):

$$ds^2 = -a^2(\eta)(1 + 2\Psi)d\eta^2 + a^2(\eta)(1 + 2\Phi)\delta_{ij}dx^i dx^j . \tag{10.8}$$

Inside a potential well, Ψ is negative. In order to be convinced of this one has just to think about the Newtonian limit and realise that 2Ψ is the Newtonian gravitational potential, hence negative. So, since $\Theta + \Psi$ stays constant, we have that:

$$\Theta(\eta_*, \mathbf{x}_*, \hat{p}) + \Psi(\eta_*, \mathbf{x}_*) = \Theta(\eta_0, \mathbf{x}_0, \hat{p}) + \Psi(\eta_0, \mathbf{x}_0) . \tag{10.9}$$

where on the left hand side we have chosen the quantities at recombination whereas on the right hand side we have chosen the present time. Note that $\mathbf{x}_0$, is where our laboratory (the CMB experiment) is, i.e. Earth, and as such is fixed. Therefore, since we can only detect photons on our past light-cone, and those from CMB comes from a fixed comoving distance $r_* = \eta_0 - \eta_*$, we have that:

$$\mathbf{x}_* = \mathbf{x}_0 - r_*\hat{p} . \tag{10.10}$$

Note that $\hat{p}$ is the photon direction and so it is opposite to the direction of the line of sight $\hat{n} = -\hat{p}$. So, the only independent variables are 2, the components of $\hat{p}$. They become just a single one, μ, because of the way in which we factorise the azimuthal dependence (and assuming axial symmetry).

The potential $\Psi(\eta_0, \mathbf{x}_0)$ is usually neglected, or incorporated in the potential at recombination, since it is not detectable. As it is well known, classically only potential differences are physically meaningful. The above equation then tells us that:

$$\Theta(\eta_*, \mathbf{x}_0 - r_*\hat{p}, \hat{p}) + \Psi(\eta_*, \mathbf{x}_0 - r_*\hat{p}) = \Theta(\eta_0, \mathbf{x}_0, \hat{p}) , \tag{10.11}$$

i.e. the observed temperature fluctuation (on the right hand side) accounts for the energy loss due to climbing out the potential well or falling down a potential hill. This is the so-called **Sachs-Wolfe effect** (Sachs and Wolfe 1967). Writing the above equation in Fourier modes, we have:

$$\int \frac{d^3\mathbf{k}}{(2\pi)^3} \Theta(\eta_0, \mathbf{k}, \hat{p}) e^{i\mathbf{k}\cdot\mathbf{x}_0} = \int \frac{d^3\mathbf{k}}{(2\pi)^3} \left[\Theta(\eta_*, \mathbf{k}, \hat{p}) + \Psi(\eta_*, \mathbf{k})\right] e^{i\mathbf{k}\cdot(\mathbf{x}_0 - r_*\hat{p})} .$$
(10.12)

We can set now $\mathbf{x}_0 = 0$, without losing of generality, and manifest the dependences as k and μ, the former since we normalise to the scalar primordial mode $\alpha(\mathbf{k})$, cf. Eq. (7.64), and the latter since we are considering axisymmetric scalar perturbations. Hence we have for the Fourier modes:

$$\Theta(\eta_0, k, \mu) = \left[\Theta(\eta_*, k, \mu) + \Psi(\eta_*, k)\right] e^{-ik\mu r_*} .$$
(10.13)

Using the partial wave expansion, we get:

$$\Theta_\ell(\eta_0, k) = \frac{1}{(-i)^\ell} \int_{-1}^{1} \frac{d\mu}{2} \mathcal{P}_\ell(\mu) \left[\Theta(\eta_*, k, \mu) + \Psi(\eta_*, k)\right] e^{-ik\mu r_*} ,$$
(10.14)

and using the relation:

$$\int_{-1}^{1} \frac{d\mu}{2} \mathcal{P}_\ell(\mu) e^{-ik\mu r_*} = (-i)^\ell j_\ell(kr_*) ,$$
(10.15)

which can be obtained by inverting the expansion of Eq. (5.75), we can write:

$$\Theta_\ell(\eta_0, k) = \Psi(\eta_*, k) j_\ell(kr_*) + \frac{1}{(-i)^\ell} \int_{-1}^{1} \frac{d\mu}{2} \mathcal{P}_\ell(\mu) \Theta(\eta_*, k, \mu) e^{-ik\mu r_*} .$$
(10.16)

Using again the partial wave expansion, we can write the above formula as:

$$\Theta_\ell(\eta_0, k) = \Psi(\eta_*, k) j_\ell(kr_*)$$
$$+ \frac{1}{(-i)^\ell} \sum_{\ell'} (-i)^{\ell'} (2\ell' + 1) \Theta_{\ell'}(\eta_*, k) \int_{-1}^{1} \frac{d\mu}{2} \mathcal{P}_\ell(\mu) \mathcal{P}_{\ell'}(\mu) e^{-ik\mu r_*} .$$
(10.17)

We shall see later that, because of tight-coupling, the monopole and the dipole contribute the most at recombination. Hence, we can write, truncating the summation at $\ell' = 1$:

$$\Theta_\ell(\eta_0, k) = (\Theta_0 + \Psi)(\eta_*, k) j_\ell(kr_*) + \frac{3\Theta_1(\eta_*, k)}{(-i)^{\ell-1}} \int_{-1}^{1} \frac{d\mu}{2} \mathcal{P}_\ell(\mu) \mu e^{-ik\mu r_*} .$$
(10.18)

The integral can be performed as follows:

$$\int_{-1}^{1} \frac{d\mu}{2} \mathcal{P}_\ell(\mu)\mu e^{-ik\mu r_*} = i \frac{d}{d(kr_*)} \int_{-1}^{1} \frac{d\mu}{2} \mathcal{P}_\ell(\mu) e^{-ik\mu r_*} = \frac{1}{i^{\ell-1}} \frac{d}{d(kr_*)} j_\ell(kr_*) \ .$$

(10.19)

The same technique can be used, in principle, to calculate the integral for any ℓ': for each power of μ one gains a derivative of the spherical Bessel function. Recalling the formula (Abramowitz and Stegun 1972)[1]:

$$\frac{dj_\ell(x)}{dx} = j_{\ell-1}(x) - \frac{\ell+1}{x} j_\ell(x) \ ,$$

(10.20)

we can write:

$$\Theta_\ell(\eta_0, k) = (\Theta_0 + \Psi)\,(\eta_*, k)\, j_\ell(kr_*)$$
$$+3\Theta_1(\eta_*, k)\left[j_{\ell-1}(kr_*) - \frac{\ell+1}{kr_*} j_\ell(kr_*) \right] \ .$$

(10.21)

So, the spherical Bessel functions that we have mentioned in Chap. 9 start to appear. We have obtained the above free-streaming solution neglecting the potentials derivatives in Eq. (10.7). Taking them into account is not difficult, since an additional piece containing the integration of the potential derivatives would appear in Eq. (10.13):

$$\Theta(\eta_0, k, \mu) = [\Theta(\eta_*, k, \mu) + \Psi(\eta_*, k)]\, e^{-ik\mu r_*} + \int_{\eta_*}^{\eta_0} d\eta\,(\Psi' - \Phi')(\eta, k) e^{-ik\mu(\eta_0-\eta)} \ .$$

(10.22)

The exponential factor in the integral comes from the Fourier transform of the potentials and from considering:

$$\mathbf{x} = \mathbf{x}_0 - (\eta_0 - \eta)\hat{p} \ ,$$

(10.23)

at any given time η along the photon trajectory (this is the "line of sight", in practice). Performing again the expansion in partial waves, we get:

$$\Theta_\ell(\eta_0, k) = (\Theta_0 + \Psi)\,(\eta_*, k)\, j_\ell(kr_*)$$
$$+3\Theta_1(\eta_*, k)\left[j_{\ell-1}(kr_*) - \frac{\ell+1}{kr_*} j_\ell(kr_*) \right]$$
$$+ \int_{\eta_*}^{\eta_0} d\eta\,[\Psi'(\eta, k) - \Phi'(\eta, k)] j_\ell(kr) \ ,$$

(10.24)

where

$$r \equiv \eta_0 - \eta \ .$$

(10.25)

As we are going to see, the first two terms of the above formula contains the **primary anisotropies** of the CMB, which are the **acoustic oscillations** and the **Doppler effect**. The $\Psi(\eta_*, k)$ contribution in the first term is, as we have already anticipated,

---

[1]The website http://functions.wolfram.com is also very useful.

Fig. 10.2 Evolution of the spherical Bessel function $j_{10}^2(x)$

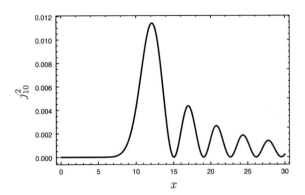

the **Sachs-Wolfe effect**. The last term is the **Integrated Sachs-Wolfe** (ISW) effect (Sachs and Wolfe 1967) and contributes only when the gravitational potentials are time-varying. This happens, as we have seen in Chap. 9, when radiation and DE are relevant. For this reason the ISW effect is usually separated in the early-times one, due to a small presence of radiation still at decoupling, and in the late-times one, due to DE.

Once we know all the contributions of the above formula, we can use Eq. (7.81) and provide the prediction on the $C_{TT,\ell}^S$ spectrum.

The presence of the spherical Bessel function is interesting for two reasons, which we display in Fig. 10.2 for the arbitrary choice $\ell = 10$.

We have chosen to plot the squared spherical Bessel function because it is the relevant window function when computing the C_ℓ's, as we shall see briefly. First, the maximum value is attained roughly when $x \approx \ell$ and for $x < \ell$ the spherical Bessel function is practically vanishing. Therefore, for a given multipole ℓ the scale which contribute most for the observed anisotropy is:

$$k \approx \frac{\ell}{\eta_0 - \eta_*} . \tag{10.26}$$

We have anticipated this already in Chap. 9. The second reason of interest is that the spherical Bessel function goes to zero for large x. This means that scales such that $kr_* \gg 1$ do not contribute to the observed anisotropy. Physically, this is an effect due to the free-streaming phase for which, on very small scales, hot and cold photons mix up destroying thus the anisotropy.

We have thus seen that the predicted anisotropy today is given by formula Eq. (10.24). We have now to justify the fact of considering only the monopole and the dipole at recombination. We shall commence in the next section discussing very large scales.

In principle, Eq. (10.24) has the very same form for neutrinos, but with an initial conformal time η_i which is well anterior to η_*, since neutrinos do not interact and therefore they only free-stream (at least for temperatures of the primordial plasma below 1 MeV).

10.2 Anisotropies on Large Scales

On large scales, i.e. $k\eta \ll 1$, the relevant equations are those of Chap. 6, which we report here:

$$\delta'_\gamma = -4\Phi' , \quad \delta'_\nu = -4\Phi' , \quad \delta'_c = -3\Phi' , \quad \delta'_b = -3\Phi' , \tag{10.27}$$

i.e. only the monopoles are relevant. Since we want to describe CMB, let us focus on the photon density contrast, which can be written as:

$$\delta_\gamma(k, \eta) = 4\Theta_0(k, \eta) , \tag{10.28}$$

introducing the monopole of the temperature fluctuation. The equation $\Theta'_0 = -\Phi'$ can be immediately integrated, obtaining:

$$\Theta_0(k, \eta) = -\Phi(k, \eta) + C_\gamma(k) . \tag{10.29}$$

For the adiabatic primordial mode, the only which we are going to consider, we know from Eq. (6.94) that $C_\gamma(k) = \Phi_P(k) - \Psi_P(k)/2$ and thus:

$$\Theta_0(k, \eta) = -\Phi(k, \eta) + \Phi_P(k) - \frac{1}{2}\Psi_P(k) . \tag{10.30}$$

As we know from Chap. 9, we can consider the gravitational potentials to be equal in modulus and on large scales $\Phi(k, \eta)$ is independent of time and since recombination $\eta_* \gg \eta_{eq}$ takes place well after radiation-matter equality, we know that $\Phi(k, \eta_*) = 9\Phi_P(k)/10$, i.e. the value of the gravitational potential drops of 10% in passing through radiation-matter domination. Therefore:

$$\Theta_0(k, \eta_*) = \frac{3}{5}\Phi_P(k) = \frac{2}{3}\Phi(k, \eta_*) = -\frac{2}{3}\Psi(k, \eta_*) . \tag{10.31}$$

As we saw earlier in Eq. (10.24), the observed anisotropy is not $\Theta_0(k, \eta_*)$ but $\Theta_0(k, \eta_*) + \Psi(k, \eta_*)$, because of the gravitational redshift. Again, this is the Sachs-Wolfe effect, amounting to a shift in the photons frequency when they decouple from the baryonic plasma depending whether they are in a well or hill of the gravitational potential. So, we have from Eq. (10.31) that:

$$(\Theta_0 + \Psi)(k, \eta_*) = \frac{1}{3}\Psi(k, \eta_*) . \tag{10.32}$$

On the other hand, for δ_c we know that

$$\delta_c(k, \eta) = -3\Phi(k, \eta) + \frac{9\Phi_P(k)}{2} , \tag{10.33}$$

again assuming adiabatic primordial modes. Using again $\Phi(k, \eta_*) = 9\Phi_P(k)/10$, we get:

$$\delta_c(k, \eta_*) = 2\Phi(k, \eta_*) = -2\Psi(k, \eta_*) . \tag{10.34}$$

The fluctuations in CDM contribute more in generating the potential wells than photons, a factor 2 against a factor $-2/3$. Combining the two equations:

$$\boxed{(\Theta_0 + \Psi)(k, \eta_*) = -\frac{\delta_c(k, \eta_*)}{6}} \tag{10.35}$$

This result tells us that on large scales colder spots represent larger overdensities, a counter-intuitive result. One expects hotter photons the deeper the well is and in fact this is the case with just $\Theta_0(k, \eta_*)$, since we have:

$$\Theta_0(k, \eta_*) = -\frac{2}{3}\Psi(k, \eta_*) = \frac{\delta_c(k, \eta_*)}{3} , \tag{10.36}$$

i.e. the larger the CDM overdensity, the larger the well and $\Theta_0(k, \eta_*)$ are. However, photons' response to the gravitational potential is only a factor $-2/3$ whereas the gravitational redshift adds a Ψ contribution, changing thus the sign of the observed anisotropy. In the limit of $\delta_c \to -1$, one gets $(\Theta_0 + \Psi)(k, \eta_*) \to 1/6$, so cosmic voids correspond to hot spots!

The results found here are valid only on large scales, i.e. for $k\eta_* \ll 1$, scales much larger than the horizon at recombination, which has an angular size of approximatively 1 degree. Moreover, they also depend on the choice of initial conditions. We have opted for the adiabatic ones, as usual.

Exercise 10.2 Reproduce the above argument for the other primordial modes.

Let us use the theoretical prediction on the $C^S_{TT,\ell}$ given in Eq. (7.81) together with the first contribution only from Eq. (10.24). The latter approximation is justified by the fact that we are considering large scales, hence the dipole contribution is negligible and the ISW effect is vanishing because the potentials are constant. Since:

$$(\Theta_0 + \Psi)(k, \eta_*) = \frac{1}{3}\Psi(k, \eta_*) = -\frac{1}{3}\Phi(k, \eta_*) = -\frac{3}{10}\Phi_P(k) = -\frac{1}{5}\mathcal{R}(k) , \tag{10.37}$$

the transfer function is just the constant $-1/5$ (recall that we are neglecting the neutrino fraction R_ν) and thus the angular power spectrum is:

$$C^S_{TT,\ell}(\text{SW}) = \frac{4\pi}{25} \int_0^\infty \frac{dk}{k} \Delta^2_{\mathcal{R}}(k) j^2_\ell(k\eta_0) , \tag{10.38}$$

since $\eta_* \ll \eta_0$. Note that $k\eta_* \ll k\eta_0$ and we have seen in Fig. 10.2 that the spherical Bessel function contributes the most about $k\eta_0 \approx \ell$. Thus, for small ℓ, i.e. large angular scales, $k\eta_0$ is small and $k\eta_*$ is very small, where in fact $| (\Theta_0 + \Psi) (k, \eta_*)|^2$ is constant. In other words, the above approximation is valid for small ℓ, typically $\ell \lesssim 30$.

In the above integral we can look at $j_\ell^2(k\eta_0)$ as a very peaked window function and approximate it as:

$$C_{TT,\ell}^S(\mathrm{SW}) \approx \frac{4\pi}{25} \Delta_{\mathcal{R}}^2(\ell/\eta_0) \int_0^\infty \frac{dk}{k} j_\ell^2(k\eta_0) \,. \tag{10.39}$$

Using the result:

$$\int_0^\infty \frac{dx}{x} j_\ell^2(x) = \frac{1}{2\ell(\ell+1)} \,, \tag{10.40}$$

we have then:

$$\frac{\ell(\ell+1)C_{TT,\ell}^S(\mathrm{SW})}{2\pi} \approx \frac{1}{25} \Delta_{\mathcal{R}}^2(\ell/\eta_0) \,. \tag{10.41}$$

Hence, for a scale-invariant spectrum $n_S = 1$ the combination $\ell(\ell+1)C_{TT,\ell}^S(\mathrm{SW})$ is constant and it is called **Sachs-Wolfe plateau**. This also explains why CMB power spectra are usually presented with the $\ell(\ell+1)$ normalisation, as in Fig. 10.1.

If $n_S \neq 1$, then $\ell(\ell+1)C_{TT,\ell}^S(\mathrm{SW})$ is proportional to ℓ^{n_S-1}, i.e. the **primordial tilt** in the power spectrum leaves its mark in a tilted plateau for small ℓ.

10.3 Tight-Coupling and Acoustic Oscillations

We have seen that in order to determine the prediction on the present time $C_{TT,\ell}$'s we need to know what happens at recombination. We devote this section to such purpose, showing that the monopole and the dipole contribute the most.

Let us recover here the hierarchy of Boltzmann equations for the Θ_ℓ's (not taking into account polarisation) that we have derived in Chap. 5:

$$(2\ell+1)\Theta_\ell' + k[(\ell+1)\Theta_{\ell+1} - \ell\Theta_{\ell-1}] = (2\ell+1)\tau'\Theta_\ell \,, \quad (\ell > 2) \,, \tag{10.42}$$
$$10\Theta_2' + 2k(3\Theta_3 - 2\Theta_1) = 10\tau'\Theta_2 - \tau'\Pi \,, \tag{10.43}$$
$$3\Theta_1' + k(2\Theta_2 - \Theta_0) = k\Psi + \tau'(3\Theta_1 - V_b) \,, \tag{10.44}$$
$$\Theta_0' + k\Theta_1 = -\Phi' \,, \tag{10.45}$$

where recall that $\delta_\gamma = 4\Theta_0$ and $3\Theta_1 = V_\gamma$. The best way to deal with these equations is to solve them numerically by using Boltzmann codes such as CAMB or CLASS, but in this way the physics behind the $C_{TT,\ell}$'s remains hidden or unclear. For this reason we attack these equations in an approximate fashion, but analytically.

We take the limit $-\tau' \gg \mathcal{H}$, which is called **tight-coupling** (TC) approximation. This limit physically means that the Thomson scattering rate between photons and electrons is much larger than the Hubble rate until recombination and then drops abruptly since the free electron fraction X_e goes to zero very rapidly, as we have seen when studying thermal history in Chap. 3. We shall first consider the case of **sudden recombination**, where all the photons last scatter at the same time. It is a fair approximation, though unrealistic.

Exercise 10.3 From the definition of the optical depth:

$$\tau \equiv \int_{\eta}^{\eta_0} d\eta' \, n_e \sigma_T a \, , \tag{10.46}$$

show that $\tau \propto 1/\eta^3$ when matter dominates and $\tau \propto 1/\eta$ when radiation dominates.

We can be more quantitative and write:

$$-\tau' = n_e \sigma_T a = n_b \sigma_T a = \frac{\rho_b}{m_b} \sigma_T a \, , \tag{10.47}$$

where we have used the definition of τ and assumed to be in an epoch before recombination, so that we can approximate n_e with n_b, since all the electrons are free.

Exercise 10.4 Introducing the baryon density parameter and using $m_b = 1$ GeV, the mass of the proton, show that:

$$-\tau' \approx 1.46 \times 10^{-19} \frac{\Omega_{b0} h^2}{a^2} \, \text{s}^{-1} \, . \tag{10.48}$$

Now we need to compare this scattering rate with the Hubble rate, in order to check the goodness of the TC approximation. Assuming matter-domination and using the conformal time Friedmann equation (this because τ' is derived with respect to the conformal time), we have:

$$\mathcal{H} = H_0 \sqrt{\Omega_{m0}} a^{-1/2} \approx 3.33 \times 10^{-18} h \sqrt{\Omega_{m0}} \, a^{-1/2} \, \text{s}^{-1} \, . \tag{10.49}$$

Therefore, the ratio:

$$\frac{-\tau'}{\mathcal{H}} = 0.044 \, \frac{\Omega_{b0} h^2}{\sqrt{\Omega_{m0} h^2}} a^{-3/2} \, , \tag{10.50}$$

diverges for $a \to 0$ as expected (though the formula should be generalised to the case of radiation-domination), so if it is sufficiently big at recombination then the

TC approximation would be reliable. Substituting the *Planck* values $\Omega_{b0}h^2 = 0.022$ and $\Omega_{m0}h^2 = 0.12$ one gets at recombination, i.e. for $a = 10^{-3}$:

$$\frac{-\tau'}{\mathcal{H}} \approx 10^2 . \tag{10.51}$$

This means that the scattering rate is much larger than the Hubble rate even at recombination as long as there are free electrons around and thus we are going to use the tight-coupling approximation with reliability.

Let us see in detail how the TC limit works. Let us compare in the hierarchy for $\ell \geq 2$ the terms Θ_ℓ' and $k\Theta_\ell$ with $\tau'\Theta_\ell$, which have all the same dimensions of inverse time. There are two physical time scales in our problem, one is given by the expansion rate and the other by the scattering rate, hence

$$\Theta_\ell' \propto \mathcal{H}\Theta_\ell, \tau'\Theta_\ell , \tag{10.52}$$

from a dimensional analysis. However, the mode for which $\Theta_\ell' \propto \tau'\Theta_\ell$ implies that $\Theta_\ell \propto \exp\tau$ and hence diverges at early times, which is unacceptable for a small fluctuation. We then dismiss this mode as unphysical and take into account just that for which $\Theta_\ell' \propto \mathcal{H}\Theta_\ell$, which is small compared to $\tau'\Theta_\ell$.

Now, let us inspect the ratio

$$\frac{-\tau'}{k} . \tag{10.53}$$

This is the number of collisions which take place on a scale $1/k$. Hence, this number is very large, provided that we consider sufficiently large scales, i.e. small k. If the scale is too small, i.e. large k, then the TC approximation does not work well and we must take into account the multipole moments for $\ell \geq 2$. We will see this when investigating the **diffusion damping** or **Silk damping** effect.

From the above analysis, for sufficiently large scales we can conclude then that $\Theta_\ell \approx 0$ for $\ell \geq 2$. Sufficiently large means much larger than the mean free path $-1/\tau'$ which is approximately of the order of 10 Mpc at recombination. This number can be computed from Eq. (10.48) and is a comoving scale; the physical one is divided by a factor a thousand and so it is 10 kpc.

Finally, note that $\Theta_\ell \sim \tau'/k\Theta_{\ell-1}$. Therefore, considering smaller and smaller scales makes necessary to include higher and higher order multipoles.

Eliminating all the multipoles $\ell \geq 2$, the relevant equations are just the following two:

$$\Theta_0' + k\Theta_1 = -\Phi' , \tag{10.54}$$
$$3\Theta_1' - k\Theta_0 = k\Psi + \tau'(3\Theta_1 - V_b) , \tag{10.55}$$

i.e. the TC approximation allows us to treat photons as a fluid until recombination. Note the coupling to baryons via the baryon velocity V_b. Thus, we need also the equations for baryons:

$$\delta_b' + kV_b = -3\Phi' , \tag{10.56}$$

$$V_b' + \mathcal{H}V_b = k\Psi + \frac{\tau'}{R}(V_b - 3\Theta_1) , \tag{10.57}$$

where we have introduced $R \equiv 3\rho_b/4\rho_\gamma$, i.e. the baryon density to photon density ratio. This number can be cast as:

$$R = \frac{3\Omega_{b0}}{4\Omega_{\gamma 0}}a \approx 600a , \tag{10.58}$$

using the usual values and it grows from zero at early times to $R_* \approx 0.6$ at recombination. So it is small, but not that negligible. Let us rewrite the velocity equation for baryons in the following way:

$$V_b = 3\Theta_1 + \frac{R}{\tau'}\left(V_b' + \mathcal{H}V_b - k\Psi\right) . \tag{10.59}$$

We can solve this equation via successive approximation, exploiting the fact that $R < 1$ before recombination. That is, assume the expansion:

$$V_b = V_b^{(0)} + RV_b^{(1)} + R^2 V_b^{(2)} + \cdots . \tag{10.60}$$

The solution for $R = 0$ simply gives $V_b^{(0)} = 3\Theta_1$, which we have used in Chap. 6 in order to investigate the primordial modes. This solution is reliable well before recombination, say at $a = 10^{-7}$ for example, because $R \approx 6 \times 10^{-5}$ there, but it is not satisfactory at recombination and we shall take into account the first order in R in the above expansion.

10.3.1   The Acoustic Peaks for $R = 0$

Let us start with the simple case of $R = 0$, which amounts to neglect baryons.

Exercise 10.5 Combine the photon Eqs. (10.54)–(10.55) with the zeroth-order TC condition $V_b = 3\Theta_1$ and find the following second-order equation for Θ_0:

$$\Theta_0'' + \frac{k^2}{3}\Theta_0 = -\frac{k^2\Psi}{3} - \Phi'' . \tag{10.61}$$

We have here already the first fundamental piece of physics of the CMB. This is the equation of motion of a driven harmonic oscillator where instead of the position we have the monopole of the temperature fluctuation and the driving force is given by the gravitational potential. This equation describe **acoustic oscillations** of the baryon-photon fluid until recombination. After recombination we expect to observe these fluctuations in the $C_{TT,\ell}$'s, using the free-streaming formula (10.24), and in fact we do, cf. Fig. 10.1.

Note that these oscillations are in the baryon-photon fluid and therefore affect also baryons. We therefore expect to see oscillations in the baryon distribution after recombination, called **baryon acoustic oscillations** (BAO), and detected by Eisenstein and collaborators in 2005 (Eisenstein et al. 2005). The BAO are the manifestation of a special length, the sound horizon at recombination, in the correlation function of galaxies which appears as a bump, i.e. an excess probability. In the Fourier space, i.e. for the power spectrum, a given scale is represented with various oscillations. We have already encountered BAO in Chap. 9. BAO and weak gravitational lensing are among the main observables on which current and future experiments (such as *Euclid* and *LSST*) are based.

Exercise 10.6 Combine Eqs. (10.56) and (10.57) and the TC condition $V_b = 3\Theta_1$ and find the following equation for δ_b:

$$\delta_b' = 3\Theta_0' . \tag{10.62}$$

Hence, the same oscillatory solution of Θ_0 holds true for δ_b.

Now, consider the fact that close to recombination CDM is already dominating and thus the potentials are equal and constant at all scales. We get:

$$\Theta_0'' + \frac{k^2}{3}\Theta_0 = -\frac{k^2\Psi}{3} . \tag{10.63}$$

This equation can be put in the following form:

$$(\Theta_0 + \Psi)'' + \frac{k^2}{3}(\Theta_0 + \Psi) = 0 , \tag{10.64}$$

where we have used the constancy of Ψ. Note how the observed temperature fluctuation, used in Eq. (10.24), has appeared. The solution is:

$$(\Theta_0 + \Psi)(\eta, k) = A(k)\sin\left(\frac{k\eta}{\sqrt{3}}\right) + B(k)\cos\left(\frac{k\eta}{\sqrt{3}}\right) , \tag{10.65}$$

with the driving potential, i.e. CDM, providing just an offset for the oscillations. At recombination we have

$$(\Theta_0 + \Psi)(\eta_*, k) = A(k)\sin\left(\frac{k\eta_*}{\sqrt{3}}\right) + B(k)\cos\left(\frac{k\eta_*}{\sqrt{3}}\right) , \tag{10.66}$$

with peaks and valleys in the temperature fluctuations given by this combination of sine and cosine, therefore dependent on the functions $A(k)$ and $B(k)$. Inserting these formula into Eq. (10.24) in order to compute the $\Theta_\ell(\eta_0, k)$ (the anisotropies today) and then into Eq. (7.81) in order to compute the $C_{TT,\ell}$'s, we are able to explain the **acoustic oscillations** feature of the CMB TT spectrum, of Fig. 10.1.

The functions $A(k)$ and $B(k)$ are determined by the initial condition, i.e. for $k\eta_* \ll 1$:

$$(\Theta_0 + \Psi)(k\eta_* \ll 1) \sim A(k)\frac{k\eta_*}{\sqrt{3}} + B(k) \ . \tag{10.67}$$

Hence, if we choose adiabatic modes, we must put $A(k) = 0$. So, considering different initial conditions changes the position of the acoustic peaks and observation allows to test the choice made. As we saw in Chap. 6, Planck limits the presence of isocurvature modes to a few percent. With $A(k) = 0$, i.e. for adiabatic perturbations, using the large-scale solution that we found in Eq. (10.37), we have:

$$(\Theta_0 + \Psi)(\eta_*, k) = -\frac{1}{5}\mathcal{R}(k)T(k)\cos\left(\frac{k\eta_*}{\sqrt{3}}\right) \ , \tag{10.68}$$

where $T(k)$ is the transfer function of $\Theta_0 + \Psi$. We did not calculate it in Chap. 9, but it can be shown that it is limited to a range 0.4-2, approximately. See Mukhanov (2005).

The extrema of the effective temperature fluctuations are thus given by:

$$\frac{k\eta_*}{\sqrt{3}} = n\pi \ , \qquad (n = 1, 2, \dots) \ , \tag{10.69}$$

where the odd values provide peaks, corresponding to the highest temperature fluctuations and thus to scales at which photons are maximally compressed and hot, whereas the even values provide throats, corresponding to the lowest temperature fluctuations and thus to scales at which photons are maximally rarefied and cold. In the spectrum, cf. Fig. 10.1, only peaks appear because of the quadratic nature of the $C_{TT,\ell}$'s as functions of the Θ_ℓ's, but it should be clear that the first and the third peaks are compressional.

From Eqs. (10.54) and (10.68), we can determine easily the dipole contribution:

$$\Theta_1(\eta_*, k) = -\frac{\Theta_0'(\eta_*, k)}{k} = -\frac{1}{5\sqrt{3}}\mathcal{R}(k)T(k)\sin\left(\frac{k\eta_*}{\sqrt{3}}\right) \ , \tag{10.70}$$

where we are still continuing in keeping the potentials constant. Substituting this equation and Eq. (10.68) into Eq. (10.24) and then into Eq. (7.81) in order to compute the angular power spectrum, we get:

$$C_{TT,\ell} = \frac{4\pi}{25} \int_0^\infty \frac{dk}{k} \Delta_{\mathcal{R}}^2(k) \left[\cos\left(\frac{k\eta_*}{\sqrt{3}}\right) j_\ell(k\eta_0) + \sqrt{3} \sin\left(\frac{k\eta_*}{\sqrt{3}}\right) \frac{d j_\ell(k\eta_0)}{d(k\eta_0)} \right]^2 ,$$

$$(10.71)$$

where the derivative of the spherical Bessel function is given in Eq. (10.20). We have put $T(k) = 1$ here for simplicity.

We can manipulate analytically this integral following the technique used in Mukhanov (2004, 2005). In these references, baryon loading and diffusion damping are taken into account but here we just tackle a simpler case.

The idea is to avoid the oscillatory nature of the Bessel function and of the trigonometric ones (which are also problematic from a numerical perspective) by approximating $j_\ell(x)$ as follows, for large ℓ:

$$j_\ell(x) \approx \begin{cases} 0, & (x < \ell), \\ \frac{1}{\sqrt{x}(x^2-\ell^2)^{1/4}} \cos\left[\sqrt{x^2-\ell^2} - \ell \arccos(\ell/x) - \pi/4\right], & (x > \ell). \end{cases}$$

$$(10.72)$$

This approximation is identical either for $j_\ell(x)$ and for $j_{\ell-1}(x)$, since we are assuming ℓ to be large. Hence, when we deal with the derivative of the spherical Bessel function in Eq. (10.71), we can factorise a $j_\ell^2(x)$ and we can approximate the squared cosine coming from the above approximation with its average, i.e. a factor $1/2$. We thus have the following integration:

$$C_{TT,\ell} = \frac{2\pi\Delta_{\mathcal{R}}^2}{25} \int_{\ell/\eta_0}^\infty \frac{dk}{k^2 \eta_0 \sqrt{(k\eta_0)^2 - \ell^2}} \left[\cos\left(\frac{k\eta_*}{\sqrt{3}}\right) + \sqrt{3}\left(1 - \frac{\ell}{k\eta_0}\right) \sin\left(\frac{k\eta_*}{\sqrt{3}}\right) \right]^2 ,$$

$$(10.73)$$

where we have already assumed a scale-invariant spectrum, for simplicity. Using now the variable

$$x \equiv \frac{k\eta_0}{\ell} ,$$

$$(10.74)$$

we can write:

$$\ell^2 C_{TT,\ell} = \frac{2\pi\Delta_{\mathcal{R}}^2}{25} \int_1^\infty \frac{dx}{x^2\sqrt{x^2-1}} \left[\cos(\ell\varrho x) + \sqrt{3}\frac{x-1}{x} \sin(\ell\varrho x) \right]^2 , \quad (10.75)$$

where note the appearance of the factor ℓ^2 on the left hand side and we have defined the quantity:

$$\varrho \equiv \frac{\eta_*}{\sqrt{3}\eta_0} .$$

$$(10.76)$$

Now, developing the square and using the trigonometric formulae:

$$\cos^2\alpha = \frac{1 + \cos 2\alpha}{2} , \quad \sin^2\alpha = \frac{1 - \cos 2\alpha}{2} , \quad 2\sin\alpha\cos\alpha = \sin 2\alpha ,$$

$$(10.77)$$

we can write:

$$\ell^2 C_{TT,\ell} = \frac{2\pi \Delta_{\mathcal{R}}^2(k)}{25} \int_1^\infty \frac{dx}{x^2 \sqrt{x^2-1}}$$

$$\left[\frac{x^2 + 3(x-1)^2}{2x^2} + \frac{x^2 - 3(x-1)^2}{2x^2} \cos(2\ell\varrho x) + \frac{\sqrt{3}(x-1)}{x} \sin(2\ell\varrho x) \right] .$$

$$(10.78)$$

Now, let us treat separately the three integrands. The first, non-oscillatory one is simplest one:

$$N \equiv \int_1^\infty \frac{dx}{x^2 \sqrt{x^2-1}} \frac{x^2 + 3(x-1)^2}{2x^2} = 3\left(1 - \frac{\pi}{4}\right) , \qquad (10.79)$$

but also the less interesting. The oscillatory ones can be dealt with following Mukhanov (2005). Define:

$$O_1 \equiv \int_1^\infty \frac{dx}{\sqrt{x-1}} \frac{x^2 - 3(x-1)^2}{2x^4 \sqrt{x+1}} \cos(2\ell\varrho x) , \qquad (10.80)$$

then solving the problem in Mukhanov (2005, p. 383), we can use the formula:

$$\int_1^\infty \frac{dx}{\sqrt{x-1}} f(x) \cos(bx) \approx f(1) \sqrt{\frac{\pi}{b}} \cos(b + \pi/4) , \qquad (10.81)$$

for large values of b and a slowly varying $f(x)$. A similar result holds true also for the sine function. Using this formula we have then:

$$O_1 = \frac{1}{2\sqrt{2}} \sqrt{\frac{\pi}{2\ell\varrho}} \cos(2\ell\varrho + \pi/4) , \qquad (10.82)$$

whereas for the integral containing the sine:

$$O_2 \equiv \int_1^\infty \frac{dx}{\sqrt{x-1}} \frac{\sqrt{3}(x-1)}{x^3 \sqrt{x+1}} \sin(2\ell\varrho x) \approx 0 , \qquad (10.83)$$

since $f(1) = 0$ here. The contribution O_2 comes from the cross product between the monopole and the dipole terms and it is usually neglected in the calculations. We have explicitly shown why here. Gathering the N and O_1 contributions, we plot the sum $N + O_1$ in Fig. 10.3.

In order to make this plot, we have used $a \propto \eta^2$, since we are in the matter-dominated epoch, and thus we have evaluated ϱ as follows:

$$\varrho = \frac{\eta_*}{\sqrt{3}\eta_0} = \frac{1}{\sqrt{3(1+z_*)}} = \frac{1}{\sqrt{3000}} \approx 0.0183 . \qquad (10.84)$$

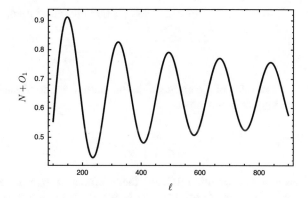

Fig. 10.3 Sum of the N and O_1 contributions

The agreement between the plots of Figs. 10.1 and 10.3 is poor but at least we have understood how the acoustic oscillations free-stream until today and are seen in the CMB TT power spectrum. There are several feature missing in Fig. 10.3: there are too many peaks, their relative height diminishes too slowly and the overall trend does not decay as in Fig. 10.1. The reason is that we have neglected baryons and diffusion damping, which we are going to tackle in the next sections.

10.3.2 Baryon Loading

The oscillations in Eq. (10.64) take place with frequency $k/\sqrt{3}$, i.e. as if the speed of sound was $1/\sqrt{3}$, i.e. the speed of sound of a pure photon fluid. We have been too radical in assuming $V_b = 3\Theta_1$ in the equation for baryons. In fact we saw that this assumption is equivalent to say that $R = 0$, i.e. the baryon density is negligible with respect to the photon one. That is why photons do not feel baryons at all and baryons fluctuations oscillate in the same way as photons do.

We now take into account R up to first-order. If we consider $V_b^{(0)} = 3\Theta_1$ substituted in Eq. (10.59) we get up to order R:

$$V_b = 3\Theta_1 + \frac{R}{\tau'}\left(3\Theta_1' + 3\mathcal{H}\Theta_1 - k\Psi\right) . \tag{10.85}$$

Exercise 10.7 Combine the above equation and Eqs. (10.54)–(10.55) in order to find the following second-order equation for Θ_0:

$$\Theta_0'' + \mathcal{H}\frac{R}{1+R}\Theta_0' + \frac{k^2}{3(1+R)}\Theta_0 = -\frac{k^2\Psi}{3} - \Phi'' - \mathcal{H}\frac{R}{1+R}\Phi' . \tag{10.86}$$

Now the speed of sound, i.e. the quantity multiplying k^2, has been reduced:

$$c_s^2 = \frac{1}{3(1 + R)} \ . \tag{10.87}$$

The extrema of the temperature fluctuation at recombination are now expected to be slightly changed, since:

$$\frac{k\eta_*}{\sqrt{3(1 + R)}} = n\pi \ , \qquad (n = 1, 2, \dots) \ , \tag{10.88}$$

is now the condition defining them. Moreover, baryons are also responsible for the damping term $\mathcal{H}R\Theta_0'/(1 + R)$, hence we also expect the extrema to have less and less amplitude. These features translate, once free-streamed,[2] in a relative suppression of the second peak with respect to the first one, as seen in Fig. 10.1.

This effect is due to the **baryon loading** and it is also called **baryon drag**. Physically, baryons are heavy and prevent the oscillations in Θ_0 to be symmetric, favouring compression over rarefaction. Since $R \propto a$, then we have that:

$$R' = \mathcal{H}R \ . \tag{10.89}$$

Let us write Eq. (10.86) in the following form:

$$\left(\frac{d^2}{d\eta^2} + \frac{R'}{1 + R}\frac{d}{d\eta} + k^2 c_s^2\right)(\Theta_0 + \Phi) = \frac{k^2}{3}\left(\frac{\Phi}{1 + R} - \Psi\right) \ . \tag{10.90}$$

The above equation cannot be solved analytically, but we can use a semi-analytic approximation, provided by Hu and Sugiyama (1996). Let us employ the WKB method and use the following ansatz:

$$(\Theta_0 + \Phi)(\eta, k) = A(\eta)e^{i B(\eta,k)} \ , \tag{10.91}$$

where $A(\eta)$ and $B(\eta, k)$ are functions to be determined via Eq. (10.90).

Exercise 10.8 Substitute this ansatz into the homogenous part of Eq. (10.90) and find the following couple of equations, by separately equating the real and imaginary parts to zero:

$$-A(B')^2 + A'' + \frac{R'}{1 + R}A' + k^2 c_s^2 A = 0 \ , \tag{10.92}$$

$$2B'A' + AB'' + \frac{R'}{1 + R}AB' = 0 \ . \tag{10.93}$$

---

[2]To "free-stream" means to calculate the $C_{TT,\ell}$'s weighting the solution at recombination with the spherical Bessel function of Eq. (10.24).

In the first equation, let us neglect the second and the third term with respect to the first one. That is, the oscillations provide almost at any time (except at the extrema) a much larger derivative than that of the amplitude or R. Then, the first equation is readily solved as:

$$B(\eta, k) = k \int_0^\eta c_s(\eta')d\eta' \equiv kr_s(\eta) \qquad (10.94)$$

where in the last step we have defined the **sound horizon**, i.e. the conformal distance travelled by a sound wave propagating in the baryon photon fluid. When evaluated at recombination, $r_s(\eta_*) = 150$ Mpc and this scale is fundamental for BAO, making them **standard rulers**.

Exercise 10.9 Determine now $A(\eta)$. Show that the above equations, together with the found solution for $B(\eta, k)$, can be cast as:

$$\frac{A'}{A} = -\frac{1}{4}\frac{R'}{1+R} , \qquad (10.95)$$

which gives:

$$A(\eta) = (1 + R)^{-1/4} . \qquad (10.96)$$

The general, approximate, solution of the homogeneous equation is then:

$$(\Theta_0 + \Phi)(\eta, k) = \frac{1}{(1+R)^{1/4}}[C(k)\sin(kr_s) + D(k)\cos(kr_s)] \qquad (10.97)$$

The condition $|A'|, R' \ll |B'|$, which was employed in order to find the above solution, can be checked as follows:

$$\frac{R'}{4(1+R)^{5/4}}, R' \ll \frac{k}{\sqrt{3(1+R)}} , \qquad (10.98)$$

which essentially amounts to say that:

$$k \gg R' , \qquad (10.99)$$

i.e. the solution found is good on sufficiently small scales. Since $R' = \mathcal{H}R \sim R/\eta$, we must have that $k\eta \gg R$. Since R is pretty small, being at most $R_* \approx 0.6$ at recombination, this condition means any scale at early times, but sub-horizon scales at recombination.

Equation (10.97) gives us the general solution of the homogeneous part of Eq. (10.90). In order to find the general solution of the full equation we need to find a particular solution of Eq. (10.90). This can be obtained via Green's functions method, which we recall in Chap. 12. Let us define, in order to keep a more compact notation, the independent solutions of the homogeneous equation that we have just found in Eq. (10.97) as follows:

$$S_1(\eta, k) \equiv \frac{1}{(1 + R)^{1/4}} \sin(kr_s) , \qquad S_2(\eta, k) \equiv \frac{1}{(1 + R)^{1/4}} \cos(kr_s) . \quad (10.100)$$

Taking into account the non-homogeneous term, the general solution of Eq. (10.90) is:

$$(\Theta_0 + \Phi)(\eta, k) = C(k)S_1 + D(k)S_2 + \frac{k^2}{3} \int_0^\eta d\eta' \left[\frac{\Phi(\eta')}{1 + R} - \Psi(\eta') \right] G(\eta, \eta') , \quad (10.101)$$

where $G(\eta, \eta')$ is the Green's function.

Exercise 10.10 As done in Chap. 12, cf. Eq. (12.121), determine the Green's function:

$$G(\eta, \eta') = \frac{S_1(\eta')S_2(\eta) - S_1(\eta)S_2(\eta')}{W(\eta')} , \quad (10.102)$$

using the homogeneous solution. Show that:

$$G(\eta, \eta') = \frac{1}{\sqrt{1 + R)}} \frac{\sin[kr_s(\eta') - kr_s(\eta)]}{W(\eta')} , \quad (10.103)$$

and

$$W(\eta') = -\frac{1}{\sqrt{3}(1 + R)} . \quad (10.104)$$

We are omitting the k-dependence for simplicity.

With the above results we can write:

$$(\Theta_0 + \Phi)(\eta, k) = C(k)S_1 + D(k)S_2$$
$$+ \frac{k}{\sqrt{3}} \int_0^\eta d\eta' \left[\frac{\Phi(\eta')}{1 + R(\eta')} - \Psi(\eta') \right] \sqrt{1 + R(\eta')} \sin[kr_s(\eta) - kr_s(\eta')] . \quad (10.105)$$

For the primordial modes, in the limit $k\eta \to 0$, one gets at the dominant order:

$$(\Theta_0 + \Phi)(0, k) = D(k) . \quad (10.106)$$

Hence, it is the adiabatic mode which multiplies the cosine. Since sine and cosine have a $\pi/2$ phase difference, the effect of different initial conditions is to change the scales for which the effective temperature fluctuations is maximum or minimum and hence the positions of the peaks in the $C_{TT,\ell}$'s.

In the adiabatic case, we have:

$$(\Theta_0 + \Phi)(\eta, k) = (\Theta_0 + \Phi)(0, k) \frac{\cos[kr_s(\eta)]}{(1 + R)^{1/4}}$$

$$+ \frac{k}{\sqrt{3}} \int_0^\eta d\eta' \left[\frac{\Phi(\eta')}{1 + R} - \Psi(\eta') \right] \sqrt{1 + R} \sin[kr_s(\eta) - kr_s(\eta')] \qquad (10.107)$$

This is the semi-analytic (semi because the integral has to be performed numerically) formula of Hu and Sugiyama (1996).

The above solution (10.107) can also be used for baryons. Indeed, combining Eq. (10.56) with Eq. (10.85) and then with Eq. (10.54) we get:

$$\delta'_b = 3\Theta'_0 + \frac{3R}{\tau'} \left[\Theta''_0 + \Phi'' + \mathcal{H}(\Theta'_0 + \Phi') + \frac{k^2\Psi}{3} \right] . \qquad (10.108)$$

Eliminating the second derivative by means of the differential equation (10.90), we have:

$$\delta'_b = 3\Theta'_0 + \frac{R}{\tau'(1 + R)} \left[-k^2\Theta_0 + 3\mathcal{H}(\Theta'_0 + \Phi') \right] . \qquad (10.109)$$

Just to make a rough estimative, let us neglect the second contribution (which is divided by τ' anyway which is much larger than $\mathcal{H}$ and also than k, for suitable scales) and use the homogeneous part of Eq. (10.107). It is straightforward then to integrate δ'_b and obtain at recombination:

$$\delta_b(\eta_*, k) \propto \cos[kr_s(\eta_*)] = \cos\left[2\pi \frac{r_s(\eta_*)}{\lambda} \right] . \qquad (10.110)$$

So the scale $r_s(\eta_*) \approx 150$ Mpc is relevant for baryons, too. Indeed, at about this scale the matter power spectrum display the BAO feature, as we saw in Chap. 9.

10.4 Diffusion Damping

In order to understand what happens to the $C_{TT,\ell}$'s when ℓ grows larger and larger we need to take into account smaller and smaller scales, because of the relation $\ell \approx k\eta_0$. As discussed earlier, for larger and larger k the ratio $-\tau'/k$ becomes smaller and smaller and so the TC approximation must be relaxed.

In this section then we investigate what happens to the temperature fluctuations when the quadrupole moment Θ_2 is taken into account. Since this analysis accounts for the behavior of very small scales which entered the horizon deep into the radiation-dominated epoch, we can neglect the gravitational potentials since these, as we saw in Chap. 9, rapidly decay.

Moreover, being deep into the radiation-dominated epoch, we can also neglect R and thus take $3\Theta_1 = V_b$. Neglecting also polarisation, we have the following set of three equations for Θ_0, Θ_1 and Θ_2:

$$\Theta_0' + k\Theta_1 = 0 , \tag{10.111}$$

$$3\Theta_1' + 2k\Theta_2 - k\Theta_0 = 0 , \tag{10.112}$$

$$10\Theta_2' - 4k\Theta_1 = 9\tau'\Theta_2 . \tag{10.113}$$

In the last equation we can neglect Θ_2' with respect $\tau'\Theta_2$, as we already did earlier, and then find:

$$\Theta_2 = -\frac{4k}{9\tau'}\Theta_1 . \tag{10.114}$$

The minus sign might ring some alarm, but recall that τ' is always negative by definition.

Exercise 10.11 Combine the above condition with the remaining equations in order to find a closed equation for Θ_0:

$$\boxed{\Theta_0'' + \left(-\frac{8k^2}{27\tau'}\right)\Theta_0' + \frac{k^2}{3}\Theta_0 = 0} \tag{10.115}$$

This is the equation for an harmonic oscillator that we have already found earlier in Eq. (10.64), only that now there appears a damping term which is relevant on small scales, i.e. when $k \sim -\tau'$. Baryons also provide a damping term, cf. Eq. (10.86), but they are irrelevant in the present case since we set $R = 0$.

This damping term here depends on Θ_2 and is time-dependent. Let us consider it constant and assume a solution of the type $\Theta_0 \propto \exp(i\omega\eta)$. Substituting this ansatz in the equation, we find:

$$-\omega^2 + \left(-\frac{8k^2}{27\tau'}\right)i\omega + \frac{k^2}{3} = 0 . \tag{10.116}$$

The frequency must have an imaginary part, which accounts for the damping, thus let us stipulate:

$$\omega = \omega_R + i\omega_I , \tag{10.117}$$

Exercise 10.12 Substitute this ansatz in the equation and find:

$$\omega_R = \frac{k}{\sqrt{3}}, \qquad \omega_I = -\frac{4k^2}{27\tau'} . \tag{10.118}$$

Hence, we can write the general solution for Θ_0 as:

$$\Theta_0 \propto e^{ik\eta/\sqrt{3}} e^{-k^2/k_{\text{Silk}}^2} , \tag{10.119}$$

where we have introduced the comoving diffusion length, the **Silk length**, as

$$\lambda_{\text{Silk}}^2 = \frac{1}{k_{\text{Silk}}^2} \equiv -\frac{4\eta}{27\tau'} . \tag{10.120}$$

What does the diffusion length physically represent? It is the comoving distance travelled by a photon in a time η, but taking into account the collisions which it is suffering, i.e. its diffusion. Let us see this in some more detail.

Since $-\tau'$ is the scattering rate, i.e. how many collisions take place per unit conformal time, then $-1/\tau'$ is the average conformal time between 2 consecutive collisions, which for a photon is also the average comoving distance between two collision, i.e. the mean free path.

Now, we have:

$$\lambda_{\text{Silk}}^2 \propto -\frac{\eta}{\tau'} \propto \lambda_{\text{MFP}} \eta , \tag{10.121}$$

where we have used the comoving mean free path, λ_{MFP}. Now, multiply and divide by λ_{MFP} and take the square root:

$$\lambda_{\text{Silk}} \propto \lambda_{\text{MFP}} \sqrt{\frac{\eta}{\lambda_{\text{MFP}}}} , \tag{10.122}$$

Under the square root we have the comoving distance η divided by the photon comoving mean free path. This gives us the average number of collision N which the photons experience up to the time η and hence:

$$\lambda_{\text{Silk}} \propto \sqrt{N} \lambda_{\text{MFP}} , \tag{10.123}$$

which is the typical relation for diffusion. Below this scales λ_{Silk} all fluctuations are suppressed because photons cannot agglomerate since they escape away. This effect is known as **Silk damping** (Silk 1967). Therefore, the behaviour of the C_l's for large l's, as seen in Fig. 10.1, is also decaying, though not exactly as in the above solution since this has to be free-streamed first.

We can do a more detailed calculation of the damping scale as follows. Let us neglect the gravitational potential and the $\ell \geq 3$ multipoles as before, but let us deal with more care of baryons and take into account polarisation. From Eq. (10.59) we have:

$$V_b = 3\Theta_1 + \frac{R}{\tau'} \left(V_b' + \mathcal{H} V_b \right) , \qquad (10.124)$$

and the six equations for the monopole, dipole and quadrupole of the temperature fluctuations and polarisation:

$$\Theta_0' + k\Theta_1 = 0 , \qquad (10.125)$$

$$3\Theta_1' + 2k\Theta_2 - k\Theta_0 = \tau'(3\Theta_1 - V_b) , \qquad (10.126)$$

$$10\Theta_2' - 4k\Theta_1 = 9\tau'\Theta_2 - \tau'\Theta_{P0} - \tau'\Theta_{P2} , \qquad (10.127)$$

$$2\Theta_{P0}' + 2k\Theta_{P1} = \tau'\Theta_{P0} - \tau'\Theta_{P2} - \tau'\Theta_2 , \qquad (10.128)$$

$$3\Theta_{P1}' + 2k\Theta_{P2} - k\Theta_{P0} = 3\tau'\Theta_{P1} , \qquad (10.129)$$

$$10\Theta_{P2}' - 4k\Theta_{P1} = 9\tau'\Theta_{P2} - \tau'\Theta_{P0} - \tau'\Theta_2 . \qquad (10.130)$$

Now, assuming a solution of the type $\exp(i \int \omega d\eta)$ for all the above 7 variables and also assuming that $\omega \gg \mathcal{H}$, we have:

$$V_b = \frac{3\Theta_1}{1 + Ri\omega\eta_c} , \qquad (10.131)$$

where we have defined $\eta_c \equiv -1/\tau'$ as the the average conformal time between 2 consecutive collisions. We have thus a closed system for Θ_0, Θ_1, Θ_2, Θ_{P0}, Θ_{P1} and Θ_{P2}:

$$i\omega\Theta_0 + k\Theta_1 = 0 , \qquad (10.132)$$

$$-k\Theta_0 + 3i\omega\Theta_1 \left(1 + \frac{R}{1 + Ri\omega\eta_c} \right) + 2k\Theta_2 = 0 , \qquad (10.133)$$

$$-4k\eta_c\Theta_1 + (10i\omega\eta_c + 9)\Theta_2 - \Theta_{P0} - \Theta_{P2} = 0 , \qquad (10.134)$$

$$-\Theta_2 + (2i\omega\eta_c + 1)\Theta_{P0} + 2k\eta_c\Theta_{P1} - \Theta_{P2} = 0 , \qquad (10.135)$$

$$-k\Theta_{P0} + 3(i\omega\eta_c + 1)\Theta_{P1} + 2k\eta_c\Theta_{P2} = 0 , \qquad (10.136)$$

$$-\Theta_2 - \Theta_{P0} - 4k\eta_c\Theta_{P1} + (10i\omega\eta_c + 9)\Theta_{P2} = 0 . \qquad (10.137)$$

We have already arranged the variables in order for the system matrix to appear clearly. The determinant of this matrix, in order to have a non trivial solution, must be zero. Considering the limit $\omega\eta_c \ll 1$, and keeping the first-order only in $\omega\eta_c$ we get:

$$\frac{k^2}{3} - \omega^2(1 + R) + \frac{2i}{30}\omega\eta_c \left[37k^2 - 285(1 + R)\omega^2 + 15\omega^2 R^2 \right] = 0 . \qquad (10.138)$$

In order to solve for ω, let us again employ the smallness of $\omega\eta_c$ and stipulate that:

$$\omega = \omega_0 + \delta\omega , \tag{10.139}$$

where $\delta\omega$ is a small correction. From the above equation is then straightforward to obtain:

$$\frac{k^2}{3} - \omega_0^2(1 + R) = 0 , \tag{10.140}$$

$$-2\omega_0\delta\omega(1 + R) + \frac{2i}{30}\omega_0\eta_c\left[37k^2 - 285(1 + R)\omega_0^2 + 15\omega_0^2 R^2\right] = 0 . \tag{10.141}$$

The first equation gives the result that we have already encountered:

$$\boxed{\omega_0^2 = \frac{k^2}{3(1 + R)} = k^2 c_s^2} \tag{10.142}$$

which, substituted in the second equation, gives us:

$$\boxed{\delta\omega = \frac{i\eta_c k^2}{6(1 + R)}\left[\frac{16}{15} + \frac{R^2}{1 + R}\right]} \tag{10.143}$$

This result was obtained for the first time by Kaiser (1983). See also the derivation of Weinberg (2008).

Therefore, the evolution of the multipoles is proportional to the following factor:

$$\exp\left(i\int\omega d\eta\right) = e^{ikr_s(\eta)}e^{-k^2/k_{\text{Silk}}^2} , \tag{10.144}$$

where

$$\boxed{\frac{1}{k_{\text{Silk}}^2} \equiv -\int_0^\eta d\eta' \frac{1}{6\tau'(1 + R)}\left(\frac{16}{15} + \frac{R^2}{1 + R}\right)} \tag{10.145}$$

From the best fit values of the parameter of the ΛCDM model we have:

$$\boxed{d_{\text{Silk}} = 0.0066 \text{ Mpc}} \tag{10.146}$$

10.5   Line-of-Sight Integration

The approximate solutions found earlier are based on the TC limit, which allows us to take into account just the monopole and the dipole until recombination and then to better understand the physics behind the CMB anisotropies. On the other hand, observation demands more precise calculations to be compared with and therefore, at the end, numerical computation and codes such as CLASS are needed. Even so, there is a more efficient way of computing predictions on the CMB anisotropies than dealing directly with the hierarchy of Boltzmann equation and that is to formally integrate along the photon past light-cone according to a semi-analytic technique called **line-of-sight integration**, due to Seljak and Zaldarriaga (1996), and which was the basis for the CMBFAST code.[3]

Recall the photon Boltzmann equations (5.114) and (5.115):

$$\Theta' + ik\mu\Theta = -\Phi' - ik\mu\Psi - \tau'\left[\Theta_0 - \Theta - i\mu V_b - \frac{1}{2}\mathcal{P}_2(\mu)\Pi\right] ,$$
(10.147)

$$\Theta'_P + ik\mu\Theta_P = -\tau'\left[-\Theta_P + \frac{1}{2}[1 - \mathcal{P}_2(\mu)]\Pi\right] ,$$
(10.148)

where $\Pi = \Theta_2 + \Theta_{P2} + \Theta_{P0}$. Let us rewrite them as follows:

$$\Theta' + (ik\mu - \tau')\Theta = -\Phi' - ik\mu\Psi - \tau'\left[\Theta_0 - i\mu V_b - \frac{1}{2}\mathcal{P}_2(\mu)\Pi\right] \equiv \mathcal{S}(\eta, k, \mu) ,$$
(10.149)

$$\Theta'_P + (ik\mu - \tau')\Theta_P = -\frac{\tau'}{2}[1 - \mathcal{P}_2(\mu)]\Pi \equiv \mathcal{S}_P(\eta, k, \mu) ,$$
(10.150)

where we have introduced two source functions on the right hand sides. Note that the dependence in on k and not on $\mathbf{k} = k\hat{z}$ because we are considering the equations for the transfer functions. Afterwards, before performing the anti-Fourier transform, we must rotate back $\hat{k}$ in a generic direction.

Let us write the left hand sides as follows:

$$\Theta' + (ik\mu - \tau')\Theta = e^{-ik\mu\eta + \tau}\frac{d}{d\eta}\left(\Theta\, e^{ik\mu\eta - \tau}\right) ,$$
(10.151)

with a similar expression for Θ_P. Substituting these into the Boltzmann equations and integrating formally from a certain initial $\eta_i \to 0$ to today η_0, we get:

---

[3]https://lambda.gsfc.nasa.gov/toolbox/tb_cmbfast_ov.cfm.

$$\Theta(\eta_0)e^{-\tau(\eta_0)} = \Theta(\eta_i)e^{ik\mu(\eta_i-\eta_0)-\tau(\eta_i)} + \int_{\eta_i}^{\eta_0} d\eta \, e^{ik\mu(\eta-\eta_0)-\tau(\eta)}S(\eta, k, \mu) \,,$$

$$(10.152)$$

$$\Theta_P(\eta_0)e^{-\tau(\eta_0)} = \Theta_P(\eta_i)e^{ik\mu(\eta_i-\eta_0)-\tau(\eta_i)} + \int_{\eta_i}^{\eta_0} d\eta \, e^{ik\mu(\eta-\eta_0)-\tau(\eta)}S_P(\eta, k, \mu) \,,$$

$$(10.153)$$

Now recall the definition of the optical depth:

$$\tau \equiv \int_{\eta}^{\eta_0} d\eta' \, n_e \sigma_T a \,. \tag{10.154}$$

It is clear then that $\tau(\eta_0) = 0$ and, since $\eta_i \to 0$ is deep into the radiation-dominated epoch, then $\tau \propto 1/\eta$ is very large and we can neglect $\exp[-\tau(\eta_i)]$. Therefore, we are left with

$$\Theta(\eta_0, k, \mu) = \int_0^{\eta_0} d\eta \, e^{ik\mu(\eta-\eta_0)-\tau(\eta)}S(\eta, k, \mu) \,, \tag{10.155}$$

$$\Theta_P(\eta_0, k, \mu) = \int_0^{\eta_0} d\eta \, e^{ik\mu(\eta-\eta_0)-\tau(\eta)}S_P(\eta, k, \mu) \,. \tag{10.156}$$

where we have already implemented the limit $\eta_i \to 0$. Now we calculate the Θ_ℓ's inverting the Legendre expansion as done in Eq. (5.113) and obtain:

$$\Theta_\ell(\eta_0, k) = \frac{1}{(-i)^\ell} \int_{-1}^{1} \frac{d\mu}{2} \, \mathcal{P}_\ell(\mu) \int_0^{\eta_0} d\eta \, e^{ik\mu(\eta-\eta_0)-\tau(\eta)}S(k, \eta, \mu) \,,$$

$$(10.157)$$

$$\Theta_{P\ell}(\eta_0, k) = \frac{1}{(-i)^\ell} \int_{-1}^{1} \frac{d\mu}{2} \, \mathcal{P}_\ell(\mu) \int_0^{\eta_0} d\eta \, e^{ik\mu(\eta-\eta_0)-\tau(\eta)}S_P(k, \eta, \mu) \,,$$

$$(10.158)$$

The source terms have μ-dependent contributions (up to μ^2) that we can handle integrating by parts. Take for example the $-ik\mu\Psi$ contribution of $S(k, \eta, \mu)$. Let I_Ψ be its integral, which can be rewritten as follows:

$$I_\Psi \equiv -\int_0^{\eta_0} d\eta \, ik\mu\Psi e^{ik\mu(\eta-\eta_0)-\tau(\eta)} = -\int_0^{\eta_0} d\eta \, \Psi e^{-\tau(\eta)} \frac{d}{d\eta}\left[e^{ik\mu(\eta-\eta_0)}\right] \,,$$

$$(10.159)$$

and now it is easy to integrate by parts and obtain:

$$I_\Psi = -\left. \Psi e^{-\tau(\eta)} e^{ik\mu(\eta-\eta_0)}\right|_0^{\eta_0} + \int_0^{\eta_0} d\eta \, e^{ik\mu(\eta-\eta_0)} \frac{d}{d\eta}\left[\Psi e^{-\tau(\eta)}\right] \,. \tag{10.160}$$

The first contribution gives $-\Psi(\eta_0)$, i.e. the gravitational potential evaluated at present time. This is just an undetectable offset that we incorporate into the definition of $\Theta_\ell(\eta_0, k)$, as the observed anisotropy, like we did at the beginning of this chapter when dealing with the free-streaming solution.

Exercise 10.13 Take care of the term containing μ^2, in $\mathcal{P}_2(\mu)$. Show that:

$$\int_0^{\eta_0} d\eta \, \tau' \mu^2 \Pi e^{ik\mu(\eta-\eta_0)-\tau(\eta)} = -\frac{1}{k^2} \int_0^{\eta_0} d\eta \, e^{ik\mu(\eta-\eta_0)} \frac{d^2}{d\eta^2} \left[\tau' \Pi e^{-\tau(\eta)} \right] .$$

$$(10.161)$$

Combining all the terms treated with integration by parts, we get:

$$\Theta_\ell(k, \eta_0) = \frac{1}{(-i)^\ell} \int_{-1}^1 \frac{d\mu}{2} \, \mathcal{P}_\ell(\mu) \int_0^{\eta_0} d\eta \, e^{ik\mu(\eta-\eta_0)}$$
$$\left[-\left(\Phi' + \tau' \Theta_0 + \frac{\tau' \Pi}{4} \right) e^{-\tau} + \left(\Psi e^{-\tau} - \frac{\tau' V_b e^{-\tau}}{k} \right)' - \frac{3}{4k^2} (\tau' \Pi e^{-\tau})'' \right] ,$$

$$(10.162)$$

$$\Theta_{P\ell}(k, \eta_0) = -\frac{3}{4(-i)^\ell} \int_{-1}^1 \frac{d\mu}{2} \, \mathcal{P}_\ell(\mu) \int_0^{\eta_0} d\eta \, e^{ik\mu(\eta-\eta_0)} \left[\tau' \Pi e^{-\tau} + \frac{1}{k^2} (\tau' \Pi e^{-\tau})'' \right] .$$

$$(10.163)$$

Using now the relation of Eq. (10.15), we can cast the above equations as:

$$\Theta_\ell(\eta_0, k) = \int_0^{\eta_0} d\eta \, S(\eta, k) j_\ell [k(\eta_0 - \eta)] , \qquad (10.164)$$

$$\Theta_{P\ell}(\eta_0, k) = \int_0^{\eta_0} d\eta \, S_P(\eta, k) j_\ell [k(\eta_0 - \eta)] . \qquad (10.165)$$

with

$$S(\eta, k) \equiv (\Psi' - \Phi') e^{-\tau} + g \left(\Theta_0 + \frac{\Pi}{4} + \Psi \right) + \frac{1}{k} (g V_b)' + \frac{3}{4k^2} (g\Pi)'' ,$$

$$(10.166)$$

$$S_P(\eta, k) \equiv \frac{3}{4} g\Pi + \frac{3}{4k^2} (g\Pi)'' ,$$

$$(10.167)$$

where we have introduced the **visibility function**:

$$\boxed{g(\eta) \equiv -\tau' e^{-\tau}} \qquad (10.168)$$

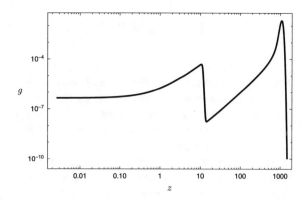

Fig. 10.4 Visibility function g as function of the redshift from the numerical calculation performed with CLASS for the standard model

---

Exercise 10.14 Show that the visibility function is normalised to unity, i.e.

$$\int_0^{\eta_0} d\eta \, g(\eta) = 1 \,. \tag{10.169}$$

The visibility function represents the Poissonian probability that a photon is last scattered at a time η. It is very peaked at a time that we define as the one of recombination, i.e. at $\eta = \eta_*$, because for $\eta > \eta_*$ it is basically zero, since $\tau' = 0$. Before recombination, in the radiation-dominated epoch, we saw that $-\tau' \propto 1/\eta^2$ and thus $\tau \propto 1/\eta$ and $g \propto \exp(-1/\eta)/\eta^2$, i.e. it goes to zero exponentially fast.

In Fig. 10.4 we plot the numerical calculation of the visibility function performed with CLASS for the standard model. Note the peak at about $z = 1000$, which has always been our reference for the recombination redshift. Note also another peak at about $z = 10$, representing the epoch of **reionisation**. Until now we have used the peakedness of the visibility function as if it were a Dirac delta $\delta(\eta - \eta_*)$, i.e. we have made the **sudden recombination approximation**. From Fig. 10.4 we can appreciate that it is a good approximation (mind the logarithmic scale there). As usual, in cosmology but not only, the calculations get more and more complicated and impossible to do analytically the more precision we demand.

Inserting the source terms (10.166) and (10.167) in the expressions for $\Theta_\ell(\eta_0, k)$ and $\Theta_{P\ell}(\eta_0, k)$ in the expression of Θ_ℓ and integrating by parts, we get:

$$\Theta_\ell(k, \eta_0) = \int_0^{\eta_0} d\eta \, g \left(\Theta_0 + \Psi + \frac{\Pi}{4} \right) j_\ell \left[k(\eta_0 - \eta) \right]$$

$$- \int_0^{\eta_0} d\eta \frac{g V_b}{k} \frac{d}{d\eta} j_\ell \left[k(\eta_0 - \eta) \right] + \int_0^{\eta_0} d\eta \frac{3g\Pi}{4k^2} \frac{d^2}{d\eta^2} j_\ell \left[k(\eta_0 - \eta) \right]$$

$$+ \int_0^{\eta_0} d\eta \, e^{-\tau} (\Psi' - \Phi') j_\ell \left[k(\eta_0 - \eta) \right] \,, \tag{10.170}$$

$$\Theta_{P\ell}(k, \eta_0) = \int_0^{\eta_0} d\eta \, \frac{3g\Pi}{4} \, j_\ell \left[k(\eta_0 - \eta)\right] + \int_0^{\eta_0} d\eta \frac{3g\Pi}{4k^2} \frac{d^2}{d\eta^2} j_\ell \left[k(\eta_0 - \eta)\right] \, .$$

$$(10.171)$$

Assuming the visibility function to be a Dirac delta $\delta(\eta - \eta_*)$, i.e. the sudden recombination mentioned earlier, and neglecting Π, we recover formula (10.24). Note that neglecting Π no polarisation is present. Indeed, from the above equation we see that a non-zero quadrupole moment of the photon distribution at recombination is essential in order to have polarisation.

The above equations still need the Boltzmann hierarchy in order to be integrated, but just up to $\ell = 4$ (because Θ_2 and Θ_4 moments are contained in the equation for Θ_3') and hence are much more convenient from the computational point of view.

The partial wave expansion of Θ given in Eq. (7.77):

$$\Theta(k, \mu) = \sum_\ell (-i)^\ell (2\ell + 1) \mathcal{P}_\ell(\mu) \Theta_\ell(k) \, , \qquad (10.172)$$

and that we have used in the above calculations is valid as long as $\hat{k} = \hat{z}$. Now we have to rotate it in a general direction before performing the Fourier anti-transform. The task is simple because the temperature fluctuation is a scalar. Therefore:

$$\Theta(k, \hat{k} \cdot \hat{p}) = \sum_\ell (-i)^\ell (2\ell + 1) \mathcal{P}_\ell(\hat{k} \cdot \hat{p}) \Theta_\ell(k) \, . \qquad (10.173)$$

The same is not true for Θ_P, since the Stokes parameters are not scalars.

Using the definition of $a_{T,\ell m}$ given in Eq. (7.73) we can then write:

$$a_{T,\ell m}^S = \int d^2\hat{n} \, Y_\ell^{m*}(\hat{n}) \sum_l (-i)^\ell (2\ell + 1) \int \frac{d^3 k}{(2\pi)^3} \mathcal{P}_\ell(\hat{k} \cdot \hat{p}) \alpha(\mathbf{k}) \Theta_\ell(k) \, .$$

$$(10.174)$$

The integration is over $d^2\hat{n}$, hence we must change $\hat{p} \to \hat{n} = -\hat{p}$ in the Legendre polynomial. This gives an extra $(-1)^\ell$ factor, due to the parity of the Legendre polynomials, and then using the addition theorem we obtain:

$$a_{T,\ell m}^S = \int d^2\hat{n} \, Y_\ell^{m*}(\hat{n}) \sum_l i^\ell (2\ell + 1) \int \frac{d^3 k}{(2\pi)^3} \alpha(\mathbf{k}) \frac{4\pi}{2\ell' + 1}$$

$$\sum_{m'=-\ell'}^{\ell'} Y_{\ell'}^{*m'}(\hat{k}) Y_{\ell'}^{m'}(\hat{n}) \Theta_\ell(k) \, . \qquad (10.175)$$

Now the integration over the whole solid angle can be performed and the orthonormality of the spherical harmonics can be employed, obtaining thus:

$$a_{T,\ell m}^S = 4\pi i^\ell \int \frac{d^3\mathbf{k}}{(2\pi)^3} Y_\ell^{m*}(\hat{k})\alpha(\mathbf{k})\Theta_\ell(k) \qquad (10.176)$$

This formula, together with Eq. (10.170) allows us to explicitly calculate the scalar contribution to the $a_{T,\ell m}$'s. Earlier, we have focused on the $C_{TT,\ell}$'s only, for which the calculations are simpler because there is no need of performing a spatial rotation, but we need to know the explicit form of the $a_{T,\ell m}$'s in order to compute the TE correlation spectrum of Eq. (7.87).

10.6 Finite Thickness Effect and Reionization

In this section we discuss two more effects that influence the CMB spectrum, namely the finite thickness effect and the reionisation. The first one is related to the fact that the visibility function g in Fig. 10.4 is very peaked but it is not a Dirac delta. In other words, CMB photons do not last scatter all at once at η_* but during a finite amount of time say $\Delta\eta_*$. This is the **finite thickness effect**. Physically, on scales smaller than the thickness $\Delta\eta_*$ we expect fluctuations to be washed out because they are averaged over a finite amount of time. This is similar to what is called **Landau damping** (although the latter arises from a spread in frequency and not in time). It may seem that Landau damping is a small effect, but actually is of the same order of Silk damping and therefore it must be taken into account.

Let us take advantage of this investigation and derive the form of the visibility function here. The question is: what is the probability that a photon last scatter during some sufficiently small interval between the instants η and $\eta + \Delta\eta$? The time interval $\Delta\eta$ is sufficiently small so that only one collision can take place into it. The attentive reader has noticed that this is the same requirement we make when we derive the Poisson distribution, cf. Chap. 12, which in fact rules the statistics of e.g. scattering process.

So, we divide the time interval $\eta_0 - \eta$ in many, i.e. $N \equiv (\eta_0 - \eta)/\Delta\eta$, intervals and write the probability as:

$$\Delta P = \frac{\Delta\eta}{\eta_c(\eta)}\left[1 - \frac{\Delta\eta}{\eta_c(\eta_1)}\right]\left[1 - \frac{\Delta\eta}{\eta_c(\eta_2)}\right]\cdots\left[1 - \frac{\Delta\eta}{\eta_c(\eta_N)}\right], \qquad (10.177)$$

where recall that $\eta_c \equiv -1/\tau'$ is the average time between two consecutive collisions and it is time-dependent. We have chosen the time interval $\Delta\eta$ small enough in order for $\Delta\eta/\eta_c$ to be the probability of having one scattering during its duration and hence $1 - \Delta\eta/\eta_c$ being that of having no scattering. Now, in the limit $\Delta\eta \to 0$ we can write:

$$dP = \frac{d\eta}{\eta_c}\exp\left(-\int_\eta^{\eta_0}\frac{d\eta}{\eta_c}\right) = -\tau'\exp[-\tau(\eta)]\,d\eta = g(\eta)d\eta\,, \qquad (10.178)$$

i.e. the visibility function defined in Eq. (10.168) appears. As we have anticipated, the maximum of the visibility function occurs in a time that we dub η_*, i.e. the recombination time if we make the assumption of sudden recombination. From the condition for the extrema of a function:

$$g'(\eta) = -\tau'' e^{-\tau} + (\tau')^2 e^{-\tau} = 0 \,, \tag{10.179}$$

we get:

$$- \tau'' = (\tau')^2 \,, \tag{10.180}$$

as the condition which defines η_*. Therefore, employing the definition of the optical depth, we get:

$$(n_e \sigma_T a)'_* = -(n_e \sigma_T a)^2_* \,, \tag{10.181}$$

where the derivative is evaluated at η_*, as well as the function on the right hand side. Now, let us write the free-electron number density as $n_e = X_e n_b$, i.e. introducing the free-electron fraction and the baryon number density and recall from Boltzmann equation, cf. Chap. 3, that:

$$X'_e = -\frac{1.44 \times 10^4}{z} \mathcal{H} X_e \,. \tag{10.182}$$

Evaluating the above equation at η_* and combining it with the extrema condition for the visibility function, we get:

$$X_e(\eta_*) \approx \frac{1.44 \times 10^4}{z_*(n_b \sigma_T a)_*} \mathcal{H}(\eta_*) \equiv \frac{K}{(n_b \sigma_T a)_*} \mathcal{H}(\eta_*) \,, \tag{10.183}$$

and from this we get the recombination redshift $z_* \approx 1050$.

Let us approximate the visibility function by expanding it about its maximum:

$$g(\eta) = \exp[\ln(-\tau') - \tau] \approx \exp\left[-\frac{1}{2}[\tau - \ln(-\tau')]''_*(\eta - \eta_*)^2\right] \,, \tag{10.184}$$

i.e. we have a Gaussian function. Now we determine the second derivative, and hence the variance of the distribution by using Boltzmann equation and employing the following approximation: we consider only the first derivative of X_e to be different from zero. All the other quantities are approximately constant indeed since recombination takes place quite rapidly. So, we can write:

$$\left(\tau' - \frac{\tau''}{\tau'}\right)'_* = \left(-n_b X_e \sigma_T a - \frac{n_b X'_e \sigma_T a}{n_b X_e \sigma_T a}\right)'_* \,. \tag{10.185}$$

Now, within our approximation X'_e/X_e is constant, and thus:

$$\left(\tau' - \frac{\tau''}{\tau'}\right)'_* = -(n_{\rm b} X'_e \sigma_{\rm T} a)_* = (n_{\rm b} X_e \sigma_{\rm T} a)^2_* = K^2 \mathcal{H}(\eta_*)^2 \ . \tag{10.186}$$

Hence the visibility function can be approximated as a Gaussian function:

$$g(\eta) \approx \frac{K\mathcal{H}_*}{\sqrt{2\pi}} \left[-\frac{1}{2}(K\mathcal{H}_*)^2(\eta - \eta_*)^2 \right] , \tag{10.187}$$

with variance $1/(K\mathcal{H}_*)$.

When we substitute this approximation of the visibility function in the line-of-sight integrals of Eqs. (10.170) and (10.171) we can extract the spherical Bessel functions as $j_l[k(\eta_0 - \eta_*)]$, since η_0 is much larger than any conformal time about recombination, and the other integrals are oscillating functions of $kr_s(\eta)$, as we saw in Eq. (10.107). We can approximate this about η_*, as follows:

$$kr_s(\eta) = k \int_0^\eta d\eta' \frac{1}{\sqrt{3(1+R)}} \approx kr_s(\eta_*) + \frac{k}{\sqrt{3(1+R_*)}}(\eta - \eta_*) \ . \tag{10.188}$$

Hence, we have finally a conformal time Gaussian integral of the following type:

$$\int_{-\infty}^\infty d\eta \exp\left[-\frac{1}{2}(K\mathcal{H}_*)^2(\eta - \eta_*)^2 + \frac{ik}{\sqrt{3(1+R_*)}}(\eta - \eta_*) \right] . \tag{10.189}$$

Now, using the formula for the Gaussian integral:

$$\int_{-\infty}^\infty e^{-ax^2+bx} dx = \sqrt{\frac{\pi}{a}} e^{b^2/4a} \ , \tag{10.190}$$

we can conclude that the Θ_ℓ's get an additional damping factor of the following form:

$$\exp\left[-\frac{k^2}{6(1+R_*)(K\mathcal{H}_*)^2} \right] , \tag{10.191}$$

so a new damping scale appears:

$$\boxed{d^2_{\rm Landau} = \frac{1}{k^2_{\rm Landau}} \equiv \frac{1}{6(1+R_*)(K\mathcal{H}_*)^2}} \tag{10.192}$$

which, following Weinberg (2008), we call **Landau damping scale**. For the ΛCDM model best fit parameters we have:

$$\boxed{d_{\rm Landau} \approx 0.0048 \text{ Mpc}} \tag{10.193}$$

Now let us turn to reionisation. At a redshift of about 10 hydrogen gets ionised again by the ultraviolet radiation of the first structures. Hence, the free-electron fraction grows again increasing the probability of a CMB photon to be scattered again, cf. the extra bump in the visibility function in Fig. 10.4. Following the calculation done above in order to obtain the visibility function, we know that the probability for a photon not to be scattered from reionisation until today is:

$$\exp[-\tau(\eta_{\text{reion}})] ,\qquad (10.194)$$

and of course the one for being scattered is 1 minus the above quantity, which we compute now:

$$\tau(\eta_{\text{reion}}) = \int_{\eta_{\text{reion}}}^{\eta_0} d\eta \, n_e \sigma_{\text{T}} a = \sigma_{\text{T}} \int_0^{z_{\text{reion}}} \frac{dz}{(1+z)^2 \mathcal{H}} n_e .\qquad (10.195)$$

For the free-electron number density we can write:

$$n_e = 0.88 n_{\text{b}} X_e = 0.88 \frac{3 H_0^2}{8 \pi m_{\text{b}}} \Omega_{\text{b0}} (1+z)^3 ,\qquad (10.196)$$

where the factor 0.88 is due to the fact that not all the baryons are electrons or protons, but there are also neutrons in Helium nuclei. Assuming matter-domination and instantaneous reionisation, i.e.

$$\mathcal{H}^2 = H_0^2 \Omega_{\text{m0}} (1+z)^3 ,\qquad (10.197)$$

and $X_e = 1$ for $z < z_{\text{reion}}$, we get:

$$\tau(z_{\text{reion}}) \approx 0.04 \frac{\Omega_{\text{b0}} h^2}{\sqrt{\Omega_{\text{m0}} h^2}} z_{\text{reion}}^{3/2} .\qquad (10.198)$$

Hence, the probability for a CMB photon not to be scattered for reionisation taking place at redshift $z_{\text{reion}} = 10$ is about 0.99, i.e. very high. Those photons which are scattered are mixed up, hence the correlation in their temperature is destroyed. So, the effect of reionisation on the $C_{TT,\ell}$'s is simply to weigh them by a factor $\exp(-2\tau_{\text{reion}})$, the factor 2 appearing because the spectrum is a quadratic function of the temperature fluctuations.

10.7   Cosmological Parameters Determination

In this section we discuss how the CMB TT spectrum, i.e. the $C_{TT,\ell}$'s, are sensitive to the cosmological parameters. We have learned in this chapter about many quantities which are of relevance in forming the shape of the spectrum but we have not actually derived an analytic, approximated formula in order to see this explicitly. These can be found in Mukhanov (2005) and Weinberg (2008). Here instead we plot with CLASS various spectra for varying parameters and discuss the physics behind the changes.

Note that, for the standard Λ model, 6 of the overall parameters are usually left free and constrained by observation:

1. The amplitude of the primordial power spectrum: A_S;
2. The primordial tilt: n_S;
3. The baryonic abundance: $\Omega_{b0}h^2$;
4. The CDM abundance: $\Omega_{c0}h^2$;
5. The reionization epoch: z_{reion};
6. The sound horizon at recombination: $r_s(\eta_*)$, which is related to the Hubble constant value H_0.

The other parameters can be derived by these ones. In particular, the amount of radiation is already well known by measuring the CMB temperature and so the amount of Λ and curvature is determined via the positions of the peaks, which depend on $r_s(\eta_*)$, which in turn depends on the baryon content.

In Figs. 10.5 and 10.6 we start to show the numerical calculation of CMB TT power spectrum decomposed in the contributions discussed in this chapter. See also Wands et al. (2016). We consider the ΛCDM as fiducial model.

In Fig. 10.7 we show what happens to CMB TT the spectrum for $\Omega_{b0}h^2 = 0.010, 0.014, 0.018, 0.022, 0.026, 0.030, 0.034$. Taking the first peak height as reference, the larger the value of $\Omega_{b0}h^2$ is, the higher the peak is. When we vary one of the density parameters, since their sum must be equal to one that means that also something else must vary. In this case we have chosen to vary Ω_Λ.

Why so? We have seen that baryons loading makes compression favoured over rarefaction and hence the first and the third peaks are higher for higher values of

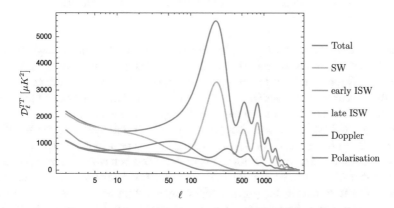

Fig. 10.5 Total CMB TT power spectrum (blue line) computed with CLASS and decomposed in the physically different contributions: Sachs-Wolfe effect (yellow line), early-times ISW effect (green line), late-times ISW effect (red line), Doppler effect (purple line), and polarisation (brown line)

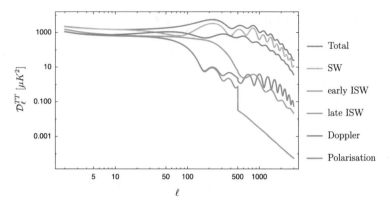

Fig. 10.6 Same as Fig. 10.5 but in logarithmic scale, in order to better distinguish the weakest contributions

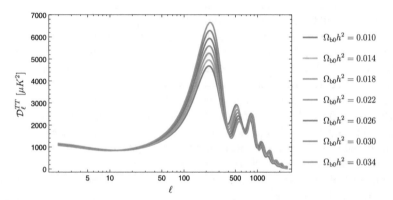

Fig. 10.7 CMB TT power spectrum computed with CLASS and varying $\Omega_{b0}h^2$. From the lowest first peak to the highest: $\Omega_{b0}h^2 = 0.010, 0.014, 0.018, 0.022, 0.026, 0.030, 0.034$

$\Omega_{b0}h^2$, but the second one is lower. In other words, the peaks relative height is very sensitive to the baryon content. The position of the first peak does not change much because it is most sensitive to the spatial curvature and this has been fixed to zero. Finally, the curves for larger $\Omega_{b0}h^2$, as we commented, have less Ω_Λ and therefore less ISW effect. For these reason they are slightly lower for small ℓ.

In Fig. 10.8 we show what happens to CMB TT the spectrum for $\Omega_{c0}h^2 = 0.09, 0.10, 0.11, 0.12, 0.13, 0.14, 0.15$. Taking the first peak height as reference, the larger the value of $\Omega_{c0}h^2$ is, the lower the peak is. This behaviour is the opposite of the one that we found by varying $\Omega_{b0}h^2$. Mostly CDM intervenes through the SW effect since it dominates the gravitational potential Ψ at recombination. The first peak is affected more because it corresponds to large scales, basically the horizon at recombination, and there the transfer function is approximately unit, meaning that $-\Psi$ is as large as possible. The subsequent peaks correspond to scales which entered the horizon much earlier and therefore the CDM influence there is weak.

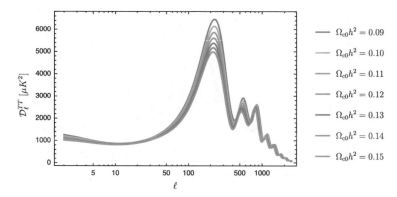

Fig. 10.8 CMB TT power spectrum computed with CLASS and varying $\Omega_{c0}h^2$. From the highest first peak to the lowest: $\Omega_{c0}h^2 = 0.09, 0.10, 0.11, 0.12, 0.13, 0.14, 0.15$

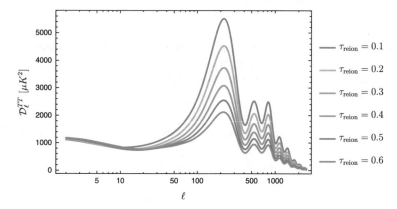

Fig. 10.9 CMB TT power spectrum computed with CLASS and varying τ_{reion}. From the highest first peak to the lowest: $\tau_{\text{reion}} = 0.1, 0.2, 0.3, 0.4, 0.5, 0.6, 0.7$

In this case also we have chosen to vary Ω_Λ in order to keep the total density budget. Indeed, the more CDM, the less Λ and the less ISW effect, as expected.

In Fig. 10.9 we show what happens to CMB TT the spectrum for $\tau_{\text{reion}} = 0.1, 0.2, 0.3, 0.4, 0.5, 0.6, 0.7$. As we have commented in the previous section, the overall effect of reionisation is simple because it happens very lately: a damping of the order $\exp(-2\tau_{\text{reion}})$ for multipoles larger than a certain ℓ_{reion} which we infer to be about 10 from the plots in Fig. 10.9.

From Fig. 10.10 we can appreciate how the CMB TT power spectrum is affected by the spatial geometry of the universe. From the leftmost spectrum to the rightmost one $\Omega_{K0} = -0.2, -0.1, 0, 0.1, 0.2$. Hence, the position of the first peak is of great importance in order to determine whether our universe is closed or open. Note that the flat case is a limiting value which we cannot determine observationally, because

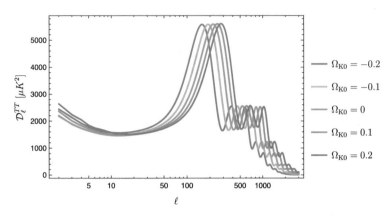

Fig. 10.10 CMB TT power spectrum computed with CLASS and varying Ω_{K0}. From the left to the right: $\Omega_{K0} = -0.2, -0.1, 0, 0.1, 0.2$

of the experimental error; we can only conclude that observation is consistent with $\Omega_{K0} = 0$, i.e. this value is not ruled out.

As we saw in Eq. (10.107), the length scale associated to the acoustic peaks is the sound horizon at recombination:

$$r_s(\eta_*) = \int_0^{\eta_*} c_s d\eta \, , \tag{10.199}$$

where the speed of sound of the baryon-photon plasma is given by Eq. (10.87):

$$c_s^2 = \frac{1}{3(1 + R)} = \frac{4\Omega_{\gamma 0}}{3(4\Omega_{\gamma 0} + 3\Omega_{b0}a)} \, . \tag{10.200}$$

The physical sound horizon is given by:

$$r_s^{\text{phys}}(z_*) = \int_0^{t_*} c_s(t)dt = \int_{z_*}^{\infty} dz \frac{c_s(z)}{H(z)(1 + z)} \, , \tag{10.201}$$

i.e. integrating the lookback time. We need the physical quantity in order to relate it with the angular-diameter distance to recombination:

$$d_A(z_*) = \frac{1}{(1 + z_*)} \int_0^{z_*} \frac{dz}{H(z)} \, , \tag{10.202}$$

and thus estimate the multipole corresponding to the first peak:

$$\ell_{1st} \approx \frac{1}{\theta_{1st}} = \frac{d_A(z_*)}{r_s^{\text{phys}}(z_*)} \, . \tag{10.203}$$

Let us approximate the physical sound horizon by assuming c_s constant and a matter-dominated universe. We have thus:

$$r_s^{\text{phys}}(z_*) \approx \frac{c_s}{H_0\sqrt{\Omega_{m0}}} \int_{z_*}^{\infty} \frac{dz}{(1+z)^{5/2}} = \frac{2c_s}{3H_0\sqrt{\Omega_{m0}}} \frac{1}{(1+z_*)^{3/2}} \,, \qquad (10.204)$$

and for the angular-diameter distance we also assume a matter plus Λ universe:

$$d_A(z_*) = \frac{1}{H_0(1+z_*)} \int_0^{z_*} \frac{dz}{\sqrt{\Omega_{m0}(1+z)^3 + (1-\Omega_{m0})}} \,. \qquad (10.205)$$

Exercise 10.15 Show that $d_A(z_*)$ can be approximated as:

$$d_A(z_*) \approx \frac{2}{7H_0(1+z_*)\sqrt{\Omega_{m0}}}(9 - 2\Omega_{m0}^3) \,. \qquad (10.206)$$

Hence, we have:

$$\ell_{1\text{st}} \approx 0.74\sqrt{1+z_*}(9 - 2\Omega_{m0}^3) \approx 220 \,, \qquad (10.207)$$

which clearly shows how the position of the first peak changes as function of the total matter content.

In Fig. 10.11 we show how the initial conditions dramatically affect the CMB TT power spectrum and how the adiabatic ones are favoured by observation (when comparing with the data points of Fig. 10.1).

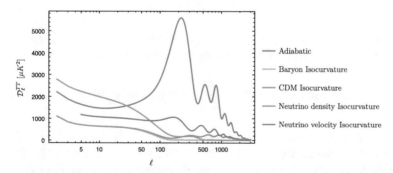

Fig. 10.11 CMB TT power spectrum computed with CLASS and varying initial conditions: adiabatic (blue line), baryon isocurvature (yellow line), CDM isocurvature (green line), neutrino density isocurvature (red line), neutrino velocity isocurvature (purple line)

10.8   Tensor Contribution to the CMB TT Correlation

Tensor perturbations also contribute to generate temperature anisotropies, as we see in Eq. (5.138), which we report here after renormalising to the primordial mode $\beta(\mathbf{k}, \lambda)$, cf. Eq. (7.101):

$$
\left(\frac{\partial}{\partial \eta} + ik\mu - \tau'\right) \Theta^{(T)}(\eta, k, \mu) + \frac{h'^T}{2} =
$$

$$
-\tau' \left[\frac{3}{70}\Theta_4^{(T)} + \frac{1}{7}\Theta_2^{(T)} + \frac{1}{10}\Theta_0^{(T)} - \frac{3}{70}\Theta_{P4}^{(T)} + \frac{6}{7}\Theta_{P2}^{(T)} - \frac{3}{5}\Theta_{P0}^{(T)}\right]
$$

$$
\equiv -\tau' \mathcal{S}^T(\eta, k) , \qquad (10.208)
$$

$$
\left(\frac{\partial}{\partial \eta} + ik\mu - \tau'\right) \Theta_P^{(T)}(\eta, k, \mu) = \tau' \mathcal{S}^T(\eta, k) , \qquad (10.209)
$$

The label λ representing the two possible states of helicity is absent because of the renormalisation with $\beta(\mathbf{k}, \lambda)$. It represents the fact that the evolution of the two helicities is the same.

The line-of-sight solutions of the above equations are the following:

$$
\Theta^{(T)}(\eta_0, k, \mu) = \int_0^{\eta_0} d\eta \, e^{ik\mu(\eta-\eta_0)-\tau} \left[-h'^T/2 - \tau' \mathcal{S}^T(\eta, k)\right] , \qquad (10.210)
$$

$$
\Theta_P^{(T)}(\eta_0, k, \mu) = \int_0^{\eta_0} d\eta \, e^{ik\mu(\eta-\eta_0)-\tau} \tau' \mathcal{S}^T(\eta, k) . \qquad (10.211)
$$

We now focus on $\Theta^{(T)}(\eta_0, k, \mu)$. Defining:

$$
S^T(\eta, k) \equiv e^{-\tau} \left[-h'^T/2 - \tau' \mathcal{S}^T(\eta, k)\right] , \qquad (10.212)
$$

and using Eqs. (5.131) and (5.32), the tensor contribution to the temperature fluctuation is made up of the sum of the following two contributions:

$$
f_\lambda(k\hat{z}, \hat{p}) \equiv 4\sqrt{\frac{\pi}{15}} Y_2^\lambda(\hat{p}) \int_0^{\eta_0} d\eta \, S^T(\eta, k) e^{-i\mu kr(\eta)} , \qquad (10.213)
$$

where $r(\eta) \equiv \eta_0 - \eta$ and where we stress that the result holds true for $\hat{k} = \hat{z}$ since this was the condition under which we derived the Boltzmann equation for photons. We cannot yet sum over λ because we have to include $\beta(\mathbf{k}, \lambda)$ first. For this reason, we shall work on $f_\lambda(k\hat{z}, \hat{p})$.

In order to investigate temperature fluctuations in the sky, we need to anti-transform $\Theta^{(T)}(\mathbf{k}, \hat{p})$ in order to employ the usual expansion:

$$
\Theta^{(T)}(\hat{n}) = \sum_{\ell m} a_{T,\ell m}^T Y_\ell^m(\hat{n}) , \quad a_{T,\ell m}^T = \int d^2\hat{n} \, Y_\ell^{m*}(\hat{n}) \Theta^{(T)}(\hat{n}) . \qquad (10.214)
$$

So, let us proceed as follows. We trade $\hat{p}$ for the line-of-sight $\hat{n} = -\hat{p}$ and use the expansion of a plane wave in spherical harmonics:

$$e^{i\hat{k}\cdot\hat{n}kr} = \sum_{LM} i^L Y_L^{M*}(\hat{k}) Y_L^M(\hat{n}) j_L(kr) , \tag{10.215}$$

in Eq. (10.213).

Exercise 10.16 First of all, since $\hat{k} = \hat{z}$ then show that:

$$Y_L^{M*}(\hat{k}) = Y_L^{M*}(\hat{z}) = \delta_{M0}\sqrt{\frac{2L+1}{4\pi}} , \tag{10.216}$$

i.e. for $\theta = 0$ (representing the $\hat{z}$ direction) the spherical harmonics are non-vanishing only if $M = 0$.

Therefore, we can write:

$$f_\lambda(k\hat{z}, \hat{n}) = \frac{2}{\sqrt{15}} Y_2^\lambda(\hat{n}) \sum_L i^L \sqrt{2L+1} Y_L^0(\hat{n}) \int_0^{\eta_0} d\eta \, S^T(\eta, k) j_L(kr) . \tag{10.217}$$

The idea is now to perform a rotation in order to put $\hat{k}$ in a generic direction. But then also $\hat{n}$ rotates and therefore we need to know how a spherical harmonics behaves under rotations. In order to deal with just one spherical harmonic we take advantage of the following decomposition:

$$Y_2^{\pm 2}(\hat{n}) Y_L^0(\hat{n}) = \sqrt{\frac{5(2L+1)}{4\pi}} \sum_{L'} \sqrt{2L'+1}$$
$$\begin{pmatrix} L & 2 & L' \\ 0 & \pm 2 & \mp 2 \end{pmatrix} \begin{pmatrix} L & 2 & L' \\ 0 & 0 & 0 \end{pmatrix} Y_{L'}^{\pm 2}(\hat{n}) , \tag{10.218}$$

where we have employed the **Wigner $3j$-symbols**, which are coefficients appearing in the quantum theory of angular momentum, when we combine two angular momenta and we want to write the state of total angular momentum as a linear combination on the basis of the tensor product of the two combined angular momenta. They are an alternative to the (perhaps more commonly used) Clebsch-Gordan coefficient See e.g. Landau and Lifshits (1991) and Weinberg (2015).

This expansion allows us to deal with just one spherical harmonics. Now we take advantage of the properties of the spherical harmonics under spatial rotation, i.e.

$$Y_\ell^m(R\hat{n}) = \sum_{m'=-\ell}^{\ell} D_{m'm}^{(\ell)}(R^{-1})Y_\ell^{m'}(\hat{n}) \qquad (10.219)$$

where the $D_{m'm}^{(\ell)}$ are the elements of the **Wigner D-matrix**. See Landau and Lifshits (1991) for more detail. The above R is a generic rotation. Of course, we are interested in a $R(\hat{k})$ rotation which brings $\hat{k}$ in a generic direction. Hence, we can write:

$$f_\lambda(\mathbf{k},\hat{n}) = \sum_L i^L \frac{2L+1}{\sqrt{3\pi}} \sum_{L'} \sqrt{2L'+1} \begin{pmatrix} L & 2 & L' \\ 0 & \lambda & -\lambda \end{pmatrix}$$

$$\begin{pmatrix} L & 2 & L' \\ 0 & 0 & 0 \end{pmatrix} \sum_{m'} D_{m'\lambda}^{(L')}[R(\hat{k})]Y_{L'}^{m'}(\hat{n}) \int_0^{\eta_0} d\eta\, S^T(\eta,k)j_L(kr) . \qquad (10.220)$$

Here we have dubbed $R\hat{n}$ the original line of sight and $\hat{n}$ the resulting one after the rotation.

Now we can perform the Fourier anti-transform. Let us multiply $f_\lambda(\mathbf{k},\hat{n})$ by $\beta(\mathbf{k},\lambda)$ and $Y_\ell^{m*}(\hat{n})$ and integrate over $d^2\hat{n}$ in order to obtain the $a_{T,\ell m}^T$'s. We obtain:

$$a_{\ell m,\pm 2}^T = \sum_L i^L \frac{2L+1}{\sqrt{3\pi}} \sqrt{2\ell+1} \begin{pmatrix} L & 2 & \ell \\ 0 & \pm 2 & \mp 2 \end{pmatrix}$$

$$\begin{pmatrix} L & 2 & \ell \\ 0 & 0 & 0 \end{pmatrix} \int \frac{d^3\mathbf{k}}{(2\pi)^3} D_{m\pm 2}^{(\ell)}[R(\hat{k})]\beta(\mathbf{k},\pm 2) \int_0^{\eta_0} d\eta\, S^T(\eta,k)j_L(kr) . \quad (10.221)$$

We have used here the orthonormality relation of the spherical harmonics and distinguished the contributions from different helicities. Of course $a_{\ell m} = a_{\ell m,+2} + a_{\ell m,-2}$.

It is now time to compute the $3j$ symbols and to perform the summation over L. A general formula for those was obtained in Racah (1942), but we can read their expression from Landau and Lifshits (1991). We then have the only following non-vanishing occurrences:

$$\begin{pmatrix} \ell & 2 & \ell \\ 0 & 0 & 0 \end{pmatrix} = (-1)^{\ell+1} \sqrt{\frac{\ell(\ell+1)}{(2\ell-1)(2\ell+1)(2\ell+3)}} , \qquad (10.222)$$

$$\begin{pmatrix} \ell+2 & 2 & \ell \\ 0 & 0 & 0 \end{pmatrix} = (-1)^\ell \sqrt{\frac{3(\ell+1)(\ell+2)}{2(2\ell+1)(2\ell+3)(2\ell+5)}} , \qquad (10.223)$$

$$\begin{pmatrix} \ell-2 & 2 & \ell \\ 0 & 0 & 0 \end{pmatrix} = (-1)^\ell \sqrt{\frac{3\ell(\ell-1)}{2(2\ell-3)(2\ell-1)(2\ell+1)}} . \qquad (10.224)$$

In particular, there is no contribution coming from $L = \ell \pm 1$. The other three relevant (i.e. not considering those for $L = \ell \pm 1$ which are non-vanishing in this case) symbols are:

$$\begin{pmatrix} \ell & 2 & \ell \\ 0 & \pm2 & \mp2 \end{pmatrix} = (-1)^\ell \sqrt{\frac{3(\ell-1)(\ell+2)}{2(2\ell-1)(2\ell+1)(2\ell+3)}} \,, \tag{10.225}$$

$$\begin{pmatrix} \ell+2 & 2 & \ell \\ 0 & \pm2 & \mp2 \end{pmatrix} = (-1)^\ell \frac{1}{2}\sqrt{\frac{(\ell-1)\ell}{(2\ell+1)(2\ell+3)(2\ell+5)}} \,, \tag{10.226}$$

$$\begin{pmatrix} \ell-2 & 2 & \ell \\ 0 & \pm2 & \mp2 \end{pmatrix} = (-1)^\ell \frac{1}{2}\sqrt{\frac{(\ell+1)(\ell+2)}{(2\ell-3)(2\ell-1)(2\ell+1)}} \,, \tag{10.227}$$

Exercise 10.17 Derive the above expressions for the relevant Wigner $3j$ symbols given in Landau and Lifshits (1991) and put them in Eq. (10.221). Show that:

$$a_{T,\ell m,\pm2}^{T} = -i^\ell \sqrt{\frac{(2\ell+1)(\ell+2)!}{8\pi(\ell-2)!}} \int \frac{d^3k}{(2\pi)^3} D_{m\pm2}^{(\ell)}[R(\hat{k})]\beta(\mathbf{k},\pm2) \int_0^{\eta_0} d\eta\, S^T(\eta,k)$$

$$\left[\frac{j_{\ell-2}(kr)}{(2\ell-1)(2\ell+1)} + \frac{2j_\ell(kr)}{(2\ell-1)(2\ell+3)} + \frac{j_{\ell+2}(kr)}{(2\ell+1)(2\ell+3)}\right] . \tag{10.228}$$

Recall that $r = r(\eta) \equiv \eta_0 - \eta$.

Exercise 10.18 Show that, using the recurrence relation (Abramowitz and Stegun 1972):

$$\frac{j_\ell(x)}{x} = \frac{j_{\ell-1}(x) + j_{\ell+1}(x)}{2\ell+1} \,, \tag{10.229}$$

we can write:

$$a_{T,\ell m}^{T} = -i^\ell \sqrt{\frac{(2\ell+1)(\ell+2)!}{8\pi(\ell-2)!}} \sum_{\lambda=\pm2} \int \frac{d^3k}{(2\pi)^3} D_{m,\lambda}^{(\ell)}[R(\hat{k})]\beta(\mathbf{k},\lambda)$$

$$\int_0^{\eta_0} d\eta\, S^T(\eta,k)\frac{j_\ell(kr)}{(kr)^2} \,. \tag{10.230}$$

The Wigner D-matrix can be related to the spin-weighted spherical harmonics as follows:

$$D_{m,\pm2}^{(\ell)}(\hat{k}) = \sqrt{\frac{4\pi}{2\ell+1}} \,_{\pm2}Y_\ell^{-m}(\hat{k}) = \sqrt{\frac{4\pi}{2\ell+1}} \,_{\mp2}Y_\ell^{m*}(\hat{k}) \,, \tag{10.231}$$

so we have:

$$a_{T,\ell m}^T = -i^\ell \sqrt{\frac{(\ell+2)!}{2(\ell-2)!}} \sum_{\lambda=\pm 2} \int \frac{d^3 k}{(2\pi)^3} \lambda Y_\ell^{m*}(\hat{k}) \beta(\mathbf{k}, \lambda) \int_0^{\eta_0} d\eta \, S^T(\eta, k) \frac{j_\ell(kr)}{(kr)^2}$$

(10.232)

This is our main result of this section. It is not surprising that $Y_\ell^{m*}(\hat{k})$ eventually appeared, being GW a spin-2 field.

In order to compute the tensor contribution to the $C_{TT,\ell}$'s, we perform the ensemble average:

$$\langle a_{T,\ell m}^T a_{T,\ell' m'}^{T*} \rangle = C_{TT,\ell}^T \delta_{\ell\ell'} \delta_{mm'} .$$

(10.233)

Exercise 10.19 Assuming Gaussian perturbations, using Eq. (7.102) and the orthogonality property of the Wigner D-matrices or the spin-weighted spherical harmonics:

$$\int d^2\hat{k} \, D_{m,\pm 2}^{(\ell)}[R(\hat{k})] D_{m',\pm 2}^{(\ell')*}[R(\hat{k})] = \frac{4\pi}{2\ell+1} \delta_{\ell\ell'} \delta_{mm'} ,$$

(10.234)

show that:

$$C_{TT,\ell}^T = \frac{(\ell+2)!}{4\pi(\ell-2)!} \int_0^\infty \frac{dk}{k} \Delta_h^2(k) \left| \int_0^{\eta_0} d\eta \, S^T(\eta, k) \frac{j_\ell(kr)}{(kr)^2} \right|^2$$

(10.235)

Note that a factor 2 arises because of the two polarisation states.

The above result was originally obtained in Abbott and Wise (1984) (though not exactly in the same way and final form).

The main difficulty we faced in computing the $a_{T,\ell m}^T$ was the spatial rotation which brought $\hat{k}$ in a generic direction. This can be avoided if we calculate straightaway $C_{TT,\ell}^T$ because it is rotationally invariant. Note that no correlation exists between scalar and tensor modes. In fact if we compute:

$$\langle a_{T,\ell m}^T a_{T,\ell' m'}^{S*} \rangle ,$$

(10.236)

we would get zero, mathematically because of the integral:

$$\int d^2\hat{k} \, _2Y_\ell^m(\hat{k}) Y_{\ell'}^{m'}(\hat{k}) = 0 ,$$

(10.237)

between a spin-2 spherical harmonic and a spin-0 one. Physically, because we know that at the linear order scalar and tensor perturbations do not couple.

We can again approximate this angular power spectrum for large values of ℓ as follows. First, $S^T(\eta, k)$ contains the derivative of h, which is maximum when a mode enters the horizon, for $k\eta \approx 1$, being almost zero elsewhere. Therefore, assuming instantaneous recombination, we can write:

$$C_{TT,\ell}^T = \frac{(\ell-1)\ell(\ell+1)(\ell+2)}{4\pi} \int_0^\infty \frac{dk}{k} \Delta_h^2(k) \frac{j_\ell^2(k\eta_0)}{(k\eta_0)^4} . \tag{10.238}$$

Defining the new variable $x \equiv k\eta_0$ and introducing the primordial tensor power spectrum we get:

$$C_{TT,\ell}^T \propto \frac{(\ell-1)\ell(\ell+1)(\ell+2)}{4\pi} \int_0^\infty dx\, x^{n_T-5} j_\ell^2(x) . \tag{10.239}$$

The integral can be performed exactly:

$$\int_0^\infty dx\, x^{n_T-5} = \frac{\sqrt{\pi}}{2} \frac{\Gamma[1-(n_T-4)/2]\Gamma[(n_T-4)/2+\ell]}{(4-n_T)\Gamma[1/2-(n_T-4)/2]\Gamma[\ell+2-(n_T-4)/2]}, \tag{10.240}$$

but in the case of $n_T = 0$, a scale-invariant primordial tensor spectrum, we get:

$$\frac{\ell(\ell+1)C_{TT,\ell}^T}{2\pi} \propto \frac{\ell(\ell+1)}{(\ell-2)(\ell+3)} . \tag{10.241}$$

The behaviour of the tensor contribution to the TT power spectrum is thus very different from the one coming from scalar perturbations. In Figs. 10.12 and 10.13 we display the numerical calculations done with CLASS of the total (solid line), scalar (dashed line) and tensor (dotted line) angular power spectra.

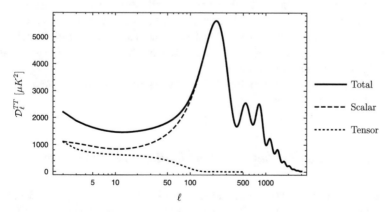

Fig. 10.12 Numerical calculations done with CLASS of the total (solid line), scalar (dashed line) and tensor (dotted line) angular power spectra

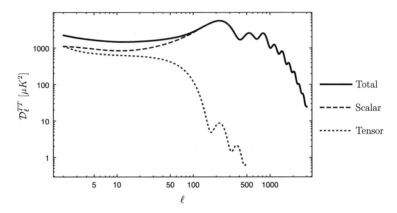

Fig. 10.13 Same as Fig. 10.12 but using a logarithmic scale

The tensor contribution is practically irrelevant on very small angular scale (i.e. large ℓ) and on large angular scales they can be as large as 10% of the total. Typically then one can give upper limits on $C^T_{TT,\ell}/C^S_{TT,\ell}$ for small multipoles ($\ell = 2$ or $\ell = 10$) and this ratio is proportional to A_T/A_S and therefore on the parameter r, the tensor-to-scalar ratio. From Eq. (8.176) we saw that $r < 0.1$. In order to determine this constraint one has also to use polarisation data since with these we are able to disentangle the $A_S \exp(-2\tau_{\text{reion}})$ dependence coming from the scalar contribution only to the temperature power spectrum.

10.9   Polarisation

In this section we address CMB polarisation. Recall that before recombination polarisation is also erased because of tight-coupling. Polarisation is generated thanks to the fact that recombination does not take place instantaneously, so the finite-thickness effect is indeed important. Moreover, since Thomson scattering is axially-symmetric, circular polarisation is not produced.

In Sect. 12.7 we recall the main terminology regarding polarisation and in particular the Stokes parameters.

10.9.1   Scalar Perturbations Contribution to Polarisation

Now, let us focus on scalar perturbations only and write down from Eq. (5.115) the line of sight solution for the combination $Q + iU$. Since we have chosen a reference frame in which $\hat{k} = \hat{z}$, there is no U polarisation. This can be also seen from the fact that $\mathcal{B}^0 = 0$. Hence, we shall again perform a rotation in order to compute the $a_{P,\ell m}$.

We have called Θ_P the Stokes parameter Q in the $\hat{k} = \hat{z}$ frame. So, let us work on its line-of-sight solution.

Exercise 10.20 Show that:

$$\Theta_P(k\hat{z}, \hat{n}) = \frac{3}{2}\sqrt{\frac{8\pi}{15}} {}_2Y_2^0(\hat{n}) \int_0^{\eta_0} d\eta \, e^{-i\mu k r} S_P^S(\eta, k) , \tag{10.242}$$

where we have defined a new source term:

$$S_P^S(\eta, k) \equiv g(\eta)\Pi(\eta, k) . \tag{10.243}$$

We could have written ${}_{-2}Y_2^0(\hat{n})$ instead of ${}_2Y_2^0(\hat{n})$, since they are equal. However, we are going to deal with $Q + iU$ first. The above equation can written as:

$$\Theta_P(k\hat{z}, \hat{n}) = \sqrt{\frac{9}{30}} {}_2Y_2^0(\hat{p}) \sum_L i^L \sqrt{2L + 1} Y_L^0(\hat{n}) \int_0^{\eta_0} d\eta \, S_P^S(\eta, k) j_L(kr) , \tag{10.244}$$

where again $r \equiv \eta_0 - \eta$ and we have used the well-known-by-now expansion of a plane wave into spherical harmonics plus the fact that $\hat{k} = \hat{z}$.

Now, as in Eq. (10.218) we can write the product of spherical harmonics as follows:

$${}_2Y_2^0(\hat{n})Y_L^0(\hat{n}) = \sqrt{\frac{5(2L + 1)}{4\pi}} \sum_{L'} \sqrt{2L' + 1}$$
$$\begin{pmatrix} L & 2 & L' \\ 0 & -2 & +2 \end{pmatrix} \begin{pmatrix} L & 2 & L' \\ 0 & 0 & 0 \end{pmatrix} {}_2Y_{L'}^0(\hat{n}) , \tag{10.245}$$

and thus obtain:

$$(Q + iU)^S(\hat{n}) = \sqrt{\frac{3}{8\pi}} \sum_L i^L (2L + 1) \sum_{L'} \sqrt{2L' + 1} \begin{pmatrix} L & 2 & L' \\ 0 & -2 & 2 \end{pmatrix}$$
$$\begin{pmatrix} L & 2 & L' \\ 0 & 0 & 0 \end{pmatrix} \sum_{m'} {}_2Y_{L'}^{m'}(\hat{n}) \int \frac{d^3k}{(2\pi)^3} D_{m'0}^{(L')}(\hat{k})\alpha(\mathbf{k}) \int_0^{\eta_0} d\eta \, S_P^S(\eta, k) j_L(kr) , \tag{10.246}$$

where we have already considered the rotation which brings $\hat{k}$ in a generic direction.

Now, from the expansion:

$$(Q + iU)^S(\hat{n}) = \sum_{\ell m} a_{P,\ell m}^S \, {}_2Y_\ell^m(\hat{n}) , \tag{10.247}$$

we are able to calculate the coefficients $a^S_{P,\ell m}$ by taking advantage of the orthonormality of the spin-2 spherical harmonics. We can therefore write:

$$a^S_{P,\ell m} = \sqrt{\frac{3}{8\pi}} \sum_L i^L (2L+1)\sqrt{2\ell+1} \begin{pmatrix} L & 2 & \ell \\ 0 & -2 & 2 \end{pmatrix}$$

$$\begin{pmatrix} L & 2 & \ell \\ 0 & 0 & 0 \end{pmatrix} \int \frac{d^3 k}{(2\pi)^3} D^{(\ell)}_{m0}(\hat{k})\alpha(\mathbf{k}) \int_0^{\eta_0} d\eta\, S^S_P(\eta,k) j_L(kr) \;. \tag{10.248}$$

Remarkably, the sum over L can be performed in the very same way we did for the $a^T_{T,\ell m}$, since the $3j$ symbols are the same. Therefore, we have:

$$a^S_{P,\ell m} = -\frac{3i^\ell}{8}\sqrt{\frac{(2\ell+1)(\ell+2)!}{\pi(\ell-2)!}} \int \frac{d^3 k}{(2\pi)^3} D^{(\ell)}_{m0}(\hat{k})\alpha(\mathbf{k})$$

$$\int_0^{\eta_0} d\eta\, S^S_P(\eta,k)\frac{j_\ell(kr)}{(kr)^2} \;, \tag{10.249}$$

and using

$$D^{(\ell)}_{m0}(\hat{k}) = \sqrt{\frac{4\pi}{2\ell+1}} Y^{-m}_\ell(\hat{k}) = \sqrt{\frac{4\pi}{2\ell+1}} Y^{m*}_\ell(\hat{k}) \;, \tag{10.250}$$

we can write:

$$\boxed{a^S_{P,\ell m} = -\frac{3i^\ell}{4}\sqrt{\frac{(\ell+2)!}{(\ell-2)!}} \int \frac{d^3 k}{(2\pi)^3} Y^{m*}_\ell(\hat{k})\alpha(\mathbf{k}) \int_0^{\eta_0} d\eta\, S^S_P(\eta,k)\frac{j_\ell(kr)}{(kr)^2}}$$

$$\tag{10.251}$$

The expansion for $(Q - iU)^S(\hat{n})$ can be obtained by complex conjugation, i.e.

$$(Q - iU)(\hat{n}) = \sum_{\ell m} a^*_{P,\ell m}\, {}_2Y^{m*}_\ell(\hat{n}) = \sum_{\ell m} a^*_{P,\ell m}\, {}_{-2}Y^{-m}_\ell(\hat{n})$$

$$= \sum_{\ell m} a^*_{P,\ell,-m}\, {}_{-2}Y^m_\ell(\hat{n}) \;. \tag{10.252}$$

There is no reality condition here holding true for the $a_{P,\ell m}$ as the one holding true for the $a_{T,\ell m}$, because $Q + iU$ is not real and is not a scalar. It is thus convenient to define the following combinations:

$$a_{E,\ell m} \equiv -(a_{P,\ell m} + a^*_{P,\ell,-m})/2 \;, \quad a_{B,\ell m} \equiv i(a_{P,\ell m} - a^*_{P,\ell,-m})/2 \;, \tag{10.253}$$

because the first has parity $(-1)^\ell$ whereas the second $(-1)^{\ell+1}$. Thus $Q \pm iU$ can be expanded as:

$$(Q \pm iU)(\hat{n}) = \sum_{\ell m} (-a_{E,\ell m} \mp i a_{B,\ell m}) \, {}_2Y_\ell^m(\hat{n}) . \qquad (10.254)$$

Now, if we compute $a_{P,\ell m}^{S*}$, we obtain:

$$a_{P,\ell m}^{S*} = -\frac{3(-i)^\ell}{4} \sqrt{\frac{(\ell+2)!}{(\ell-2)!}} \int \frac{d^3\mathbf{k}}{(2\pi)^3} Y_\ell^{-m*}(\hat{k}) \alpha(-\mathbf{k})$$

$$\int_0^{\eta_0} d\eta \, S_P^S(\eta, k) \frac{j_\ell(kr)}{(kr)^2} , \qquad (10.255)$$

since $\alpha(\mathbf{k})^* = \alpha(-\mathbf{k})$ because of the reality condition of the power spectrum. Changing the integration variable to $\mathbf{k}$ and using the parity of the spherical harmonic:

$$Y_\ell^{-m*}(-\hat{k}) = (-1)^\ell Y_\ell^{-m*}(\hat{k}) , \qquad (10.256)$$

we can finally conclude that:

$$a_{P,\ell m}^S = a_{P,\ell,-m}^{S*} , \qquad (10.257)$$

and therefore scalar perturbations only affect the E-mode, i.e.

$$a_{E,\ell m}^S = -a_{P,\ell m}^S , \qquad a_{B,\ell m}^S = 0 . \qquad (10.258)$$

This means that, if the B-mode was detected, it would be a clear indication of the existence of primordial gravitational waves.

From Eq. (10.251) we can then obtain the scalar contribution to the EE spectrum. Assuming adiabatic Gaussian perturbations:

$$\boxed{C_{EE,\ell}^S = \frac{9}{64\pi} \frac{(\ell+2)!}{(\ell-2)!} \int \frac{dk}{k} \Delta_\mathcal{R}^2 \left| \int_0^{\eta_0} d\eta \, S_P^S(\eta, k) \frac{j_\ell(kr)}{(kr)^2} \right|^2} \qquad (10.259)$$

Using instead Eq. (10.176) we can compute the cross-correlation TE multipole coefficients:

$$\boxed{C_{TE,\ell}^S = -\frac{3}{4} \sqrt{\frac{(\ell+2)!}{(\ell-2)!}} \int \frac{dk}{k} \Delta_\mathcal{R}^2 \Theta_\ell(k) \int_0^{\eta_0} d\eta \, S_P^S(\eta, k) \frac{j_\ell(kr)}{(kr)^2}} \qquad (10.260)$$

10.9.2 Tensor Perturbations Contribution to Polarisation

Let us now calculate the contribution to CMB polarisation coming from tensor perturbations. From Eq. (10.211) we have

$$\Theta_P^{(T)}(\eta_0, k\hat{z}, \mu) = \int_0^{\eta_0} d\eta\, e^{ik\mu(\eta-\eta_0)} S_P^T(\eta, k)\,, \qquad (10.261)$$

with

$$S_P^T(\eta, k) \equiv g(\eta)\mathcal{S}^T(\eta, k)\,. \qquad (10.262)$$

and then use Eq. (5.132) in order to write part of the tensor contribution to polarization:

$$\mathcal{Q}_\lambda^{(T)}(k\hat{z}, \hat{p}) \equiv \sqrt{\frac{8\pi}{5}} \mathcal{E}^\lambda(\hat{p}) \int_0^{\eta_0} d\eta\, e^{-i\mu kr(\eta)} S_P^T(\eta, k)\,, \qquad (10.263)$$

where $r(\eta) \equiv \eta_0 - \eta$ and where we stress that the result holds true for $\hat{k} = \hat{z}$ since this was the condition under which we derived the Boltzmann equation for photons.

In the scalar case no U contribution to polarisation is produced, so the above expression already furnishes the quantity $Q + iU$. However, the same is not true for tensor perturbation. We have thus to add the iU contribution. As we saw in Chap. 5 this is equal to $\mathcal{B}^m Q/\mathcal{E}^m$ and for this reason we have just one polarisation hierarchy. Hence, summing up we get:

$$(\mathcal{Q}_\lambda + i\mathcal{U}_\lambda)^{(T)}(k\hat{z}, \hat{n}) = \sqrt{\frac{32\pi}{5}}\, _2Y_2^\lambda(\hat{n}) \int_0^{\eta_0} d\eta\, e^{i\hat{k}\cdot\hat{n}kr(\eta)} S_P^T(\eta, k)\,. \qquad (10.264)$$

Using the usual plane-wave expansion and recalling that $\hat{k} = \hat{z}$ we get:

$$(\mathcal{Q}_\lambda + i\mathcal{U}_\lambda)^{(T)}(k\hat{z}, \hat{n}) = \sqrt{\frac{8}{5}}\, _2Y_2^\lambda(\hat{n}) \sum_L i^L \sqrt{2L+1}\, Y_L^0(\hat{n})$$
$$\int_0^{\eta_0} d\eta\, S_P^T(\eta, k)\, j_L(kr)\,. \qquad (10.265)$$

The product of the two spherical harmonics can be written via the Wigner $3j$-symbols as follows:

$$_2Y_2^{\pm 2}(\hat{n})Y_L^0(\hat{n}) = \sqrt{\frac{5(2L+1)}{4\pi}} \sum_{L'} \sqrt{2L'+1}$$
$$\begin{pmatrix} L & 2 & L' \\ 0 & \pm 2 & \mp 2 \end{pmatrix} \begin{pmatrix} L & 2 & L' \\ 0 & -2 & 2 \end{pmatrix} {}_2Y_{L'}^{\pm 2}(\hat{n})\,, \qquad (10.266)$$

and we rotate in a generic $\hat{k}$ direction the only spherical harmonic left, i.e.

$$_2Y_{L'}^{\pm 2}(R\hat{n}) = \sum_{m'=-L'}^{L'} D_{m',\pm 2}^{(L')}[R^{-1}(\hat{k})]\, _2Y_{L'}^{m'}(\hat{n})\,. \qquad (10.267)$$

Hence, we can write:

$$(\mathcal{Q}_\lambda + i\mathcal{U}_\lambda)^{(T)}(\mathbf{k}, \hat{n}) = \sqrt{\frac{2}{\pi}} \sum_L i^L (2L+1) \sum_{L'} \sqrt{2L'+1} \begin{pmatrix} L & 2 & L' \\ 0 & \lambda & -\lambda \end{pmatrix}$$

$$\begin{pmatrix} L & 2 & L' \\ 0 & -2 & 2 \end{pmatrix} \sum_{m'} D_{m'\lambda}^{(L')}[R(\hat{k})] Y_{L'}^{m'}(\hat{n}) \int_0^{\eta_0} d\eta \, S_P^T(\eta, k) j_L(kr) \, . \quad (10.268)$$

Now we can perform the Fourier anti-transform. Multiply by $\beta(\mathbf{k}, \lambda)$ and $_2Y_\ell^{m*}(\hat{n})$ and integrate over $d^2\hat{n}$ in order to obtain the $a_{P,\ell m}^T$'s. We obtain:

$$a_{P,\ell m,\pm 2}^T = \sqrt{\frac{2}{\pi}} \sum_L i^L (2L+1)\sqrt{2\ell+1} \begin{pmatrix} L & 2 & \ell \\ 0 & \pm 2 & \mp 2 \end{pmatrix}$$

$$\begin{pmatrix} L & 2 & \ell \\ 0 & -2 & 2 \end{pmatrix} \int \frac{d^3\mathbf{k}}{(2\pi)^3} D_{m\pm 2}^{(\ell)}[R(\hat{k})]\beta(\mathbf{k}, \lambda) \int_0^{\eta_0} d\eta \, S_P^T(\eta, k) j_L(kr) \, . \quad (10.269)$$

We have used here the orthonormality relation of the spin-2 spherical harmonics and distinguished the contributions of different helicity. Of course $a_{P,\ell m}^T = a_{P,\ell m,+2}^T + a_{P,\ell m,-2}^T$.

We have already computed some of the $3j$ symbols earlier, for the tensor case but now two more enter: those for $L = \ell \pm 1$. We shall see that these contributions will characterise the B-mode of polarisation. They are:

$$\begin{pmatrix} \ell+1 & \ell & 2 \\ 0 & -2 & +2 \end{pmatrix} = (-1)^{\ell+1}\sqrt{\frac{(\ell-1)}{2(2\ell+1)(2\ell+3)}} \, , \quad (10.270)$$

$$\begin{pmatrix} \ell & \ell-1 & 2 \\ -2 & 0 & +2 \end{pmatrix} = (-1)^\ell \sqrt{\frac{(\ell+2)}{2(2\ell-1)(2\ell+1)}} \, . \quad (10.271)$$

Extra care has to be used when manipulating these terms. The reason is that the $3j$ symbols gain an overall phase factor

$$(-1)^{j_1+j_2+j_3} \, , \quad (10.272)$$

where $j_{1,2,3}$ are the momenta which are being combined, each time we swap two columns or change simultaneously all the signs of the bottom row.[4] Therefore, as long as $j_1 + j_2 + j_3$ is even, no matter how many times we perform the above operations. This is the case for $L = \ell \pm 2$ or $L = \ell$. However, for $L = \ell \pm 1$ we have that

---

[4] These signs can be changed only simultaneously since the sums $m_1 + m_2 + m_3 = 0$ always. This is a selection rule.

$$L + \ell + 2 = 2(\ell + 1) \pm 1 \,, \tag{10.273}$$

which is odd and thus we have to keep track of the correct sign.

Exercise 10.21 Using the formulas for the $3j$ symbols, show that:

$$a^T_{P,\ell m,\pm 2} = \frac{-i^\ell \sqrt{2\ell+1}}{\sqrt{8\pi}} \int \frac{d^3\mathbf{k}}{(2\pi)^3} D^{(\ell)}_{m\pm 2}[R(\hat{k})]\beta(\mathbf{k},\lambda) \int_0^{\eta_0} d\eta \, S^T_P(\eta,k)$$

$$\left[\frac{(\ell+1)(\ell+2)}{(2\ell-1)(2\ell+1)} j_{\ell-2}(kr) - \frac{6(\ell-1)(\ell+2)}{(2\ell-1)(2\ell+3)} j_\ell(kr) + \frac{(\ell-1)\ell}{(2\ell+1)(2\ell+3)} j_{\ell+2}(kr) \right.$$

$$\left. \pm 2i \frac{\ell-1}{2\ell+1} j_{\ell+1}(kr) \mp 2i \frac{\ell+2}{2\ell+1} j_{\ell-1}(kr) \right] \,. $$
$$\tag{10.274}$$

Recall that $r = r(\eta) \equiv \eta_0 - \eta$.

Exercise 10.22 Show that, using the recurrence relation of Eq. (10.229) and the following ones for the derivatives:

$$j'_\ell(x) = j_{\ell-1}(x) - \frac{\ell+1}{x} j_\ell(x) \,, \quad j'_\ell(x) = \frac{\ell}{x} j_\ell(x) - j_{\ell+1}(x) \,, \tag{10.275}$$

we can write:

$$a^T_{P,\ell m} = \frac{-i^\ell \sqrt{2\ell+1}}{\sqrt{8\pi}} \sum_{\lambda=\pm 2} \int \frac{d^3\mathbf{k}}{(2\pi)^3} D^{(\ell)}_{m,\lambda}[R(\hat{k})]\beta(\mathbf{k},\lambda) \int_0^{\eta_0} d\eta \, S^T_P(\eta,k)$$

$$\left[\frac{2}{kr} j'_\ell - 2 j_\ell + \frac{2+\ell(\ell+1)}{(kr)^2} j_\ell - i\lambda \left(j'_\ell + \frac{2}{kr} j_\ell \right) \right] \,. $$
$$\tag{10.276}$$

Show that the term between square brackets is equal to the corresponding one in Weinberg (2008, p. 389). In order to make the second derivative of the spherical Bessel function to appear one must use Bessel differential equation:

$$j''_\ell + \frac{2}{x} j_\ell + \frac{1-\ell(\ell+1)}{x^2} j_\ell = 0 \,. \tag{10.277}$$

Now we are ready to investigate the reality property of $a^T_{P,\ell m}$ and discern from it the E-mode and B-mode contributions. Recall that Wigner D-matrix can be related to the spin-weighted spherical harmonics as follows:

$$D_{m,\pm 2}^{(\ell)}(\hat{k}) = \sqrt{\frac{4\pi}{2\ell+1}} \, {}_{\pm 2}Y_\ell^{-m}(\hat{k}) = \sqrt{\frac{4\pi}{2\ell+1}} \, {}_{\mp 2}Y_\ell^{m*}(\hat{k}) \ . \tag{10.278}$$

So taking the complex conjugate we find:

$$a_{P,\ell m}^{T*} = -\frac{(-i)^\ell}{\sqrt{2}} \sum_{\lambda=\pm 2} \int \frac{d^3\mathbf{k}}{(2\pi)^3} {}_{\mp 2}Y_\ell^m(\hat{k})\beta(-\mathbf{k},\lambda) \int_0^{\eta_0} d\eta \, S_P^T(\eta,k)$$
$$\left[\frac{2}{kr}j_\ell' - 2j_\ell + \frac{2+\ell(\ell+1)}{(kr)^2}j_\ell + i\lambda\left(j_\ell' + \frac{2}{kr}j_\ell\right) \right] \ . \tag{10.279}$$

Note that $\beta(\mathbf{k},\lambda)* = \beta(-\mathbf{k},\lambda)$ because of the reality condition and beware that the sign of the imaginary unit inside the square brackets has changed. Now, changing integration variable

$$\mathbf{k} \to -\mathbf{k} \ , \tag{10.280}$$

and taking advantage of the parity property and the complex conjugation property:

$$\mp 2Y_\ell^m(-\hat{k}) = (-1)^\ell {}_{\pm 2}Y_\ell^m(\hat{k}) = (-1)^\ell {}_{\mp 2}Y_\ell^{-m*}(\hat{k}) \ , \tag{10.281}$$

we get:

$$a_{P,\ell m}^{T*} = -\frac{i^\ell}{\sqrt{2}} \sum_{\lambda=\pm 2} \int \frac{d^3\mathbf{k}}{(2\pi)^3} {}_{\mp 2}Y_\ell^{-m*}(\hat{k})\beta(\mathbf{k},\lambda) \int_0^{\eta_0} d\eta \, S_P^T(\eta,k)$$
$$\left[\frac{2}{kr}j_\ell' - 2j_\ell + \frac{2+\ell(\ell+1)}{(kr)^2}j_\ell + i\lambda\left(j_\ell' + \frac{2}{kr}j_\ell\right) \right] \ . \tag{10.282}$$

This time we have not the same situation as in Eq. (10.257) because of the $i\lambda$ contribution. Hence, we can compute the E-mode:

$$a_{E,\ell m}^T = \frac{i^\ell}{\sqrt{2}} \sum_{\lambda=\pm 2} \int \frac{d^3\mathbf{k}}{(2\pi)^3} {}_{\mp 2}Y_\ell^{m*}(\hat{k})\beta(\mathbf{k},\lambda) \int_0^{\eta_0} d\eta \, S_P^T(\eta,k)$$
$$\left[\frac{2}{kr}j_\ell' - 2j_\ell + \frac{2+\ell(\ell+1)}{(kr)^2}j_\ell \right] \ . \tag{10.283}$$

and the B-mode is also present:

$$\boxed{a_{B,\ell m}^T = -\frac{i^\ell}{\sqrt{2}} \sum_{\lambda=\pm 2} \lambda \int \frac{d^3\mathbf{k}}{(2\pi)^3} {}_{\mp 2}Y_\ell^{m*}(\hat{k})\beta(\mathbf{k},\lambda) \int_0^{\eta_0} d\eta \, S_P^T(\eta,k) \left(j_\ell' + \frac{2}{kr}j_\ell \right)}$$
$$\tag{10.284}$$

Now we are in position of giving the formulas for the angular power spectra. Assuming Gaussian perturbations we have:

$$C_{EE,\ell}^T = \int \frac{dk}{4\pi k} \Delta_h^2(k) \left| \int_0^{\eta_0} d\eta\, S_P^T(\eta, k) \left[\frac{2}{kr} j_\ell' - 2 j_\ell + \frac{2 + \ell(\ell+1)}{(kr)^2} j_\ell \right] \right|^2,$$

$$(10.285)$$

and

$$C_{BB,\ell}^T = \int \frac{dk}{4\pi k} \Delta_h^2 \left| \int_0^{\eta_0} d\eta\, S_P^T(\eta, k) \left(2 j_\ell' + \frac{4}{kr} j_\ell \right) \right|^2. \qquad (10.286)$$

The cross-correlation $C_{TE,\ell}^T$, using Eq. (10.232) gives:

$$C_{TE,\ell}^T = -\sqrt{\frac{(\ell+2)!}{(\ell-2)!}} \int \frac{dk}{8\pi k} \Delta_h^2(k) \int_0^{\eta_0} d\eta\, S^T(\eta, k) \frac{j_\ell}{(kr)^2}$$

$$\int_0^{\eta_0} d\eta'\, S_P^T(\eta', k) \left[\frac{2}{kr} j_\ell' - 2 j_\ell + \frac{2 + \ell(\ell+1)}{(kr)^2} j_\ell \right]. \qquad (10.287)$$

If we try to compute the cross correlations $C_{TB,\ell}^T$ and $C_{EB,\ell}^T$ we obtain a vanishing result, as expected, because of the term λ in the sum of Eq. (10.284). In fact we get, considering for example $C_{EB,\ell}^T$:

$$C_{EB,\ell}^T = -\sum_{\lambda=\pm 2} \frac{\lambda}{2} \int \frac{dk}{4\pi k} \Delta_h^2 \int_0^{\eta_0} d\eta\, S_P^T(\eta, k) \left[\frac{2}{kr} j_\ell' - 2 j_\ell + \frac{2 + \ell(\ell+1)}{(kr)^2} j_\ell \right]$$

$$\int_0^{\eta_0} d\bar{\eta}\, S_P^T(\bar{\eta}, k) \left(2 j_\ell' + \frac{4}{kr} j_\ell \right). \qquad (10.288)$$

Now, the sum over λ is equivalent to a difference and since nothing else depends on λ the result is zero. The same happens with the correlation $C_{TB,\ell}^T$.

In Fig. 10.14 we display the 4 angular CMB power spectra which constitute a wealth of cosmological information. The only possible cross-correlation is the one between temperature and the E-mode of polarisation. The largest signal is the TT one and then, in order of decreasing power, the TE, EE and BB one. The latter is 5 orders of magnitude smaller than the TT one and it has not yet been detected. The bump it displays for small ℓ's is due to reionisation.

From Fig. 10.14 we can appreciate how small the polarisation spectra are with respect to the TT one. This is due to the fact that a quadrupole moment in the distribution of photons is needed in order to have production of polarisation. Before recombination, Thomson scattering rate is so high that photons are in nearly perfect

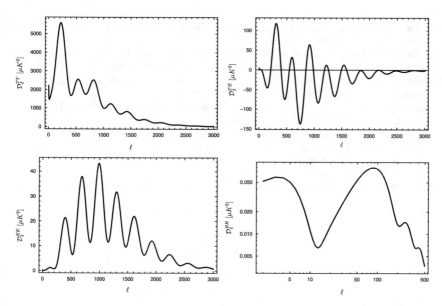

Fig. 10.14 The four angular power spectra characterising the CMB, computed with CLASS for the standard model. Top left: TT. Top right: TE. Bottom left: EE. Bottom right: BB. As for the TT spectrum, $\mathcal{D} \equiv \ell(\ell + 1)C_\ell/(2\pi)$

thermal equilibrium and any moment from the quadrupole up is washed out. After recombination, photons free stream and thus have no more chance of being polarised by Thomson scattering.

References

Abbott, L.F., Wise, M.B.: Constraints on generalized inflationary cosmologies. Nucl. Phys. B **244**, 541–548 (1984)

Abramowitz, M., Stegun, I.A.: Handbook of Mathematical Functions With Formulas, Graphs, and Mathematical Tables. Dover (1972)

Ade, P.A.R., et al.: Planck 2015 results. XIII. Cosmological parameters. Astron. Astrophys. **594**, A13 (2016)

Eisenstein, D.J., et al.: Detection of the baryon acoustic peak in the large-scale correlation function of SDSS luminous red galaxies. Astrophys. J. **633**, 560–574 (2005)

Hu, W., Sugiyama, N.: Small scale cosmological perturbations: an analytic approach. Astrophys. J. **471**, 542–570 (1996)

Kaiser, N.: Small-angle anisotropy of the microwave background radiation in the adiabatic theory. MNRAS **202**, 1169–1180 (1983)

Landau, L.D., Lifshits, E.M.: Quantum Mechanics. Volume 3 of Course of Theoretical Physics. Butterworth-Heinemann, Oxford (1991)

Ma, C.-P., Bertschinger, E.: Cosmological perturbation theory in the synchronous and conformal Newtonian gauges. Astrophys. J. **455**, 7–25 (1995)

Mukhanov, V.F.: CMB-slow, or how to estimate cosmological parameters by hand. Int. J. Theor. Phys. **43**, 623–668 (2004)

Mukhanov, V.: Physical Foundations of Cosmology. Cambridge University Press, Cambridge (2005)

Racah, G.: Theory of complex spectra. II. Phys. Rev. **62**(9–10), 438 (1942)

Sachs, R.K., Wolfe, A.M.: Perturbations of a cosmological model and angular variations of the microwave background. Astrophys. J. **147**, 73–90 (1967)

Seljak, U., Zaldarriaga, M.: A line of sight integration approach to cosmic microwave background anisotropies. Astrophys. J. **469**, 437–444 (1996)

Silk, J.: Fluctuations in the primordial fireball. Nature **215**(5106), 1155–1156 (1967)

Wands, D., Piattella, O.F., Casarini, L.: Physics of the cosmic microwave background radiation. Astrophys. Space Sci. Proc. **45**, 3–39 (2016)

Weinberg, S.: Cosmology. Oxford University Press, Oxford (2008)

Weinberg, S.: Lectures on Quantum Mechanics. Cambridge University Press, Cambridge (2015)

Chapter 11
Miscellanea

*E desse modo ele ia levando a vida, metade na repartição, sem
ser compreendido, a outra metade em casa, também sem ser
compreendido
(And in this way he took life, half in the division, without being
understood, the other half at home, also without being
understood)*

Lima Barreto, Triste fim de Policarpo Quaresma

In this Chapter we collect some extra material which extends the topics treated so far.

11.1 Bayesian Analysis Using Type Ia Supernovae Data

We discuss here how we can use type Ia supernovae in order to infer cosmological information. We do not address important issues such as the calibration of the light curves, how the distance moduli are calculated and the treatment of the systematic errors. Moreover, some knowledge of statistics and Bayesian analysis is required. The latter can be helped by e.g. Trotta (2017).

First of all, the measured quantities of interest are *fluxes*, which for historical reasons are given in a logarithmic scale called **apparent magnitude** m, defined as follows:

$$m \equiv -\frac{5}{2} \log_{10}\left(\frac{F}{F_0}\right) , \tag{11.1}$$

where F is the observed flux and F_0 is some reference flux for which $m = 0$. One should take into account that observations are made via filters which select a certain

© Springer International Publishing AG, part of Springer Nature 2018

O. Piattella, *Lecture Notes in Cosmology*, UNITEXT for Physics,
https://doi.org/10.1007/978-3-319-95570-4_11

Table 11.1 Binned type Ia supernovae data. From Betoule et al. (2014)

z_b	μ_b	z_b	μ_b	z_b	μ_b
0.010	32.9538	0.051	36.6511	0.257	40.5649
0.012	33.8790	0.060	37.1580	0.302	40.9052
0.014	33.8421	0.070	37.4301	0.355	41.4214
0.016	34.1185	0.082	37.9566	0.418	41.7909
0.019	34.5934	0.097	38.2532	0.491	42.2314
0.023	34.9390	0.114	38.6128	0.578	42.6170
0.026	35.2520	0.134	39.0678	0.679	43.0527
0.031	35.7485	0.158	39.3414	0.799	43.5041
0.037	36.0697	0.186	39.7921	0.940	43.9725
0.043	36.4345	0.218	40.1565	1.105	44.5140
				1.300	44.8218

portion of the emitted spectrum. Therefore, the measured apparent magnitude m depends on the filter which has been used. Moreover, one should also correct m by the effect of interstellar absorption in such a way that its value really depends only on how far the source is and not on the matter existing along the line of sight.

We do not take into account these important issues here and simply suppose to be given with the **bolometric magnitude**, i.e. the magnitude corresponding to the emission in the whole spectrum.

The **absolute magnitude** is the hypothetical apparent magnitude of an object as if it were at a distance of $10\,\mathrm{pc}$. Since $F \propto 1/d_L^2$, using Eq. (11.1) the absolute magnitude can be written as:

$$M \equiv -\frac{5}{2}\log_{10}\left[\frac{F}{F_0}\left(\frac{d_L}{10\,\mathrm{pc}}\right)^2\right] = m - 5\log_{10}\left(\frac{d_L}{10\,\mathrm{pc}}\right) . \tag{11.2}$$

The **distance modulus** is defined as:

$$\mu \equiv m - M = 5\log_{10}\left(\frac{d_L}{10\,\mathrm{pc}}\right) = 5\log_{10}\left(\frac{d_L}{1\,\mathrm{Mpc}}\right) + 25 , \tag{11.3}$$

where in the last equality we have introduced the Megaparsec (Mpc) as a more appropriate distance scale for cosmology.

Almost a thousand type Ia supernovae are known today and many more are expected to be detected by future experiments such as the *LSST*. For the sake of simplicity, we use now the 31 binned data of Betoule et al. (2014), which we report in Table 11.1, and the relative binned 31×31 covariant matrix $\mathbf{C}_b$, which we do not report here.

Combining Eq. (11.3) with Eq. (2.155) allows us to find the theoretical prediction on $\mu(z)$, which we can adjust to the observed binned values of Table 11.1 and thus find the best-fit values for H_0 and q_0.

Note that there are binned redshifts larger than 1 in Table 11.1 and for these certainly the expansion of Eq. (2.155) is inaccurate. However, let us not worry about this now, but focus only on how the analysis is performed.

The χ^2 is calculated as follows:

$$\chi^2(h, q_0) = \mathbf{r}^T \cdot \mathbf{C}_b^{-1} \cdot \mathbf{r} , \tag{11.4}$$

where $\mathbf{r}$ is the vector of the differences between the observed μ_b and the predicted one, i.e.

$$r_i \equiv \mu_{bi} - 5 \log_{10} \left[\frac{3000}{h} \left(z_i + \frac{1}{2}(1 - q_0)z_i^2 \right) \right] - 25 , \tag{11.5}$$

for $i = 1, \ldots, 31$. The μ_{bi} and z_i are those of Table 11.1. Note that we have expressed H_0 as $100 \, h \, \mathrm{km \, s^{-1} \, Mpc^{-1}}$ and $c = 3 \times 10^5 \, \mathrm{km \, s^{-1}}$.

Through the χ^2 we define the **likelihood**:

$$\mathcal{L}(h, q_0) = N e^{-\chi^2(h,q_0)/2} , \tag{11.6}$$

where N is a normalisation constant. The likelihood represents the probability of having a dataset given a cosmological model. We are interested in the contrary, i.e. in the probability of having a certain cosmological model given a dataset. This is called **posterior probability**. If there is no *a priori* reason for which some values of the parameters are preferred with respect to others, then the likelihood and the posterior probability are equal.

Let us see what happens in our case. What we do is the following: we set up a 100×100 grid in the parameter space (h, q_0) with $0.65 < h < 0.75$ and $-0.7 < q_0 < 0.2$. We choose these values because we already know the results, but in general one has to explore the parameter space. Also the 100×100 grid is purely arbitrary as one may choose a finer one in order to improve precision.

For each point of the grid we compute the χ^2 of Eq. (11.4) and thus construct the likelihood as a function in the parameter space.[1] Its maximum corresponds to the minimum χ^2 which in turn corresponds to the best fit values for the parameters. In our case, we find:

$$\chi^2_{\min} = 37.94 , \quad h^{(\mathrm{bf})} = 0.69 , \quad q_0^{(\mathrm{bf})} = -0.18 , \tag{11.7}$$

where the superscript *bf* means "best fit", of course. Note the negative best fit value of q_0, which corresponds to an accelerated expansion.

---

[1] Mathematica codes which perform this task are available on my personal webpage http://ofp.cosmo-ufes.org.

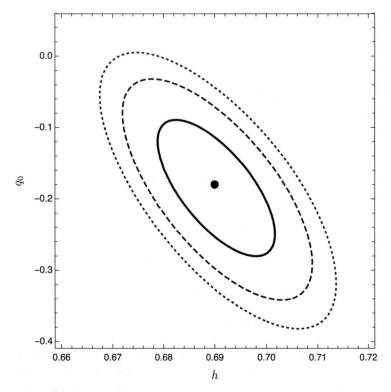

Fig. 11.1 Contour plots at 68, 95 and 99% confidence level (from the inner to the outer contour). The big dot represent the best fit values of Eq. (11.7)

We could in principle plot the likelihood as function of the parameters as a 3D plot, but it is more convenient and visually clearer to plot contours, i.e. horizontal slices of the likelihood, where horizontal means parallel to the (h, q_0) plane. This is what we have in Fig. 11.1.

In Fig. 11.1 there are three contour plots, which correspond to 68, 95 and 99% **confidence level**. What are these? Starting from the maximum value of the likelihood (represented by the big dot in Fig. 11.1) we slice its graph with a horizontal plane such that the volume enclosed is 68, 95 and 99%. If the contours are very close one to each other then it means that the likelihood is very peaked and thus the measure has been very precise.

Therefore, from Fig. 11.1 we can say that q_0 is negative with almost 99% of confidence and this is remarkable.

What happens if there are more than two parameters? In this case the likelihood cannot be plotted, nor its 2-dimensional contours. Then, one performs **marginalisation**, i.e. one integrates the likelihood with respect to all the parameters but two. In our case we did not need to do this because we had two parameters since the begin-

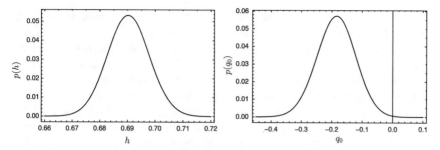

Fig. 11.2 Normalised probability distribution functions for h and q_0

ning, but we can marginalise over one and get the probability distribution function (PDF) for the other. This is done in Fig. 11.2.

Now we can do something similar to the calculation of the contour plots based on the confidence levels. For each one of the PDF in Fig. 11.2, starting from the maximum and going down, we calculate the values of the parameters for which the area encompassed is 68, 95 and 99%. these values represent the uncertainty over the best fit values of our parameters. In our case we have:

$$h = 0.69^{+0.015}_{-0.015}, \qquad q_0 = -0.18^{+0.12}_{-0.13}, \qquad (11.8)$$

at 95% confidence level.

Both the PDF's for h and q_0 are beautifully symmetric. This happens because of the special way h and q_0 enter in the χ^2, i.e. the latter is quadratic in h and q_0 and therefore the PDF for each one of these parameters is a Gaussian function. This can also be seen in the contour plots of Fig. 11.1, which are ellipses. On the other hand, if a parameter enters in a more complicated way into the χ^2, then the contour plots might be very asymmetric (typically they are "banana" shaped).

Exercise 11.1 Reproduce the analysis of this section developing a suitable numerical code.

11.2 Doing Statistics in the Sky

In Chap. 7 we have tackled the stochastic character of cosmological perturbations from a theoretical perspective. Here we offer a simplistic approach to how statistical methods are applied observationally. More detailed and comprehensive treatments can be found e.g. in Peebles (1980), Bonometto (2008).

As we learn in the very first year of our Physics course, determining the value of some physical quantity is no trivial task. The true value that we hope to find is only an abstraction, being our measurements imperfect i.e. characterised by an error that we try to harness with some mathematical tools. Statistics is one of them and tells us for example that the more we repeat our measurements the closer to a certain value the average goes. This certain value might be the true value, or not if systematic errors are present. In other words, we might have a very precise but inaccurate experiment. The approach of repeating experiments and measurements in order to extract informations about an underlying pattern is called **frequentist**. The question is: how do we apply that machinery to the universe, being this only one?

We want to learn something about gravity by observing how structures (galaxies) are distributed in the universe. We need statistics in order for our results to be meaningful and thankfully there are many galaxies. Imagine a big volume V which contains N galaxies. The galaxy number density is $n = N/V$. This, for example, represents the volume of a certain survey. Getting information on the volume V in itself is not meaningful, because we have just a single realisation of it. Therefore, let us consider small spheres of radius R and volume

$$V_R = \frac{4\pi}{3} R^3 \, , \tag{11.9}$$

with, of course, $V_R \ll V$. Now, inside these spheres we would expect a number $\bar{N}_R \equiv n V_R$ of galaxies if these were randomly distributed, i.e. distributed according to a Poisson distribution. But actually this is not the case, because of gravity. The latter is attractive and therefore it is more probable to find a galaxy closer to a big cluster rather than to a smaller one. Therefore, studying the large scale structure of the universe amounts to study the properties of gravity on large scales.

Now, let us count the galaxies in each one of the sphere that we have constructed. We have different numbers depending on the sphere considered, say $N_R(\mathbf{x})$ because the sphere has centre in $\mathbf{x}$. The number density thus becomes

$$n_R(\mathbf{x}) = \frac{N_R(\mathbf{x})}{V_R(\mathbf{x})} \, . \tag{11.10}$$

It is not really a function of a continuous variable, because in practice $\mathbf{x}$ gets only a countable amount of values, as many as the centres of the spheres chosen. Formally, we have:

$$\langle n_R(\mathbf{x}) \rangle = \lim_{v \to \infty} \frac{1}{v} \sum_{i=1}^{v} n_R(\mathbf{x}_i) = n \, . \tag{11.11}$$

For $\mathbf{x}$ being a continuous variable, we would have an integral:

$$\langle n_R(\mathbf{x}) \rangle = \frac{1}{V} \int_V d^3\mathbf{x} \, n_R(\mathbf{x}) = n \, . \tag{11.12}$$

So, we have divided the initial big volume V into small spherical regions V_R in order to gain statistics, i.e. in order to define averages. In particular, the **mass variance**

$$\sigma_R^2 = \frac{\langle (N_R(\mathbf{x}) - \bar{N}_R)^2 \rangle}{\bar{N}_R^2} \tag{11.13}$$

is a very important indicator, as we shall see in a moment.

Exercise 11.2 Assume that all galaxies have the same mass m_g. Show that from Eq. (11.13) we get

$$\sigma_R^2 = \frac{\langle (\rho_R(\mathbf{x}) - \rho)^2 \rangle}{\rho^2} = \left\langle \left(\frac{\delta \rho_R(\mathbf{x})}{\rho} \right)^2 \right\rangle = \langle \delta_R(\mathbf{x})^2 \rangle \,, \tag{11.14}$$

where $\rho_R(\mathbf{x}) \equiv n_R(\mathbf{x})m_g$ is the mass-density inside a sphere of radius R, centered in $\mathbf{x}$ and $\rho = nm_g$, i.e. it is the background density, averaged over all V.

The important point shown by the above exercise is that the mass variance is related to the averaged smoothed squared density contrast. From the theory developed in these lecture notes we know how to get $\delta(\mathbf{x})$. The point now is to learn how to smooth it and for this purpose we need to use filters.

11.2.1 Top Hat and Gaussian Filters

The density contrast field smoothed over a sphere of radius R can be formally written as:

$$\delta_R(\mathbf{x}) = \int_{V_R(\mathbf{x})} d^3\mathbf{u} \, \delta(\mathbf{u}) \,. \tag{11.15}$$

Note that the $\mathbf{x}$ dependence of $\delta_R(\mathbf{x})$ comes from where we have centred the sphere. We can transform the above integral into an equivalent one over the whole space, which we need in order to use the Fourier transform:

$$\delta_R(\mathbf{x}) = \int_{V_R(\mathbf{x})} d^3\mathbf{u} \, \delta(\mathbf{u}) = \int d^3\mathbf{u} \, W_R(|\mathbf{x} - \mathbf{u}|)\delta(\mathbf{u}) \,, \tag{11.16}$$

where

$$W_R(y) = \begin{cases} 1/V_R & \text{for } y < R \,, \\ 0 & \text{for } y > R \,, \end{cases} \tag{11.17}$$

is the **Top Hat filter**, or window function (hence the symbol W for denoting it). It is a simple trick to exclude all the information content beyond a certain scale R. There is no reason for this to depend from the direction, therefore the argument of W_R is the modulus $|\mathbf{x} - \mathbf{u}|$. A filter is a function that must not add or subtract any information, therefore it must be normalised to unity:

$$\int d^3\mathbf{y}\, W_R(y) = 1 \ . \tag{11.18}$$

In general, the functional form of a filter is:

$$W_R(y) = \frac{1}{V_R} w(y/R) \ , \tag{11.19}$$

where $w(y/R)$ is a generic function which goes to zero rapidly as y/R grows. In the case of the Top Hat filter, for example, w is a Heaviside function. One can also have a **Gaussian filter**:

$$W_R(y) = (2\pi R^2)^{-3/2} e^{-y^2/(2R^2)} \ . \tag{11.20}$$

Equation (11.16) is a convolution between the density contrast and the filter, hence its Fourier transform is the product between the two Fourier transforms:

$$\delta_R(\mathbf{k}) = \delta(\mathbf{k}) \tilde{W}(kR) \ , \tag{11.21}$$

where we have already guessed the dependence kR for the Fourier transformed filter since W_R is a function of y/R.

Exercise 11.3 Calculate the Fourier transform of the Top Hat filter. Show that:

$$\tilde{W}(kR) = \frac{3}{(kR)^3} \left[\sin(kR) - (kR)\cos(kR) \right] \ . \tag{11.22}$$

Of course, the Fourier transform of the Gaussian filter is still a Gaussian. The Top Hat filter is very effective in the configuration space since it cuts out all the scales above a given one, i.e. $W_R = 0$ if $y/R > 1$ or $kR < 2\pi$. However, its Fourier transform scales as $1/(kR)^3$, so it is not cut out drastically. This gives rise to spurious effects and therefore sometimes it might be safer to use the Gaussian filter, for which one has the same cutoff also for the Fourier transform.

So, in Eq. (11.21) the left hand side is the smoothed density contrast, which on should compare with observation. On the right hand side we have the full density contrast and the window function. Actually, there is another observational limitation. We have integrated over the whole space, together with a window function, in order to smooth out but often a survey does not cover the full sky. Certainly not in depth (we cannot observe up to infinite redshift) and also not the full celestial sphere. There is a **mask** say $f(\mathbf{x})$ by which the effective density contrast that we are going to use is

$$\delta_{\text{eff}}(\mathbf{x}) = f(\mathbf{x})\delta(\mathbf{x}) \ . \tag{11.23}$$

Hence, its Fourier transform is the convolution of the two Fourier transforms:

$$\delta_{\text{eff}}(\mathbf{k}) = (\tilde{f} * \delta)(\mathbf{k}) = \int d^3k' \, \tilde{f}(\mathbf{k}')\delta(\mathbf{k} - \mathbf{k}') \,. \tag{11.24}$$

This convolution has important effects on the final predictions of a model. In fact, suppose that we have a theory providing a density contrast which oscillates and then we expect of course an oscillating power spectrum. This would be ruled out if compared directly with the data, but since the result of the convolution smooths out oscillations, it might happen that the model still remains viable. See e.g. Moffat and Toth (2011) for a discussion on this point.

11.2.2 Sampling and Shot Noise

We can formally define the following quantity:

$$n(\mathbf{x}) \equiv \lim_{R \to 0} n_R(\mathbf{x}) \,, \tag{11.25}$$

as the number density field. On the other hand, we do not observe such field, but only a finite amount of galaxies. Hence, observationally $n(\mathbf{x})$ is realised by

$$n(\mathbf{x}) = \sum_{i=1}^{N} \delta^{(3)}(\mathbf{x} - \mathbf{x}_i) \,, \tag{11.26}$$

i.e. abusing (since it is a distribution) of the Dirac delta of the positions of the galaxies $\mathbf{x}_i$. Avoiding any abuse we could write:

$$n(\mathbf{x}) = \sum_{i=1}^{N} \frac{1}{a^3 \pi^{3/2}} e^{-|\mathbf{x} - \mathbf{x}_i|^2/a^2} \,, \tag{11.27}$$

for $a \to 0$, i.e. using a representation of the Dirac delta.

Technically, when integrating Eq. (11.26) over a certain volume V we should obtain the number of galaxies there contained. This is formally achieved by using the Top Hat filter:

$$N = \int_V dV \, n(\mathbf{x}) = \sum_{i=1}^{N} \int dV \, w(\mathbf{x})\delta^{(3)}(\mathbf{x} - \mathbf{x}_i) = \sum_{i=1}^{N} w(\mathbf{x}_i) \,, \tag{11.28}$$

where $w(\mathbf{x}) = V W(\mathbf{x})$ and $W(\mathbf{x})$ is equal to $1/V$ when $\mathbf{x}$ is inside the volume V and zero otherwise (this ensures the normalisation to unity). In the last sum above $w(\mathbf{x}_i) = 1$ only when $\mathbf{x}_i$ is inside the volume V and hence the equation holds true.

Now let us keep our sampling volume V, imagining that is the volume probed by a certain survey. As we commented earlier, the effective density contrast field is then given by Eq. (11.23). The difference now is that we are constructing the observed density contrast field whereas in that equation it was our theoretical prediction. The density field is:

$$\rho(\mathbf{x}) = \sum_{i=1}^{N} m_i \delta^{(3)}(\mathbf{x} - \mathbf{x}_i) , \tag{11.29}$$

where m_i is the mass of the ith galaxy. The effective, or sampled, density contrast is

$$\delta_s(\mathbf{x}) = \left(\frac{\rho(\mathbf{x})}{\rho} - 1 \right) w(\mathbf{x}) = \left(\frac{\sum_{i=1}^{N} m_i \delta^{(3)}(\mathbf{x} - \mathbf{x}_i)}{\sum_{i=1}^{N} m_i / V} - 1 \right) w(\mathbf{x}) . \tag{11.30}$$

where ρ is the background density, which is equal to $\sum_{i=1}^{N} m_i / V$ in the sampled volume. Assuming the same mass for each galaxy, one obtains:

$$\delta_s(\mathbf{x}) = \left(\frac{V}{N} \sum_{i=1}^{N} \delta^{(3)}(\mathbf{x} - \mathbf{x}_i) - 1 \right) w(\mathbf{x}) . \tag{11.31}$$

Exercise 11.4 Show that the Fourier transform of the density contrast is then:

$$\delta_s(\mathbf{k}) = \frac{1}{N} \sum_{i=1}^{N} w(\mathbf{x}_i) e^{-i\mathbf{k}\cdot\mathbf{x}_i} - \tilde{W}(\mathbf{k}) , \tag{11.32}$$

where $\tilde{W}(\mathbf{k})$ is the Fourier transform of the Top Hat filter, computed in Eq. (11.22).

When we compute the power spectrum from the above realisation, and we suppose a Gaussian distribution, we get for the mode $\mathbf{k}$:

$$\langle \delta_s(\mathbf{k}) \delta_s^*(\mathbf{k}) \rangle = \frac{1}{N^2} \sum_{i,j=1}^{N} \langle w(\mathbf{x}_i) w(\mathbf{x}_j) \rangle e^{-i\mathbf{k}\cdot(\mathbf{x}_i - \mathbf{x}_j)} - \frac{\tilde{W}(\mathbf{k})}{N} \sum_{i=1}^{N} \langle w(\mathbf{x}_i) \rangle e^{-i\mathbf{k}\cdot\mathbf{x}_i}$$

$$- \frac{\tilde{W}(\mathbf{k})}{N} \sum_{i=1}^{N} \langle w(\mathbf{x}_i) \rangle e^{i\mathbf{k}\cdot\mathbf{x}_i} + \tilde{W}(\mathbf{k})^2 , \tag{11.33}$$

where the average is an ensemble average. Since by definition:

$$\langle \delta_s(\mathbf{k}) \rangle = 0 , \tag{11.34}$$

we have that:

$$\frac{1}{N} \sum_{i=1}^{N} \langle w(\mathbf{x}_i) \rangle e^{-i\mathbf{k} \cdot \mathbf{x}_i} = \tilde{W}(\mathbf{k}) \; . \tag{11.35}$$

Therefore:

$$\langle \delta_s(\mathbf{k}) \delta_s^*(\mathbf{k}) \rangle = \frac{1}{N^2} \sum_{i,j=1}^{N} \langle w(\mathbf{x}_i) w(\mathbf{x}_j) \rangle e^{-i\mathbf{k} \cdot (\mathbf{x}_i - \mathbf{x}_j)} - \tilde{W}(\mathbf{k})^2 \; . \tag{11.36}$$

When $i = j$ there is a contribution above of the form:

$$\frac{1}{N^2} \sum_{i=1}^{N} \langle w(\mathbf{x}_i)^2 \rangle = \frac{1}{N} \; . \tag{11.37}$$

This is an error (a variance) on the Fourier mode $\mathbf{k}$ which does not depend on the wavenumber and is called **shot noise**. It comes from the fact that we are mapping a continuous density field with a discrete distribution, or sampling. It is the Poissonian part of the distribution of galaxies, which must be subtracted in order to obtain the true spectrum which is given by the correlations $\langle w(\mathbf{x}_i) w(\mathbf{x}_j) \rangle$.

Multiplying by V for dimensional reasons, we have thus the true power spectrum:

$$P(\mathbf{k}) = \frac{V}{N^2} \sum_{i \neq j}^{N} \langle w(\mathbf{x}_i) w(\mathbf{x}_j) \rangle e^{-i\mathbf{k} \cdot (\mathbf{x}_i - \mathbf{x}_j)} - V \tilde{W}(\mathbf{k})^2 \; , \tag{11.38}$$

and the noise spectrum:

$$P_n = \frac{V}{N}. \tag{11.39}$$

11.2.3 Correlation Function

How do we measure $P(k)$ and put the data on a plot? All we observe are galaxies in the sky. Consider again the very large volume V upon which we have worked until now, containing N galaxies. So, the mean number density is $n = N/V$. Assuming the same mass m_g for every galaxy, $\rho = N m_g / V$ is the mean density. Consider two small volumes δV_1 and δV_2, centered in $\mathbf{x}_1$ and $\mathbf{x}_2$ respectively and much smaller than V. If the distribution of galaxies was random, i.e. Poissonian, one would expect $\delta N_1 = n \delta V_1$. Deviations from this randomness are due to gravity and are encoded in the **correlation function** $\xi(\mathbf{x}_1, \mathbf{x}_2)$:

$$\langle \delta N_1(\mathbf{x}_1) \delta N_2(\mathbf{x}_2) \rangle = n^2 \delta V_1 \delta V_2 \left[1 + \xi(\mathbf{x}_1, \mathbf{x}_2) \right] \; . \tag{11.40}$$

We have already met the correlation function in Chap. 7, in Eq. (7.11). There it was defined through the ensemble average, whereas here the average is made over all the couples of volumes and clearly these must be chosen large enough in order to contain many galaxies but sufficiently small in order to identify many of them in V. Usually one assume that

$$\xi(\mathbf{x}_1, \mathbf{x}_2) = \xi(r) \,, \tag{11.41}$$

where $r = |\mathbf{x}_1 - \mathbf{x}_2|$, i.e. the correlation function depends only on the distance of the volumes, not on their positions or along which direction they are aligned. This is again statistical homogeneity and isotropy. Dividing the above equation by $\delta V_1 \delta V_2$, one gets

$$\langle \delta n_1(\mathbf{x}) \delta n_2(\mathbf{x} + \mathbf{r}) \rangle = n^2 \left[1 + \xi(r) \right] \,, \tag{11.42}$$

where now the average can be interpreted as the integration over $\mathbf{x}$.

Exercise 11.5 Show that the above equation can be written as:

$$\langle \delta(\mathbf{x}) \delta(\mathbf{x} + \mathbf{r}) \rangle = \xi(r) \,, \tag{11.43}$$

and this gives a direct relation between the density contrast and the correlation function. Compare with Eq. (7.40).

The correlation function measures the galaxy clustering. From observation we have the following empirical formula:

$$\xi(r) = \left(\frac{r_0}{r} \right)^\gamma \,, \tag{11.44}$$

where $r_0 = 5.5 \, h^{-1}$ Mpc and $\gamma \approx 1.77$ (Peebles 1980). The mass variance can be related to the correlation function as follows:

$$\sigma_R^2 = G(\gamma) \xi(R) \,, \tag{11.45}$$

where $G(\gamma) \approx 2$. We can argue that the non-linear regime of cosmological becomes dominant when $\sigma_R^2 = 1$ and from the above formulae one can find that this happens at $R = 8 \, h^{-1}$ Mpc. Hence, this is why σ_8 is so often used in cosmology.

The mass variance can be written as:

$$\sigma_R^2 = \langle \delta_R^2(\mathbf{x}) \rangle = \int \frac{d^3 k}{(2\pi)^3} \int \frac{d^3 k'}{(2\pi)^3} \, \langle \delta(\mathbf{k}) \tilde{W}(kR) \delta(\mathbf{k}') \tilde{W}(k'R) \rangle \,, \tag{11.46}$$

where we have used the Fourier transform of the smoothed density contrast, which is the product of the density contrast times the window function (or filter). Using the ensemble average we get:

$$\boxed{\sigma_R^2 = \int \frac{d^3 \mathbf{k}}{(2\pi)^3} \, P_\delta(k) \tilde{W}(kR)^2 = \int_0^\infty \frac{dk}{k} \, \Delta_\delta^2(k) \tilde{W}(kR)^2}$$

(11.47)

The above expression for the mass variance and that for the correlation function in Eq. (7.46) are very similar being the difference in the function which weighs the dimensionless power spectrum Δ^2. For the correlation function it is $\sin(kr)/(kr)$ whereas for the mass-variance is the squared Fourier transform of the filter. If we take the Top Hat filter then from Eq. (11.22) we see that $\tilde{W}(kR)^2$ decays as $(kR)^6$.

11.2.4 Bias

The bias is the deviation of the clustering behaviour of ordinary matter from the one of CDM. In general, we might suppose that structures of scale R are formed in those sites where

$$\delta_R(\mathbf{x}) > \nu \sigma_R ,$$

(11.48)

where ν is some parameter. In this way we are stating that galaxies or clusters have not exactly the same fluctuation pattern as CDM. Define the biased density field:

$$\rho_{R,\nu} = \theta[\delta_R(\mathbf{x}) - \nu \sigma_R] ,$$

(11.49)

where θ is the step function. For a Gaussian process, it can be shown that (Bonometto 2008)

$$\langle \rho_{R,\nu} \rangle \approx \frac{1}{\sqrt{2\pi}} \frac{1}{\nu} e^{-\nu^2/2} ,$$

(11.50)

which gives

$$\langle n_{R,\nu} \rangle \approx \frac{3}{(2\pi)^{3/2} R^3 \nu} e^{-\nu^2/2} .$$

(11.51)

The correlation function can be written then as

$$\xi^{(\nu, R)}(r) = e^{\nu^2 \xi^R(r)/\sigma_R^2} - 1 ,$$

(11.52)

which, for a small exponential, can be cast as

$$\xi^{(\nu, R)}(r) = \frac{\nu^2}{\sigma_R^2} \xi^R(r) \equiv b^2 \xi^R(r) ,$$

(11.53)

and in general the relation between galactic and CDM density contrast is written as:

$$\delta_g = b\delta ,$$

(11.54)

i.e. the same Eq. (9.1) that we presented in Chap. 9. As we mentioned there, typically b is treated as a parameter, though it may depend on the scale and on time.

References

Betoule, M., et al.: Improved cosmological constraints from a joint analysis of the SDSS-II and SNLS supernova samples. Astron. Astrophys. **568**, A22 (2014)

Bonometto, S.: Cosmologia and Cosmologie. Zanichelli (2008)

Moffat, J.W., Toth, V.T.: Comment on The Real Problem with MOND by Scott Dodelson (2011). arXiv:1112.1320. ArXiv e-prints

Peebles, P.J.E.: The Large-scale Structure of the Universe. Princeton university press (1980)

Trotta, R.: Bayesian Methods in Cosmology (2017)

Chapter 12
Appendices

> *Like all people who try to exhaust a subject, he exhausted his listeners*
>
> Oscar Wilde, The picture of Dorian Gray

Here are collected miscellaneous topics which are helpful in order to understand the former chapters and to make these lecture notes as self-contained as possible.

12.1 Thermal Distributions

In this section we derive the functional form of the Maxwell–Boltzmann, Fermi–Dirac and Bose–Einstein thermal distributions in the grand-canonical ensemble. See e.g. Huang (1987) for a textbook reference.

12.1.1 Derivation of the Maxwell–Boltzmann Distribution

Suppose that we have a system of N classical particles distributed in K states of different energies. Let us assume $N > K$. For each state i there are n_i particles with energy e_i and the total energy of the system is fixed and equal to E. Therefore, we have two constraints:

$$N = \sum_{i=1}^{K} n_i , \qquad E = \sum_{i=1}^{K} n_i e_i . \tag{12.1}$$

The number of microstates Ω which correspond to the macroscopical configuration of energy E is the following:

© Springer International Publishing AG, part of Springer Nature 2018
O. Piattella, *Lecture Notes in Cosmology*, UNITEXT for Physics,
https://doi.org/10.1007/978-3-319-95570-4_12

$$\Omega\left(\{n_i\}\right) = \frac{N!}{n_1!n_2!\cdots n_K!} . \tag{12.2}$$

The numerator is $N!$ because classical particles are *distinguishable*, so each of their permutations counts as a microstate. However, permutations done within the same state i do not count, hence we eliminate these possibilities dividing by $n_1!n_2!\cdots n_K!$.

Now take the logarithm of Ω and use Stirling's approximation:

$$\log\Omega = \log(N!) - \sum_{i=1}^{K}\log(n_i!) \approx N\log N - N - \sum_{i=1}^{K}(n_i\log n_i - n_i) . \tag{12.3}$$

We have to find the maximum for this expression, but taking into account the constraints (12.1). We then introduce Lagrange multipliers α and β and calculate the differential:

$$d\left[N\log N - N - \sum_{i=1}^{K}(n_i\log n_i - n_i) - \alpha\sum_{i=1}^{K}n_i - \beta\sum_{i=1}^{K}n_i e_i\right] = 0 . \tag{12.4}$$

The variables are the n_i's, thus:

$$\sum_{i=1}^{K}(-\log n_i - \alpha - \beta e_i)\,dn_i = 0 . \tag{12.5}$$

From this equation we obtain:

$$n_i = \exp(-\alpha - \beta e_i) \equiv \exp\left(-\frac{e_i - \mu}{k_{\mathrm{B}}T}\right) , \tag{12.6}$$

where in the last equality we have associated the Lagrange multiplier α as the chemical potential and $1/\beta$ as the thermal energy of the bath.

12.1.2   Derivation of the Fermi–Dirac Distribution

When calculating the Fermi–Dirac distribution two main differences with respect to the Maxwell–Boltzmann case appear: the first is that we have now quantum particles, which are *indistinguishable*, and the second is that fermions obey Pauli's exclusion principle, so there could be at most one fermion per quantum state.

Let us assume again N quantum particles and K energy states. The only difference now is that for each energy state i there are g_i sub-states, i.e. the energy states are degenerate. Let us focus on the energy state i. Here we must place n_i fermions. Because of Pauli's exclusion principle, $g_i \geq n_i$, otherwise we would have more than one fermion per state.

Exercise 12.1 In how many ways can we fit n_i indistinguishable fermions in g_i energy "slots"? Prove that the answer is:

$$\Omega_i = \frac{g_i!}{n_i!(g_i - n_i)!} \, . \tag{12.7}$$

Since combinatorics might be sometimes irksome, and after all these are lecture notes in cosmology, we offer a proof in the footnote. Try however not jumping to it right away and to work out the solution yourself.[1]

The total number of micro-states is then:

$$\Omega\left(\{n_i\}\right) = \prod_{i=1}^{K} \Omega_i = \prod_{i=1}^{K} \frac{g_i!}{n_i!(g_i - n_i)!} \, . \tag{12.8}$$

Now we proceed as before, taking the logarithm of Ω:

$$\log \Omega = \sum_{i=1}^{K} \left[\log(g_i!) - \log(n_i!) - \log((g_i - n_i)!) \right] \, , \tag{12.9}$$

using Stirling's approximation and calculating the constrained maximum:

$$\sum_{i=1}^{K} \left[-\log n_i + \log(g_i - n_i) - \alpha - \beta e_i \right] dn_i = 0 \, , \tag{12.10}$$

one obtains:

$$\log \frac{g_i - n_i}{n_i} = \alpha + \beta e_i \, , \tag{12.11}$$

and finally:

$$n_i = g_i \frac{1}{1 + \exp(\alpha + \beta e_i)} \equiv g_i \frac{1}{1 + \exp\left(\frac{e_i - \mu}{k_B T}\right)} \, , \tag{12.12}$$

with the same physical meanings for α and β as those stated for the Maxwell–Boltzmann distribution.

---

[1] We can choose g_i slots for the first fermion, $g_i - 1$ for the second and so on until finally we can choose $g_i - n_i + 1$ slots for the n_ith fermion. This gives $g_i!/(g_i - n_i)!$ and we only used Pauli's exclusion principle. Until here, then, the order of the chosen particles matter, e.g. having fermion 1 in the first slot is different from having fermion 2 in the first slot. But fermions are indistinguishable, therefore we must divide by $n_i!$ and hence the result (12.7).

12.1.3   Derivation of the Bose–Einstein Distribution

The setup for deriving the Bose–Einstein distribution is the same as the one used in the previous subsection for the Fermi–Dirac distribution except for the fact that now Pauli's exclusion principle does not apply and so the constraint $n_i \leq g_i$ does not hold true anymore. This changes the way in which we calculate Ω_i.

Exercise 12.2 Show that:

$$\Omega_i = \binom{n_i + g_i - 1}{g_i - 1} = \frac{(n_i + g_i - 1)!}{n_i!(g_i - 1)!} . \tag{12.13}$$

This is the same calculation of how many n_ith partial derivatives of a function of g_i variables there are. This is of course related to the fact that the wave-function of an ensemble of bosons is symmetric, just as partial derivatives are. Again, a proof is given in the footnote.[2]

Proceeding as we did in the previous two subsections, we find:

$$\sum_{i=1}^{K} \left[\log(n_i + g_i - 1) - \log n_i - \alpha - \beta e_i \right] dn_i = 0 , \tag{12.14}$$

and finally:

$$n_i = g_i \frac{1}{\exp(\alpha + \beta e_i) - 1} \equiv g_i \frac{1}{\exp\left(\frac{e_i - \mu}{k_\mathrm{B} T}\right) - 1} . \tag{12.15}$$

12.2   Derivation of the Poisson Distribution

Poisson distribution describes stochastic *independent* events happening randomly in time but with a certain average rate say λ. It is important for the description particle scattering or decaying processes and we have used it in Chap. 3 in order to take into account the neutron decay during BBN and also in Chap. 10 when we have discussed the visibility function.

Let $P(n; \lambda, t)$ be the probability that n events occur in a time interval t, given the rate λ. Consider a sufficiently small time interval δt, for which:

$$P(1; \lambda, \delta t) = \lambda \delta t , \qquad P(0; \lambda, \delta t) = 1 - \lambda \delta t . \tag{12.16}$$

---

[2]Imagine n_i particles and g_i slots where to fit them. These slots are separated by $g_i - 1$ walls. So, compute all the permutations among these objects, which are $(n_i + g_i - 1)!$, but do not consider the permutations among the walls $(g_i - 1)!$ and the particles, $n_i!$, because they are indistinguishable. So, we find Eq. (12.13).

Indeed, we can regard the first formula as the *definition* of λ. Therefore, calculating $P(0; \lambda, t + \delta t)$ one gets:

$$P(0; \lambda, t + \delta t) = P(0; \lambda, t)(1 - \lambda \delta t) , \tag{12.17}$$

where we have used the independence of the randomly occurring events.

Exercise 12.3 From Eq. (12.17) show that:

$$P(0; \lambda, t) = e^{-\lambda t} . \tag{12.18}$$

Now, exploiting the independence of the events, we can find a recurrence relation for $P(n; \lambda, t)$. Consider the following:

$$P(n; \lambda, t + \delta t) = P(n; \lambda, t)(1 - \lambda \delta t) + P(n - 1; \lambda, t)\lambda \delta t , \tag{12.19}$$

i.e. n events in a time interval $t + \delta t$ can either occur as n in the time interval t and none in the subsequent δt or $n - 1$ in the time interval t and just one in the subsequent δt. This is because we have chosen δt sufficiently small in order to accommodate at most one event.

From the above equation we have then:

$$\frac{dP(n; \lambda, t)}{dt} + \lambda P(n; \lambda, t) = \lambda P(n - 1; \lambda, t) . \tag{12.20}$$

Exercise 12.4 From Eq. (12.20) show that:

$$P(1; \lambda, t) = \lambda t e^{-\lambda t} . \tag{12.21}$$

By induction show that:

$$P(n; \lambda, t) = \frac{(\lambda t)^n}{n!} e^{-\lambda t} . \tag{12.22}$$

This is Poisson distribution. Another way to find it is from the binomial distribution:

$$P(N; n, p) = \binom{N}{n} p^n (1 - p)^{N-n} , \tag{12.23}$$

where N is the number of trials, n are the successful ones and p is the probability of success for a single trial.

Exercise 12.5 Assume $N \to \infty$ and $p \to 0$ such that Np stays constant. Performing these limits in Eq. (12.23) and using Stirling's approximation show that:

$$P(n, Np) = \frac{(Np)^n}{n!} e^{-Np} . \tag{12.24}$$

Defining $Np \equiv \lambda t$, we recover again Poisson distribution (12.22).

12.3 Helmholtz Theorem

We follow here Appendix B of Griffiths (2017). Let $\mathbf{F}(\mathbf{r})$ be a vector field. Let its divergence and curl be the following:

$$\nabla \cdot \mathbf{F}(\mathbf{r}) \equiv D(\mathbf{r}) , \qquad \nabla \times \mathbf{F}(\mathbf{r}) \equiv \mathbf{C}(\mathbf{r}) . \tag{12.25}$$

By construction, the divergence of the curl is vanishing:

$$\nabla \cdot [\nabla \times \mathbf{F}(\mathbf{r})] = \nabla \cdot \mathbf{C}(\mathbf{r}) = 0 , \tag{12.26}$$

hence $\mathbf{C}(\mathbf{r})$ is a solenoidal, i.e. divergenceless, vector field.

Helmholtz theorem states that $\mathbf{F}(\mathbf{r})$ can be decomposed as:

$$\mathbf{F}(\mathbf{r}) = -\nabla U(\mathbf{r}) + \nabla \times \mathbf{W}(\mathbf{r}) , \tag{12.27}$$

where:

$$U(\mathbf{r}) \equiv \frac{1}{4\pi} \int_V d^3\mathbf{r}' \frac{D(\mathbf{r}')}{|\mathbf{r} - \mathbf{r}'|} , \qquad \mathbf{W}(\mathbf{r}) = \frac{1}{4\pi} \int_V d^3\mathbf{r}' \frac{\mathbf{C}(\mathbf{r}')}{|\mathbf{r} - \mathbf{r}'|} . \tag{12.28}$$

Since the integration is over the whole space, in order for it to be well-defined $D(\mathbf{r})$ and $\mathbf{C}(\mathbf{r})$ must tend to zero more rapidly than $1/r^2$ for $r \to \infty$. This can be seen from the fact that:

$$\frac{d^3\mathbf{r}'}{|\mathbf{r} - \mathbf{r}'|} \sim dr' \, r' , \qquad \text{for } r' \to \infty , \tag{12.29}$$

and then if the two functions D and $\mathbf{C}$ scaled as $1/r'^2$ we would have a logarithmic divergence.

By taking the divergent of Eq. (12.27), remembering that the divergent of a curl is vanishing, and using Eq. (12.25) we can easily check that:

$$D(\mathbf{r}) = \nabla \cdot \mathbf{F}(\mathbf{r}) = -\nabla^2 U(\mathbf{r}) , \tag{12.30}$$

which is a Poisson equation and its solution is the first equation of Eq. (12.28).

On the other hand, by taking the curl of (12.27), remembering that the curl of the gradient is vanishing, we can check that:

$$\mathbf{C}(\mathbf{r}) = \nabla \times [\nabla \times \mathbf{W}(\mathbf{r})] = \nabla [\nabla \cdot \mathbf{W}(\mathbf{r})] - \nabla^2 \mathbf{W} . \tag{12.31}$$

The Laplacian term only gives the second equation of Eq. (12.28), but what about the $\nabla [\nabla \cdot \mathbf{W}(\mathbf{r})]$ term? Does it vanishes? Let us check that this is the case:

$$\nabla \cdot \mathbf{W}(\mathbf{r}) = \frac{1}{4\pi} \int_V d^3\mathbf{r}' \mathbf{C}(\mathbf{r}') \cdot \nabla \left(\frac{1}{|\mathbf{r} - \mathbf{r}'|} \right)$$

$$= -\frac{1}{4\pi} \int_V d^3\mathbf{r}' \mathbf{C}(\mathbf{r}') \cdot \nabla' \left(\frac{1}{|\mathbf{r} - \mathbf{r}'|} \right) =$$

$$-\frac{1}{4\pi} \int_{\partial V} d\mathbf{S}' \cdot \mathbf{C}(\mathbf{r}') \frac{1}{|\mathbf{r} - \mathbf{r}'|} + \frac{1}{4\pi} \int_V d^3\mathbf{r}' \nabla' \cdot \mathbf{C}(\mathbf{r}') \frac{1}{|\mathbf{r} - \mathbf{r}'|} . \tag{12.32}$$

In the last line, both contributions vanish. The first is a surface integral at infinity and in the second the divergence of $\mathbf{C}(\mathbf{r})$ is zero by construction.

In principle, the decomposition of Eq. (12.27) might not be unique because we could add to $\mathbf{F}(\mathbf{r})$ a vector field say $\mathbf{G}(\mathbf{r})$ which has both vanishing divergence and curl. On the other hand, it can be proved that there is no such $\mathbf{G}(\mathbf{r})$, with both vanishing divergence and curl and which goes to zero at infinity. Therefore, if $\mathbf{F}(\mathbf{r})$ goes to zero sufficiently fast at infinity then the decomposition in Eq. (12.27) is indeed unique.

12.4 Conservation of $\mathcal{R}$ on Large Scales and for Adiabatic Perturbations

In order to show the constancy of $\mathcal{R}$ on large scales and for adiabatic perturbations, let us start from Eq. (4.185) and combine its scalar part with the definition of $\mathcal{R}$ given in Eq. (4.131):

$$\mathcal{R} = \Phi + \mathcal{H} \frac{\Phi' - \mathcal{H}\Psi}{4\pi G a^2 (\rho + P)} . \tag{12.33}$$

Exercise 12.6 Differentiate Eq. (4.131) with respect to the conformal time and show that:

$$4\pi G a^2 (\rho + P)\mathcal{R}' = 4\pi G a^2 (\rho + P)\Phi' + \mathcal{H}' (\Phi' - \mathcal{H}\Psi)$$

$$+\mathcal{H} \left(\Phi'' - \mathcal{H}\Psi' - \mathcal{H}'\Psi \right) + \mathcal{H}^2 \left(\Phi' - \mathcal{H}\Psi \right) + 3\mathcal{H}^2 \frac{P'}{\rho'} \left(\Phi' - \mathcal{H}\Psi \right) ,$$

$$\tag{12.34}$$

where $\rho' = -3\mathcal{H}(\rho + P)$ has been used.

Exercise 12.7 Use the generalised Poisson equation (4.179), written as

$$3\mathcal{H}\left(\Phi' - \mathcal{H}\Psi\right) + k^2\Phi = 4\pi Ga^2\delta\rho \,, \tag{12.35}$$

and the background relation:

$$4\pi Ga^2(\rho + P) = \mathcal{H}^2 - \mathcal{H}' \,, \tag{12.36}$$

in order to cast Eq. (12.34) as follows:

$$4\pi Ga^2(\rho + P)\mathcal{R}' = \mathcal{H}\left(\Phi'' + 2\mathcal{H}\Phi' - \mathcal{H}\Psi' - 2\mathcal{H}'\Psi - \mathcal{H}^2\Psi\right)$$
$$+\mathcal{H}\frac{P'}{\rho'}\left(4\pi Ga^2\delta\rho - k^2\Phi\right) \,. \tag{12.37}$$

The first term on the right hand side can be simplified using the Einstein equation (4.189), so that we have:

$$4\pi Ga^2(\rho + P)\mathcal{R}' = -4\pi Ga^2\mathcal{H}\delta P - \frac{k^2\mathcal{H}}{3}(\Phi + \Psi) + \mathcal{H}\frac{P'}{\rho'}\left(4\pi Ga^2\delta\rho - k^2\Phi\right) \,. \tag{12.38}$$

Recalling the gauge-invariant entropy perturbation that we introduced in Eq. (4.129):

$$\Gamma \equiv \delta P - \frac{P'}{\rho'}\delta\rho \,, \tag{12.39}$$

we can finally write:

$$\boxed{\mathcal{R}' = -\mathcal{H}\frac{\Gamma}{\rho + P} - \mathcal{H}\frac{P'}{\rho'}\frac{k^2\Phi}{\mathcal{H}^2 - \mathcal{H}'} - \mathcal{H}\frac{k^2(\Phi + \Psi)}{3(\mathcal{H}^2 - \mathcal{H}')}} \tag{12.40}$$

The first term on the right-hand side vanishes if the perturbations are adiabatic, whereas the second and third terms on the right-hand side vanish on large scales, leaving thus $\mathcal{R}$ constant.

The gauge-invariant variables $\mathcal{R}$ and ζ can be related in the following way. First, using their definitions (4.131) and (4.132) in the Newtonian gauge we can write:

$$\zeta = \mathcal{R} - \mathcal{H}v + \frac{\delta\rho}{3(\rho + P)} \,. \tag{12.41}$$

Second, write down the relativistic Poisson equation (4.179) and the velocity equation (4.185) in the following form:

$$3\mathcal{H}\Phi' - 3\mathcal{H}^2\Psi + k^2\Phi = 4\pi Ga^2\delta\rho , \tag{12.42}$$
$$\Phi' - \mathcal{H}\Psi = 4\pi Ga^2 (\rho + P) v , \tag{12.43}$$

and combine them in order to give the constraint:

$$12\pi G\mathcal{H}a^2(\rho + P)v + k^2\Phi = 4\pi G\delta\rho . \tag{12.44}$$

Using this constraint, the relation in Eq. (12.41) becomes:

$$\boxed{\zeta = \mathcal{R} + \frac{k^2\Phi}{12\pi Ga^2(\rho + P)}} \tag{12.45}$$

Since $(\rho + P)a^2$ is of order $\mathcal{H}^2$, then $\zeta - \mathcal{R}$ is of order $k^2/\mathcal{H}^2$, which vanishes for large scales. This means that if $\mathcal{R}$ is conserved then ζ also is.

The result found here confirms our previous calculation ended with Eq. (6.94), in Chap. 6, when we investigated the adiabatic primordial modes.

12.5 Spherical Harmonics

In this section we briefly review spherical harmonics. These are the eigenfunctions of the Laplacian operator on the sphere or, equivalently, the eigenfunctions of the square of the angular momentum operator. They form a complete orthonormal system and therefore any function on the sphere can be expanded in a linear combination of spherical harmonics. That is why they are extensively used in cosmology to analyse CMB anisotropies in temperature and polarisation: the latter are fields (of spin zero and spin 2, respectively) on the celestial sphere. In this section we shall follow principally (Butkov 1968). However, many references exist dealing with the theory of the angular momentum and spherical harmonics, so the reader is encouraged to find the treatment which most suits them.

Exercise 12.8 Find the expression of the Laplacian operator on the sphere. The line element on the unit sphere is:

$$ds_S^2 = g_{ab}dx^a dx^b = d\theta^2 + \sin^2\theta d\phi^2 , \tag{12.46}$$

with $a, b = \theta, \phi$, and naturally we have chosen spherical coordinates. Use the formula:

$$\nabla^2 = \frac{1}{\sqrt{g}} \frac{\partial}{\partial x^a} \left(g^{ab} \sqrt{g} \frac{\partial}{\partial x^b} \right) , \tag{12.47}$$

where $\sqrt{g}$ is the determinant of the metric and find:

$$\nabla^2 = \frac{1}{\sin\theta}\frac{\partial}{\partial\theta}\left(\sin\theta\frac{\partial}{\partial\theta}\right) + \frac{1}{\sin^2\theta}\frac{\partial^2}{\partial\phi^2} . \qquad (12.48)$$

Let us then set up the eigenvalue equation:

$$\frac{1}{\sin\theta}\frac{\partial}{\partial\theta}\left[\sin\theta\frac{\partial Y(\theta,\phi)}{\partial\theta}\right] + \frac{1}{\sin^2\theta}\frac{\partial^2 Y(\theta,\phi)}{\partial\phi^2} = \lambda Y(\theta,\phi) . \qquad (12.49)$$

Assume that the θ and ϕ dependences can be factorised, i.e. let us write[3]:

$$Y(\theta,\phi) = \Theta(\theta)\Phi(\phi) , \qquad (12.50)$$

so that:

$$\frac{\sin\theta}{\Theta}\frac{d}{d\theta}\left[\sin\theta\frac{d\Theta}{d\theta}\right] - \lambda\sin^2\theta = -\frac{1}{\Phi}\frac{d^2\Phi}{d\phi^2} . \qquad (12.51)$$

Now, the left hand side depends only on θ whereas the right hand side only on ϕ. Therefore, the only possible way for them to be equal is to be equal to the same constant, which we call m^2. In this way we have now two equations:

$$\frac{d^2\Phi}{d\phi} + m^2\Phi = 0 , \qquad \sin\theta\frac{d}{d\theta}\left[\sin\theta\frac{d\Theta}{d\theta}\right] - \lambda\sin^2\theta\Theta = m^2\Theta . \qquad (12.52)$$

The first one is very simple to solve and gives us:

$$\Phi = e^{\pm im\phi} , \qquad (12.53)$$

with some normalisation which we will determine afterwards for the full $Y(\theta,\phi)$. On the other hand, Φ must be periodic since ϕ or $\phi + 2\pi$ denote the same angular position. Hence:

$$e^{im\phi} = e^{im(\phi+2\pi)} , \qquad (12.54)$$

which implies:

$$e^{i2m\pi} = 1 , \qquad (12.55)$$

and therefore m must be an integer, positive or negative, or zero. The equation for Θ can be treated as follows.

Exercise 12.9 Consider the new variable:

$$x \equiv \cos\theta . \qquad (12.56)$$

[3]The function $\Theta(\theta)$ here is not the relative temperature fluctuation and $\Phi(\phi)$ is not the Bardeen potential.

Show that the derivative with respect to θ satisfies:

$$\sin\theta\frac{d}{d\theta} = -(1-x^2)\frac{d}{dx}\ ,\tag{12.57}$$

and hence the equation for $\Theta(x)$ can be written as follows:

$$\frac{d}{dx}\left[(1-x^2)\frac{d\Theta(x)}{dx}\right] - \lambda\Theta(x) - \frac{m^2}{1-x^2}\Theta(x) = 0\ .\tag{12.58}$$

Now, we need $\Theta(x)$ to be regular at $x = \pm 1$, i.e. for $\theta = 0$ or $\theta = \pi$. A way to investigate this is to change variable:

$$z = 1 \mp x\ ,\tag{12.59}$$

in order to trade the neighbourhood of $x = \pm 1$ for that of $z = 0$.

Exercise 12.10 Obtain the new equation for $\Theta(z)$ and $z = 1 - x$:

$$z(2-z)\frac{d^2\Theta}{dz^2} + 2(1-z)\frac{d\Theta}{dz} - \lambda\Theta - \frac{m^2}{z(2-z)}\Theta = 0\ .\tag{12.60}$$

Now, using Frobenius method we look for a solution of the form:

$$\Theta(z) = z^s\sum_{n=0}^{\infty}a_n z^n\ .\tag{12.61}$$

where $s \geq 0$ in order for $\Theta(z)$ to be regular for $z \to 0$.

Exercise 12.11 Substitute the above ansatz in Eq. (12.60) and equate to zero power by power. Show that we get from the lowest power (which is z^s) that:

$$s = \pm\frac{m}{2}\ .\tag{12.62}$$

Let us stipulate that $m \geq 0$. Then we must choose the positive sign, i.e. $s = m/2$, since $z^{-m/2} \to \infty$ for $z \to 0$.

Exercise 12.12 Carry on a similar analysis for $x = -1$, by transforming variable to $z = 1 + x$. Show that again $s = m/2$.

Combining the results from the above exercises we can conclude that $\Theta(x)$ must have the following form:

$$\Theta(x) = (1 - x^2)^{m/2} f(x) , \qquad (12.63)$$

where $f(x)$ is an analytic function, non-vanishing for $x = \pm 1$. We have also learnt that $\Theta(x)$ does vanish for $x \pm 1$, except when $m = 0$. Now, let us find an equation for f.

Exercise 12.13 Substituting Eq. (12.63) into (12.58) show that:

$$(1 - x^2)\frac{d^2 f}{dx^2} - 2x(m + 1)\frac{df}{dx} - (\lambda + m + m^2)f = 0 . \qquad (12.64)$$

We now prove that $\lambda = -\ell(\ell + 1)$, with ℓ integer such that $\ell \geq m$. This result will stem out of the requirement of regularity of f. Adopting again Frobenius method, let us stipulate that:

$$f(x) = x^s \sum_{n=0}^{\infty} a_n x^n . \qquad (12.65)$$

Exercise 12.14 Substitute the above ansatz into Eq. (12.64) and find the following relation:

$$\sum_{n=0}^{\infty} a_n(n + s)(n + s - 1)x^{n+s-2} =$$

$$\sum_{n=0}^{\infty} a_n \left[m(m + 1) + \lambda + 2(m + 1)(n + s) + (n + s)(n + s - 1) \right] x^{n+s} . \qquad (12.66)$$

The two series can be combined starting from $n = 2$.

Exercise 12.15 Show that:

$$a_0 s(s - 1)x^{s-2} + a_1 s(s + 1)x^{s-1} + \sum_{n=2}^{\infty} a_n(n + s)(n + s - 1)x^{n+s-2} =$$

$$\sum_{n=2}^{\infty} a_{n-2} \left[m(m + 1) + \lambda + 2(m + 1)(n + s - 2) + (n + s - 2)(n + s - 3) \right] x^{n+s-2} .$$

$$(12.67)$$

Now the two series can be merged in a single one and equating to zero each power we have that either $s = 0$ or $s = 1$. These conditions lead to the following recursion relations for the series coefficients:

$$a_{n+2} = \frac{n(n-1) + \lambda + m(m+1) + 2n(m+1)}{(n+2)(n+1)} a_n , \qquad (s = 0) ,$$

(12.68)

$$a_{n+2} = \frac{n(n+1) + \lambda + m(m+1) + 2(n+1)(m+1)}{(n+3)(n+2)} a_n , \qquad (s = 1) .$$

(12.69)

The integral test of convergence fails because $a_{n+2}/a_n \to 1$ for $n \to \infty$, hence the series solution would diverge unless:

$$n(n-1) + \lambda + m(m+1) + 2n(m+1) = 0 , \quad (s = 0) , \quad (12.70)$$
$$n(n+1) + \lambda + m(m+1) + 2(n+1)(m+1) = 0 , \quad (s = 1) . \quad (12.71)$$

These are constraints on λ, which has then to assume the following form:

$$\lambda = -(m+n)(m+n+1) , \quad (s = 0) , \quad (12.72)$$
$$\lambda = -(m+n+1)(m+n+2) , \quad (s = 1) , \quad (12.73)$$

or, in general:

$$\boxed{\lambda \equiv -\ell(\ell+1)}$$

(12.74)

with ℓ integer and $\ell \geq m$, which is what we wanted to prove. This result can be obtained also in quantum mechanics, by exploiting the commutation relations of the components of the angular momentum operator, see e.g. Weinberg (2015). We have adopted here an approach based on calculus.

The above condition on λ implies that the series solution for f terminates after the $(\ell - m)$th term and thus $\Theta(x)$ is a polynomial. The polynomials thus obtained for all the possible choices of (ℓ, m) are known as **associated Legendre polynomials** and denoted as $P_\ell^m(x)$. Equation (12.58) can be thus written as:

$$\boxed{\frac{d}{dx}\left[(1-x^2)\frac{dP_\ell^m(x)}{dx}\right] + \left[\ell(\ell+1) - \frac{m^2}{1-x^2}\right]P_\ell^m(x) = 0}$$

(12.75)

and it is known as the **general Legendre equation**. Via some manipulation it is possible to obtain it from the **Legendre equation**:

$$\boxed{\frac{d}{dx}\left[(1-x^2)\frac{d\mathcal{P}_\ell(x)}{dx}\right] + \ell(\ell+1)\mathcal{P}_\ell(x) = 0}$$

(12.76)

and therefore write the associated Legendre polynomials as:

$$P_\ell^m(x) = (1 - x^2)^{m/2} \frac{d^m}{dx^m}[\mathcal{P}_\ell(x)] .$$

(12.77)

Using Rodrigues' formula:

$$\mathcal{P}_\ell(x) = \frac{1}{2^\ell \ell!} \frac{d^\ell}{dx^\ell}[(x^2 - 1)^\ell] ,$$

(12.78)

we have then

$$P_\ell^m(x) = \frac{(1 - x^2)^{m/2}}{2^\ell \ell!} \frac{d^{\ell+m}}{dx^{\ell+m}}[(x^2 - 1)^\ell] ,$$

(12.79)

and this formula allows us to extend the definition of $P_\ell^m(x)$ also for negative values of m, such that $\ell \geq |m|$ or $-\ell \leq m \leq \ell$, as we are used from the theory of the angular momentum in quantum mechanics. In these notes we have adopted the following convention:

$$P_\ell^{-m} = \frac{(\ell - m)!}{(\ell + m)!} P_\ell^m .$$

(12.80)

Another, perhaps more standard convention includes a $(-1)^m$ factor, but we choose to omit it in order to have the following property for the spherical harmonics under complex conjugation:

$$\boxed{Y_\ell^{m*}(\theta, \phi) = Y_\ell^{-m}(\theta, \phi)}$$

(12.81)

We can now finally write down the explicit formula for the spherical harmonics:

$$\boxed{Y_\ell^m(\theta, \phi) = \sqrt{\frac{(2\ell + 1)(\ell - m)!}{4\pi(\ell + m)!}} P_\ell^m(\cos\theta)e^{im\phi}}$$

(12.82)

where the normalisation has been chosen in order to have

$$\boxed{\int_0^\pi d\theta \, \sin\theta \int_0^{2\pi} d\phi \, Y_\ell^m(\theta, \phi)Y_{\ell'}^{m'*}(\theta, \phi) = \delta_{\ell\ell'}\delta_{mm'}}$$

(12.83)

which is probably the single most important relation about spherical harmonics because it tells us that they form an **orthonormal system**. This is also complete, i.e. the spherical harmonics satisfy the following **completeness relation**:

$$\sum_{\ell=0}^\infty \sum_{m=-\ell}^\ell Y_\ell^m(\theta, \phi)Y_\ell^{m*}(\theta', \phi') = \frac{1}{\sin\theta}\delta(\theta - \theta')\delta(\phi - \phi') ,$$

(12.84)

Therefore, any square-integrable function $f(\theta, \phi)$ can be expanded as:

$$f(\theta, \phi) = \sum_{\ell=0}^{\infty} \sum_{m=-\ell}^{\ell} a_{\ell m} Y_\ell^m(\theta, \phi) , \qquad (12.85)$$

in an unique way (meaning that the $a_{\ell m}$'s are unique for that function). The $\delta_{mm'}$ in the orthonormality can be easily understood from the $d\phi$ integration, since:

$$\int_0^{2\pi} d\phi \, e^{i(m-m')\phi} = \left. \frac{e^{i(m-m')\phi}}{i(m-m')} \right|_0^{2\pi} = 0 , \qquad (12.86)$$

for $m \neq m'$, leaving only the $m = m'$ possibility and the 2π value of the integral. The $\delta_{\ell\ell'}$ part of the orthonormality relation depends of course on the properties of the associated Legendre polynomials, but we do not give further detail here.

It is important to know the parity of the spherical harmonics, i.e. how $Y_\ell^m(\hat{n})$ changes under $\hat{n} \to -\hat{n}$. In terms of the angles:

$$\hat{n} \to -\hat{n} \quad \Rightarrow \quad (\theta, \phi) \to (\pi - \theta, \pi + \phi) . \qquad (12.87)$$

Since $\cos(\pi - \theta) = -\cos\theta$, spatial inversion amount to $x \to -x$ for the associated Legendre polynomial and therefore a $(-1)^{\ell+m}$ factor coming from the derivative in the formula of Eq. (12.79). The exponential $\exp(im\phi)$ instead becomes:

$$e^{im\phi} \to e^{im(\pi+\phi)} = e^{im\pi} e^{im\phi} = (-1)^m e^{im\phi} . \qquad (12.88)$$

Therefore, we finally have:

$$\boxed{Y_\ell^m(\pi - \theta, \pi + \phi) = (-1)^\ell Y_\ell^m(\theta, \phi)} \qquad (12.89)$$

A formula that we have intensively employed in these notes is the expansion of a plane wave into spherical harmonics:

$$\boxed{e^{i\mathbf{k}\cdot\mathbf{r}} = 4\pi \sum_{\ell=0}^{\infty} \sum_{m=-\ell}^{\ell} i^\ell Y_\ell^{m*}(\hat{k}) Y_\ell^m(\hat{r}) j_\ell(kr)} \qquad (12.90)$$

where the complex conjugation can be switched from one spherical harmonic to the other, since $\mathbf{k} \cdot \mathbf{r} = \mathbf{r} \cdot \mathbf{k}$. Also, we have made great use of the **addition theorem**:

$$\boxed{\mathcal{P}_\ell(\mathbf{x} \cdot \mathbf{y}) = \frac{4\pi}{2\ell+1} \sum_{m=-\ell}^{\ell} Y_\ell^{m*}(\mathbf{y}) Y_\ell^m(\mathbf{x})} \qquad (12.91)$$

We can then combine the plane wave expansion with the addition theorem and get:

$$e^{i\mathbf{k}\cdot\mathbf{r}} = \sum_{\ell=0}^{\infty} i^{\ell}(2\ell + 1)\mathcal{P}_{\ell}(\hat{k} \cdot \hat{r}) j_{\ell}(kr) \,. \tag{12.92}$$

We now turn to the spin-weighted spherical harmonics.

12.5.1 Spin-Weighted Spherical Harmonics

The usual spherical harmonics allow to expand a scalar quantity such as the temperature on the sphere. Consider a rotation R such that

$$\hat{n} \to \hat{n}' = R\hat{n} \,. \tag{12.93}$$

Since the relative temperature fluctuation is a scalar, we have that:

$$\Theta'(\hat{n}') = \Theta(\hat{n}) \,. \tag{12.94}$$

Using the expansion in spherical harmonics we can write:

$$\sum_{\ell m} a'_{\ell m} Y_{\ell}^{m}(\hat{n}') = \sum_{\ell m} a_{\ell m} Y_{\ell}^{m}(\hat{n}) \,. \tag{12.95}$$

Now we take advantage of the properties of the spherical harmonics under spatial rotation, i.e.

$$\boxed{Y_{\ell}^{m}(R\hat{n}) = \sum_{m'=-\ell}^{\ell} D_{m'm}^{(\ell)}(R^{-1}) Y_{\ell}^{m'}(\hat{n})} \tag{12.96}$$

where the $D_{m'm}^{(\ell)}$ are the elements of the **Wigner D-matrix**. See Landau and Lifshits (1991) for more detail. Hence, we can write:

$$\sum_{\ell m} a'_{\ell m} Y_{\ell}^{m}(\hat{n}') = \sum_{\ell m m'} a_{\ell m} D_{m'm}^{(\ell)}(R) Y_{\ell}^{m'}(\hat{n}') \,, \tag{12.97}$$

and readjusting the indices we finally have:

$$\boxed{a'_{\ell m} = \sum_{m'} D_{m'm}^{(\ell)}(R) a_{\ell m'}} \tag{12.98}$$

since the spherical harmonics form an orthonormal basis. This is an important relation because it tells us that the angular power spectrum is independent from the direction, i.e. it is rotationally invariant:

$$C'_\ell = \langle a'_{\ell m} a'^*_{\ell m} \rangle = \sum_{MM'} D^{(\ell)}_{mM} D^{(\ell)*}_{mM'} \langle a_{\ell M} a^{*'}_{\ell M} \rangle = C_\ell \sum_M D^{(\ell)}_{mM} D^{(\ell)*}_{mM} = C_\ell \, ,$$

$$(12.99)$$

the product of Wigner D-matrices being unity since it represents a product of one rotation with its inverse. Hence we have found that the power spectrum is a rotational invariant. This is a consequence of the fact that Θ is a scalar, but what about another function which is not? In Sect. 12.7 we shall see that indeed the Stokes parameters are not scalars, i.e. upon a rotation $\hat{n} \to \hat{n}' = R\hat{n}$ we have in general that:

$$Q'(\hat{n}') \neq Q(\hat{n}) , \qquad U'(\hat{n}') \neq U(\hat{n}) . \qquad (12.100)$$

Therefore, if we expand them on a spherical harmonics basis we would get power spectra depending on the orientation of our coordinate frame.

In order to avoid this, we must employ the **spin-weighted spherical harmonics**. These were introduced in Newman and Penrose (1966) in the context of gravitational waves, which also have polarisation. Using their notation, η is a spin-s quantity if it transforms as:

$$\eta(\hat{n}) \to \eta(\hat{n})' = e^{si\psi} \eta(\hat{n}) , \qquad (12.101)$$

under a rotation of an angle ψ about the line of sight $\hat{n}$. Then, one defines the differential operators $\eth$ and $\bar{\eth}$ as:

$$\eth\eta = -(\sin\theta)^s \left[\frac{\partial}{\partial\theta} + \frac{i}{\sin\theta} \frac{\partial}{\partial\phi} \right] \left[(\sin\theta)^{-s} \eta \right] , \qquad (12.102)$$

$$\bar{\eth}\eta = -(\sin\theta)^{-s} \left[\frac{\partial}{\partial\theta} + \frac{i}{\sin\theta} \frac{\partial}{\partial\phi} \right] \left[(\sin\theta)^{s} \eta \right] , \qquad (12.103)$$

in spherical coordinates and proves that $\eth\eta$ is a spin-$(s+1)$ quantity whereas $\bar{\eth}\eta$ is a spin-$(s-1)$ quantity. These operators are essentially covariant derivative on the sphere, as we shall see in Sect. 12.7. The spin-s spherical harmonics are thus defined as:

$$_sY^m_\ell = \sqrt{\frac{(\ell-s)!}{(\ell+s)!}} \eth^s Y^m_\ell \qquad (0 \le s \le \ell) , \qquad (12.104)$$

$$_sY^m_\ell = (-1)^s \sqrt{\frac{(\ell+s)!}{(\ell-s)!}} \bar{\eth}^{-s} Y^m_\ell \qquad (-\ell \le s \le 0) . \qquad (12.105)$$

The spin-weighted spherical harmonics have fundamental properties similar to those characterising the usual ones. First of all:

$$_0Y^m_\ell(\theta, \phi) = Y^m_\ell(\theta, \phi) , \qquad (12.106)$$

i.e. the usual spherical harmonics can be regarded as spin-0. Then, the $_sY_\ell^m$ form a complete orthonormal system, i.e. they are orthonormal

$$\int_0^\pi d\theta \, \sin\theta \int_0^{2\pi} d\phi \, _sY_\ell^m(\theta, \phi)_s Y_{\ell'}^{m'*}(\theta, \phi) = \delta_{\ell\ell'}\delta_{mm'} \qquad (12.107)$$

and complete:

$$\sum_{\ell=0}^{\infty} \sum_{m=-\ell}^{\ell} {}_sY_\ell^m(\theta, \phi)_s Y_\ell^{m*}(\theta', \phi') = \frac{1}{\sin\theta}\delta(\theta - \theta')\delta(\phi - \phi') \qquad (12.108)$$

and therefore any spin-s quantity η can be expanded as:

$$\eta(\theta, \phi) = \sum_{\ell=0}^{\infty} \sum_{m=-\ell}^{\ell} \eta_{\ell m} \, {}_sY_\ell^m(\theta, \phi) \qquad (12.109)$$

Under complex conjugation and spatial inversion the spin-weighted spherical harmonics transform as:

$$_sY_\ell^{m*} = (-1)^s {}_{-s}Y_\ell^{-m}, \qquad _sY_\ell^m(-\hat{n}) = (-1)^\ell {}_{-s}Y_\ell^m(\hat{n}), \qquad (12.110)$$

and they can be related to the Wigner D-matrix as follows:

$$D_{ms}^{(\ell)}(\phi, \theta, \psi) = \sqrt{\frac{4\pi}{2\ell+1}} {}_sY_\ell^{-m}(\theta, \phi)e^{is\psi}, \qquad (12.111)$$

where (ϕ, θ, ψ) are the Euler angles.

12.6   Method of Green's Functions

Suppose that we have a linear differential equation, which we write as follows:

$$\mathcal{L}y(x) = f(x), \qquad (12.112)$$

where $\mathcal{L}$ is a second-order linear differential operator. Suppose also that we have been able to solve the homogeneous equation $\mathcal{L}y(x) = 0$ and found the solution:

$$y_{\text{hom}}(x) = C_1 y_1(x) + C_2 y_2(x). \qquad (12.113)$$

How, knowing this, do we determine the solution to the complete, non-homogeneous equation? Recall that Green's function is defined by:

$$\mathcal{L}G(x, x') = \delta(x - x') , \tag{12.114}$$

and from $G(x, x')$ the general solution to the complete equation is readily found via convolution with the non-homogeneous term:

$$y(x) = \int_{x_i}^{x} dx' \, G(x, x')f(x') , \tag{12.115}$$

where x_i is our initial value.

Exercise 12.16 Check that the above convolution Eq. (12.115) is solution of Eq. (12.112).

So, we must determine $G(x, x')$. We can do this by noticing that it is solution of the homogeneous equation when $x \neq x'$ and thus we can write:

$$G(x, x') = \begin{cases} A_1 y_1(x) + A_2 y_2(x) & \text{for } x < x' \\ B_1 y_1(x) + B_2 y_2(x) & \text{for } x > x' \end{cases} , \tag{12.116}$$

where the A's and B's are integration constants to be determined. Since $\mathcal{L}$ is a second order linear differential equation, we want that:

- Since $\mathcal{L}G(x, x') = \delta(x - x')$, the derivative of $G(x, x')$ must be discontinuous in $x = x'$, i.e. it must be a θ function.
- On the other hand, $G(x, x')$ is continuous in $x = x'$.

So, with these two conditions, we impose the following matching:

$$\begin{cases} A_1 y_1(x') + A_2 y_2(x') = B_1 y_1(x') + B_2 y_2(x') \\ B_1 y_1'(x') + B_2 y_2'(x') - A_1 y_1'(x') - A_2 y_2'(x') = 1 \end{cases} . \tag{12.117}$$

Note that whatever the gap of the discontinuity might be, we can always normalise it to unity by renormalising the integration constants. The above system can be cast in the following form:

$$\begin{cases} (B_1 - A_1)y_1(x') + (B_2 - A_2)y_2(x') = 0 \\ (B_1 - A_1)y_1'(x') + (B_2 - A_2)y_2'(x') = 1 \end{cases} . \tag{12.118}$$

The unknowns of this system are the differences among the integration constants and the determinant of the system is:

$$W(x') = (y_1 y_2' - y_1' y_2)(x') , \qquad (12.119)$$

i.e. the Wronskian, which is certainly different from zero because y_1 and y_2 are independent solutions. Therefore, we use Cramer's rule and find:

$$B_1 - A_1 = \frac{-y_2(x')}{W(x')} , \qquad B_2 - A_2 = \frac{y_1(x')}{W(x')} . \qquad (12.120)$$

We have freedom of choosing two of the four constants. Therefore, choosing $A_1 = A_2 = 0$:

$$G(x, x') = \begin{cases} 0 & \text{for } x < x' \\ \frac{-y_2(x')y_1(x)+y_1(x')y_2(x)}{W(x')} & \text{for } x > x' \end{cases} , \qquad (12.121)$$

and so this is our Green function which, in convolution with $f(x')$ gives us the complete solution to our non-homogenous differential equation:

$$\boxed{y(x) = \int_{x_i}^{x} dx' \, \frac{-y_2(x')y_1(x) + y_1(x')y_2(x)}{W(x')} f(x')} \qquad (12.122)$$

12.7 Polarisation

In this section we recall the terminology regarding the polarisation of an electromagnetic wave. The standard reference is Jackson (1998), but see also Griffiths (2017).

12.7.1 Electromagnetic Waves

Let us start from Maxwell equations in vacuum and in Minkowski space:

$$\nabla \cdot \mathbf{E} = 0 , \quad \nabla \times \mathbf{E} = -\frac{\partial \mathbf{B}}{\partial t} , \quad \nabla \cdot \mathbf{B} = 0 , \quad \nabla \times \mathbf{B} = \frac{1}{c^2} \frac{\partial \mathbf{E}}{\partial t} . \qquad (12.123)$$

Exercise 12.17 Combine Maxwell equations in order to find the wave equations:

$$\Box \mathbf{E} = \frac{1}{c^2} \frac{\partial^2 \mathbf{E}}{\partial t^2} - \nabla^2 \mathbf{E} = 0 , \quad \Box \mathbf{B} = \frac{1}{c^2} \frac{\partial^2 \mathbf{B}}{\partial t^2} - \nabla^2 \mathbf{B} = 0 . \qquad (12.124)$$

The general solutions of the wave equations are thus:

$$\mathbf{E}(t, \mathbf{x}) = \mathbf{E}_0 f(\hat{p} \cdot \mathbf{x} - ct) , \quad \mathbf{B}(t, \mathbf{x}) = \mathbf{B}_0 g(\hat{p}' \cdot \mathbf{x} - ct) , \qquad (12.125)$$

where $\hat{p}$ and $\hat{p}'$ are in principle distinct directions of propagation, f and g two distinct functions and $\mathbf{E}_0$ and $\mathbf{B}_0$ are the amplitudes of the wave describing the electric field and of the one describing the magnetic field.

Now we can constrain all this freedom using indeed the very Maxwell equations. We have:

$$\nabla \cdot \mathbf{E} = 0 \quad \Rightarrow \quad \mathbf{E}_0 \cdot \hat{p} f' = 0 \quad \Rightarrow \quad \boxed{\mathbf{E}_0 \cdot \hat{p} = 0} ,$$
$$(12.126)$$

$$\nabla \times \mathbf{E} = -\frac{\partial \mathbf{B}}{\partial t} \quad \Rightarrow \quad \hat{p} \times \mathbf{E}_0 f' = -\frac{\partial \mathbf{B}}{\partial t} \quad \Rightarrow \quad \hat{p} \times \mathbf{E}_0 f' = \mathbf{B}_0 g'c ,$$
$$(12.127)$$

where f' and g' are the derivatives of f and g with respect to their arguments. From the first equation we learn that the electric field oscillates perpendicularly to the direction of its propagation.

Exercise 12.18 Show from the other couple of Maxwell equations that:

$$\nabla \cdot \mathbf{B} = 0 \quad \Rightarrow \quad \boxed{\mathbf{B}_0 \cdot \hat{p}' = 0} , \qquad (12.128)$$

$$\nabla \times \mathbf{B} = \frac{1}{c^2} \frac{\partial \mathbf{E}}{\partial t} \quad \Rightarrow \quad \hat{p}' \times \mathbf{B}_0 g' = -\frac{1}{c} \mathbf{E}_0 f' , \qquad (12.129)$$

from which we learn that also the magnetic field oscillates perpendicularly to the direction of its propagation. For now this is distinct from the one of the electric field.

Now, since $\hat{p} \cdot (\hat{p} \times \mathbf{E}_0) = 0$, we can conclude that $\hat{p} \times \mathbf{B}_0 = 0$, and therefore that the magnetic field is perpendicular to both the directions of propagation. On the other hand:

$$\hat{p}' \times (\hat{p} \times \mathbf{E}_0) f' = \hat{p}' \times \mathbf{B}_0 g'c = -\mathbf{E}_0 f' . \qquad (12.130)$$

Hence, being f' not identically vanishing otherwise we would not have a wave, we can conclude that:

$$\hat{p}' \times (\hat{p} \times \mathbf{E}_0) = -\mathbf{E}_0 . \qquad (12.131)$$

Exercise 12.19 Use the rule for the triple cross product and show that:

$$\hat{p}(\hat{p} \cdot \mathbf{E}_0) - \mathbf{E}_0(\hat{p} \cdot \hat{p}') = -\mathbf{E}_0 . \qquad (12.132)$$

Therefore, being $\hat{p} \cdot \mathbf{E}_0 = 0$ and $\mathbf{E}_0$ non-vanishing, we must conclude that:

$$\hat{p} \cdot \hat{p}' = 1 \quad \Rightarrow \quad \boxed{\hat{p} = \hat{p}'} \,, \tag{12.133}$$

since both are unit vectors and therefore the scalar product equal to unity means that the angle between them must be vanishing. Thus, we have proven that electric and magnetic field propagate in the same direction. Now we prove that $f = g$:

$$\frac{|\mathbf{E}_0|}{|\mathbf{B}_0|c} = \frac{g'}{f'} \,. \tag{12.134}$$

In principle we should ask $f' \neq 0$ in order to make meaningful the above expression, but since the left hand side is time-independent, so the right hand side must be. Therefore:

$$g = \frac{|\mathbf{E}_0|}{|\mathbf{B}_0|c} f + K \,, \tag{12.135}$$

where K is an integration constant. Hence, we can write:

$$\mathbf{E} = \mathbf{E}_0 f \,, \qquad \mathbf{B} = \mathbf{B}_0 \left(\frac{|\mathbf{E}_0|}{|\mathbf{B}_0|c} f + K \right) \,. \tag{12.136}$$

Without losing any generality we can take $|\mathbf{E}_0| = |\mathbf{B}_0|$ and $K = 0$, and still we have a solution to Maxwell equations. Hence:

$$\boxed{\mathbf{B} = \frac{1}{c} \hat{p} \times \mathbf{E}} \tag{12.137}$$

For this reason, the polarisation properties of an electromagnetic wave can be analysed with respect to the electric field only.

12.7.2 Polarisation Ellipse and Stokes Parameters

Now we focus on a monochromatic electromagnetic plane wave, propagating in the direction $\hat{p} = \hat{z}$. In complex notation the electric field can be written as:

$$\mathbf{E}(t, \mathbf{x}) = \mathbf{E}_0 e^{i(z-ct)} = \begin{pmatrix} E_x \\ E_y \\ 0 \end{pmatrix} e^{i(z-ct)} \,, \tag{12.138}$$

where the two-dimensional vector is called the **Jones vector**. It is often useful to work with complex fields in electrodynamics, but here it will not be necessary so let us instead work with:

$$\mathbf{E}(t, \mathbf{x}) = \begin{bmatrix} E_x(t, \mathbf{x}) \\ E_y(t, \mathbf{x}) \\ 0 \end{bmatrix} = \begin{bmatrix} A_x \cos(z - ct + \phi_x) \\ A_y \cos(z - ct + \phi_y) \\ 0 \end{bmatrix}. \tag{12.139}$$

The wave equation that we have solved in the previous subsection is for a vector quantity, the electric field. Hence, there are actually three wave equations, one for each component. However, we have seen that the electromagnetic wave is transversal, therefore one of the component can always be put to zero, as we have done above by choosing $\hat{p} = \hat{z}$. We have then two second-order differential equations and thus we expect 4 integration constants, which we have introduced above as the amplitudes A_x and A_y and the phases ϕ_x and ϕ_y. From these 4 quantities depends the polarisation state of the wave.

We can always redefine our clock such that:

$$\mathbf{E} = \begin{bmatrix} E_x(t, \mathbf{x}) \\ E_y(t, \mathbf{x}) \\ 0 \end{bmatrix} = \begin{bmatrix} A_x \cos(z - ct) \\ A_y \cos(z - ct + \beta) \\ 0 \end{bmatrix}, \tag{12.140}$$

where $\beta \equiv \phi_y - \phi_x$ is the relative phase, which is indeed the relevant physical quantity. So we have actually 3 independent parameters which describe polarisation. When $\beta = 0$, the polarisation is purely linear, whereas when $\beta = \pi/2$ it is purely circular. This can be seen in general by studying the time-evolution of the field in the $E_y - E_x$ plane. In fact, we have:

$$\boxed{\frac{E_x^2}{A_x^2} + \frac{E_y^2}{A_y^2} - \frac{2 E_x E_y}{A_x A_y} \cos \beta = \sin^2 \beta} \tag{12.141}$$

This is the equation of an ellipse, the **polarisation ellipse**, rotated by an angle α in the $E_y - E_x$ plane.

Exercise 12.20 Defining:

$$A_x = A \cos \theta, \qquad A_y = A \sin \theta, \tag{12.142}$$

show that:

$$\tan 2\alpha = \tan 2\theta \cos \beta, \tag{12.143}$$

and that the semi-major and semi-minor axis of the polarisation ellipse are given by:

$$a^2 = \frac{A^2}{2} \left(1 + \sqrt{1 - \sin^2 2\theta \sin^2 \beta} \right), \tag{12.144}$$

$$b^2 = \frac{A^2}{2} \left(1 - \sqrt{1 - \sin^2 2\theta \sin^2 \beta} \right). \tag{12.145}$$

When $\beta = 0$ we have $b = 0$, i.e. the ellipse degenerates in a straight line, titled by an angle α with respect to the E_x axis. In this case the wave is purely linearly polarised. On the other hand, when $\beta = \pi/2$ the ellipse degenerates in a circle, and thus we have a purely circularly polarised wave.

The **Stokes parameters** are defined as follows in terms of the polarisation ellipse parameters:

$$I \equiv a^2 + b^2 = A^2 ,$$
$$(12.146)$$

$$Q \equiv (a^2 - b^2) \cos 2\alpha = A^2 \sqrt{1 - \sin^2 2\theta \sin^2 \beta} \cos 2\alpha = A^2 \cos 2\theta ,$$
$$(12.147)$$

$$U \equiv (a^2 - b^2) \sin 2\alpha = A^2 \sqrt{1 - \sin^2 2\theta \sin^2 \beta} \sin 2\alpha = A^2 \sin 2\theta \cos \beta ,$$
$$(12.148)$$

$$V \equiv 2abh = (A^2 \sin 2\theta \sin \beta)h ,$$
$$(12.149)$$

where $h = \pm 1$ simply establishes the direction of rotation, clockwise ($h = -1$) or anti-clockwise ($h = 1$). Note that the Stokes parameter I is the intensity of the wave.

Pure Q polarisation. If $U = V = 0$, then $\beta = 0$ and therefore we have pure linear polarisation. Since $\sin 2\alpha = 0$ then either $\alpha = 0$ or $\alpha = \pi/2$. In the former case the electric field oscillates along the x-axis and $Q = A^2 = I > 0$. If $\alpha = \pi/2$ then the electric field oscillates along the y-axis and $Q = -A^2 = -I < 0$.

Pure U polarisation. If $Q = V = 0$, then $\beta = 0$ and therefore we have again pure linear polarisation. This time $\cos 2\alpha = 0$ then either $\alpha = \pi/4$ or $\alpha = 3\pi/4$. In the former case the electric field oscillates along the $y = x$ line and $Q = A^2 = I > 0$. If $\alpha = 3\pi/4$ then the electric field oscillates along the $y = -x$ line and $Q = -A^2 = -I < 0$.

Pure V polarisation. If $Q = U = 0$, then $\theta = \pi/4$ and $\beta = \pi/2$ and therefore we have pure circular polarisation. We have $V = A^2 h = \pm A^2 = \pm I$, with the sign determined by the direction of rotation.

As we have seen, the polarisation is fully determined by 3 parameters. These means that the 4 Stokes parameters are not independent.

Exercise 12.21 Show that the Stokes parameters satisfy the following constraint:

$$\boxed{Q^2 + U^2 + V^2 = I^2}$$
$$(12.150)$$

Other definitions of the Stokes parameters are the following. Using the complex electric field:

$$\mathbf{E}(t, \mathbf{x}) = \begin{pmatrix} E_x \\ E_y \\ 0 \end{pmatrix} e^{i(z-ct)} , \qquad (12.151)$$

they are defined as:

$$I \equiv |E_x|^2 + |E_y|^2 , \quad Q \equiv |E_x|^2 - |E_y|^2 , \qquad (12.152)$$
$$U \equiv 2\mathrm{Re}(E_x E_y^*) , \quad V \equiv -2\mathrm{Im}(E_x E_y^*) . \qquad (12.153)$$

We have considered so far a monochromatic wave, possibly made up of many waves but all coherent. On the other hand, if we consider the superposition of incoherent waves then we can write a generic time-dependence for the electric field:

$$\mathbf{E}(t, \mathbf{x}) = \begin{bmatrix} E_x(t, \mathbf{x}) \\ E_y(t, \mathbf{x}) \\ 0 \end{bmatrix} , \qquad (12.154)$$

and therefore the Stokes parameters are defined as expectation values:

$$I \equiv \langle E_x^2 \rangle + \langle E_y^2 \rangle , \quad Q \equiv \langle E_x^2 \rangle - \langle E_y^2 \rangle , \qquad (12.155)$$
$$U \equiv 2\langle E_x E_y \rangle \cos \beta , \quad V \equiv 2\langle E_x E_y \rangle \sin \beta , \qquad (12.156)$$

as for example time-averages:

$$\langle X \rangle = \frac{1}{T} \int_0^T dt \, X(t) . \qquad (12.157)$$

In this case the electric field is a random variable and we can just define polarisation through averages. Moreover, in this case one has:

$$\boxed{Q^2 + U^2 + V^2 \le I^2} \qquad (12.158)$$

since each Stokes parameter is defined as the sum of the corresponding Stokes parameters of all the waves composing the total one. See e.g. Chandrasekhar (1960).

Exercise 12.22 Show that the three definitions of Stokes parameters given above are in fact equivalent for a single monochromatic sinusoidal wave.

Now, suppose to make successive anti-clockwise rotations of $\pi/4$ in the polarisation plane. What happens is the following:

$$Q \to U \to -Q \to -U \to Q . \qquad (12.159)$$

That is, after a half rotation about the propagation direction we recover the initial polarisation state, as if we applied an identity operator. This suggests us that we are dealing with a spin-2 field. Indeed, let us introduce the intensity matrix for an electromagnetic wave propagating along $-\hat{z}$:

$$
J_{ij}(-\hat{z}) = \frac{1}{2}
\begin{pmatrix}
I(\hat{z}) + Q(\hat{z}) & U(\hat{z}) - iV(\hat{z}) & 0 \\
U(\hat{z}) + iV(\hat{z}) & I(\hat{z}) - Q(\hat{z}) & 0 \\
0 & 0 & 0
\end{pmatrix},
\tag{12.160}
$$

where the factor $1/2$ serves in order to reproduce the correct result that the trace $J_{ii} = I$. This intensity matrix reminds us of h_{ij}^T, the tensor perturbation of the metric, defined in Eq. (4.198).

Exercise 12.23 Show that applying the a rotation of an angle θ about the axis $\hat{z}$ we have that:

$$
I \to I,
\tag{12.161}
$$

$$
Q \pm iU \to e^{\pm 2i\theta}(Q \pm iU),
\tag{12.162}
$$

$$
V \to V.
\tag{12.163}
$$

The different sign with respect to the GW case is due to the fact that the rotation is made about the propagation direction $-\hat{z}$. Hence, a rotation by θ about $-\hat{z}$ corresponds to a rotation by $-\theta$ about $\hat{z}$, which is the line-of-sight direction.

The matrix defined on the two-dimensional subspace orthogonal to the propagation direction:

$$
\rho_{ij}(-\hat{z}) = \frac{1}{2}
\begin{pmatrix}
I(\hat{z}) + Q(\hat{z}) & U(\hat{z}) - iV(\hat{z}) \\
U(\hat{z}) + iV(\hat{z}) & I(\hat{z}) - Q(\hat{z})
\end{pmatrix},
\tag{12.164}
$$

is called **photon density matrix** and can be suggestively put in the following form:

$$
\rho = \frac{1}{2}(I\mathbf{1} + Q\sigma_3 + U\sigma_1 + V\sigma_2),
\tag{12.165}
$$

where $\mathbf{1}$ is the 2×2 identity matrix and the $\sigma_{1,2,3}$ are the Pauli matrices. It might seem strange that the latter could be associated to a photon, since they are usually employed for the description of spin-$1/2$ particles whereas the photon is a spin-1 boson. However, since the photon is massless it is characterised by only two spin states ± 1 (or helicities), just as any spin-$1/2$ particle, massive or massless, is characterised by two spin states $\pm 1/2$.

The polarisation vectors introduced in Eq. (4.213) can be defined equivalently here. Along the $\hat{z}$ axis they are simply:

$$e_\pm(\hat{z}) = (1, \pm i, 0)/\sqrt{2} \,, \tag{12.166}$$

and the Stokes parameters are defined as:

$$Q(\hat{z}) \pm i U(\hat{z}) = 2e_{\pm,i}(\hat{z})e_{\pm,j}(\hat{z})J_{ij}(-\hat{z}) \,. \tag{12.167}$$

For a generic direction of propagation, say $-\hat{n}$, we define:

$$Q(\hat{n}) \pm i U(\hat{n}) = 2e_{\pm,i}(\hat{n})e_{\pm,j}(\hat{n})J_{ij}(-\hat{n}) \,, \tag{12.168}$$

where in spherical coordinates:

$$\hat{n} = (\sin\theta\cos\phi, \sin\theta\sin\phi, \cos\theta) \,, \tag{12.169}$$

and the basis vectors which form the polarisation vectors can be chosen as:

$$\hat{\theta} = (\cos\theta\cos\phi, \cos\theta\sin\phi, -\sin\theta) \,, \qquad \hat{\phi} = (-\sin\phi, \cos\phi, 0) \,. \tag{12.170}$$

Hence:

$$e_\pm(\hat{n}) = \frac{\hat{\theta} \pm i\hat{\phi}}{\sqrt{2}} \,. \tag{12.171}$$

As we have already discussed, $Q \pm iU$ are spin-2 fields and thus are expanded as:

$$(Q + iU)(\hat{n}) = \sum_{\ell m} a_{P,\ell m} \,_2 Y_\ell^m(\hat{n}) \,. \tag{12.172}$$

Through the polarisation vectors we can characterise the operator $\eth$ as

$$\boxed{\eth^s = (-1)^s 2^{s/2} e_{+,i_1}(\hat{n}) \cdots e_{+,i_s}(\hat{n}) \tilde{\nabla}_{i_1} \cdots \tilde{\nabla}_{i_s}} \tag{12.173}$$

where

$$\tilde{\nabla} = \hat{\theta}\frac{\partial}{\partial\theta} + \frac{\hat{\phi}}{\sin\theta}\frac{\partial}{\partial\phi} \,, \tag{12.174}$$

is the covariant derivative on the sphere.

Exercise 12.24 Show that the definition of Eq. (12.102) and formula (12.173) are identical for $s = 1$ and $s = 2$.

Fig. 12.1 Scattering plane.
The unit vectors $\hat{y}'$ and $\hat{y}$ are
equal and going out from the
page orthogonally

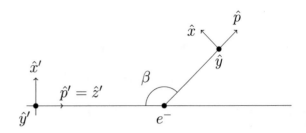

12.8 Thomson Scattering

In this section we work out the collisional term of Eq. (5.104). We follow the calculations of Chandrasekhar (1960), but take advantage of the properties of spherical harmonics, as done in Hu and White (1997). See also Kosowsky (1996).

Consider an incoming wave propagating in the direction $\hat{p}'$ and scattered in the direction $\hat{p}$. The plane in which $\hat{p}'$ and $\hat{p}$ lie is called **scattering plane**.

Choosing the simple geometry of Fig. 12.1, the incoming electric field can be decomposed as follows:

$$\mathbf{E}' = E'_x \hat{x}' + E'_y \hat{y}' , \qquad (12.175)$$

i.e. with no $\hat{z}'$ component, since this is the direction of propagation. When the electric field interacts with the electron, this starts to oscillate and to emit electromagnetic waves in almost all directions. The "almost" is make quantitative by **Larmor formula**, for which the irradiated power per solid angle is:

$$\frac{dP}{d\Omega} = \frac{e^2}{4\pi c^3} \left| \hat{p} \times (\hat{p} \times \mathbf{a}) \right|^2 , \qquad (12.176)$$

where $\mathbf{a}$ is the electron acceleration. See e.g. Jackson (1998) and note that here we are using Gaussian units for which $[F] = [e^2/r^2]$, i.e. the force can be dimensionally expressed as squared Coulomb divided by a squared length. For a particular outgoing polarisation say $\hat{\epsilon}$, we make the scalar product with $\hat{\epsilon}$ in the square modulus of the above formula and obtain:

$$\frac{dP}{d\Omega} = \frac{e^2}{4\pi c^3} \left| \hat{\epsilon} \cdot \mathbf{a} \right|^2 . \qquad (12.177)$$

The acceleration of the electron is produced by the incident electric field $\mathbf{E}'$:

$$\mathbf{a} = \frac{e}{m} \mathbf{E}' , \qquad (12.178)$$

so that:

$$\frac{dP}{d\Omega} = \frac{e^4}{4\pi m^2 c^3} \left| \hat{\epsilon} \cdot \mathbf{E}' \right|^2 . \qquad (12.179)$$

Writing now $\mathbf{E}' = \hat{\epsilon}'|\mathbf{E}'|$, we get:

$$\frac{dP}{d\Omega} = \frac{c}{4\pi}|\mathbf{E}'|^2 \left(\frac{e^2}{mc^2}\right)^2 |\hat{\epsilon} \cdot \hat{\epsilon}'|^2 . \tag{12.180}$$

The contribution $c|\mathbf{E}'|^2/(4\pi)$ is the Poynting vector and hence the flux of the incoming wave, leaving:

$$\boxed{\frac{d\sigma}{d\Omega} = \left(\frac{e^2}{mc^2}\right)^2 |\hat{\epsilon} \cdot \hat{\epsilon}'|^2 \equiv \frac{3\sigma_T}{8\pi}|\hat{\epsilon} \cdot \hat{\epsilon}'|^2} \tag{12.181}$$

as the differential cross section, where of course σ_T is the **Thomson cross section**. The above derivation is based on Larmor formula and thus it is purely classical and valid only for a non-relativistic movement of the electron and for photon energies much smaller than the electron mass. If these conditions are not met, one should employ the Klein–Nishina formula, see e.g. Weinberg (2005).

Let us assume $\sigma_T = 1$, for simplicity. We shall recover the correct dimensions only at the end of our derivation via the scattering rate $-\tau' = n_e\sigma_T a$. The electric field scattered in the direction $\hat{p}$ is of the form:

$$\mathbf{E} = A\left[\hat{p} \times (\hat{p} \times \mathbf{E}')\right] , \tag{12.182}$$

where $A^2 = 3/(8\pi)$. Let us use now the geometry in the scattering plane of Fig. 12.1 and calculate the double cross product.

Exercise 12.25 Show that:

$$\mathbf{E} = -A\left(\hat{x}'E_x' \cos^2\beta + \hat{y}'E_y' + \hat{z}'E_x' \sin\beta \cos\beta\right) . \tag{12.183}$$

Since $\hat{y} = \hat{y}'$, we can already conclude that:

$$E_y = -AE_y' , \tag{12.184}$$

i.e. the electric field contribution perpendicular to the scattering plane gains no angular dependence.

In order to calculate E_x, which is the contribution of electric field parallel to the scattering plane, we need to know $\hat{x}$. One has:

$$\hat{x} = \hat{y} \times \hat{p} , \tag{12.185}$$

and

$$\hat{p} = (\sin\beta, 0, -\cos\beta) . \tag{12.186}$$

Hence:

$$\hat{x} = -\hat{x}' \cos\beta - \hat{z}' \sin\beta . \tag{12.187}$$

Therefore we have:

$$E_x = \mathbf{E} \cdot \hat{x} = A E_x' \cos\beta . \tag{12.188}$$

We are now in the position of relating the Stokes parameter of the outgoing wave to those of the incoming one:

$$I = \langle E_x^2 \rangle + \langle E_y^2 \rangle = A^2 \langle E_x'^2 \rangle \cos^2\beta + A^2 \langle E_y'^2 \rangle \equiv A^2 I_x' \cos^2\beta + A^2 I_y' , \tag{12.189}$$

$$Q = \langle E_x^2 \rangle - \langle E_y^2 \rangle = A^2 I_x' \cos^2\beta - A^2 I_y' , \tag{12.190}$$

$$U = 2\langle E_x E_y \rangle \cos\beta = A^2 U' \cos\beta , \tag{12.191}$$

$$V = 2\langle E_x E_y \rangle \sin\beta = A^2 V' \cos\beta . \tag{12.192}$$

Using now the definitions:

$$I' = \langle E_x'^2 \rangle + \langle E_{y'}^2 \rangle , \tag{12.193}$$

$$Q' = \langle E_x'^2 \rangle - \langle E_{y'}^2 \rangle , \tag{12.194}$$

we have the complete transformation:

$$\begin{pmatrix} I \\ Q \\ U \\ V \end{pmatrix} = \frac{3}{8\pi} \begin{pmatrix} \frac{1+\cos^2\beta}{2} & -\frac{\sin^2\beta}{2} & 0 & 0 \\ -\frac{\sin^2\beta}{2} & \frac{1+\cos^2\beta}{2} & 0 & 0 \\ 0 & 0 & \cos\beta & 0 \\ 0 & 0 & 0 & \cos\beta \end{pmatrix} \begin{pmatrix} I' \\ Q' \\ U' \\ V' \end{pmatrix} . \tag{12.195}$$

These Stokes parameters have dimension of intensity. We now introduce Θ, the temperature fluctuation instead of I and suitably redefine all the other Stokes parameters. Moreover, we introduce the combination $Q \pm iU$, since we know how it transforms under a rotation in the polarisation plane, cf. Eq. (12.161). We shall need this rotation in a moment. We have then:

$$\begin{pmatrix} \Theta \\ Q + iU \\ Q - iU \end{pmatrix} = \frac{3}{16\pi} \begin{pmatrix} 1 + \cos^2\beta & -\frac{\sin^2\beta}{2} & -\frac{\sin^2\beta}{2} \\ -\sin^2\beta & \frac{(1+\cos\beta)^2}{2} & \frac{(1-\cos\beta)^2}{2} \\ -\sin^2\beta & \frac{(1-\cos\beta)^2}{2} & \frac{(1+\cos\beta)^2}{2} \end{pmatrix} \begin{pmatrix} \Theta' \\ Q' + iU' \\ Q' - iU' \end{pmatrix} , \tag{12.196}$$

and V does not mix up with the other Stokes parameter, so we consider it separately:

Fig. 12.2 Scattering geometry

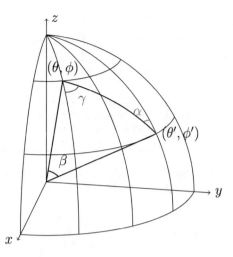

$$V = \frac{3\cos\beta}{16\pi} V' . \tag{12.197}$$

Recall that the above transformations hold true in the scattering plane and the primed quantities depend on the incoming direction $\hat{p}'$ whereas the unprimed quantities from $\hat{p}$.

We now have to transform to a generic reference frame in which the incident wave comes from a direction (θ', ϕ') and the outgoing wave goes along a direction (θ, ϕ). This calculation is performed in Chandrasekhar (1960) but we follow Hu and White (1997) because it is faster and takes advantage of the properties of the spherical harmonics which we have seen earlier. In particular, we refer to Fig. 12.2.

We need to perform a rotation $R(-\alpha)$ in order to pass from the laboratory frame to the scattering frame. Recall that the rotation is performed in the polarisation plane, i.e. about $-\hat{p}'$. Hence it is a clockwise rotation and for this reason we have $-\alpha$. Then we apply the above matrix say $S(\beta)$ in order to obtain the outgoing Stokes parameters. Finally, we apply another rotation $R(-\gamma)$ in order to obtain the Stokes parameters in the laboratory frame. This rotation seems to be anti-clockwise because it is in the direction opposite to $R(-\alpha)$. However, notice that $\hat{p}$ is now outgoing and thus the polarisation plane is reflected, giving then $R(-\gamma)$ instead of $R(\gamma)$.

Exercise 12.26 Using the transformation of Eq. (12.161) and the fact that Θ is invariant under rotation in the polarisation plane, show that:
selectfont

$$R(-\gamma)S(\beta)R(-\alpha) = \frac{3}{16\pi} \begin{pmatrix} 1+\cos^2\beta & -\frac{\sin^2\beta}{2}e^{-2i\alpha} & -\frac{\sin^2\beta}{2}e^{2i\alpha} \\ -e^{-2i\gamma}\sin^2\beta & e^{-2i\gamma}\frac{(1+\cos\beta)^2}{2}e^{2i\alpha} & e^{-2i\gamma}\frac{(1-\cos\beta)^2}{2}e^{2i\alpha} \\ -e^{2i\gamma}\sin^2\beta & e^{2i\gamma}\frac{(1-\cos\beta)^2}{2}e^{-2i\alpha} & e^{2i\gamma}\frac{(1+\cos\beta)^2}{2}e^{2i\alpha} \end{pmatrix},$$

$$(12.198)$$

Since also V is invariant under rotation in the polarisation plane, Eq. (12.200) is valid even in the laboratory frame.

Exercise 12.27 Using the definitions of spherical harmonics given in Table 5.1, write the above matrix as:

$$R(\gamma)S(\beta)R(-\alpha) = \frac{1}{8\pi}\sqrt{\frac{4\pi}{5}} \begin{pmatrix} Y_2^0 + 2\sqrt{5}Y_0^0 & -\sqrt{3/2}Y_2^{-2} & -\sqrt{3/2}Y_2^2 \\ -\sqrt{6}e^{-2i\gamma}{}_2Y_2^0 & 3e^{-2i\gamma}{}_2Y_2^{-2} & 3e^{-2i\gamma}{}_2Y_2^2 \\ -\sqrt{6}e^{2i\gamma}{}_{-2}Y_2^0 & 3e^{2i\gamma}{}_{-2}Y_2^{-2} & 3e^{2i\gamma}{}_{-2}Y_2^2 \end{pmatrix}.$$

$$(12.199)$$

The spherical harmonics inside this matrix are function of (β, α). For V show that:

$$V = \frac{1}{4}\sqrt{\frac{3}{4\pi}}Y_1^0(\beta, \alpha)V'.$$

$$(12.200)$$

Now, using the addition theorem:

$$\sum_{m=-\ell}^{\ell} {}_{s_1}Y_\ell^{m*}(\theta', \phi')\,{}_{s_2}Y_\ell^m(\theta, \phi) = \frac{2\ell+1}{4\pi}\,{}_{s_2}Y_\ell^{-s_1}e^{-is_2\gamma},$$

$$(12.201)$$

we can introduce the matrix:

$$P^{(m)}(\hat{p}, \hat{p}') = \begin{pmatrix} Y_2^{m*}Y_2^m & -\sqrt{3/2}\,{}_2Y_2^{m*}Y_2^m & -\sqrt{3/2}\,{}_{-2}Y_2^{m*}Y_2^m \\ -\sqrt{6}Y_2^{m*}{}_2Y_2^m & 3{}_2Y_2^{m*}{}_2Y_2^m & 3{}_{-2}Y_2^{m*}{}_2Y_2^m \\ -\sqrt{6}Y_2^{m*}{}_{-2}Y_2^m & 3{}_2Y_2^{m*}{}_{-2}Y_2^m & 3{}_{-2}Y_2^{m*}{}_{-2}Y_2^m \end{pmatrix}, \quad (12.202)$$

where the complex conjugates spherical harmonics depend on $\hat{p}'$ whereas the others depend on $\hat{p}$. Hence, the scattered Stokes parameters can then be written as:

$$\begin{pmatrix} \Theta \\ Q+iU \\ Q-iU \end{pmatrix} = \frac{1}{4\pi}\begin{pmatrix} \Theta' \\ 0 \\ 0 \end{pmatrix} + \frac{1}{10}\sum_{m=-2}^{2} P^{(m)}(\hat{p}, \hat{p}')\begin{pmatrix} \Theta' \\ Q'+iU' \\ Q'-iU' \end{pmatrix}, \quad (12.203)$$

whereas for V we can write:

$$V(\hat{p}) = \frac{1}{4}\sum_{m=-1}^{1} Y_1^m(\hat{p})Y_1^{m*}(\hat{p}')V(\hat{p}').$$

$$(12.204)$$

Integrating over all the incoming directions and multiplying by the Thomson scattering rate:

$$- \tau' = n_e \sigma_T a ,$$ (12.205)

we obtain part of the collisional terms for the Boltzmann equation (5.104). This is the scattering rate into photons with momentum in the direction $\hat{p}$, which we have taken as the reference direction. There is also a contribution which takes into account the scattering of photons into other directions different from $\hat{p}$. This generates the term:

$$\tau' \begin{pmatrix} \Theta \\ Q+iU \\ Q-iU \end{pmatrix} .$$ (12.206)

Finally, all the calculations performed here assume the electron fluid at rest. Performing a boost, we get the Doppler shift term $-\tau' \hat{p} \cdot \mathbf{v}_b$. Indeed, the photon energy in the new frame is given by:

$$\tilde{p} = p(1 - \mathbf{v}_b \cdot \hat{p}) ,$$ (12.207)

upon a boost with velocity $\mathbf{v}_b$. The Lorentz factor on the right hand side is unity since $|\mathbf{v}_b|^2$ is negligible, being a first-order quantity. The perturbed distribution $\mathcal{F}_\gamma$, being a scalar, transforms as:

$$\mathcal{F}_\gamma(p) = \tilde{\mathcal{F}}_\gamma[p(1 - \mathbf{v}_b \cdot \hat{p})] ,$$ (12.208)

and thus, developing the distribution function up to first-order, we get:

$$\mathcal{F}_\gamma(p) = \tilde{\mathcal{F}}_\gamma(p) - \frac{\partial \bar{f}_\gamma}{\partial p} p \mathbf{v}_b \cdot \hat{p} ,$$ (12.209)

and using Eq. (5.80) we have that:

$$\Theta = \tilde{\Theta} + \mathbf{v}_b \cdot \hat{p} .$$ (12.210)

We have thus fully recovered the collisional term of Eq. (5.104).

Finally, consider initially unpolarised photons. Then, upon scattering:

$$\begin{pmatrix} \Theta \\ Q+iU \\ Q-iU \end{pmatrix} (\hat{p}) = \begin{pmatrix} \int \frac{d^2\hat{p}'}{4\pi} \Theta(\hat{p}') \\ 0 \\ 0 \end{pmatrix}$$

$$+ \frac{1}{10} \sum_{m=-2}^{2} \begin{pmatrix} Y_2^m \\ -\sqrt{6}_2 Y_2^m \\ -\sqrt{6}_{-2} Y_2^m \end{pmatrix} (\hat{p}) \int d^2\hat{p}' Y_2^{m*}(\hat{p}') \Theta(\hat{p}') .$$ (12.211)

Expanding the relative temperature fluctuation in spherical harmonics:

$$\Theta(\hat{p}') = \sum_{\ell m} a_{\ell m} Y_\ell^m(\hat{p}') , \tag{12.212}$$

due to the orthogonality properties of the latter, polarisation will be produced only if the incident $\Theta(\hat{p}')$ has a quadrupole moment.

References

Butkov, E.: Mathematical Physics. Addison-Wesley Publishing Company, Incorporated (1968)

Chandrasekhar, S.: Radiative Transfer. Dover, New York (1960)

Griffiths, D.J.: Introduction to Electrodynamics. Cambridge University Press, Cambridge (2017)

Hu, W., White, M.J.: CMB anisotropies: total angular momentum method. Phys. Rev. D **56**, 596–615 (1997)

Huang, K.: Statistical Mechanics, 2nd edn. Wiley-VCH, New York (1987)

Jackson, J.D.: Classical Electrodynamics, 3rd edn. Wiley-VCH, New York (1998)

Kosowsky, A.: Cosmic microwave background polarization. Ann. Phys. **246**, 49–85 (1996)

Landau, L.D., Lifshits, E.M.: Quantum Mechanics. Course of Theoretical Physics, vol. 3. Butterworth-Heinemann, Oxford (1991)

Newman, E.T., Penrose, R.: Note on the Bondi-Metzner-Sachs group. J. Math. Phys. **7**, 863–870 (1966)

Weinberg, S.: The Quantum Theory of Fields. Vol. 1: Foundations. Cambridge University Press, Cambridge (2005)

Weinberg, S.: Lectures on Quantum Mechanics. Cambridge University Press, Cambridge (2015)

Index

A

Absolute magnitude, 366
Acceleration equation, 28
Acoustic oscillations, 320
Adiabatic speed of sound, 138, 279
Age of the universe, 30, 31
Alternatives to Λ, 11
Angular diameter distance, 51
Anisotropic stress, 126, 147
Anthropic principle, 11
Apparent magnitude, 365
Associated Legendre polynomials, 391
Axion, 6

B

Bardeen's potentials, 137
Baryogenesis, 58
Baryon Acoustic Oscillations (BAO), 288, 321
Baryon feedback, 12
Baryon loading, 326
Baryons, 34
    Boltzmann equation, 186
    density parameter, 35
Baryon-to-photon ratio, 58, 91
Bias, 273, 377
Big-Bang, 13
Big Bang Nucleosynthesis (BBN)
    temperature, 90
Bispectrum, 221
    reduced, 222
Bolometric magnitude, 366
Boltzmann equation
    collisional term, 84
    collisionless, 82
    coupled to FLRW metric, 82

force term, 160
    scalar perturbations, 162
    tensor perturbations, 162
    vector perturbations, 163
moments, 83
non-relativistic case, 78
perturbation, 158
relativistic case, 81
Bose–Einstein distribution, 65, 382
Bouncing cosmology, 235
Bulk viscosity, 126
Bullet cluster, 5

C

Chemical equilibrium, 56, 90
Christoffel symbols
    perturbation, 119
CMBFAST, 334
Cold Dark Matter (CDM), 34
    density parameter, 35
    perturbed Boltzmann equation, 164
    small-scale anomalies, 12
Comoving coordinates, 21
Comoving curvature perturbation, 138
    conservation, 387
Comoving distance, 45
Comoving momentum, 64
Comoving-gauge density perturbation, 138
Confidence level, 368
Conformal Hubble factor, 30
Conformal time, 22
Conformal transformation, 267
Continuity equation, 34, 83
    perturbation, 165
    using thermodynamics laws, 56
Copernican principle, 13, 227

© Springer International Publishing AG, part of Springer Nature 2018
O. Piattella, *Lecture Notes in Cosmology*, UNITEXT for Physics,
https://doi.org/10.1007/978-3-319-95570-4

Core/Cusp problem, 12
Correlation function, 214, 375
Cosmic coincidence, 11
Cosmic Microwave Background (CMB)
  acoustic oscillations, 313
  anomalies, 13, 227
  $C_{TT,\ell}^{S}$ spectrum, 226
  $C_{BB,\ell}^{T}$ spectrum, 362
  $C_{EE,\ell}^{S}$ spectrum, 357
  $C_{EE,\ell}^{T}$ spectrum, 361
  $C_{TE,\ell}^{S}$ spectrum, 357
  $C_{TE,\ell}^{T}$ spectrum, 362
  $C_{TT,\ell}^{T}$ spectrum, 352
  doppler effect, 313
  integrated Sachs-Wolfe effect, 314
  large scale anisotropies, 315
  observation, 7
  Planck TT spectrum, 309
  polarisation spectra, 227
  power spectrum, 225
  primary anisotropies, 313
  spectral distortions, 67
  temperature, 69
  temperature fluctuations expansion, 224
Cosmic time, 21
Cosmic variance, 216
  angular power spectrum, 228
  CMB power spectrum, 229
  power spectrum, 221
Cosmography, 46
Cosmological constant
  density parameter, 35
  problem, 11
Cosmological observations, 7
Cosmological principle, 19
Coulomb scattering, 187
Critical density, 32
Cross section
  thermally averaged, 89

D
Dark Energy (DE), 4
Dark Energy Survey (DES), 9
Dark Matter (DM), 4
  Cold Dark Matter (CDM), 6
  Hot Dark Matter (HDM), 7
  Warm Dark Matter (WDM), 6
Dark matter decoupling, 58
Dark matter searches, 10
Deceleration parameter, 31
Density contrast

definition, 127
Density parameter
  closure relation, 33
Density parameter Ω, 32
de Sitter universe, 38
  deceleration parameter, 39
Deuterium bottleneck, 91
Diffusion damping, 319, 329
Distance modulus, 366
Distribution function, 60
  energy density, 62
  energy-momentum tensor, 62
  particle number density, 62
  perturbation, 130, 158
  pressure, 62
Dust, 34
Dust-dominated universe
  scale factor solution, 42

E
Effective number of relativistic degrees of
  freedom, 96, 110
Effective speed of sound, 138
Einstein-de Sitter universe, 41
Einstein equations
  scalar perturbations, 145
  tensor perturbations, 149
  vector perturbations, 152
Einstein frame, 268
Einstein static universe, 37
Einstein tensor
  perturbation, 124
Electron-positron annihilation, 59
Electroweak phase transition, 58
Energy momentum tensor
  imperfect fluid, 126
  kinetic theory, 130
  perturbations, 127
Ensemble, 213
Ensemble average, 213
Ensemble variance, 215
Entropy density, 65
Entropy modes, 196
Entropy perturbation, 138
Equation of state, 34
Equivalence wavenumber, 282
Ergodic theorem, 216, 231
Etherington's distance duality, 51
Euclid (experiment), 9
Euler equation
  perturbation, 166
Event horizon, 47

Evolution of perturbation
  Λ-dominated epoch, 281
  matter-dominated epoch, 287
  Mészáros equation, 295
  radiation-dominated epoch, 293
  super-horizon, 278

F
Fermi–Dirac distribution, 65, 380, 381
Fine structure constant, 60
Fine tuning, 236
Finite thickness effect, 339
Flatness problem, 32, 235
FLRW metric, 21
  Christoffel symbols, 25, 121
  light-cone structure, 23
  perturbation, 118
  perturbed Christoffel symbols, 122
  perturbed Einstein tensor, 125
  perturbed Ricci scalar, 125
  perturbed Ricci tensor, 123
  Ricci tensor, 28
Fourier transform, 142
Four-velocity
  perturbations, 129
Free electron fraction, 100
Free-streaming
  solution, 313
Freeze out, 57
$f(R)$ gravity, 268
Friedmann equation, 28

G
Gauge, 118
  problem, 131
Gauge transformation, 132
  energy-momentum tensor, 133
  metric, 133
Gaussian filter, 372
Grand Unified Theory (GUT), 58
Gravitational waves
  equation, 150
  helicity, 150
  observatories, 9
  sum over helicities, 231
Green's function, 397
Growth factor, 287

H
Heat transfer, 126
Helium mass fraction, 99

Helmholtz theorem, 384
Horizon crossing, 195
Horizon problem, 238
Hot Big Bang
  thermal history, 57
Hot dark matter, 35
Hubble constant, 2, 30
Hubble parameter, 22
Hubble radius, 2, 25
Hubble's law, 1
  derivation, 45

I
IceCube, 9
Inflation, 236
  e-folds number, 237
  energy scale, 238
  horizon problem, 238
  inflaton, 240
  Lyth bound, 246
  maximum number of e-folds, 242
  Planck constraints, 265
  power-law, 266
  slow-roll, 240
  Starobinsky model, 267
  trans-Planckian problem, 243
Interacting dark matter, 13
Interaction rate, 56
Isocurvature modes, 196
Isometry, 133

J
Jerk parameter, 46
Jordan frame, 267

K
Killing equations, 133
Kinetic equilibrium, 56
Kodama–Sasaki equation, 278

L
ΛCDM model, 7, 35
  age of the universe, 36
  scale factor solution, 44
Landau damping, 339
Landau damping scale, 341
Larmor formula, 406
Last scattering surface, 100
Legendre polynomials
  expansion, 168
  orthogonality relation, 169

recurrence relation, 169
Legendre polynomials expansion, 179
Leptogenesis, 58
Likelihood, 367
Linearised Einstein equations, 131
Line-of-sight, 311
    unit vector, 224
Line-of-sight integration, 334
Liouville operator, 78
Liouville theorem, 78
    proof, 79
Lithium problem, 13
Lookback time, 45
Lukash variable, 138
Luminosity distance, 49

M
Marginalisation, 369
Mass variance, 371
Matter-radiation equality, 77
Maximally symmetric space, 19
Maxwell–Boltzmann distribution, 380
Maxwell equations, 398
Method of characteristics, 80
Milne universe, 44
Missing satellites problem, 12
Mukhanov-Sasaki equation, 260
Multiverse, 11

N
Neutralino, 6
Neutrino decoupling, 58
Neutrinos
    density parameter, 35
    effective number of families N_{eff}, 73
    energy density, 70
    fraction, 200
    isocurvature velocity mode, 199
    mass constraints, 74
    massive neutrinos density parameter, 76
    perturbed Boltzmann equation, 167
    relative neutrino heat flux, 199
    right-handed, 70
    sterile, 70
    temperature, 72
    temperature equations hierarchy, 169
Neutron abundance, 95
Newtonian cosmology, 18
Newtonian gauge, 141
Non-Gaussianity, 221
    local type, 222

Planck constraint on f_{NL}, 222
Normal mode decomposition, 142
Number density
    equilibrium, 86

O
Olbers's paradox, 3
Open problems in cosmology, 13
Optical depth, 176

P
Partial wave expansion, 168, 179
Particle horizon, 47
Particle number conservation, 83
Particles
    thermal production, 6
Particles in a thermal bath, 75
Perfect fluid, 29
Periodic boundary conditions, 61
Phase-space
    volume of the fundamental cell, 61
Photon density matrix, 404
Photons
    decoupling from electrons, 100
    density parameter, 35, 69
    energy density, 67
    free-streaming, 310
    number density, 69
    perturbed Boltzmann equation
        collisional term, 173
        scalar perturbations, 179
        tensor perturbations, 182
        vector perturbations, 184
    relative temperature fluctuations, 172
Poisson distribution, 384
Polarisation ellipse, 401
Polarisation vectors, 152, 404
Power spectrum, 212, 218
    amplitude, 262
    dimensionless, 219
    scale invariant, 253
Present time t_0, 31
Primordial modes, 195
    adiabatic, 201
    baryon density isocurvature, 206
    Cold Dark Matter density isocurvature,
        206
    neutrino density isocurvature, 205
    neutrino velocity isocurvature, 207
    Planck constraints, 209
Primordial plasma, 57
Primordial tilt, 317

Proper distance, 45
Proper momentum, 26, 64, 130
Proper radius, 22

Q
Quantum-to-classical transition, 248

R
Radiation, 35
Radiation-dominated universe
    scale factor solution, 40
Radiation plus dust universe
    scale factor solution, 43
Random field, 213
    Gaussian, 219
Reality condition, 217, 225
Recombination, 100
Redshift, 10, 27
    photometric, 10
    spectroscopic, 10
Redshift drift, 31
Reheating, 247
Reionisation, 105, 337
Relativistic Poisson equation, 146
Ricci scalar
    perturbation, 124
Ricci tensor
    definition, 123
    perturbation, 123
Riemann ζ function, 68
Rodrigues' formula, 392

S
Sachs-Wolfe effect, 312
Sachs-Wolfe plateau, 317
Saha equation, 90
Scalar field
    Klein–Gordon equation, 241
Scalar-Vector-Tensor decomposition, 134
Scale factor, 21, 27
See-saw mechanism, 70
Shot noise, 274, 375
σ_8, 376
Silk damping, 319
Silk length, 331
Size of the visible universe, 31
Sloan Digital Sky Survey (SDSS), 8
Slow-roll
    conditions, 241
    parameters, 242
Snap parameter, 46

Spatial curvature density, 33
Spectral index
    running, 262
    scalar, 262
    tensor, 255
Spherical harmonics, 224
    addition theorem, 393
    completeness relation, 392, 396
    complex conjugation, 392
    normalisation, 392
    orthonormality, 392, 396
    parity, 393
    plane wave expansion, 170, 393
    spin-weighted, 395
Spurious gauge modes, 137, 141
Standard candle, 48
Standard ruler, 50, 327
Statistical homogeneity, 214
Statistical isotropy, 214
Stefan–Boltzmann law, 69
Sterile neutrino, 6
Stewart-Walker lemma, 134
Stochastic initial conditions, 223
Stokes parameters, 402
    Boltzmann equation, 177
Structure formation
    bottom-up scenario, 286
    top-down scenario, 286
Sudden recombination, 318, 337
Synchronicity problem, 36
Synchronous gauge, 140

T
Tensions in cosmology, 13
Tensor perturbation, 136
Tensor-to-scalar ratio, 264
Thermal distributions, 379
Thermal relic, 106
Thomson scattering
    cross section, 60, 407
    rate, 104
Tight-coupling, 197, 318
Too big to fail problem, 12
Top Hat filter, 371
Top quark, 58
Transfer function, 211, 223
    BBKS, 299
    gravitational waves, 307
    matter, 298

V
Vector perturbation, 135

Visibility function, 336

W
Warm dark matter, 12
Wigner D-matrix, 350, 394

Wigner $3j$-symbols, 349
WIMP, 6
    miracle, 113
    WIMPless miracle, 113
Wronskian, 397

Printed in the United States
By Bookmasters